NEW HORIZONS OF QUANTUM CHEMISTRY

ACADÉMIE INTERNATIONALE
DES SCIENCES MOLÉCULAIRES QUANTIQUES

INTERNATIONAL ACADEMY
OF QUANTUM MOLECULAR SCIENCE

NEW HORIZONS OF QUANTUM CHEMISTRY

PROCEEDINGS OF THE FOURTH INTERNATIONAL CONGRESS OF QUANTUM CHEMISTRY HELD AT UPPSALA, SWEDEN, JUNE 14–19, 1982

Edited by

PER-OLOV LÖWDIN
Department of Quantum Chemistry, Uppsala University, Uppsala, Sweden

and

BERNARD PULLMAN
Institut de Biologie Physico-Chimique, Paris, France

D. REIDEL PUBLISHING COMPANY
DORDRECHT : HOLLAND / BOSTON : U.S.A.
LONDON : ENGLAND

Library of Congress Cataloging in Publication Data

CIP

International Congress of Quantum Chemistry (4th : 1982 : Uppsala, Sweden)
New horizons of quantum chemistry.

Held under the auspices of the International Academy of Quantum Molecular Science.
Includes index.
1. Quantum chemistry–Congresses. I. Löwdin, Per Olov, 1916- . II. Pullman, Bernard, 1919- . III. International Academy of Quantum Molecular Science. IV. Title.
QD462.A1I57 1982 541.2'8 82-23029
ISBN-13: 978-94-009-7952-9 e-ISBN-13: 978-94-009-7950-5
DOI: 10.1007/978-94-009-7950-5

Published by D. Reidel Publishing Company,
P.O. Box 17, 3300 AA Dordrecht, Holland.

Sold and distributed in the U.S.A. and Canada
by Kluwer Boston Inc.,
190 Old Derby Street, Hingham, MA 02043, U.S.A.

In all other countries, sold and distributed
by Kluwer Academic Publishers Group,
P.O. Box 322, 3300 AH Dordrecht, Holland.

D. Reidel Publishing Company is a member of the Kluwer Group.

TABLE OF CONTENTS

SYMPOSIUM III. TIME DEPENDENT PHENOMENA

Chairman: K. Fukui

SYMPOSIUM IV. COMPUTATIONAL QUANTUM CHEMISTRY

Chairman: J. A. Pople

SYMPOSIUM V. A. SURFACES AND CATALYSIS

Chairman: M. Simonetta

SYMPOSIUM V. B. THEORY OF LARGE MOLECULES AND SOLIDS

Chairman: J. Koutecký

SYMPOSIUM VI. QUANTUM BIOLOGY AND QUANTUM PHARMACOLOGY

Chairman: B. Pullman

PREFACE

The Fourth International Congress in Quantum Chemistry under the auspices of the International Academy of Molecular Quantum Science in Menton, France was arranged at Uppsala University, Uppsala, Sweden, during the period June 14 - 19, 1982, in close collaboration with the University of Florida. The previous congresses were held in Menton 1973, New Orleans 1976, and Kyoto 1979, and the 1985 congress is tentatively planned to be held in the province of Quebec, Canada.

The Congress consisted of six symposia in various areas of quantum chemistry, solid-state theory, and quantum biology. The meeting was attended by about 450 scientists from 45 different nations, and a total of more than 300 scientific papers were presented. Even the poster contributions were given some plenary time.

These proceedings contain the text of the plenary lectures as well as the chairmen's introductions, whereas the contributed papers will be published in the International Journal of Quantum Chemistry, (John Wiley & Sons, New York) in the regular January - April 1983 issues.

The organizers would like to express their gratitude for moral and financial support from the following Swedish authorities: the Department of Education, the Nobel Committee in Physics, the Nobel Committee in Chemistry, the Swedish Natural Science Research Council, and Uppsala University. They are further indebted to a number of members of the Uppsala Quantum Chemistry Group and the Florida Quantum Theory Project for their skillful work in organizing the conference. They would also like to thank the D. Reidel Publishing Company for the valuable cooperation which made these proceedings possible.

Per-Olov Löwdin
Editor

Jean-Louis Calais　　　　Osvaldo Goscinski

Quantum Chemistry Group
Uppsala University, Box 518
Uppsala, S-751 20 Sweden

P.-O. Löwdin and B. Pullman (eds.), New Horizons of Quantum Chemistry, ix.

INTRODUCTION TO THE FIRST SYMPOSIUM

Professor Raymond Daudel, Honorary President of
The International Academy of Quantum Molecular Science

To be the first chairman of the fourth international congress of quantum chemistry is really a great privilege

Perhaps I owe this honour to the fact that I have been the president of the first of this set of important meetings organized by the international academy of quantum molecular science.

This is why I know the tremendous amount of delicate work which is needed to prepare such a congress and I must present my warmest thanks to Professor Löwdin and collaborators for their excellent organization.

I am sure that this congress will be a success as every body knows that the Uppsala group is an out standing expert is the field of international meetings.

In this short introduction I do not want to duplicate what our brilliant lecturers have to say.

I shall present only two general remarks on quantum mechanical methods.

Interpretations of quantum mechanics :

The first one is concerned with the interpretations of quantum mechanics, a field still open for discussions.
I want to recall the original interpretation of Louis de Broglie which is essentially made of two parts :

a) the double solution theory
b) the thermodynamics of the isolated particule (1)

Louis de Broglie consider two kinds of waves. The classical one Ψ solution of the Schrodinger equation :

$$H \Psi = - \frac{h}{2\eta i} \frac{\partial \Psi}{\partial t}$$

P.-O. Löwdin and B. Pullman (eds.), New Horizons of Quantum Chemistry, 1–4.

As well known this wave is interprated as a probability wave. It permits to calculate the probability of finding the particle in a small given volume at a given instant. It is a subjective wave which vanishes when we know where is the particle.

But Louis de Broglie admits the existence of a second wave U which would be an objective one directly related with the structure of the particle which is considered as an object possessing a rather large extension in the space.

The particle would present a kind of nucleus, a hard core if you want in which the wave U becomes very large. In this area the wave U possesses a kind of singularity.

The wave U should be a solution of the equation :

$$HU + k(U) = -\frac{h}{2\pi i}\frac{\partial U}{\partial t}$$

where k (U) is a non linear term which is only important in the singularity region.

As a consequence, except for that region the two waves Ψ and U are proportional and have the same phase ϕ.

Theoretically it becomes possible to calculate the trajectory of the singularity and its velocity is given by the following formula :

$$\vec{v} = -\frac{1}{m}\,\text{grad}\ \phi$$

where m is the mass of the particle.

To go further it is necessary to admit that a particle is always in interaction with the remaining part of the Universe, with which it exchanges energy and momentum.

By using this hypothesis it is possible to demonstrate that the density probability of finding the particle singularity at a point M is given by the relation :

$$\rho\ (M) = |\Psi\ (M)|^2$$

Therefore is we consider the double hole Young experiment each photon is able to react simultaneouxly with the two holes, in such a way that each photon goes throuth only one hole, but that its trajectory depends on both of them.

Starting from that description of a particle it is possible to associate to this isolated particle certain thermodynamical properties by using a formalism introduced by Boltzmann to establish a bridge between mechanics and thermodynamics.

Boltzmann has shown that the heat amount δQ necessary to produce a change δA of the action of molecules possessing a periodical movement is given by

$$\delta Q = \nu \delta A$$

if ν represents the frequency of the movement. But furthermore :

$$\delta Q = T\delta S.$$

if δS denotes the entropie change.

Therefore it appears that the temperature T of an isolate particle must be proportional to its frequency :

$$T = C\nu$$

The constant C will be taken such that :

$$h\nu = kT$$

Finally the temperature is given by the relation :

$$T = \frac{h\nu}{k}$$

Following a similar way it is also possible to express the entropy S of the particle as a function of its action.

By using these expressions it is possible to show that the entropy is a maximum along a possible trajectory of the singularity associated with a particle.

Relation between the formalism of the wave mechanics and the formalism of the statistical mechanics.

I shall now present a remark related to a field in which the Uppsala group as made very important contribution : the analogy between the formalisms of wave mechanics and statistical mechanics.

The Liouwille equation :

$$i\frac{h}{2\pi}\frac{d}{dt} = \rho(t) = [H,\rho(t)]$$

is much more difficult to integrate that the Schrödinger equation :

$$i\frac{h}{2\pi}\frac{d}{dt}|\psi\rangle = H|\psi\rangle$$

but it is possible to transform the Liouville equation into an other equation possessing the same structure as the Schrodinger equation if $\rho(t)$ is considered to be a vector in an algebra including superopetators.

Let us call A the algebra of operators O of the classical Hilbert space. With each operator O we associate a superoperator |O) scalar product of two of them being defined by the equation :

$$(|O),|P)) = \mathrm{Tr}\ (O^{+}.P)$$

in which Tr denotes the trace and + the hermitian conjugate.

Consider know a basis of superopetators $\{|\Omega_k)\}$ and the structural constant $\sigma^{\tau}_{\mu\lambda}$ of the Lie algebra generated by the Ω_k. These constants obey the equation :

$$[\Omega^{+}_{\mu}, \Omega_{\lambda}] = \sum_{\tau} \sigma^{\tau}_{\mu\lambda}\ \Omega_{\tau}$$

Starting from this relation we can define an id-operator O-associated with each operator O by the equation :

$$(\Omega_{\mu}|O\_|\Omega_{\lambda}) = \sum_{\tau} \sigma^{\tau}_{\mu\lambda}\ (\Omega_{\tau}|O)$$

in which the superoperator $(\Omega_{\mu}|$ is the one associated with $|\Omega_{\lambda})$ in the dual of A.

If finally H_ is the id-operator associated with the hamiltonian the Liouville equation takes the form :

$$i\frac{h}{2\pi}|\frac{d\rho(t)}{dt}) = H\_|\rho(t)$$

which is exactly similar to the Schrödinger equation.

This procedure has made possible the solution of important problems of molecular physics in particular various topics related to the density operator description of excitons in molecular aggregates[1].

REFERENCES

(1) C. ASLANGUL and P. KOTTIS
Compte rendus de l'Académie des Sciences, 278, 33, (1974)
Phys. Rev. B 10, 4364, (1974)
Phys. Rev. B 13, 5544, (1976)
Phys. Rev. B 18, 4462, (1981)
Phys. Rev. B 24, 4805, (1981)

CLASSICAL IMPLICATIONS OF THE QUANTUM STATISTICAL INTERPRETATION

Gérard G. Emch
Departments of Mathematics and of Physics
The University of Rochester, Rochester, NY 14627, USA

Regarding Quantum Mechanics as the fundamental theory we derive, from its underlying geometry, a complete version of Classical Mechanics, including: the interpretation of the phase space as the manifold of classical states; the Wigner-Moyal correspondance rule, now obtained from first principles; the Jordan and Lie structures of the algebra of classical observables, given by the point-wise multiplication and the Poisson bracket of functions on the cotangent bundle of the configuration manifold. The geometric formulation of the Dirac problem followed here allows to extend the solution which we propose to cases where the configuration space is an homogeneous Riemannian manifold. In case this manifold is not flat, the theory provides a natural distinction between the momenta and the Darboux coordinates relative to the symplectic form associated to the Poisson bracket. For instance, the momentum map is explicitly derived for the case where the configuration space is a Riemannian manifold of constant negative curvature. Finally, the connection between our approach and the general geometric quantization programme is discussed.

Asides from our emphasis of the geometric underpinnings of our approach, the essential ingredient of the theory which we present is a quantum definition of classical expectations; the latter include for instance: canonical states encountered in Statistical Mechanics, and the coherent states of Schroedinger and Perelomov, as well as a generalization of these to mixture states.

Details are being published separately as :

Prequantization and KMS Structures, Intern.Journ.Theor.Phys.20(1981);
Geometric Dequantization and the Correspondance Principle, Intern.Journ. Theor.Phys. (to appear);
Quantum and Classical Mechanics on Homogeneous Riemannian Manifolds, Journ. Math. Phys. (to appear).

P.-O. Löwdin and B. Pullman (eds.), New Horizons of Quantum Chemistry, 5.

PROPAGATORS, SCATTERING THEORY AND CHEMICAL REACTIONS.

Jan Linderberg
Department of Chemistry
Aarhus University
Aarhus, Denmark

ABSTRACT: Some recent developments in the theory of chemical reaction rates are reviewed. The conceptual advantage of hyperspherical coordinates is stressed.

INTRODUCTION.

Two points are made in this paper: (i) hyperspherical coordinates can be invoked to provide a mapping of the general molecular reactive scattering process onto states on a sphere, and (ii) density matrices are convenient tools for the description of non equilibrium situations.

None of these points is original. Fano has recently advocated the conceptual advantages of hyperspherical coordinates [1] and he has also long been a champion of density matrix descriptions [2]. A measure of novelty is attributed to the mapping, in the second section of the paper, of binary collision channels on points on a sphere. The use of hyperspherical coordinates offers a somewhat different view on the interpretation of reaction rates. It has been shown that even unmolecular rearrangement reactions can be interpreted as transitions between points on a sphere [3] and that the transition rates are conveniently calculated in terms of a theory advanced by Lax [4] for radiative processes.

Density matrices have been prominent here at Uppsala since Löwdin's landmark paper [5] and have been brought to the fore again through the trace algebra formulation of quantum theory [6]. Analysis of scattering experiments in terms of density matrices is standard practice [7] and the information theoretical analysis introduced by Bernstein and Levine [8] has provided exciting new concepts [9].

The second section of this paper is devoted to the hyperspherical coordinates and some mappings which are brought out in such a frame. The third section discusses the transition matrix and the transition rates in realistic systems. An example is discussed in the fourth section and a few conclusive remarks make up the last part.

P.-O. Löwdin and B. Pullman (eds.), New Horizons of Quantum Chemistry, 7–14.

HYPERSPHERICAL COORDINATES.

Fano [1] presents a fine argument for the unified picture one obtains of scattering processes in terms of hyperspherical coordinates and calculations have shown their operational effectiveness [10]. They seem to have appeared in the detailed studies of the helium atom [11]. Delves [12] resurrected them in the nuclear context while Smith [13] provides the first general study for an arbitrary number of particles. The three-particle problem has received particular attention [1,10] and Johnson [14] has reviewed the separation of variables in detail. Some consequences of symmetry and phase problems have been analyzed by Mead and Truhlar [15].

We consider a set of point masses, $\{\underline{r}_j, m_j | j=1,2,..N\}$, and define the center of mass as

$$M = \Sigma_j m_j, \; M\underline{G} = \Sigma_j m_j \underline{r}_j . \tag{1}$$

The hyperspherical radius q is then defined through the relation

$$Mq^2 = |\Sigma_j m_j (\underline{r}_j - G)^2|^{\frac{1}{2}} \tag{2}$$

It is invariant with respect to translation and rotation of the reference frame and is further invariant under orthogonal transformations of the set of mass weighted coordinates $\{\underline{U}_j = \underline{r}_j \sqrt{m_j/M}\}$ which leave $\underline{G}$ constant. One such transformation is offered by the Householder form:

$$\underline{U}'_j = \underline{U}_j - 2(\Sigma_k \underline{U}_k C_k) C_j , \tag{3}$$

where $\{C_j\}$ is a unit vector. We may require that $\underline{U}'_N = \underline{G}$ and find then that

$$2C_N^2 = 1 - \sqrt{m_N/M} , \tag{4}$$

$$2C_j C_N = - \sqrt{m_j/M} , \tag{5}$$

which determines the transformation. The q-coordinate is then the length of a 3N-3 component vector:

$$q = |\Sigma_{j=1}^{N-1} (\underline{U}'_j)^2|^{\frac{1}{2}} \tag{6}$$

and the hyperspherical radius concept refers to this feature.

Further substitutions can be made in several ways as examplified by Smith [13]. In order to take advantage of the invariance of q and other internal features in the system under space notations we arrange the components of $\{\underline{U}'_j\}$ as a 3x(N-1) matrix

$$X = \begin{bmatrix} U'_{11} & U'_{12} & \cdots & U'_{1N-1} \\ U'_{21} & U'_{23} & \cdots & U'_{2N-1} \\ U'_{31} & U'_{32} & \cdots & U'_{3N-1} \end{bmatrix} \tag{7}$$

and use a polar decomposition to write

$$X = D(\alpha\beta\gamma) \begin{pmatrix} q \sin\theta \cos\psi & 0 & 0 \\ 0 & q \sin\theta \sin\psi & 0 \\ 0 & 0 & q \cos\theta \end{pmatrix} Y \tag{8}$$

where Y is another 3x(N-1) matrix. The rows of Y form 3 orthonormal vectors and $D(\alpha\beta\gamma)$ denotes a 3x3 orthogonal notation matrix parametrized through Euler angles. The internal coordinate system is such that the tensor of moments of inertia is diagonal with elements

$$I_1 = Mq^2(1 - \sin^2\theta \cos^2\psi), \tag{9}$$

$$I_2 = Mq^2(1 - \sin^2\theta \sin^2\psi), \tag{10}$$

$$I_3 = Mq^2(1 - \cos^2\theta). \tag{11}$$

The range of the variables and the explicit parametrization of Y will not be of concern here. Neither will we be interested in the explicit form of the kinetic energy operator nor the volume element in configuration space.

The form (8) is now of the main interest. A system, such as the one considered here, may evolve in time so that it is meaningful to talk about two separate subsystems which are separated by a large distance. These conditions impose the relations that I_1 and I_2 are much smaller than I_3 and the limiting situation of infinite separation requires that the parameter θ approaches zero. This implies in turn that only the third row of Y will be explicitly characterizing the particular separation we are dealing with. It is a vector in N-1 dimensions with components that are the projection of the $\{\underline{U}_j'\}$ on the vector between the two subsystems.

Particular simplifications occur when we consider a four particle system so that N-1 = 3. Each asymptotic channel of separation into two subsystems corresponds then to points on a unit sphere in three dimensions. A map of the various possibilities is given in Figure 1.

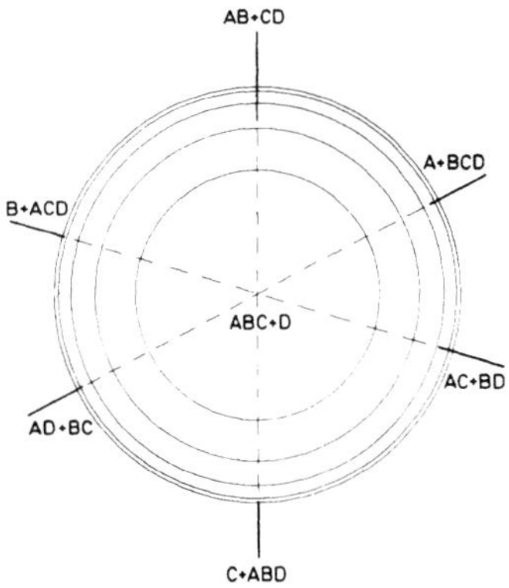

Fig. 1. Map of asymptotic hyperangle directions for a four particle system.

The relative directions are determined by the mass ratios of the particles. Three particle systems give asymptotic states in terms of direction in a plane [13].

A four particle system offers the simplest picture of a general chemical reaction where two bonds are broken and two new ones are formed. Theoretical models for such reactions are based on a picture of the electronic structure constructed from four orbitals [16] and the analysis is based on orbital energy correlation diagrams. It has recently been shown [3] that the essential features of such a diagram can be mapped on a sphere as well. The four orbital energies are conveniently parametrized in the form [17]

$$\begin{aligned} \varepsilon_1 &= t + x - y - z, \\ \varepsilon_2 &= t - x + y - z, \\ \varepsilon_3 &= t - x - y + z, \\ \varepsilon_4 &= t + x + y + z. \end{aligned} \tag{12}$$

The mean energy t and the second moment of the spectrum, $x^2 + y^2 + z^2$, are irrelevant for describing the degeneracies and the relative position of the levels, except for the four-fold degeneracy. The direction of the vector (x,y,z) is thus the property to investigate. There are eight directions or points on the unit sphere corresponding to three-fold degeneracy which are joined by linesegments where two-fold degeneracies occur. The remaining set of points consists of disjoint subsets where no degeneracy exists. A particular conformation of the molecule can then be mapped onto the point which describes its orbital energy spectrum and a distortion of the molecule will result in a motion of this point and it passes into another subset we expect a "critical" intermediate where an orbital degeneracy should result.

The orbital energy mapping is a rather indirect one and a more direct porcedure may have a quantitative content. Rearrangement reactions in larger molecules do not cause a separation of the system into fragments and in order to deal with such situations we develop a more general mapping of conformations onto points on a sphere. The nuclear coordinates of a molecule can be transformed as before so that the internal coordinates are those of Eq. (8). Two different conformations give two sets of parameters. The Euler angles are unimportant for the structure of the molecule and we define a new matrix

$$X' = D^{-1}X. \tag{13}$$

Each conformation has a hyperspherical radius

$$q_S^2 = \mathrm{Tr}\; X_S'(X_S')^T \tag{14}$$

and we may now define a scalar product as

$$\mathrm{Tr}\; X_R'(X_S')^T = q_R q_S \cos\omega_{RS} , \tag{15}$$

which gives relative directions between different conformations. Many rearrangement reactions cause only a moderate change in q but will give some value larger than zero for ω_{RS}. Three different conformations give three relative angles and these may be used to define a spherical triangle. Thus we have the possibility of mapping a reactant state, an intermediate state, and a product state on points on a sphere [3].

TRANSITION RATES.

Formal scattering theory [18] gives an expression for the transition rate from an initial asymptotic state to a final state in terms of the transition matrix element $\langle f|T(E)|i\rangle$ at the energy E which is the same for both states. A reaction rate is then obtained by rule of summing over final states and averaging over initial states [4]. The use of hyperspherical coordinates admits a characterization of initial and final states in terms of an asymptotic wave number and a discrete channel index labelling the asymptotic hyperangular states which form an enumerable set. The energy is related to the wave number as

$$E = \varepsilon_n + (\hbar k_n)^2/2M , \qquad (16)$$

where ε_n is the internal energy in channel n.

The discretization of the hyperangular spectrum does not mean that the levels are well separated. There will be a large number of states within a small energy range as soon as there is an excess energy above the ground state. We may consider solving for the transition matrix elements in terms of an adiabatic expansion of the eigenstates of the Hamiltonian where the basis diagonalizes the hyperangular Hamiltonian which it obtains when the hyperradial kinetic energy term is eliminated. This will give adiabatic energy curves $\{\varepsilon_n(q)\}$ and a set of coupled channel equations involving nonadiabatic coupling terms $\{A_{nm}(q)\}$. The curves will be similar to conventional electronic energy curves for diatomic molecules but they will be very closely spaced and can be expected to show numerous crossings or avoided crossings. An experiment will generally be unable to select either the initial or final channel totally and we need then to develop effective summation procedures. It is then that density matrices and their determination enter.

We follow Levine [19] and write the rate for a transition

$$(k_n,n) \rightarrow (k_m,m)$$

as

$$R(m \leftarrow n) = |\langle k_m,m|T(E^+)|k_n,n\rangle|^2 \times (2\pi M k_m V h^{-3}) , \qquad (17)$$

where V is the appropriate volume and E^+ denotes the limit as E approaches the real axis from above. There may generally be a mixture of initial states and it may either be a statistical equilibrium mixture or something else. Thermal equilibrium gives a normal Boltzmann distribution in energy while other distributions can be determined from the specifics of the initial situations through the information theoretical approach [9]. Similarly, there is an efficiency matrix [20] associated with the detection of the final state, which also is given by information theory. We obtain then the effective rate as the sum

$$R = \Sigma\ p_m R(m \leftarrow n) W_n \ . \tag{18}$$

where the weights W_n refer to the initial situation and the coefficients p_m to the detection apparatus.

The distribution with the diagonal elements p_m should generally be a microcanonical one consistent with a particular final state energy while the initial density matrix shows some energy spread. Lax [4] developed a general way of managing sums of the kind [18] when the number of degrees of freedom is large as they are in many chemical reaction problems. He concluded that when the elementary transition rate is such that it is meaningful to consider, the sum over all pairs of channels which have the same rate, then the sum of the product $p_m W_n$ over these pairs might be considered as a combined weight with a normal probability distribution. His argument also shows that the width of the distribution in a thermal equilibrium system becomes linear in temperature.

The general arguments lead to the expected result that the total rate can be written as

$$R = A \exp(-E_a/kT) \tag{19}$$

with A as the representative value of the elementary rates $R(m \leftarrow n)$ between the selected states.

PHOTOCHEMICAL REARRANGEMENT.

An example where the previously advanced arguments have been applied is the photochemical rearrangement of 6b,7a-Dihydrocycloprop[a]acenaphthylene (R) to phenalene (P) via the intermediate triplet 1,3-perinaphthadiyl biradical [3].

Fig. 2.

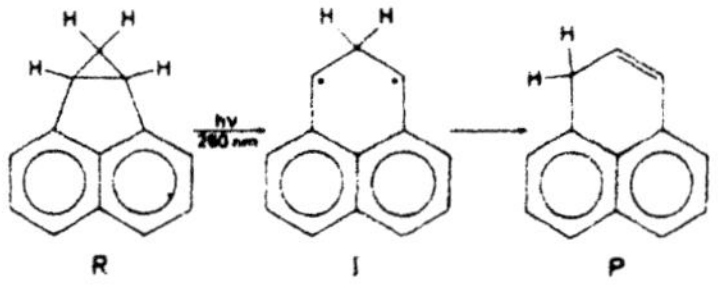

The three conformers show little change in hyperspherical radius and the relative angles are such that W(R,I) = 7.6° and W(I,P) = 24,8°. Deuteration at the apical positions changes these values to W(R,I) = 7.9° and W(I,P) = 18.7°. Both the transition from the excited singlet in the reactant to the intermediate and from the intermediate to reactant ground state singlet depend upon the spin-orbit coupling in the electronic system. The hyperangular displacement is substantially larger in the second step and this results in a lower rate. The surprising result that the deuteration decreases the displacement angle gives a larger intrinsic rate for this species as was found in the experiments by Michl and coworkers [21]. Further analysis is required to ascertain that the observed activation energies can be estimated. Numerical values for the preexponential factor were estimated in good agreement with experiment [3].

REMARKS.

The title of this paper conforms with what was the original plan for the lecture. Propagators have not been dealt with explicitly but the calculation of transition matrix elements is intimitely connected to the use of propagators. Approximate propagators and t-matrix elements may be obtained in the many channel scattering problem from a particular integral equation formulation suggested last year [22].

There is reason to believe that progress towards reliable estimates of chemical reaction rates can be based on quantum mechanics and that a conceptual development can result from a further analysis of the many channel scattering equations. The particular advantages offered by the hyperspherical description deserves attention in the future development.

ACKNOWLEDGEMENTS.

It is a privilege to be able to address this Congress and I am grateful to the International Academy of Quantum Molecular Science for the invitation and the support. The paper was written during leave of absence at the Quantum Chemistry Group, Uppsala University on a fellowship from the Nordic Council, and thanks are due to Per-Olov Löwdin and other members of the Group for hospitality extended to me.

REFERENCES.

1. Fano, U.: 1981, Phys. Rev. A24, p. 2402.
2. Fano, U.: 1957, Rev. Mod. Phys. 29, p. 74.
3. Linderberg, J.: 1982, 15th Jerusalem Symposium, D. Reidel, p. 325.
4. Lax, M.: 1952, J. Chem. Phys. 20, p. 1752.
5. Löwdin, P.O.: 1955, Phys. Rev. 97, p. 1474.
6. Löwdin, P.O.: 1977, Int. J. Quant. Chem. 12, Suppl. 1, p. 197. 1982, Int. J. Quant. Chem. 21, p. 275.
7. Blum, K.: *Density Matrices* (Plenum Press, New York 1981).
8. Bernstein, R.B. and Levine, R.D.: 1972, J. Chem. Phys. 57, p. 434.
9. Levine, R.D.: 1978, Am. Rev. Phys. Chem. 29, p. 59.
10. Kuppermann, A., Kaye, J.A., and Dwyer, J.P.: 1980, Chem. Phys. Letters 74, p. 257.
Hauke, G., Manz, J., and Römelt, J.: 1980, J. Chem. Phys. 73, p. 5040.
11. Gronwall, T.H.: 1937, Phys. Rev. 51, p. 655.
Bartlett, J.H. Jr.: 1937, Phys. Rev. 51, p. 661.
Kemble, E.C.: *The Fundamental Principles of Quantum Mechanics* (McGraw-Hill, New York 1937) p. 208 ff.
Morse, P.M. and Feshbach, H.: *Methods of Theoretical Physics* (McGraw-Hill, New York 1953) p. 1730 ff.
12. Delves, L.M.: 1960, Nucl. Phys. Rev. 120, p. 1058.
13. Smith, F.T.: 1960, Phys. Rev. 120, p. 1058.

14. Johnsson, B.R.: 1980, J. Chem. Phys. 73, p. 5051.
15. Mead, C.A. and Truhlar, D.G.: 1969, J. Chem. Phys. 70, p. 2284.
16. Woodward, R.B. and Hoffman, R.: 1969, Angewandte Chemie (Intern. Ad.), 8, p. 781.
17. Linderberg, J.: 1981, Int. J. Quant. Chem. 19, p. 237.
18. Roman, P.: *Advanced Quantum Theory* (Addison-Wesley, Reading 1965) p. 284.
19. Levine, R.D.: *Quantum Mechanics of Molecular Rate Processes* (Oxford, 1969) p. 115.
20. Ref. 18, p. 100.
21. Michl, J., Döhnert, D. and Penn, J.: Private communication.
22. Linderberg, J.: 1981. Int. J. Quant. Chem. S15, p. 559.

DEGENERATE PERTURBATION THEORY†

Baird Brandow
Los Alamos National Laboratory
University of California
Los Alamos, New Mexico 87545 USA

ABSTRACT. The algebraic structure of degenerate Rayleigh-Schroedinger perturbation theory is reviewed. There are a number of different but equivalent algorithms which generate this perturbation series; we argue that the frequent need to carry out infinite-order partial summations selects one of these algorithms as the most efficient. Recent developments include coupled-cluster formulations for open shells, a new diagrammatic representation, and the concept of incomplete model subspaces. These subjects are reviewed, as well as some applications.

1. INTRODUCTION

Degenerate perturbation theory, in the Rayleigh-Schroedinger form which exhibits the linked-cluster feature, is being used in quantum chemistry in several different ways. On the one hand it provides a general ab initio theory of effective Hamiltonians. It should, therefore, be able to shed light on the various semi-empirical Hamiltonians employed by quantum chemists, such as π-electron Hamiltonians, the Slater-Condon parametrization for atomic spectra, MINDO etc., and some which have more of a solid-state flavor, such as crystal-field and ligand-field models, and also Heisenberg-like spin Hamiltonians. Of course this formalism also applies when one is only interested in the ground state, for cases where accidental degeneracy leads to several strongly-mixed configurations. Examples are the beryllium atom and molecular potential-surface calculations, particularly at large internuclear separations and for difficult cases such as the N_2 molecule. And even though nondegenerate techniques may often be adequate for the ground states of closed-shell molecules, the excitation spectra of these molecules require some form of open-shell (degenerate or quasi-degenerate) treatment. Another useful feature of the full many-body formalism is that of "core-valence separation." This feature is useful even in cases where degeneracy is of little or no interest, such as for the calculation of ionization potentials and electron affinities. Still another feature, which so far has been largely overlooked by quantum chemists, is that the general formalism

†Work supported by US Department of Energy.

P.-O. Löwdin and B. Pullman (eds.), New Horizons of Quantum Chemistry, 15–29.

also provides linked-cluster prescriptions for electronic properties other than energy, i.e., for the calculation of expectation values and transition matrix elements of a general physical operator.

Much of this talk will be tutorial, reviewing material which should be well known to the practitioners in this field. I'm going to focus primarily on algebraic aspects, because I have found that confusion or misunderstanding is frequently rooted in these aspects, and the use of diagrams would tend to obscure these issues. For further details and references, I recommend a recent survey article.[1] I will also review several recent formal developments.

Since I am not a quantum chemist I will not attempt to discuss calculations in any detail; I will simply mention some applications that serve to illustrate various pedagogical points. I hereby apologize to investigators whose work is not mentioned.

2. THE BASIC FORMAL STRUCTURE

The simplest way to introduce the concept of an effective Hamiltonian is by "partitioning the Hamiltonian," as proposed by Löwdin.[2] One partitions the Hilbert space into two disjoint subspaces, P and Q, and writes the Schroedinger equation in block-matrix form,

$$\begin{pmatrix} H_{PP} & H_{PQ} \\ H_{QP} & H_{QQ} \end{pmatrix} \begin{pmatrix} \Psi_P \\ \Psi_Q \end{pmatrix} = E \begin{pmatrix} \Psi_P \\ \Psi_Q \end{pmatrix} \quad . \tag{2.1}$$

The (vector) variable Ψ_Q is then eliminated to express the Schroedinger equation in terms of Ψ_P alone,

$$\left[H_{PP} + H_{PQ}\,(E-H_{QQ})^{-1} H_{QP}\right]\Psi_P = E\,\Psi_P \quad . \tag{2.2}$$

This has the form of an effective Hamiltonian (hereafter denoted by $\mathcal{H}$) acting on the model subspace P.

We observe, however, that (2.2) has several defects which must be resolved before one can claim a practical formalism for many-body systems. First, this $\mathcal{H}$ depends explicitly on E. But this E is unknown, and furthermore, it generally takes on different values for different eigenstates. This negates the goal of fully determining $\mathcal{H}$ before any secular diagonalization to obtain the desired eigenstates. Secondly, we note that the Ψ_P's are generally not mutually orthogonal, since they are merely projections of the orthogonal state vectors onto the subspace P. This means that the present $\mathcal{H}$ is, in effect, a non-Hermitean matrix. (It's eigenvalues E remain real, however.) Finally, the form (2.2) is too abstract for most applications, and takes no account of many-body aspects.

The first step towards a practical many-body formalism is to introduce a perturbative analog of (2.2), namely, the degenerate version of

Brillouin-Wigner perturbation theory.[3] One selects an H_0 and $V = H-H_0$, and uses the set of H_0 eigenstates

$$H_o \Phi_i = E_i \Phi_i \tag{2.3}$$

to define the two subspaces. That is,

$$\begin{aligned}\Psi &= \sum_i a_i \Phi_i \\ &= \sum_{i\varepsilon P} a_i \Phi_i + \sum_{i\varepsilon Q} a_i \Phi_i \equiv P\Psi + Q\Psi \quad .\end{aligned} \tag{2.4}$$

(We use P and Q interchangeably as labels and as projection operators.) One easily obtains the secular equation

$$\left[H_o + P\mathcal{V}(E_\alpha)\right] A_\alpha = E_\alpha A_\alpha \quad , \tag{2.5}$$

where α labels the total eigenstate, A_α is the column vector of amplitudes a_i within $P\Psi_\alpha$, and the energy-dependent effective interaction matrix P is determined by

$$\begin{aligned}\mathcal{V} &= VP + V \frac{Q}{E-H_o} \mathcal{V} \\ &= V \sum_{n=0}^{\infty} \left(\frac{Q}{E-H_o} V\right)^n P \quad .\end{aligned} \tag{2.6}$$

In effect, we have simply cast (2.2) into a perturbative form. This is an important step, because it allows the use of diagrams to help visualize the many-body aspects, and thereby to organize the practical details of calculation. Unfortunately, however, the explicit energy dependence of this $\mathcal{H} = P[H_o + \mathcal{V}(E)]P$ is still a serious problem, especially for many-body systems where this leads to unlinked diagrams and lack of size-consistency.

This problem can be resolved[1,3] just as in the nondegenerate (closed shell) case, by finding a Rayleigh-Schroedinger analog, $\mathcal{W}$, for the effective interaction operator $P\mathcal{V}$. An easy way to do this is to first assume exact degeneracy of H_o within the model space, whereby H_oA can be replaced by E_oA. (This restriction is not essential; it can easily be removed later.) Equation (2.5) then reduces to

$$P\mathcal{V}(E_\alpha)A_\alpha = \Delta E_\alpha A_\alpha \quad , \tag{2.7}$$

where $E_\alpha = E_o + \Delta E_\alpha$. A Taylor-series expansion for $\mathcal{W}$ can be developed by repeated use of this result, but it is easier to proceed in an implicit manner. Assuming that $\mathcal{W}$ exists, whereby $\mathcal{H} = (H_oP + \mathcal{W})$ and

$$\mathcal{W} A_\alpha = \Delta E_\alpha A_\alpha \quad , \tag{2.8}$$

the desired expansion can be expressed formally as

$$\mathcal{W} = \sum_{n=0}^{\infty} \mathcal{V}_n [-\mathcal{W}]^n \quad , \tag{2.9}$$

where

$$\mathcal{V}_n = \frac{(-1)^n}{n!} \left(\frac{d}{dE_o}\right)^n P \mathcal{V}(E_o) \quad . \tag{2.10}$$

The Rayleigh-Schroedinger expansion for $\mathcal{W}$, i.e. the ordinary power series in V, can now be obtained by first solving (2.9) recursively, in terms of matrix products of the various $\mathcal{V}_n$'s, and then replacing the latter by their respective perturbation series as deduced from (2.6) and (2.10). Through third order, the result is

$$\begin{aligned}\mathcal{W} &= \mathcal{V}_o + \mathcal{V}_1 [-\mathcal{V}_o] + \mathcal{V}_2 [-\mathcal{V}_o]^2 \\ &\quad + \mathcal{V}_1 [-\mathcal{V}_1][-\mathcal{V}_o] + \mathcal{O}[\mathcal{V}_n]^4 \\ &= PVP + PV\frac{Q}{e}VP + PV\frac{Q}{e}V\frac{Q}{e}VP \\ &\quad + PV\left(\frac{Q}{e}\right)^2 V(-PVP) + \mathcal{O}(V^4) \quad ,\end{aligned} \tag{2.11}$$

where e stands for $E_o - H_o$. In the resulting many-body diagrams, each matrix-multiplication (and the associated minus sign) between successive $\mathcal{V}_n$'s is represented by a "fold."[3,4] Thus, study of (2.8)-(2.11) should dispel the mystery of folded diagrams.

The order of the various $\mathcal{V}_n$'s in (2.11) is quite important, due to (2.8), which has the consequence that $\mathcal{W}$ is not Hermitean. This is to be expected, of course, because the eigenvectors A_α (of $\mathcal{H} = H_oP + \mathcal{W}$) are still formally identical to the non-orthogonal Ψ_p^α's of (2.2). On the other hand, it is often desirable to have a Hermitean form for $\mathcal{H}$. This is particulary important when one is seeking an <u>ab initio</u> replacement for some semiempirical effective Hamiltonian, since the latter are almost always assumed Hermitean. We simply remark that there is a rigorous "stretching" (non-unitary) transformation which replaces the effective interaction matrix $\mathcal{W}$ by a Hermitean analog $\mathcal{K}$,

$$\mathcal{K} = \frac{1}{2}\left[(I+\Theta)^{\frac{1}{2}}(H_o+\mathcal{W})(I+\Theta)^{-\frac{1}{2}} + h.c.\right] - H_o \quad , \tag{2.12}$$

where the P-space matrix Θ also has a well-defined perturbation expansion of Rayleigh-Schroedinger type. Expansion of $(I+\Theta)^{\pm\frac{1}{2}}$ in binomial series thus provides a rigorous Rayleigh-Schroedinger expansion for $\mathcal{K}$. (The indicated treatment of H_o is necessary only for the quasidegenerate case, where $H_oA \neq E_oA$.) On the other hand, the apparently naive approximation

$$\mathcal{K} \approx \tfrac{1}{2}(\mathcal{W} + \mathcal{W}^{\dagger}) \tag{2.13}$$

has often been used, and we believe that this should be sufficient for essentially all practical purposes. The lack of Hermiticity in $\mathcal{W}$ has been found to be fairly weak, in all applications to date that we are aware of, provided the calculation is otherwise reasonably converged. [In any event, corrections are straightforward to obtain from (2.12).]

Finally, we want to emphasize that this formalism is "complete" in the sense that expectation values and transition matrix elements of an arbitrary operator $\mathcal{O}$ can also be obtained entirely in terms of model-space quantities. Introducing the effective operator matrix

$$\mathcal{M}(\mathcal{O}) = (I + \Theta)^{-\frac{1}{2}}(\Omega^{\dagger}\mathcal{O}\,\Omega)(I + \Theta)^{-\frac{1}{2}} \quad , \tag{2.14}$$

where Ω is the wave operator (for intermediate normalization) defined by $\Psi_\alpha = \Omega|P\Psi_\alpha\rangle$, we obtain the identity

$$\langle\hat{A}_\alpha|\mathcal{M}(\mathcal{O})|\hat{A}_\beta\rangle = \frac{\langle\Psi_\alpha|\mathcal{O}|\Psi_\beta\rangle}{\left(\langle\Psi_\alpha|\Psi_\alpha\rangle\langle\Psi_\beta|\Psi_\beta\rangle\right)^{\frac{1}{2}}} \quad . \tag{2.15}$$

The $\hat{A}_\alpha$'s are the orthonormal eigenvectors of the Hermitean operator $\mathcal{H} = (H_o P + \mathcal{K})$. This effective operator technique has been quite useful in nuclear physics.[5] An obvious use in quantum chemistry would be for the study of electromagnetic transition elements, but I am not aware of any such applications to date.

For many-body systems and "complete" model subspaces (see below), we have shown[3] that each $\langle\Phi_i|$, $|\Phi_j\rangle$ element of the matrices $\mathcal{W}$, $\mathcal{K}$, and $\mathcal{M}(\mathcal{O})$ has a fully linked Rayleigh-Schroedinger-type expansion. Diagrammatic aspects are reviewed in Refs. 1, 4, and 5.

3. ALTERNATIVE FORMULATIONS AND THE PROBLEM OF CONVERGENCE

A few of the initial exploratory applications of degenerate perturbation theory to chemical systems have shown remarkably good convergence. Unfortunately, however, these results are almost certainly not representative for general chemical systems, since they were obtained for small, few-electron systems where the strong shell structure provides unusually large energy denominators. Indeed, a number of other applications have found convergence to be a serious problem, where acceptable results could be obtained only by some combination of careful choices for H_o and P and/or some infinite-order partial summation. This is consistent with general experience in nuclear physics, where many of the interesting applications have required some form of infinite-order partial summation.[5,6] I therefore think it important, for the sake of future progress in this field, to recognize that some form of partial summation will often be necessary.

I also want to emphasize that the variety of possible and potentiall useful partial summations is considerably greater than what most quantum chemists have recognized so far. Most chemical applications to date have been restricted to the simplest class of summations, namely those of the denominator-shift type. As an example of a more complex possibility, let me suggest that RPA-type summation may be necessary to deal accurately with σ-polarization effects within effective π-electron Hamiltonians. Good sources for techniques that have been employed in nuclear physics may be found in Refs. 5 and 6.

I now want to mesh these remarks about partial summation with comments about the plethora of alternative and seemingly different versions of degenerate Rayleigh-Schroedinger perturbation theory that can be found in the literature. It is a curious and reassuring fact that almost all of these alternative formulations, for Hermitean quasi-degenerate Rayleigh-Schroedinger perturbation theory, lead to identical algebraic expressions for the nth order perturbation terms. That is, they are merely different algorithms for generating a single, uniquely defined, perturbation expansion. This result is not as trivial as it may first appear, because, for a given model space P, the various formulations for $\mathcal{H}$ could conceivably differ by unitary transformations withi P. For a summary of these various alternatives and their formal connections, see Refs. 1 and 7.

If one could always be content with, say, a third-order calculation, then any of these equivalent formulations would be quite satisfactory. Performing infinite-order summations is a very different matter, however, since this requires considerable insight about the general structure of the series in high orders. An important corollary is that the "renormalized" expansion (the expansion which remains after performing the partial summation) should also be fairly transparent, so that the remaining perturbative corrections can be studied without ambiguity or danger of double-counting, bearing in mind also that further partial-summation may sometimes be desirable. In short, the formalism should provide an efficient and flexible bookkeeping system.

From this perspective, we believe that the version presented in Section 2 will prove to be most useful in the long run, for the following reason. The exact Hermitean Rayleigh-Schroedinger result was obtained there by means of three successive power-series expansions, as indicated in (2.6), (2.9), and (2.12). The first of these has an obvious geometric series character, and likewise for the second, since this amounts to expanding ΔE out of all the Brillouin-Wigner energy denominators $(E_o + \Delta E - H_o)^{-1}$. The practical consequence is that these expansions offer a wide variety of opportuntites for partial summation. In contrast, the expansion (2.12) has a far more complex character, but on the other hand we have argued that this difficult step can be avoided by means of the trivial recipe (2.13). The virtue of the above approach is, thus, that it clearly separates these "easy" and "difficult but avoidable" aspects. The other equivalent algorithms are all more opaque because they have these aspects mixed together.

Several of the alternative formalisms deserve some further comment. First, the "unitary transformation" or "canonical transformation" method introduced long ago by Van Vleck[8] and Kemble.[9] This is by far the oldest general formulation of effective Hamiltonian theory; it is well-known and has been applied in all branches of quantum mechanics, from quantum chemistry to elementary particle theory. The basic idea is to carry out a unitary transformation of H, such that $\tilde{H} = U^{-1}HU$ becomes block-diagonal in the P and Q subspaces. The desired effective Hamiltonian is then $\mathcal{H} = P\tilde{H}P$. In practice this becomes a perturbative procedure, since one uses $H = H_o + \lambda V$ and

$$U = e^G = \exp(\lambda G_1 + \lambda^2 G_2 + \ldots) \quad . \tag{3.1}$$

Also, P and Q are defined by H_o, just as in (2.4). One expands $U^{-1}HU$, collects terms of each order in λ, and then applies the block-diagonal condition $P\tilde{H}Q = (Q\tilde{H}P)^\dagger = 0$ (together with $G_n^\dagger = -G_n$) successively to each order, to determine the G_n's. (An indeterminary in this procedure is removed by imposing the Kemble subsidiary conditions $PG_nP = 0$, $QG_nQ = 0$.)

Quite remarkably, the $\mathcal{H}$ from this procedure turns out to be identical to that from (2.12), in every order of perturbation theory. This result has, in fact, been proven in two independent ways, as discussed in Ref. 1. (The first recognition of this important result was by Klein.[10]) This result is important because the conceptual simplicity of the transformation approach is quite appealing, and it has thus been pursued for many-electron systems by a number of quantum chemists.[11] In practice, however, this approach turns out to be exceedingly tedious and opaque, and it is therefore one of the worst methods from the standpoint of infinite partial summations. (In every case, the explicit results in Ref. 11 are restricted to low orders in V.) This is partly for the reason given above, but it is also because the algebraic series for U does not turn out to have an exponential character (see below),[1] in contrast to the assumed representation (3.1). The latter (exponential) disadvantage has been avoided in an ingenious iterative algorithm of Kvasnicka and Holubec,[12] but their method is still quite tedious in practice. Nevertheless, their method must also be fully equivalent to (2.12), because they have incorporated the Jørgensen condition[1] $(PUP)^\dagger = PUP$. The latter plays a role equivalent to the Kemble subsidiary conditions mentioned above.

Readers should be warned that when an alternative version of the Van Vleck method is used, based on a succession of unitary transformations,

$$U' = U_1U_2U_3\ldots = e^{\lambda G_1} e^{\lambda^2 G_2} e^{\lambda^3 G_3} \ldots \quad , \tag{3.2}$$

the result is not equivalent to the other methods discussed above. In fact, $U'-U \sim V^3$ and $\mathcal{H}'-\mathcal{H} \sim V^4$. The "generalized Van Vleck" method of Kirtman (see below) is also different.

For completeness, we should also mention the approach of Hirschfeld and Silverstone and their collaborators.[13] This is notable because its philosophy and formal characteristics are entirely different from all of the methods referred to above. Their basic idea is to develop each of the eigenvalues and eigenfunctions as a strict Taylor series in V, starti from the solution of the first-order effective Hamiltonian, $P(H_o + V)P$. (This assumes that all degeneracy is lifted in first order; otherwise the procedure becomes more complicated.) In other words, they diagonalize first and then perturb. This approach does not lead to an effective Hamiltonian. Furthermore, in a test calculation[14] this procedure gave lower accuracy than when the corresponding effective Hamiltonian was calculated to the same order and then diagonalized. We shall say no more about this approach. There is still another formulation, due to Kirtman, Certain, and Hirschfelder,[15] which falls somewhere in between Van Vleck-Kemble and Hirschfelder-Silverstone. This has recently been cast into diagrammatic form, but it is not fully linked.[16] Its chief virtue is that it satisfies a "2n + 1 rule."

4. RECENT FORMAL DEVELOPMENTS

There are three recent formal developments that are quite interesting and potentially important. These are (a) the coupled-cluster approa for open-shell systems, (b) an alternative diagrammatic representation, and (c) the concept of incomplete model spaces.

The coupled-cluster method originated from the observation of Hubbard[17] and Hugenholtz[18] that the exact (perturbation-theoretic) wavefunction of a closed-shell system can be expressed in the form

$$\Psi = e^{\hat{S}}\Phi_o \quad , \tag{4.1}$$

$$\hat{S} = \sum_{n=1}^{N} \hat{S}_n \quad , \tag{4.2}$$

where N is the total number of particles. Each $\hat{S}_n$ represents the sum of all linked but open perturbation diagrams which lead to the creation of n particle-hole pairs.[4] The caret symbol is a reminder that $\hat{S}_n$ is a second-quantized operator, with a creation or annihilation operator attached at the end of each outgoing particle or hole line. Coester and Kümmel[19] noted that by inserting this form of Ψ into the Schroedinger equation, one obtains a set of coupled, nonlinear, and inhomogeneous equations for the various "cluster" component amplitudes.[4] They suggeste that, after some suitable (physically motivated) truncations, one should be able to solve the resulting equations directly, without resorting to a perturbation expansion. In practice, this provides an elegant and convenient way to carry out a number of the partial summation schemes of diagrammatic perturbation theory. (It does not automatically handle all such schemes, however; this method and perturbation theory complement each other by suggesting somewhat different approximations.[4]) There is also some interesting evidence that some approximations which seem "simpl and "natural" in the context of this approach are well-matched to the quantitative characteristics of correlations in electronic systems.[20]

It was obviously desirable to extend this approach to open-shell systems, and a number of attempts have been made.[21] Although some very sophisticated nuclear structure calculations have been done with this approach,[22] until a few years ago none of the available formulations were clear, concise, and general. A beautifully clean formulation was finally presented by Lindgren,[23] based on two key ingredients: By elementary manipulation of the Schroedinger equation he obtained the operator identity

$$[\Omega, H_0] = V\Omega - \Omega V\Omega \quad , \tag{4.3}$$

where $\Omega = \Omega P$ and $\Psi_\alpha = \Omega|P\Psi_\alpha\rangle$. He also noted that the form (4.1) is inadequate for open-shell systems, and should be replaced by

$$\Omega = \{e^{\hat{S}}\} = \sum_{r=0}^{\infty} \frac{1}{r!}\{\hat{S}^r\} \quad , \tag{4.4}$$

where { } indicates normal-ordering of the various creation and annihilation operators. Failure to do this would lead to many spurious terms,[1] and thereby greatly complicate the formalism. (This point was also recognized by Kümmel and coworkers,[24] but was not clearly stated.)

It is easy to see that systematic use of (4.3) as a recursion formula will generate the Rayleigh-Schroedinger expansion for Ω, and thus for $\mathcal{W} = PV\Omega$, the effective interaction matrix of (2.8)-(2.11). Following this procedure, Lindgren was able to prove by induction that the perturbation-theoretic Ω does indeed have the form (4.4), where each term in $\hat{S}$ is fully connected and is also "open," in the sense of always leading to states in the Q subspace. (It then follows that $\mathcal{W} = PV\Omega$ is fully linked.) In a similar manner, Lindgren then obtained a formal equation for $\hat{S}$ itself, which translates into the desired inhomogeneous and nonlinear set of equations for the cluster amplitudes within $\hat{S}$. This provides an open-shell analog of the coupled-cluster technique. In view of the discussion in Section 2, we argue that the resulting $\mathcal{W}$ may be simply hermitized by means of (2.13). On the other hand, Lindgren states without any detailed justification that the individual matrix elements of $\mathcal{W}$ need not separately conserve the numbers of valence particle and valence hole lines. (See his Figs. 9 and 13.) This feature is very likely wrong, in view of the discussion below about incomplete model spaces.

We now digress for a moment in order to avoid some potential confusion. In commenting on the Van Vleck-Kemble method in Section 3 we remarked that the perturbation expansion of U does not turn out to have an exponential character, whereby the ansatz $U = e^G$ leads to opaque and tedious algebra. This may seem inconsistent with (4.4), but there is really no conflict here. The wave operator Ω is <u>non-unitary</u> [in fact, the connection is $UP = \Omega P(P\Omega^{\dagger}\Omega P)^{-\frac{1}{2}}$], and furthermore $\hat{S}$ is a second-quantized operator. An elementary (non-many-body) quantum system can always be regarded formally as a "one body" system, for which $e^{\hat{S}}$ reduces to 1 + $\hat{S}$. In short, (4.4) is a statement about many-body aspects, and not about the basic algebraic structure.

An alternative to the usual diagrammatic representation[3] for $\mathcal{W}$ has been introduced by Hose and Kaldor.[25] They express the full correlated wavefunction in the form

$$\Psi_\alpha = \sum_{i\varepsilon P} a_{i\alpha}\, \Omega_i \Phi_i \quad , \tag{4.5}$$

where the diagrams of $\Omega_i \Phi_i$ utilize Φ_i as a "vacuum state," in complete analogy to the usual treatment of the closed-shell case, (4.1). They point out that this leads to fewer diagrams, and also to an automatic cancellation of some terms which appear in both "folded" and "ordinary" diagrams of the usual representation. On the other hand, the detailed rules for deciding which orbitals each internal line is to be summed over are more complicated and tedious to implement. The relation of the diagrams to (2.6)-(2.11) is also less obvious, and we therefore suspect that the range of possibilities for partial summation is reduced. (Their claim of superiority from the partial-summation standpoint is groundless, in view of the many sophisticated nuclear applications.[5,6]) For low orders of perturbation theory, however, the choice of diagrams is really only a matter of taste.

Quite recently, Jeziorski and Monkhorst[26] have developed a coupled-cluster analog of the Hose-Kaldor diagrammatic form, using $\Omega_i = e^{\hat{S}(i)}$ within (4.5). Since $\hat{S}(i)$ acts only on Φ_i, its behavior is similar to that of the closed-shell case (4.1), and the explicit normal-ordering of (4.4) is not required. They also employ (4.3), so their analysis has some parallels with that of Lindgren.[23] (Their severe criticism of Ref. 23 is due to a misunderstanding; Lindgren's operational procedure is unambiguous, and does, in fact, determine the desired fully-linked $\hat{S}_n$ amplitudes. The existence of the latter was established by an independent argument, as outlined above.) If and when this method is applied in practice, the relative merits of this new representation for the correlations should become much clearer.

Another novel feature of Hose and Kaldor[25] is their clear specification of the concept of an incomplete model subspace, and their recognition that this leads to a special type of unlinked diagrams. The concept is best explained by example.[25,26] Consider the calculation of a bound but excited ^1S state of the neutral He atom. Here the natural and obvious model subspace consists of "one-particle, one-hole" states with respect to the $1s^2$ ground configuration, in this case the pair of determinants $1s\alpha ns\beta$ and $1s\beta ns\alpha$. On the other hand, the "complete" model subspace, for which the existence of completely linked expansions is rigorous, would also include the $1s^2$ and ns^2 determinants, since this involves all the ways of distributing two electrons among the 1s and ns shells. But here the use of such a complete model subspace is undesirable, for two reasons

(a) it requires calculation of a larger number of $\mathcal{W}$ (or $\mathcal{K}$) matrix elements, and (b) the wide spread of unperturbed model-state energies, may lead to serious convergence problems. (In the present example all ns^2 states lie in the Auger continuum, for n > 1.) Hose and Kaldor emphasize that these motivations for using an incomplete model subspace will arise whenever one considers excited states of a closed-shell system. A further difficulty is that a complete model space for the general case will involve linked many-electron interactions (effective three-electron interactions, etc.).

I thought I had solved this problem back in 1967, by allowing for model spaces P in which each Φ_i has n' valence holes in addition to n valence particles (see pg. 794 of Ref. 3, where the numbers n and n' are required to be the same for every state Φ_i in P; the numbers n, n' are "conserved" by $\mathcal{W}$). But I failed to recognize that a special type of unlinked diagram can arise whenever valence particles and valence holes occur simultaneously within P. Examples are shown in Fig. 1, in my diagrammatic representation, for the simplest case where n' = n = 1. Hose and Kaldor argue that this special class of unlinked diagrams does no harm, because their effect is still size-consistent. Unfortunately, this is not so, as can be seen by considering a system with many molecules which are well separated from each other. As usual, each linked piece can be summed independently over the entire system. (The necessary corrections for cases where separate linked pieces share common orbital indices, at a common level, are represented by additional diagrams; the latter are all fully linked.)

I have now analyzed these diagrams a bit further, and find that all possible relative "time" orders of the separate linked parts (see Fig. 1) can be combined by means of the factorization theorem,[27] giving a net contribution to $\mathcal{W}$ of the form $M_T M_B (E_B^{-1} - E_T^{-1})$, where T and B refer to the top and bottom pairs respectively. For exact degeneracy of the unperturbed particle-hole pair energies, i.e. for $E_T = E_B$, these diagrams cancel identically and may therefore be ignored. Otherwise, in the quasi-degenerate case, their contribution to $\mathcal{W}$ has anti-Hermitean character. But the sum of all such one-particle one-hole linked pieces can be made

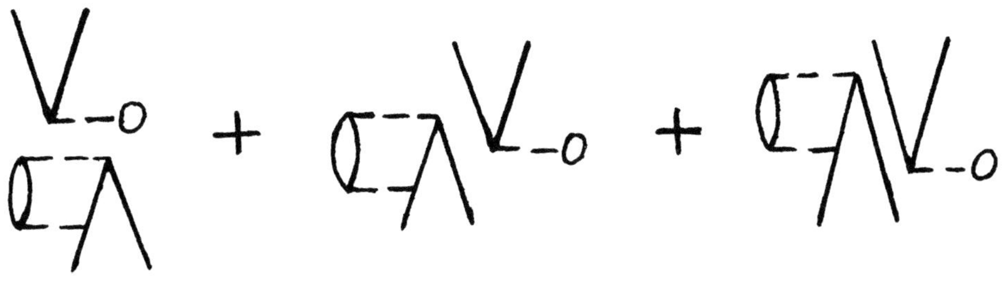

Fig. 1. Example of unlinked diagrams resulting from use of an incomplete model subspace.

to vanish, by using orbitals which satisfy the appropriate generalized Brillouin condition. Unfortunately, however, this does *not* fully resolve the problem, because the foregoing statements apply *only* to the $\mathcal{Y}_o$ terms in the development (2.11) for $\mathcal{W}$. Unlinked contributions continue to contribute to the higher $\mathcal{Y}_n$'s, and likewise to the Θ of (2.12). These residual unlinked terms are undoubtedly small effects, but they are definitely an embarrassment. I regret that at this point I do not see a clean and general formal resolution of this problem. It is important to note, however, that this difficulty arises *only* for excited states whose overall symmetry is the same as the ground state (e.g., excited ^1S states of He); in all other cases these unlinked terms must vanish identically, at least after forming the linear combinations required by symmetry. This is reassuring, because the present example of "particle-hole excitations leads to the well-known RPA approximation (as demonstrated in detail on pg. 806 of Ref. 3) and its higher-order generalizations.[28] In macroscopic systems, for which the RPA has had many successful applications, the present unlinked terms are all eliminated by momentum conservation.

5. SOME APPLICATIONS

The first quantum-chemical applications of the open-shell linked cluster expansion were by Kaldor.[29] These papers are quite pedagogical, and provide a nice introduction for newcomers to this field.

One of my favorite types of applications is to the problem of effective π-electron Hamiltonians.[1] For the relatively simple case of ethylene, with only two π electrons, the present formalism provides rigorous prescriptions for the α, β, and γ parameters, as well as for the "σ core" energy. This is significant, because some perturbation calculations[30] obtained resonable *total* correlation energies, for the relevant low-lying eigenstates of ethylene, but failed to properly sort out which correlatio contributions should be assigned to which "empirical" parameters. They thereby failed to provide *direct ab initio* calculations of these parameters (as contrasted to phenomenological fitting of the total eigenenergies). This was due to not using a fully-linked formalism, and/or not using localized instead of molecular π orbitals. Robb and coworkers have recently presented an ethylene calculation which does have these essentia features.[31] Another noteworthy feature is that the so-called "intruder state" problem[4] was encountered here, and a practical method of dealing with this was developed.[32]

Freed and coworkers have done a series of very detailed calculations on the effective valence-shell interactions for light atoms,[33] for a 3d transition metal,[34] and for the hydrides CH, NH, and OH.[35,36] (The latte calculations[35,36] are directed towards a better understanding of semi-empirical theories such as MINDO.) These papers all emphasize the very important role played by effective *many-electron* interaction terms (linke contributions involving *more than two* valence electrons); the latter are absolutely essential for obtaining reasonable agreement with experiment. There is, however, a reward for calculating these additional "nonclassical" terms--the same set of ab initio interaction parameters provides good ground and low excited energies for a number of different ionization states of the same atom (all the way from F^- to F^{+7}, for example). This is just what the formalism says should happen, but the explicit demonstration is a remarkable tour-de-force. In contrast, semi-empirical schemes generally involve only one and two-body effective interaction terms. This means that the latter parameters must be implicitly including effects of the "true" effective many-electron ("nonclassical") interactions. A number of the above papers[31,33-36] have examined the "renormalizing" role of these nonclassical terms. This is a murky but very important subject that will surely require much further work.

Potential curves of both ground and excited states have been calculated.[36,37]

A degenerate application of Lindgren's coupled-cluster approach is described in Ref. 38.

Finally, we wish to comment on applications to ionization potentials and electron affinities, where there is effectively only one valence particle (or valence hole). Here the core-valence-separation aspect is crucial, while degeneracy plays only a very minor role. Several published calculations[39] have used the degenerate linked-cluster expansion, arguing that the "energy independent" aspect of this formalism is an advantage as compared to the energy-dependent self-energy operator $\Sigma(E)$ provided by the standard Green's function theory. We disagree. We have shown in detail[40] that the standard Green's function prescription is equivalent to a partial summation of higher-order terms within our Rayleigh-Schroedinger expansion. Therefore, we consider use of the "Dyson equation" to be a more efficient method for accurate calculation of ionization potentials and electron affinities. (This method can, of course, be implemented either perturbatively or non-perturbatively.) Our type of expansion becomes advantageous only when there is more than one valence particle or hole, where the need for a fully-linked formalism takes precedence.

ACKNOWLEDGMENTS

In preparing this review I have benefitted from discussions with a number of people, especially R. Bartlett, G. Hose, V. Hubac, H. Monkhorst, J. Oddershede, M. Sheppard, and H. Silverstone.

REFERENCES

1. B. H. Brandow, Int. J. Quantum Chem. 15, 207 (1979).
2. P.-O. Löwdin, J. Chem. Phys. 19, 1396 (1951); Int. J. Quantum. Chem. 2, 867 (1968) and references therein; see also M. Lax, Phys. Rev. 79, 200A (1950).
3. B. H. Brandow, Rev. Mod. Phys. 39, 771 (1967).
4. B. H. Brandow, Adv. Quantum Chem. 10, 187 (1977).
5. P. J. Ellis and E. Osnes, Rev. Mod. Phys. 49, 777 (1977).

6. B. R. Barrett and M. W. Kirson, Adv. Nuc. Phys. 6, 219 (1973); T. T. S. Kuo, Ann. Rev. Nucl. Sci. 24, 101 (1974); "Proceedings of the International Conference on Effective Interactions and Operators in Nuclei," B. R. Barrett, ed., Lecture Notes in Physics, Vol. 40 (Springer-Verlag, Berlin, 1975); J. Shurpin, D. Strottman, T. T. S. Kuo, M. Conze, and P. Manakos, Phys. Lett. 69B, 395 (1977); see also Ref. 22.
7. B. H. Brandow, in Barrett (1975), Ref. 6.
8. J. H. Van Vleck, Phys. Rev. 33, 467 (1929); O. M. Jordahl, Phys. Rev. 45, 87 (1934).
9. E. C. Kemble, "The Fundamental Principles of Quantum Mechanics" (McGraw-Hill, New York, 1937), p. 394.
10. D. J. Klein, J. Chem. Phys. 61, 786 (1974).
11. H. Primas, Rev. Mod. Phys. 35, 710 (1963); R. A. Harris, J. Chem. Phys. 47, 3967 & 3972 (1967) and 48, 3600 (1968); S. Fischer, Int. J. Quantum Chem. Symp. 3, 651 (1970); P. Westhaus, ibid. 7, 463 (1973); P. Westhaus, E. G. Bradford, and D. Hall, J. Chem. Phys. 62, 1607 (1975); S. Iwata and K. F. Freed, ibid. 65, 1071 (1976); I. Shavitt and L. T. Redmon, ibid. 73, 5711 (1980); L. T. Redmon and R. J. Bartlett, ibid. 76, 1938 (1982); see also Ref. 10;
12. V. Kvasnicka and A. Holubec, Chem. Phys. Lett. 32, 489 (1975).
13. J. O. Hirschfelder, Int. J. Quantum Chem. 3, 731 (1969); H. J. Silverstone, J. Chem. Phys. 54, 2325 (1971); J. O. Hirschfelder and P. R. Certain, ibid. 60, 1118 (1974); H. J. Silverstone and R. K. Moats, Phys. Rev. A 23, 1645 (1981).
14. Z. Gershgorn and I. Shavitt, Int. J. Quantum Chem. 2, 751 (1968).
15. B. Kirtman, J. Chem. Phys. 49, 3890 and 3895 (1968); P. R. Certain and J. O. Hirschfelder, ibid. 52, 5977 and 53, 2992 (1970); J. O. Hirschfelder, Chem. Phys. Lett. 54, 1 (1978); also Shavitt and Redmon, Ref. 11.
16. B. Kirtman, J. Chem. Phys. 75,798 (1981); also comments in Redmon and Bartlett, Ref. 10.
17. J. Hubbard, Proc. R. Soc. (London) A240, 539 (1957).
18. N. M. Hugenholtz, Physica 23, 481 (1957).
19. F. Coester, Nucl. Phys. 7, 421 (1958); F. Coester and H. Kümmel, Nucl. Phys. 17, 477 (1960).
20. J. Paldus, J. Cizek, and I. Shavitt, Phys. Rev. A 5, 50 (1972); D. L. Freeman, Phys. Rev. B 15, 5512 (1977).
21. The first such attempt was by F. Coester, in "Lectures in Theoretica Physics," Vol. 11B, K. T. Mahanthappa and W. E. Brittin, eds. (Gordc and Breach, New York, 1969), p. 157. See also Ref. 22, and further references in Ref. 24.
22. K. Emrich, J. G. Zabolitzky, and K. H. Lührmann, Phys. Rev. C 16, 1650 (1977); J. Zabolitzky and W. Ey, Nucl. Phys. A238, 507 (1979).
23. I. Lindgren, Int. J. Quantum Chem. Symp. 12, 33 (1978).
24. R. Offerman, W. Ey, and H. Kümmel, Nucl. Phys. A273, 349 (1976); R. Offerman, Nucl. Phys. A273, 368 (1976); W. Ey, Nucl. Phys. A296, 189 (1978).
25. G. Hose and U. Kaldor, J. Phys. B 12, 3827 (1979); Phys. Scripta 21, 357 (1980); J. Chem. Phys. 62, 469 (1981).

26. B. Jeziorski and H. Monkhorst, Phys. Rev. A 24, 1668 (1981).
27. L. M. Frantz and R. L. Mills, Nucl. Phys. 15, 16 (1960).
28. The diagrams included in various RPA extensions have been carefully analyzed by J. Oddershede and P. Jørgensen, J. Chem. Phys. 66, 1541 (1977); see also E. S. Nielsen, P. Jørgensen, and J. Oddershede, ibid. 73, 6238 (1980).
29. U. Kaldor, Phys. Rev. Lett. 31, 1338 (1973); J. Chem. Phys. 62, 4634 (1975); ibid. 63, 2199 (1975); P. S. Stern and U. Kaldor, ibid. 64, 2002 (1976); U. Kaldor, J. Comput. Phys. 20, 432 (1976).
30. K. Tanaka, Int. J. Quantum Chem. 6, 1087 (1972); S. Iwata and K. F. Freed, J. Chem. Phys. 61, 1500 (1974) and Chem. Phys. Lett. 28, 176 (1974); P. Westhaus and M. Moghtaderi, J. Chem. Phys. 72, 4174 (1980).
31. H. Baker, D. Hegarty, and M. A. Robb, Mol. Phys. 41, 653 (1980).
32. D. Hegarty and M. A. Robb, Mol. Phys. 37, 1455 (1979); H. Baker, M. A. Robb, and Z. Slattery, ibid. 44, 1035 (1981). (Compare with Shurpin et al., Ref. 6.)
33. D. L. Yeager, H. Sun, K. F. Freed, and M. F. Herman, Chem. Phys. Lett. 57, 490 (1978); H. Sun, K. F. Freed, J. Chem. Phys. 72, 4158 (1980); M. G. Sheppard and K. F. Freed, ibid. 75, 4525 (1981) and Int. J. Quantum Chem. Symp. 15, 21 (1981) and Chem. Phys. Lett. 82, 235 (1981).
34. Y. S. Lee, H. Sun, M. G. Sheppard, and K. F. Freed, J. Chem. Phys. 73, 1472 (1980).
35. K. F. Freed and H. Sun, Israel J. Chem. 19, 99 (1980); H. Sun, M. G. Sheppard, K. F. Freed, and M. F. Herman, Chem. Phys. Lett. 77, 555 (1981); S. Iwata and K. F. Freed, ibid. 38, 425 (1976).
36. H. Sun and K. F. Freed, Chem. Phys. Lett. 78, 531 (1981).
37. D. W. Davies and G. J. R. Jones, Chem. Phys. Lett. 81, 279 (1981).
38. S. Salomonson, I. Lindgren, and A.-M. Martensson, Phys. Scripta 21, 351 (1980).
39. I. Hubac, V. Kvasnicka, and A.Holubec, Chem. Phys. Lett. 23, 381 (1973); I. Hubac and M. Urban, Theoret. Chim. Acta 45, 185 (1977); S. Prime and M. A. Robb, Chem. Phys. Lett. 47, 527 (1977); S. Yamamoto and A. Saika, ibid. 78, 316 (1981).
40. B. H. Brandow, Ann. Phys. (NY) 64, 21 (1971); see Appendix C.

COUPLED CLUSTER APPROACHES TO MANY-ELECTRON CORRELATION PROBLEM

Josef Paldus
Department of Applied Mathematics, Department of Chemistry and Guelph-Waterloo Center for Graduate Work in Chemistry, Waterloo Campus, University of Waterloo, Waterloo, Ontario, Canada
N2L 3G1

ABSTRACT. The origins, basic ideas and methodological developments of the coupled cluster approach (CCA) to the atomic and molecular many-electron correlation problem are briefly outlined. A rudimentary derivation of basic concepts of the CCA is attempted exploiting only very general algebraic concepts and the idea of an effective Hamiltonian. The relationship with other approaches to the correlation problem is pointed out. An extension of the CCA to quasi-degenerate cases is briefly contemplated and a generalization of the Davidson correction to a multi-reference case is suggested.

1. INTRODUCTORY REMARKS

During the past decade there has been a remarkable development in approaches to the atomic and molecular electron correlation problem. While the variational methods, namely the configuration interaction (CI) and the multi-configuration self-consistent field (MC-SCF) approaches, are today most often exploited in applications, the developments in methods based on the cluster ansatz for an exact wavefunction, and in their implementation at the ab-initio level, have been equally impressive. We cannot possibly do justice to all these developments in the space available to us here. Our choice of topics is naturally biased by our own interests and we would like to apologize beforehand for any omissions or incomplete coverage. In particular, we shall pay little attention to various approximate coupled pair approaches which received excellent coverage at the previous Congress by Meyer [1] and which were so successfully exploited and reviewed by Ahlrichs [2], Hurley [3], Kutzelnigg [4], Meyer [5,6] and others [7].

We also note that a number of reviews of various aspects of the coupled cluster approach (CCA) are currently available [8-13] and that the subject is also covered in several monographs [3,14]. We shall thus restrict ourselves to a brief sketch of methodologically important stages in the development of the CCA. We shall then present a very simple elucidation of the basic structure of this approach, which will enable

P.-O. Löwdin and B. Pullman (eds.), New Horizons of Quantum Chemistry, 31–60.

us to better understand its relationship with other approaches to the correlation problem, in particular with the variational CI approaches and the partitioning technique of Löwdin [15,16]. This comparison will not only show certain approaches in a new light, but it may also enable an exploitation of various efficient techniques, which are currently used in large scale CI methods, such as the unitary group approach [17-22], in the CCA.

2. ORIGINS OF THE CCA

With the advances in the development of various independent particle models (IPM's), particularly of the analytic SCF approach [23], the key importance of the many-electron correlation problem for a correct description of various physical properties and chemical phenomena became apparent. Similar stimuli came from nuclear and solid state problems. This renewed interest in the many-fermion correlation problem resulted in a number of key developments during the late fifties and early sixties

In the many-electron case, the main impetus was provided by the work of Löwdin, who formulated the density matrix theory [24], clearly delineated the correlation problem, elucidated and introduced different variants of the IPM [25], and laid the ground work for modern perturbatio theory [15,16,26] together with a very ingenious methodology for obtainin upper and lower bounds of eigenvalues of Hermitean operators [26,27]. In our opinion the latter techniques have preceded their time and will have a very important role to play in future developments.

Another methodological advance was provided by the mathematical techniques of quantum field theory, which led to a proof of the cancellation of unlinked terms, discovered by Brueckner [28], in all orders of perturbation theory [29-31]. These techniques also permitted the summation of certain many-body perturbation theory (MBPT) contributio to infinite order. Thus, by summing the most important ring diagram contributions, Gell-Mann and Brueckner [32] showed the existence of the logarithmic singularity in the electron gas correlation energy at high densities, while Brueckner's t-matrix theory [33] was shown [29] to originate in the sum over the ladder diagrams. These developments were later extended by Bethe [33], Bloch [34], Brandow [35] and others [36-38] Soon afterwards, the MBPT was exploited by Kelly [39] in the study of the atomic correlation problem.

The idea of the cluster expansion, characterized by the exponential ansatz in terms of pair or higher order potentials, was first introduced in classical statistical mechanics by Ursell and Mayer [40]. The quantum analogue was later developed by Kahn and Uhlenbeck [41] in quantum statistical mechanics. The usefulness of a similar expansion in the many-fermion correlation problem first became apparent from the work of Hubbard [30] who showed that a non-degenerate ground state wavefunction can be expressed as an exponential of connected diagram contributions. The idea to exploit this form of the wavefunction in lieu of the usual linear expansion, which is characteristic of the Ritz variational approach [42], was first suggested in the nuclear many-body problem by Coster and Kümmel [43]. Almost at the same time, this idea was independently exploited in the atomic and molecular correlation problem by Sinanoğlu [44]. Similar ideas were also pursued by Nesbet [45]. In both of these approaches the most important pair cluster contributions were treated as non-interacting.

One of the main attractions of the MBPT, as well as of the coupled cluster ansatz, is the proper dependence of the energy on the particle number, which is absolutely crucial in studies of extended systems, such as the nuclear matter or an electron gas. The desirability of the cluster ansatz for the exact state $|\Psi_0\rangle$,

$$|\Psi_0\rangle = \exp(T)|\Phi_0\rangle, \quad \langle\Psi_0|\Phi_0\rangle = 1, \tag{1}$$

where $|\Phi_0\rangle$ describes the same state in the IPM, in studies of various dissociative or reactive chemical phenomena, was clearly pointed out by Primas [46], who observed that this ansatz provides a convenient mapping of the multiplicative wave function structure into the addive cluster operator structure. Thus, designating the ground states of the two non-interacting species as $|\Psi_A\rangle$ and $|\Psi_B\rangle$, we have

$$|\Psi_{A+B}\rangle = |\Psi_A\rangle|\Psi_B\rangle = \exp(T_A)\exp(T_B)|\Phi_A\rangle|\Phi_B\rangle = \exp(T_A+T_B)|\Phi_{A+B}\rangle, \tag{2}$$

assuming that $[T_A, T_B]_- = 0$. Since the energy depends linearly on the cluster components, we can write

$$E_{A+B} = E(T_A+T_B) = E(T_A) + E(T_B) = E_A + E_B. \tag{3}$$

This particular feature was the main reason for the recent resurgence of the CCA's and is now referred to as the size extensive (or size consistent, cf. [13]) property of the theory.

In the middle of the sixties, the desirability of the cluster ansatz for the wavefunction was well recognized. While the general system of equations for the cluster components was proposed by Coster and Kümmel [43], and an approximate uncoupled treatment of pair clusters was given by Sinanoğlu [44], no explicit general equations for the cluster components were available. This was first accomplished by Čížek [47] who formulated a general time-independent diagrammatic technique which enabled an easy derivation of exact explicit equations for the cluster components to the desired order. A similar technique was later developed in the nuclear CCA [48]. Using this technique, one represents the Hamiltonian H by the usual Goldstone [29] or Hugenholtz [31] diagrams and, similarly for the i-particle cluster components T_i,

$$T = \sum_i T_i. \tag{4}$$

Substituting the cluster ansatz (1) into the Schrödinger equation

$$H_N|\Psi_0\rangle = \Delta E_0|\Psi_0\rangle, \tag{5}$$

which we write for convenience in the N-product form (cf., [49])

$$H_N \equiv H-\langle H\rangle, \; \Delta E_0 = E_0-\langle H\rangle, \; \langle H\rangle \equiv \langle\Phi_0|H|\Phi_0\rangle, \tag{6}$$

we can easily eliminate the disconnected cluster contributions obtaining

$$[H_N\exp(T)]_c|\Phi_0\rangle = \Delta E_0|\Phi_0\rangle, \tag{7}$$

where the subscript c designates the connected part and $|\Phi_0\rangle$ is assumed to represent the HF state. Projecting onto $|\Phi_0\rangle$ we get the correlation energy

$$\Delta E_0 = \langle[H_N\exp(T)]_c\rangle, \tag{8}$$

while projecting onto the complementary subspace spanned by excited configurations $|\Phi^*\rangle$ we get the desired equations for the cluster components T_i

$$\langle\Phi^*|[H_N\exp(T)]_c|\Phi_0\rangle = 0. \tag{9}$$

The explicit form of these equations in terms of the basic integrals in the chosen one-electron orbital basis is most easily obtained using the diagrammatic technique [47,49]. Čížek has given the explicit form of these equations for the most important case of pair clusters $T \simeq T_2$, which he referred to as the coupled pair many electron theory (CPMET) [47]. For the closed shell spin-independent H he has also given the spin-adapted form of these equations using a simple rule assigning the factor of two to each closed loop of oriented lines [47].

In contrast with the linear variational approaches which invariably lead to an eigvalue problem, the CCA requires the solution of an energy independent, non-linear algebraic system of the type

$$a_i + \sum_j b_{ij}t_j + \sum_{j\le k} c_{ijk}t_jt_k + \dots = 0, \tag{10}$$

where t_i designate the unknown cluster components and the coefficients a_i, b_{ij}, etc. are given through the one- and two-electron integrals defining the model Hamiltonian used. In the case of the CPMET, t_i represent pair cluster components and Eq. (10) contains at most the bilinear term. When appropriate normalization is chosen, the correlation energy is given simply as

$$\Delta E_0 = \sum_i a_i t_i. \tag{11}$$

By a straightforward iteration of Eqs. (10) we can show which MBPT diagrams are accounted for in this approach. It is easily found that both the ring and the ladder type diagrams, as well as their mixture, will occur. Thus, the CCA equations may be regarded as recursive formulas for the simultaneous generation and evaluation of the most important MBPT diagrams to infinite order. When only linear terms are considered in the CPMET approach (the so called L-CPMET), so that only biexcited configurations will intervene, the solution of the resulting linear system (10) is equivalent to the infinite order perturbation theory with intermediate biexcited configurations, so called D-MBPT(∞) [50]. We shall later see this and other relationships in more detail. We now only mention that the CCA equations may be also derived in a number of other ways, the most elementary being a straightforward derivation starting from the CI formulation [51]. This derivation shows very clearly the basic requirement for the validity of the CPMET, namely the negligibility of the connected vs. the disconnected tetraexcited contributions,

$$T_4 \ll \tfrac{1}{2}T_2^2, \tag{12}$$

which was first stated by Sinanoğlu [44] for non-metallic systems. It was later found that to a good approximation this inequality also holds for small metallic-like systems [52]. However, it will likely break down for large metallic-like systems [53].

3. FURTHER DEVELOPMENTS

3.1 Extended CPMET Approaches

The pair clusters and their products give the most important contribution to the correlation energy of non-degenerate closed-shell ground states, at least for small and medium sized systems. However, the singly and triply excited clusters may also give non-negligible contributions (the former will vanish exactly if maximum overlap or Brueckner orbitals are used). Explicit equations for such an extended CPMET (E-CPMET) in which T_1, T_2, and T_3 clusters are considered have been given and tested on the minimum basis set BH_3 model [54].

In agreement with a simple PT argument it was shown [54] that in contrast to the tetra-excited clusters, Eq. (12), the main contribution from the triexcited clusters is due to the connected component T_3. An overall agreement with the corresponding CI result in the case of the BH_3 model is indicated graphically in Fig. 1.

More recently, the triexcited clusters have been incorporated into the efficient coupled cluster routines based on spin-orbital formulation [55] (see Sec. 3.5). Very recently, the role of triexcited clusters was examined by Kvasnička [56] for several diatomics and BeH_2 molecule. In addition to the diagrams considered in Ref. 54, he also considered the interaction terms yielding the triexcited clusters from the $\frac{1}{2}T_2^2$ clusters. However, these terms contribute for the first time in the 5th order to the energy in contrast to the 4th order contribution from T_3 and $\frac{1}{2}T_2^2$. Indeed, his results indicate [56] that these rather complex terms are negligible.

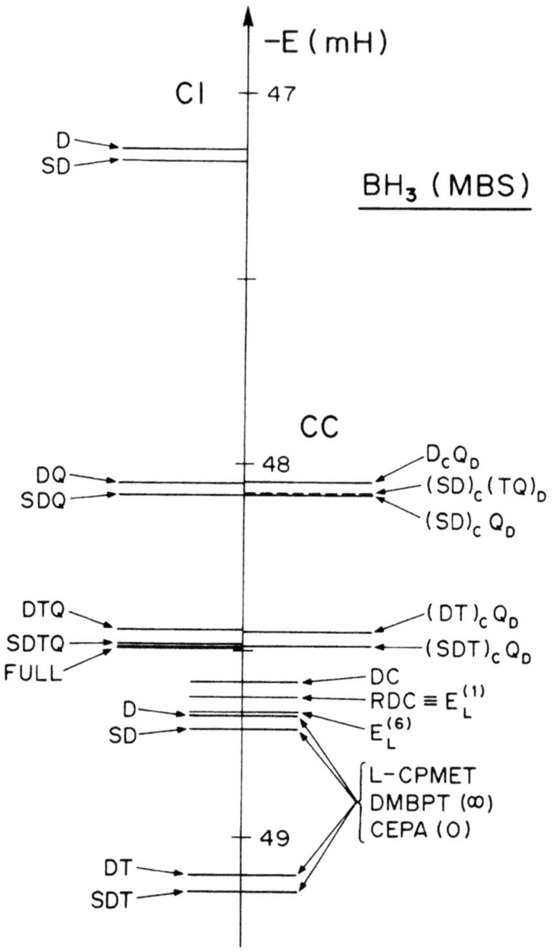

Figure 1. A comparison of CI and coupled cluster (CC) ground state correlation energies for the minimum basis set (MBS) model of the BH_3 molecule [54]. The singly (S), doubly (D), triply (T) and quadruply (Q) excited configurations or connected (subscript C) and disconnected (subscript D) clusters included in the calculation are shown for each level. Corresponding L-CPMET results are also shown, together with the Davidson (DC) and renormalized Davidson (RDC) corrections as well as the value given by Eq. (68) for k = 6.

Very recently, Purvis and Bartlett have presented and programmed the CCA equations involving all possible T_1 clusters (including $T_1^4/4!$)

in order to avoid the restriction to HF orbitals [57].

3.2 Symmetry Adapted Cluster Approaches

Assuming for simplicity that $T_1 = 0$, the biexcited component of the exact wavefunction $|\Psi_0\rangle$ can be written

$$T_2|\Phi_0\rangle = \sum_\mu t_\mu^{(2)}|\Phi_\mu^{(2)}\rangle = \sum_\mu c_\mu^{(2)}|\Phi_\mu^{(2)}\rangle, \tag{13}$$

where $t_\mu^{(2)}$ are pair cluster components and $c_\mu^{(2)}$ biexcited CI coefficients associated with biexcited configurations $|\Phi_\mu^{(2)}\rangle$, assuming intermediate normalization for $|\Psi_0\rangle$.

It can be shown that Čížek's spin-adapted closed shell CPMET formalism, obtained by assigning the factor of 2 to each closed loop of oriented lines in Goldstone diagrams [47], corresponds to the use of (unnormalized) spin bonded functions in the expansion (13). A much more complicated situation arises when triexcited clusters are considered since a straightforward spin adaptation leads not only to non-orthogonal but in fact to a linearly dependent N-electron set [54].

In order to obtain the spin-adapted formalism related to an orthonormal N-electron basis it is convenient to exploit the graphical methods of spin algebras [58,59]. Using this formalism we have formulated orthogonally spin-adapted version of both the CPMET [60] and ECPMET formalisms [61]. This formalism contains the minimal number of cluster components, is simply related to the spin-adapted CI expansion and, most importantly, yields considerably sparser matrices in actual computations.

The same technique can also be exploited to obtain LS adapted (E)CPMET formalism for atomic systems [62], which we have exploited in the study of Be atom [63,64]. The orthogonally spin-adapted version of CPMET [60] was cast into the SCEP form [6] by Chiles and Dykstra [65], who exploited this formalism in several interesting applications. The idea of the spin adapted cluster formalism, although from a different viewpoint, has been also extensively studied by Hirao and Nakatsuji [66,67] (see Sec. 6).

3.3 Approximate Approaches

As we have indicated in Sec. 1, we shall not examine the various CEPA-type approaches which result when only EPV (exclusion principle violating) diagrams of certain type are considered in the CPMET approach [cf., 2-4]. It is worthwhile to mention, however, that even when all the diagrams associated with the T_2^2 clusters are considered, a substantial approximate cancellation occurs [63,64,68,69]. In fact, the contribution of the first three diagrams in the orthogonally spin-adapted version (cf. Fig. 3(c) of Ref. 60) almost exactly cancels so that practically all the effect of pair coupling is due to the remaining two diagrams. We have studied very carefully the additivity and cancellations of individual diagrams for the Be atom [63] and the overall effect for the H_4 model systems [68]. A similar observation has been made by Chiles and Dykstra [70] who employed much larger bases and exploited this approximate CPMET version in a number of computations [71].

3.4 Properties

In principle, once the cluster components have been determined, one can use the cluster wavefunction to calculate any property as a mean value of its operator. However, in this case, the expansion for the mean value expression does not terminate and a truncation at a certain power of the cluster components must be applied. Such a procedure, though not very satisfactory, has been outlined in the second Ref. 47. It was later observed by Monkhorst [72] that a computationally much more satisfactory procedure can be obtained by exploiting the Hellmann-Feynman theorem. With this approach the property calculation requires the solution of a non-homogeneous system of linear equations, whose coefficient matrix is the same for any property calculated and only its right hand side is property dependent. The same procedure can also be used to calculate the time-dependent properties such as the dynamic polarizabilities, in which case the diagonal of the coefficient matrix contains the frequency of an incident field. For the frequency values which correspond to excitation energies of the system the coefficient matrix becomes singular. These singularities and thus excitation energies are easily determined by diagonalizing the coefficient matrix.

We have recently examined [73] a diagrammatic formulation of this approach which revealed its close relationship to both the perturbative Green function approach [74] and to our earlier generalization of the coupled cluster approach to excited and ionized states of closed shell systems [75], (see also Sec. 6) which can be regarded as a hermitized version of Monkhorst's [72] approach. We have shown that considering the singly and doubly excited cluster components for the examined state, in addition to the ground state pair clusters which are assumed to be known from the ground state calculation, will take into account all the third order RPA-type diagrams and practically all other third order Green-function approach diagrams [74] except certain self energy terms involving intermediate triply excited configurations. Unfortunately, Monkhorst's approach [72] has never been applied to even model systems and it would be highly desirable to carry out such a test. Our diagrammatic reformulation [73] of this approach should prove useful since it will provide a spin-adapted version of this formalism, which is particularly desirable if one is interested in excitation energies.

It is remarkable that in spite of the great diversity in basic formulations, there is a close proximity between Monkhorst's [72] and our CC approaches [75] to excitation energies. The advantage of Monkhorst's version is its greater simplicity which is achieved by dropping the requirement of hermiticity of the effective Hamiltonian matrix. Its further advantage is that the same coefficient matrix can be used for other property calculations. On the other hand, while

Monkhorst's approach is restricted to systems with a fixed particle number, our approach is equally applicable to ionized states thus enabling us to obtain various attachment and detachment energies. Both approaches would deserve more testing on actual systems.

3.5 Computational Algorithms and Applications

Efficient programs designed for large scale ab-initio applications and based on spin-nonadapted spin-orbital version of the CPMET, which enable to use the UHF orbitals, have been developed by Battelle's [12,13,76] and Pople's [77] groups. These algorithms and their numerous applications have been thoroughly reviewed by Bartlett [13]. A number of ab-initio applications have been also carried out in Australia [78].

Purvis and Bartlett presented recently a new algorithm for the solution of large algebraic systems encountered in the CCA, called the RLE (reduced linear equation) method [79]. From a conceptual viewpoint, this method is very similar to that proposed by Pople et al. [80], but leads to a much faster convergence. We have shown recently [81] that both methods are in fact different versions of the conjugate gradient method [82], which we have shown to be very appropriate for the CCA [83,84]. The faster convergence of the RLE method is due to the fact that it corresponds to the preconditioned conjugate gradient methods. This preconditioning is essential for the fast convergence if the studied system has a large spectral radius, as is the case with CPMET systems.

4. RUDIMENTS OF THE CCA

We shall present a very elementary motivation and derivation of the coupled cluster formalism. We will not aim to obtain detailed explicit equations at the one-electron level, but only general equations at the N-electron level. An obvious shortcoming of such a formulation is that the resulting expressions may contain disconnected cluster contributions, which would otherwise cancel out at the (spin) orbital level. On the other hand this formulation shows the motivation for the cluster ansatz and its relationship with other approaches in a new light.

We shall exploit the concept of an effective or model Hamiltonian H_{eff}, which is defined on a small model subspace V_0 of the actual N-particle space V and whose eigenvalues are the same as some of the low-lying eigenvalues of the exact Hamiltonian H in V. This idea originated in perturbative approaches and is also closely related with the underlying ideas of the partitioning or inner projection techniques of Löwdin [15,16,26]. The model space is usually spanned by the independent-particle model configurations involving a few important orbitals, whose energies lie in the vicinity of the Fermi level, in addition to the permanently occupied core orbitals. In most practical applications, the N-electron space V is given by the spin- and space-symmetry adapted antisymmetric component of the tensor produce space $v^{\otimes N}$, where v is spanned by a chosen AO basis.

Let $\{|\Phi_\mu>\}$ be some othonormal basis for V, which may or may not be spin and/or space symmetry adapted. In this basis, the Schrödinger equation (5) can be written as a matrix equation

$$\underset{\sim}{H}\,\underset{\sim}{C} = \underset{\sim}{C}\,\underset{\sim}{E}, \tag{14}$$

where

$$H_{\mu\nu} = <\Phi_\mu|H|\Phi_\nu>,\ C_{\mu i} = <\Phi_\mu|\Psi_i>, \tag{15}$$

and

$$\underset{\sim}{H} = ||H_{\mu\nu}||,\ \underset{\sim}{C} = ||C_{\mu i}||,\ \underset{\sim}{E} = ||E_i\delta_{ij}||. \tag{16}$$

Consider next a non-singular similarity transformation S (represented by $\underset{\sim}{S}$ in the same basis) which transforms problem (14) into an equivalent problem

$$\underset{\sim}{H}'\,\underset{\sim}{C}' = \underset{\sim}{C}'\,\underset{\sim}{E}, \tag{17}$$

where

$$\underset{\sim}{H}' = \underset{\sim}{S}^{-1}\,\underset{\sim}{H}\,\underset{\sim}{S},\ \underset{\sim}{C}' = \underset{\sim}{S}^{-1}\,\underset{\sim}{C}. \tag{18}$$

As is well known, the transformed Hamiltonian $\underset{\sim}{H}'$ is again Hermitean if and only if the transformation S is unitary, i.e., when $\underset{\sim}{S}^\dagger\underset{\sim}{S} = \underset{\sim}{S}\underset{\sim}{S}^\dagger = \underset{\sim}{1}$. However, the eigenvalues of $\underset{\sim}{H}'$ are identical with those of $\underset{\sim}{H}$ for an arbitrary non-singular $\underset{\sim}{S}$, i.e., spec($\underset{\sim}{H}$) = spec($\underset{\sim}{H}'$), where spec($\underset{\sim}{A}$) designates the spectrum of $\underset{\sim}{A}$. In order to keep our formalism as general as possible, we shall not require $\underset{\sim}{S}$ to be unitary, thus admitting non-Hermitean operators.

The identity of the spectra of $\underset{\sim}{H}$ and $\underset{\sim}{H}'$ would seem to indicate that no danger is involved in admitting a non-Hermitean $\underset{\sim}{H}'$. We shall see late however, that in actual applications a number of approximations must be introduced (e.g., the similarity transformation (18) is considered only to certain order), so that some eigenvalues of $\underset{\sim}{H}'$ could become complex. A similar problem is faced in a number of open-shell perturbation theorie in the method of complex scaling, etc. In such cases one should in fact simultaneously consider both $\underset{\sim}{H}'$ and $\underset{\sim}{H}'^\dagger$. Very interesting studies of this difficult problem were carried out at the University of Florida Quantum Theory Project [85].

In order to obtain the required model Hamiltonian $\underset{\sim}{H}_{eff} \equiv \underset{\sim}{H}'_0$, we shall search for such a transformation $\underset{\sim}{S}$ which gives $\underset{\sim}{H}'$ in the block forr

$$\underset{\sim}{H}' = \begin{Vmatrix} \underset{\sim}{H}'_0 & \underset{\sim}{D} \\ \underset{\sim}{0} & \underset{\sim}{H}'_1 \end{Vmatrix}, \tag{19}$$

where $\underset{\sim}{0}$ is the zero matrix and $\underset{\sim}{D}$ is arbitrary. Since the union of the spectra of $\underset{\sim}{H}'_0$ and $\underset{\sim}{H}'_1$, spec$(\underset{\sim}{H}'_0) \cup$ spec$(\underset{\sim}{H}'_1)$, is the same as spec$(\underset{\sim}{H})$, we shall further assume that $\underset{\sim}{S}$ is such that the low lying eigenvalues of interest are contained in spec$(\underset{\sim}{H}'_0)$. Then, instead of solving a large dimensional problem (14) we only have to solve a small dimensional problem

$$\underset{\sim}{H}'_0 \, \underset{\sim}{C}'_0 = \underset{\sim}{C}'_0 \, \underset{\sim}{E}_0 . \tag{20}$$

The eigenvalues E_{0i} of (20) are also eigenvalues of $\underset{\sim}{H}'$ with eigenvectors

$$\underset{\sim}{C}'_{(0)} \equiv \left\| \begin{matrix} \underset{\sim}{C}'_0 \\ \underset{\sim}{0} \end{matrix} \right\| . \tag{21}$$

We can thus visualize the transformation $\underset{\sim}{S}$ to "fold" the "content" of $\underset{\sim}{H}$ from the outside of the model space into the model space V_0, at least as far as a few low lying eigenvalues are concerned. We note that we can also introduce the orthogonal projectors P and Q = 1 - P onto V_0 and its orthogonal complement, respectively, and write

$$H'_0 = P\,H'\,P, \quad Q\,H'\,P = 0. \tag{22}$$

Let us next turn our attention to the desired form for $\underset{\sim}{S}$. The basic requirement is the regularity of $\underset{\sim}{S}$. The set of all non-singular linear transformations $\underset{\sim}{S}$ of a given dimension n constitute the general linear group GL(n). It is well known that any transformation belonging to this group can be expressed as an exponential of some matrix $\underset{\sim}{T}$,

$$\underset{\sim}{S} = \exp\,(\underset{\sim}{T}), \tag{23}$$

where in turn the set of n × n matrices $\underset{\sim}{T}$ constitute the corresponding Lie algebra gl(n). In particular, when $\underset{\sim}{T} = \underset{\sim}{0}$, then $\underset{\sim}{S} = \underset{\sim}{1}$, the unit matrix. From a practical viewpoint it would be desirable to choose for $\underset{\sim}{T}$ a nil-potent transformation, i.e., such that

$$\underset{\sim}{T}^k = 0, \tag{24}$$

for some k > 1, so that the exponential expansion (23) will truncate after the first k terms.

For the sake of simplicity consider first the case when we can choose the model space to be one-dimensional. Then $\underset{\sim}{H}'_0$ is a 1 × 1 matrix, or a scalar, and we can choose $\underset{\sim}{C}'_0 = \|1\|$. The corresponding eigenvector (21) then represents the reference configuration $|\Phi_0\rangle$ and the second relationship (18) gives

$$\underset{\sim}{C}'_{(0)} = \underset{\sim}{S}^{-1} \, \underset{\sim}{C}_{(0)}, \tag{25}$$

where $\underset{\sim}{C}_{(0)}$ represents the eigenvector $|\Psi_0\rangle$ of $\underset{\sim}{H}$, associated with the eigenvalue E_0, in the basis $\{|\Phi_\mu\rangle\}$. Thus

$$\mathcal{C}_{(0)} = \exp\ (\underset{\sim}{T})\ \underset{\sim}{C}'_{(0)}, \tag{26}$$

or, equivalently,

$$|\Psi_0\rangle = \exp\ (T)|\Phi_0\rangle. \tag{27}$$

We thus find a conventional cluster ansatz for an exact non-degenerate wavefunction, in which the wave operator W of Löwdin is written in the exponential form,

$$|\Psi_0\rangle = W\ |\Phi_0\rangle,\ W = \exp\ (T), \tag{28}$$

with W defined on V_0. More precisely

$$W = \exp\ (T)\ P, \tag{29}$$

with T defined on V.

In order to find out more about the structure of the cluster operator T, we classify the N-electron basis states $|\Phi_\mu\rangle$ by their excitation order relative to $|\Phi_0\rangle$, written as a superscript. Thus, $|\Phi_\mu^{(k)}\rangle$ represents a configuration which is obtained from $|\Phi_0\rangle$ by a k-fold excitation. Ordering the one-electron spinorbital basis $|C\rangle = |c\rangle|\gamma\rangle$ in such a way that the spinorbitals constituting $|\Phi_0\rangle$ precede the virtual (unoccupied) ones in $|\Phi_0\rangle$, and using the matric basis for the Lie algebra gl(n), we see immediately that each excitation may be characterized by lowering generator e_{AI}, which we can also write in the second quantized form as follows

$$e_{AI} = X_A^\dagger\ X_I, \tag{30}$$

X_C ($X_C^\dagger$) designating the annihilation (creation) operators associated with a given one-electron spinorbital basis $\{|C\rangle\}$. We designate the (spin) orbitals unoccupied in $|\Phi_0\rangle$ by the letters from the beginning of the alphabet (A,B,... for spinorbitals, a,b,... for orbitals) and those occupied in $|\Phi_0\rangle$ by I,J,... (i,j,...). Thus, generally,

$$|\Phi_\mu^{(k)}\rangle = \underbrace{(e_{AI}e_{BJ}\cdots)}_{(k)}|\Phi_0\rangle, \tag{31}$$

where μ designates the 2k spinorbital labels (A,B,... ; I,J,...). A similar relationship can also be established in the case when $|\Phi_\mu\rangle$ represents a spin-adapted basis using the appropriate U(n) or GL(n) generators

$$E_{ai} = e_{a\alpha,i\alpha} + e_{a\beta,i\beta}, \tag{32}$$

$|\alpha\rangle, |\beta\rangle$ designating the usual spin-up and -down functions. However, an analogous relationship to (31) generally yields a non-orthogonal basis (so called generator states of Matsen [86])

$$|\Phi_\mu^{(k)}\rangle = (E_{ai}E_{bj}\dots)|\Phi_0\rangle, \tag{33}$$

and appropriate linear combinations are needed to generate the desired orthogonal basis (say Gelfand-Tsetlin or Yamanouchi-Kotani bases).

Thus, analogously as in the CI linear expansion

$$|\Psi_0\rangle = C|\Phi_0\rangle = \sum_k C_k|\Phi_0\rangle,$$

$$C_0 = 1, \quad C_k|\Phi_0\rangle = \sum_\mu c_\mu^{(k)}|\Phi_\mu^{(k)}\rangle, \tag{34}$$

we get

$$T_k = (k!)^{-2} \sum_{\substack{AB\dots,\\ IJ\dots}} \langle AB\dots|t_k|IJ\dots\rangle_A E_{AI}E_{BJ}\dots, \tag{35}$$

where subscript A designates the antisymmetrized matrix elements. Similar expressions result in the spin-adapted case (cf. Refs. 60, 61) if the matrix elements are properly defined. For example, using the non-orthogonal basis (33), we can write the most important pair component as follows

$$T_2 = \tfrac{1}{2} \sum_{abij} \langle ab|t_2|ij\rangle\, E_{ai}E_{bj}. \tag{36}$$

We can also express this operator in terms of a given N-electron basis $\{|\Phi_\mu\rangle\}$ if desirable. Using, for example, the Gelfand-Tsetlin basis of the unitary group approach [17,18] we can write

$$T_2|\Phi_0\rangle = \sum_\mu t_\mu^{(2)}|\Phi_\mu^{(2)}\rangle = \sum_{ab,ij}\sum_k t_k(ab;ij)|[ab;ij]_k\rangle, \tag{37}$$

where $|[ab;ij]_k\rangle$ designates the biexcited singlet Gelfand-Tsetlin state with $k = 0,1$ indicating the lexical order for the two singlets when $a \neq b$ and $i \neq j$. (This index is dropped when only one singlet results.) Assuming that $C_1 = T_1 = 0$ (Brueckner one-electron basis) we have further that $C_2 = T_2$, i.e.

$$t_k(ab;ij) = c_k(ab;ij), \tag{38}$$

the pair cluster coefficients equal to biexcited CI coefficients assuming an intermediate normalization of $|\Psi_0\rangle$. In fact, the operator T_2 can be generally expressed in this case as follows (cf. Ref. 53)

$$T_2 = 2^{-1}\sum_a\sum_i c(aa;ii)E_{ai}^2$$

$$+ 2^{-\frac{1}{2}}\Big[\sum_a\sum_{i<j}(-)^{i+j}\, c(aa;ij)E_{ai}E_{aj} + \sum_{a<b}\sum_i c(ab;ii)E_{ai}E_{bi}\Big]$$

$$+ \sum_{a<b} \sum_{i<j} (-)^{i+j} \{c_-(ab;ij)E_{ai}E_{bj} + c_+(ab;ij)E_{aj}E_{bi}\} \tag{39}$$

where

$$c_\pm(ab;ij) = [c_1(ab;ij) \pm (1/\sqrt{3})c_2(ab;ij)]/2.$$

Thus, in the N-electron basis $\{|\Phi_\mu\rangle\}$ we can generally represent the cluster operators by nilpotent matrices. Assuming, for simplicity, that only even-number of times excited configurations occur in this basis and blocking the matrix representatives according to the excitation levels, we can express the pair cluster operator as follows

$$T_2 = \begin{Vmatrix} 0 & \underset{\sim}{0} & \underset{\sim}{0} & .. \\ \underset{\sim}{t} & \underset{\sim}{0} & \underset{\sim}{0} & .. \\ \underset{\sim}{0} & \underset{\sim}{\tau} & \underset{\sim}{0} & .. \end{Vmatrix}, \tag{40}$$

where $\underset{\sim}{t}$ is the column vector consisting of biexcited coefficients (38), while the block $\underset{\sim}{\tau}$ and similar higher excitation level blocks can be obtained using the representation (39). The matrix elements in these blocks either vanish or are given as linear combinations of biexcited cluster coefficients.

Consider first only the subspace $V^{(2)}$ of V spanned by $|\Phi_0\rangle$ and $|\Phi_\mu^{(2)}\rangle$, so that

$$\underset{\sim}{T}_2 = \begin{Vmatrix} 0 & \underset{\sim}{0} \\ \underset{\sim}{t} & \underset{\sim}{0} \end{Vmatrix}, \qquad \underset{\sim}{H}_D = \begin{Vmatrix} 0 & \underset{\sim}{a}^\dagger \\ \underset{\sim}{a} & \underset{\sim}{b} \end{Vmatrix}. \tag{41}$$

In this case clearly $\underset{\sim}{T}_2^2 = 0$ so that

$$\underset{\sim}{H}'_D = \exp(-\underset{\sim}{T}_2)\, \underset{\sim}{H}_D \exp(\underset{\sim}{T}_2) =$$

$$= (\underset{\sim}{1} - \underset{\sim}{T}_2)\, \underset{\sim}{H}_D (\underset{\sim}{1} + \underset{\sim}{T}_2) = \begin{Vmatrix} h'_{00} & \underset{\sim}{h}'_{0d} \\ \underset{\sim}{h}'_{d0} & \underset{\sim}{h}'_{dd} \end{Vmatrix}, \tag{42}$$

where

$$h'_{00} = E_L = \underset{\sim}{a}^\dagger \underset{\sim}{t} \quad \text{and} \quad \underset{\sim}{h}'_{d0} = \underset{\sim}{a} + \underset{\sim}{b}\underset{\sim}{t}. \tag{43}$$

Thus, choosing T (or S) in such a way that $\underset{\sim}{h}'_{d0} = \underset{\sim}{0}$, i.e.

$$\underset{\sim}{a} + \underset{\sim}{b}\underset{\sim}{t} = \underset{\sim}{0}, \tag{44}$$

we get

$$E_L = \underset{\sim}{a}^{\dagger}\underset{\sim}{t} = -\underset{\sim}{a}^{\dagger}\underset{\sim}{b}^{-1}\underset{\sim}{a}, \tag{45}$$

assuming that $\underset{\sim}{b}$ is invertible. This is exactly the linear version of the CPMET (L-CPMET). This approximation is clearly going to break down if the bi-excited submatrix $\underset{\sim}{b}$ of the D-CI matrix is singular.

In order to obtain a better approximation we must include higher excited intermediate states. Thus, truncating at the tetraexcited level, we can write

$$\underset{\sim}{H}_Q = \begin{Vmatrix} 0 & \underset{\sim}{a}^{\dagger} & \underset{\sim}{0} \\ \underset{\sim}{a} & \underset{\sim}{b} & \underset{\sim}{c}^{\dagger} \\ \underset{\sim}{0} & \underset{\sim}{c} & \underset{\sim}{q} \end{Vmatrix}, \qquad \underset{\sim}{T}_2 = \begin{Vmatrix} 0 & \underset{\sim}{0} & \underset{\sim}{0} \\ \underset{\sim}{t} & \underset{\sim}{0} & \underset{\sim}{0} \\ \underset{\sim}{0} & \underset{\sim}{\tau} & \underset{\sim}{0} \end{Vmatrix} \tag{46}$$

so that

$$\underset{\sim}{T}_2^2 = \begin{Vmatrix} 0 & \underset{\sim}{0} & \underset{\sim}{0} \\ \underset{\sim}{0} & \underset{\sim}{0} & \underset{\sim}{0} \\ \underset{\sim}{\tau}\underset{\sim}{t} & \underset{\sim}{0} & \underset{\sim}{0} \end{Vmatrix} \tag{47}$$

and

$$\underset{\sim}{T}_2^k = 0 \text{ for } k > 2. \tag{48}$$

Proceeding in the same way as in (42), we easily find the first column of $\underset{\sim}{H}'_Q$, which we again truncate at the biexcited level, since the number of conditions cannot exceed the number of unknowns in $\underset{\sim}{T}$. By a simple matrix multiplication we thus find

$$(\underset{\sim}{h}'_Q)_{00} = E_{CP} = \underset{\sim}{a}^{\dagger}\underset{\sim}{t}; \; (\underset{\sim}{h}'_Q)_{d0} = \underset{\sim}{a} + \underset{\sim}{b}\underset{\sim}{t} + \tfrac{1}{2}\underset{\sim}{c}^{\dagger}\underset{\sim}{\tau}\underset{\sim}{t} - \underset{\sim}{t}\underset{\sim}{a}^{\dagger}\underset{\sim}{t} = \underset{\sim}{0}. \tag{49}$$

The system of equations determining $\underset{\sim}{t}$ can thus be written as

$$\underset{\sim}{b} + \underset{\sim}{a}\underset{\sim}{t} + \tfrac{1}{2}\underset{\sim}{c}^{\dagger}\underset{\sim}{\tau}\underset{\sim}{t} = E_{CP}\underset{\sim}{t}, \tag{50}$$

and can be obtained directly from the CI equations (with $|\Psi_0\rangle$ in the intermediate normalization) when we approximate the tetraexcited coefficients by their disconnected components. This procedure was in fact exploited in Ref. 51 and shows that at the orbital level the right hand side member in Eq. (50) is cancelled by the disconnected terms arising from the third term on the left hand side, thus yielding the *energy independent* form of the CPMET equations. This cancellation is also automatically achieved when the diagrammatic formalism is used or, equivalently, when the similarity transformation (18) with (23) is expressed through the commutator series. Particularly the former approach enables one to write down directly the explicit form of the system (50) in terms of one- and two-electron integrals and individual pair cluster components. However, even though the disconnected clusters appear in Eq. (50), they cancel out, so that these equations together with the first Eq. (49) determine the size-extensive (ground state) energy as guaranteed by the cluster ansatz (23).

5. RELATIONSHIP WITH VARIATIONAL APPROACHES

We next examine the relationship of the above outlined procedure with the conventional CI approach. Using an intermediate normalization for the exact wavefunction, Eq. (1), we can write the CI chain of equations as follows

$$\sum_{i=k-2}^{k+2} \underset{\sim}{H}^{(k,i)} \underset{\sim}{c}_i = E\, \underset{\sim}{c}_k, \quad (\underset{\sim}{c}_0 = 1), \tag{51}$$

where $\underset{\sim}{H}^{(k,i)}$ designates the block of $\underset{\sim}{H}$ between k-times and i-times excited configurations and $\underset{\sim}{c}_i$ the i-times excited coefficient matrix in the CI expansion of $|\Psi_0\rangle$. In the case of the CI matrix (46) considered above, this chains takes the form

$$\begin{aligned} &\underset{\sim}{a}^{\dagger} \underset{\sim}{c}_2 = E \\ &\underset{\sim}{a} + \underset{\sim}{b}\underset{\sim}{c}_2 + \underset{\sim}{c}^{\dagger}\underset{\sim}{c}_4 = E\underset{\sim}{c}_2 \\ &\underset{\sim}{c}\underset{\sim}{c}_2 + \underset{\sim}{g}\underset{\sim}{c}_4 + \ldots = E\underset{\sim}{c}_4, \text{ etc.} \end{aligned} \tag{52}$$

This chain is usually decoupled at certain excitation level k by neglecting the coefficients $\underset{\sim}{c}_\ell$, $\ell \geq k$. Thus, setting $\underset{\sim}{c}_4 = 0$, we obtain the D-CI eigenvalue problem

$$\underset{\sim}{a}^{\dagger} \underset{\sim}{c}_2 = E_D; \quad \underset{\sim}{a} + \underset{\sim}{b}\underset{\sim}{c}_2 = E_D\underset{\sim}{c}_2 \tag{53}$$

or

$$\underset{\sim}{H}_D\, \underset{\sim}{c} = E_D\underset{\sim}{c}, \quad \underset{\sim}{c}^T = \| 1 \;\; \underset{\sim}{c}_2^T \| . \tag{54}$$

In contrast, the coupled cluster approaches enable a physically more meaningful decoupling of the chain (52). Thus, when (12) holds, we approximate $\underset{\sim}{c}_4$ by its disconnected component

$$c_4 \simeq (c_4)_D = \langle (4)|\tfrac{1}{2}T^2|0\rangle, \tag{55}$$

in which case we obtain the CPMET equations in the form (50).

We note that the L-CPMET equations (44) result when both non-linear terms (i.e. $\underset{\sim}{c}^{\dagger}\underset{\sim}{\tau}\underset{\sim}{t}$ and $E_{CP}\underset{\sim}{t}$) in (50) are neglected. In obtaining these equations from the corresponding D-CI equations we must set $\Delta E = 0$ in the second equation. By doing so we in fact consider the disconnected part of the tetraexcited contribution $\tfrac{1}{2}\underset{\sim}{c}^{\dagger}\underset{\sim}{\tau}\underset{\sim}{t}$ in (50), cancelled by $E_{CP}\underset{\sim}{t}$, and neglect only the connected part representing the coupling between pair clusters. This explains not only the size extensive behavior of the resulting energy but also its proximity to the CPMET result, as long as no quasi-degeneracy is present.

Thus, both the D-CI and the L-CPMET exploit the same information about the system which is contained in the D-CI matrix. We can obtain further insight into the character of both approximations if we use instead of basis states $|\Phi_\mu^{(2)}\rangle$ the canonical biexcited configurations which diagonalize $\underset{\sim}{b}$. Designating

$$\underset{\sim}{b}\,\underset{\sim}{v} = \underset{\sim}{v}\,\underset{\sim}{\beta}, \quad \underset{\sim}{v}^{\dagger}\,\underset{\sim}{v} = \underset{\sim}{v}\underset{\sim}{v}^{\dagger} = \underset{\sim}{1} \tag{56}$$

where $\underset{\sim}{\beta} = \| \beta_i\, \delta_{ij} \|$, we can write

$$\underset{\sim}{S} = \underset{\sim}{V} \cdot \underset{\sim}{S}_c; \quad \underset{\sim}{V} = \underset{\sim}{1} \dotplus \underset{\sim}{v}, \quad \underset{\sim}{V}^{\dagger}\underset{\sim}{V} = \underset{\sim}{V}\underset{\sim}{V}^{\dagger} = \underset{\sim}{1}. \tag{57}$$

Then

$$\underset{\sim}{H}_D' = \underset{\sim}{S}_c^{-1}\, \underset{\sim}{H}_D''\, \underset{\sim}{S}_c; \quad \underset{\sim}{H}_D'' = \underset{\sim}{V}^{\dagger}\, \underset{\sim}{H}_D \underset{\sim}{V} = \begin{Vmatrix} 0 & \underset{\sim}{\alpha}^{\dagger} \\ \underset{\sim}{\alpha} & \underset{\sim}{\beta} \end{Vmatrix}. \tag{58}$$

We get

$$E_D = \sum_i (\alpha_i)^2/(\beta_i - E_D) \tag{59}$$

and

$$E_L = \sum_i (\alpha_i)^2/\beta_i, \tag{60}$$

indicating that the D-CI result is obtained as a second order BWPT while the L-CPMET is given by the corresponding RSPT. Indeed, L-CPMET is equivalent to the infinite order RSPT in which only biexcited intermediate states are considered, i.e. to the DMBPT(∞).

It is also easily seen that the L-CPMET can be formulated variationally using the appropriate Hylleraas functional [2,83]. In fact, both the D-CI and the L-CPMET energies can be obtained by minimizing the same functional

$$E(\underset{\sim}{X}) = \underset{\sim}{X}^{\dagger}\, \underset{\sim}{H}_D\, \underset{\sim}{X} \tag{61}$$

with the constraint that $\underset{\sim}{X}^{\dagger}\underset{\sim}{X} = 1$ in the D-CI case and $\underset{\sim}{X}^T = \| 1 \;\; x^T \|$ in the L-CPMET case. Geometrically, these constraints may be interpreted as a minimization on the unit sphere in the D-CI case and on the tangent plane to this unit sphere at the reference state $|\Phi_0\rangle$ in the L-CPMET case [83]. Since, in the latter case, the variational space is linear, the size extensive property follows.

This fact can be exploited to obtain a simple relationship between the two approaches [83]. This is most easily seen by considering the inverse of $\underset{\sim}{H}_D$, which we write in the same block form

$$\underset{\sim}{H}_D^{-1} = \begin{Vmatrix} p & \underset{\sim}{r}^{\dagger} \\ \underset{\sim}{r} & \underset{\sim}{s} \end{Vmatrix}, \quad \underset{\sim}{s}^{\dagger} = \underset{\sim}{s}. \tag{62}$$

Requiring that $\underset{\sim}{H}_D \underset{\sim}{H}_D^{-1} = \underset{\sim}{H}_D^{-1} \underset{\sim}{H}_D = \underset{\sim}{1}$ yields immediately that

$$\underset{\sim}{a}^\dagger \underset{\sim}{r} = 1 \text{ and } \underset{\sim}{a}p + \underset{\sim}{b}\underset{\sim}{r} = \underset{\sim}{0}. \tag{63}$$

Multiplying the second equation by p^{-1}, we get in fact the L-CPMET equations (44) where $\underset{\sim}{t} = p^{-1} \underset{\sim}{r}$, so that the L-CPMET energy (45) can be expressed as follows

$$E_L = \underset{\sim}{a}^\dagger \underset{\sim}{t} = p^{-1}. \tag{64}$$

Using the spectral resolution for $\underset{\sim}{H}_D$,

$$(\underset{\sim}{H}_D)_{k\ell} = \sum_i u^*_{ki} \lambda_i u_{\ell i}, \tag{65}$$

we get immediately that

$$E_L = (\underset{\sim}{H}_D^{-1})^{-1}_{00} = p^{-1} = \left[\sum_i |u_{0i}|^2/\lambda_i\right]^{-1}. \tag{66}$$

The cluster components $\underset{\sim}{t}$ can be expressed similarly [83].

When $|\Phi_0\rangle$ is a good approximation to $|\Psi_0\rangle$, we have that

$$|u_{00}|^2 \gg |u_{0i}|^2, \; (i > 0). \tag{67}$$

Consequently, defining

$$E_L^{(k)} = \left[\sum_{i=0}^{k-1} |u_{0i}|^2/\lambda_i\right]^{-1}, \tag{68}$$

we can expect $E_L^{(1)}$ to provide a good approximation for E_L. In fact

$$E_L^{(1)} \equiv E_{RDC} = |u_{00}|^{-2} \lambda_0 = \lambda_0 + \lambda_0(1 - |u_{00}|^2)/|u_{00}|^2 \tag{69}$$

represents the renormalized Davidson correction [87-90], which will be very close to the Davidson correction [87]

$$E_{DC} = \lambda_0 + \lambda_0(1 - |u_{00}|^2) \tag{70}$$

when $|u_{00}|^2 \approx 1$. We also note that

$$E_L \le E_{RDC} \le E_{DC} \le E_D \tag{71}$$

and

$$E_L^{(k)} \le E_L^{(\ell)} \text{ if } k > \ell. \tag{72}$$

Thus, the Davidson correction represents an approximation to the L-CPMET correlation energy based on the first term in the expansion (68). Actual calculations show that for non-degenerate ground states, this approximation yields practically the L-CPMET energy. In Ref. 83 this fact was illustrated on a rather extensive calculation for the CO molecule (58 AO's, 7272 spin adapted singly and doubly excited configurations). Note that the same conclusions hold when the monoexcited configurations are explicitly considered along with the doubly excited ones [83]. In this case, $E_D = -0.25329$ a.u., $E_{DC} = -0.27127$ a.u., $E_{RDC} = -0.27264$ a.u. while $E_L = -0.27561$ a.u. Similar results for the minimum basis set BH_3 correlation energy are indicated graphically in Fig. 1.

Recently, it was observed by Mukherjee [91] that the L-CPMET energy expression (45) has the form of the Löwdin's bracketing function [15,16, 26], which for this particular case has been examined by Goscinski and Sinanoğlu [92]. Using the HF reference energy as an approximate upper bound to the ground state energy, the L-CPMET energy was shown to provide a corresponding lower bound. However, since the L-CPMET procedure is based on the D-CI matrix, the lower bound obtained in this way cannot be an exact lower bound. In fact, the "lower bound" property as formulated by Mukherjee [91] reduces essentially to the inequality (71) [83]. It is nevertheless interesting that in all model cases studied the L-CPMET energy does indeed bound the F-CI energy from below (cf. also Fig. 1). This is probably related to the fact that the L-CPMET overestimates the tetraexcited disconnected cluster contribution by roughly the amount corresponding to the connected triexcited contribution. Indeed, in a number of actual applications, the L-CPMET energies yield even better agreement with the experimental data than the CPMET results [93]. Only when the reference state becomes quasidegenerate does the L-CPMET energy deteriorate [63,68]. One might thus expect that taking the quasi-degeneracy into account, the theory at the L-CPMET level should provide sufficiently good correlation energies.

6. GENERALIZED CLUSTER ANSATZE

So far we have concentrated our attention on the case of a non-degenerate closed-shell ground state. Before we discuss more general cases, in particular those requiring a multiconfiguration reference space, let us consider the spin (or symmetry) adaptation problem for a single configuration reference case.

A comparison of the exact FCI and CCA wave functions requires that

$$C_k = T_k + \sum_{\mathcal{P}_i}{}' \prod_{j=1}^{i} (n_j!)^{-1} T_j^{\,n_j} \equiv T_k + D_k, \tag{73}$$

where

$$i = \sum_{j=1}^{i} j\, n_j, \quad 0 \le n_j \le i, \quad T_j^{\,0} = 1, \tag{74}$$

and $\mathcal{P}_i$ extends over all nontrivial partitions of i. For spin-independent H (or, generally, any symmetry group invariant H) the exact eigenstates must transform according to the irreps (irreducible representations) of the given symmetry group. Designating by $\mathcal{O}$ the appropriate projector (say, the projection operator onto a singlet manifold for spin-independen case) we have that $\mathcal{O}|\Psi_0\rangle = |\Psi_0\rangle$. Consequently, all CI components C_k must transform according to given irrep and in the cluster expansion the T_k's must similarly transform as follows from (73). However, applying $\mathcal{O}$ directly to the cluster ansatz (1),

$$|\Psi_0\rangle = \mathcal{O}\ \exp(T)|\Phi_0\rangle, \tag{75}$$

we see that the disconnected part D_k can consist of arbitrary cluster components as long as the correct symmetry is projected out. Thus, e.g., the disconnected components for the singlet ground state can consist of singlet, triplet, quintet, etc. clusters. Obviously, when clusters of other than the given irrep are used in the disconnected part, the CCA wave function contains more variational parameters than the corresponding truncated CI wave function.

This additional freedom in the CCA wave function is best seen when we restrict ourselves to the singlet case and $T = T_1$. It is well known (Thouless' theorem) that any non-orthogonal single ASP (Slater determinan state can be written in the form

$$|\Phi'\rangle = \exp(T_1)|\Phi_0\rangle. \tag{76}$$

T_1 is given as a linear combination of GL(n) generators. If normalizatio is to be preserved, U(n) must be used so that

$$|\Phi''\rangle = \exp(T_1-T_1^{\dagger})|\Phi_0\rangle. \tag{77}$$

The ansatz (76) or (77) can be conveniently used to derive various IPM approximations as well as generalized time-dependent HF approaches (cf. [66,94]). In the closed shell case, for example, the RHF is obtained when T_1 is restricted to its singlet components while the UHF results when arbitrary spin components of T_1 are allowed. Similarly, allowing arbitrary spin components of T_1 but projecting onto the singlets, Eq. (75), various projected schemes such as AMO, GVB, etc. result [66]. Thes possibilities have been studied in great detail by Hirao and Nakatsuji [66,67] who have also extended this idea to higher than T_1 clusters in their symmetry adapted cluster (SAC) approach.

We thus find that using only the appropriate SAC's, the cluster expansion of the exact wave function is unique. This is not the case when other irrep clusters are allowed in the disconnected part. Such wave functions can be very useful in the variational approach since they expand the dimension of the variational space employed. However, great care should be exercised when such a wavefunction is used in non-variational or approximate variational approaches.

Another interesting aspect of spin-adapted cluster expansion should be mentioned. Since at most two singlets can be formed for any doubly excited configuration associated with fixed orbital occupancy, it might seem that a number of tetraexcited configurations will be left out in the product representation T_2^2, since fourteen singlets are possible if all orbitals involved in the tetraexcitation are singly occupied. It may be shown, however, that all these different spin states will be represented through T_2^2 when the orbitals are arbitrarily permuted between both biexcitations [60]. However, this is not the case if H possesses certain spatial symmetry, say the cyclic symmetry C_k with $k \geq 10$. For example, in the case of $C_{10}H_{10}$ cyclic polyene model, there are 12 tetraexcited states for which $C_4 = T_4$, i.e., for which no disconnected component in terms of symmetry adapted T_2 clusters is possible [53].

The main advantage of the cluster expansion is clearly twofold: it guarantees the size-extensivity of the calculated energies and it enables a meaningful decoupling of the FCI chain. It is primarily the breakdown of the property (12) which invalidates the usefulness of a single reference CCA in quasi-degenerate cases. This is nicely illustrated by our model calculations for various H_4 systems [68] in which the amount of quasi-degeneracy can be continuously varied. The cluster analysis of the FCI wave function shows that in the quasi-degenerate region the T_4 component represents almost 50% of the disconnected component. The CPMET energies overestimate the exact ones by approximately the fraction of the tetraexcited contribution which corresponds to the size of the T_4 component [95]. For a similar reason [53] the CPMET breaks down for cyclic polyene models C_NH_H when $N \to \infty$ or in the strongly correlated limit ($\beta \to 0$) even for small N [96]. The remedy in each case is to extend the reference space so that the important T_4 contributions become representable by T_2 contributions relative to the additional reference configurations. This added flexibility will clearly shift the difficulties associated with connected T_4 contributions to higher excitation levels where, hopefully, they will be unimportant.

An extension of the cluster ansatz (1) to the multireference or open-shell cases is by no means a straightforward and easy task. Before we attempt to apply a simple approach used in Sec. 4 to this case, let us briefly review some of the ansatze which have been used in the literature so far. For a more extensive review we refer the reader to a recent paper by Jeziorski and Monkhorst [97].

/1/ In the special cases, a simple modification [47,75]

$$|\Psi\rangle = W \exp(T)|\Phi_0\rangle \tag{78}$$

enables one to treat various excited and ionized states. From the practical viewpoint this ansatz requires that the structure of the wave function is not changed drastically by the process considered (see also Sec. 3.4).

/2/ Some authors [98,99] attempted a straightforward generalization of the ansatz (1). The difficulties of such extensions have been discussed in Ref. 97. These difficulties can be avoided by a recursive definition which applies not only to all states of the reference space but also to all multiply ionized states [98,100]. This ansatz clearly leads to a rather complicated formalism.

/3/ In order to avoid the problems of ill-defined vacuum state for the hole-particle formalism, Lindgren [101] introduced the cluster expansion

$$|\Psi_i\rangle = N[\exp(T)]|\Phi_\mu\rangle, \quad (\mu=1,\ldots,M), \tag{79}$$

where N[···] designates the N-product with respect to a chosen closed-shell reference state (most conveniently chosen as a core configuration) The N-product ascertains appropriate disconnected diagram factorization in the open-shell case.

/4/ Most recently, Jeziorski and Monkhorst [97] suggested the expansion leading to the wavefunction representation

$$|\Psi_i\rangle = \sum_\mu c_{i\mu} \exp(T^{(\mu)})|\Phi_\mu\rangle, \tag{80}$$

where $T^{(\mu)}$ is defined relative to $|\Phi_\mu\rangle$, $\mu = 1,\ldots,M$. We shall briefly return to these expansions later.

Consider now an extension of our simple approach of Sec. 4 to the multi-configuration case, $M = \dim \mathcal{V}_0 > 1$. Just as in the case M = 1 we look for an effective Hamiltonian $\underset{\sim}{H}_0'$ defined in $\mathcal{V}_0$. Since the M eigenvectors of $\underset{\sim}{H}_0'$ and $\underset{\sim}{H}'$ are simply related, Eq. (21), we can write in analogy to (25)

$$\left\|\begin{matrix} \underset{\sim}{C}_0' \\ \underset{\sim}{0} \end{matrix}\right\| = \underset{\sim}{S}^{-1}\, \underset{\sim}{C}_{(0)}, \tag{81}$$

where $\underset{\sim}{C}_{(0)}$ is an n×M submatrix of $\underset{\sim}{C}$, Eq. (16), representing the exact states corresponding to those spanning $\mathcal{V}_0$. Multiplying (81) by $\underset{\sim}{S}$ and rewriting the matrix equation in terms of states and operators we get

$$|\Psi_i\rangle = S\sum_\mu c_{i\mu}|\Phi_\mu\rangle = \sum_\mu c_{i\mu}\; S|\Phi_\mu\rangle \equiv \sum_\mu c_{i\mu}|\Phi_\mu'\rangle \tag{82}$$

where $c_{i\mu}$ are the entries of $\underset{\sim}{C}_0'$ giving the eigenvectors of $\underset{\sim}{H}_0'$ in the basis $\{|\Phi_\mu\rangle\}$, and

$$|\Phi_\mu'\rangle \equiv S|\Phi_\mu\rangle. \tag{83}$$

We now have two basic possibilities for choosing S: (i) we can assume S to be μ independent, obtaining essentially the representations /2/, /3/, or (ii) we can assume S to be μ- dependent,

$$S = \sum_\mu \exp(T^{(\mu)})P_\mu, \tag{84}$$

obtaining the ansatz of Jeziorski and Monkhorst [97].

We cannot dwell here in any detail on the relative merits and shortcomings of these various ansatze. Let us only mention that the ansatz /3/ is simpler, yet leads to certain undesirable features such as the necessity to consider a complete reference space, reference configuration mixing and, perhaps most importantly, the independence of the cluster components associated with core-virtual excitations on the model space occupancies. However, should only the lowest energy state be of interest, this ansatz might be preferred. The attractiveness of the ansatz /4/ is its uniqueness in representing the whole manifold of states associated with the reference space. Unfortunately, the resulting general formalism [97] is rather complicated.

Relying on this latter ansatz, we can represent the $T_2^{(\mu)}$ operator for each component μ as we did earlier. Considering only the linear terms, we can approximate $\underset{\sim}{S}$ as follows

$$\underset{\sim}{S} \simeq \underset{\sim}{1} + \sum_{\mu} \underset{\sim}{T}_2^{(\mu)} \simeq \begin{Vmatrix} \underset{\sim}{1} & \underset{\sim}{0} \\ \underset{\sim}{t} & \underset{\sim}{1} \end{Vmatrix}, \qquad \underset{\sim}{S}^{-1} \simeq \begin{Vmatrix} \underset{\sim}{1} & \underset{\sim}{0} \\ -\underset{\sim}{t} & \underset{\sim}{1} \end{Vmatrix}. \tag{85}$$

Writing a similarly blocked and truncated $\underset{\sim}{H}$ as

$$\underset{\sim}{H} \simeq \begin{Vmatrix} \underset{\sim}{z} & \underset{\sim}{a}^{\dagger} \\ \underset{\sim}{a} & \underset{\sim}{b} \end{Vmatrix}, \tag{86}$$

we get

$$\underset{\sim}{h}'_{00} = \underset{\sim}{z} + \underset{\sim}{a}^{\dagger}\underset{\sim}{t}; \qquad \underset{\sim}{h}'_{d0} = \underset{\sim}{a} + \underset{\sim}{b}\underset{\sim}{t} - \underset{\sim}{t}\underset{\sim}{z} = \underset{\sim}{0}. \tag{87}$$

The last term obviously couples the L-CPMET systems of equations for individual reference state configurations. This coupling is also taken care of by a subsequent diagonalization of $\underset{\sim}{H}_0'$. Of course, no coupling would occur for diagonal $\underset{\sim}{z}$. Neglecting the coupling term in (87) leads to decoupled L-CPMET problems for individual model states. Such a decoupling occurs automatically if the reference space is truncated to a single configuration.

We can also obtain Eqs. (87) by similarly decoupling the multireference D-CI chain as in the single reference case. Using (86) in (14) and blocking $\underset{\sim}{C}$ accordingly, we get after the decoupling ($\underset{\sim}{c}_4 = 0$)

$$\underset{\sim}{z}\underset{\sim}{c}_0 + \underset{\sim}{a}^{\dagger}\underset{\sim}{c}_2 = \underset{\sim}{c}_0\underset{\sim}{E}_D; \qquad \underset{\sim}{a}\underset{\sim}{c}_0 + \underset{\sim}{b}\underset{\sim}{c}_2 = \underset{\sim}{c}_2\underset{\sim}{E}_D. \tag{88}$$

Assuming that $\underset{\sim}{c}_0$ is invertible, we multiply both relationships (88) by $\underset{\sim}{c}_0^{-1}$ from the right. Identifying $\underset{\sim}{t} = \underset{\sim}{c}_2\underset{\sim}{c}_0^{-1}$ we find

$$\underset{\sim}{z} + \underset{\sim}{a}^{\dagger}\underset{\sim}{t} = \underset{\sim}{c}_0\underset{\sim}{E}_D\underset{\sim}{c}_0^{-1} \equiv \underset{\sim}{H}_{eff}^{(D)}; \qquad \underset{\sim}{a} + \underset{\sim}{b}\underset{\sim}{t} = \underset{\sim}{t}\underset{\sim}{H}_{eff}^{(D)}. \tag{89}$$

Thus, approximating $\underset{\sim}{H}_{eff}^{(D)}$ in the second equation by $\underset{\sim}{z}$ [compare this with setting $E_D = 0$ in (53), which is equivalent with setting $\underset{\sim}{t} = \underset{\sim}{c}_2 = 0$ in the first Eq. (53)] we recover the L-CPMET system (87). We also note the similarity with the multidimensional partitioning technique [27,102].

We can also prediagonalize $\underset{\sim}{H}$ within the model space. Let us write

$$\underset{\sim}{S} = \underset{\sim}{V}\,\underset{\sim}{S}' \tag{90}$$

where

$$\underset{\sim}{V} = \underset{\sim}{v} \dotplus \underset{\sim}{1} = \begin{Vmatrix} \underset{\sim}{v} & \underset{\sim}{0} \\ \underset{\sim}{0} & \underset{\sim}{1} \end{Vmatrix}, \qquad \underset{\sim}{z}\underset{\sim}{v} = \underset{\sim}{v}\,\underset{\sim}{\alpha}_0, \quad \underset{\sim}{\alpha}_0 = \|\alpha_\mu \delta_{\mu\nu}\|. \tag{91}$$

Assuming that the degeneracy or quasi-degeneracy has been removed in this way, we can apply the non-degenerate formalism to the individual levels. We have tested this approach (called quasi L-CPMET) for the H_4 model systems [103] and cyclic polyenes [96] obtaining very encouraging results in spite of the fact that we did not use a full reference space.

Applying similar considerations as in the non-degenerate case to the approximate quasi-degenerate L-CPMET, we can formulate the following two extensions of the Davidson correction:

(A) Writing the inverse of $\underset{\sim}{H}$, Eq. (86), in the same block form and neglecting the coupling term in (87) we find that

$$\underset{\sim}{H}_{eff} \sim \underset{\sim}{p}^{-1}, \quad p_{ij} = \sum_k u_{ki}\lambda_k^{-1}u_{kj}. \tag{92}$$

Restricting the sum to the dimension of the reference space we obtain the following generalization of the Davidson correction: Using the multi-reference SD-CI eigenvectors, calculate the matrix

$$g_{ij} = \sum_{k=1}^{M} u_{ki}^{*}u_{kj}/\lambda_k, \quad (i,j \in V_0); \qquad \underset{\sim}{g} = \underset{\sim}{w}\underset{\sim}{\gamma}\underset{\sim}{w}^{\dagger}, \tag{93}$$

with $M = \dim V_0$ (or with properly shifted eigenvalues $(\lambda_k - \kappa)$, cf. [104]). Eigenvalues of $\underset{\sim}{g}$ will approximate the size-extensive energies for the system considered: $E_i' = 1/\gamma_i$.

(B) Based on the pre-diagonalized QL-CPMET, we get similarly the following rule: Diagonalize $\underset{\sim}{z}$ to obtain $\underset{\sim}{\alpha}_0$ and $\underset{\sim}{v}$, Eq. (91). For the state of interest associated with eigenvalue α_i, define the shifted SD-CI eigenvalues

$$\lambda_i' = \lambda_i - \alpha_i \tag{94}$$

and rotate the eigenvector $\underset{\sim}{u}_i$ associated with λ_i,

$$\underset{\sim}{u}'_i = \underset{\sim}{v}^{\dagger}\underset{\sim}{u}_i . \tag{95}$$

Then calculate the renormalized Davidson correction (or higher corrections $E_L^{(k)}$ if desirable) using λ'_i and u'_i, i.e.

$$(\lambda''_i)^{-1} = \sum_i |u'_{i0}|^2/\lambda'_i \approx |u'_{00}|^2/\lambda'_0 . \tag{96}$$

The desired energy is then

$$E''_i = \alpha_i + \lambda''_i . \tag{97}$$

We refer to (A) as DDC (degenerate Davidson correction) and to (B) as QDC (quasi-degenerate Davidson correction). We illustrate the performance of both corrections for the H_4 models studied earlier [68,95], together with the usual corrections, in Table I. Again, we only use the "incomplete" reference space, spanned by a single ground state configuration, since otherwise the SD-CI would be equivalent to the F-CI due to the simplicity of the model.

We see from Table I that both corrections represent a remarkable improvement and yield an excellent approximation to the exact FCI energies, even though only an incomplete reference space has been used. We note that the resulting energies should be in fact compared with DQ-CI energies rather than FCI energies. However, particularly in the quasi-degenerate region, both are very similar since the singly and triply excited contributions vanish exactly for the rectangular H_4 geometry. We also note that for similar reasons given by Mukherjee [91] the degenerate Davidson correction should provide an approximate lower bound to the multi-reference pair cluster energy.

To conclude this report we would like to note that an extension of the CCA to open shell systems and quasi-degenerate closed shell systems will require a better knowledge of the cluster structure of these states as well as a considerable computational effort and testing. Further methodological developments, particularly those directed to finding simple but reliable approximations as well as the relationship between various approaches, are also most desirable.

ACKNOWLEDGEMENTS

The author is greatly indebted to the Executive Committee of the Fourth International Congress in Quantum Chemistry at Uppsala University, Professor P.-O. Löwdin, Dr. O. Goscinski and Dr. J.-L. Calais, for the opportunity to present this plenary lecture. A continued support of author's research by NSERC is also greatly appreciated. He also wishes to acknowledge the help and collaboration of all his co-authors who have contributed so much to the developments mentioned in this paper and in particular to his long-time colleague and friend Professor Jiří Čížek,

TABLE I. Comparison of correlation energies (in mH) for various H4 models [68]: DC, RDC, DDC and QDC designate ordinary, renormalized, degenerate and quasi-degenerate Davidson corrections (see the text; all energy signs reversed). Parameter α is defined in Fig. 2 of Ref. 68.

Model	α	D-CI	DQ-CI	F-CI	L-CPMET	CPMET	DC	RDC	DDC	QDC
H4	.5	52.515	53.511	53.690	56.260	53.572	55.759	55.972	53.950	53.434
	.2	55.861	57.163	57.260	60.553	57.233	59.815	60.116	57.718	57.135
	.1	63.340	65.227	65.321	71.017	65.318	69.252	69.860	65.969	65.266
	.05	73.785	76.401	76.429	91.829	76.635	84.268	86.004	77.275	76.312
	.02	88.781	92.124	92.148	320.410	93.034	112.666	121.457	93.066	91.544
	.015	93.219	96.686	96.711	-517.543	97.900	122.607	136.137	97.689	95.978
	.005	105.552	109.188	109.196	-8.129	111.355	152.316	189.517	110.509	108.250
P4	10	53.013	54.650		56.507	54.650	56.100	56.291	55.872	55.595
	4	55.504	57.269		59.364	57.258	58.889	59.109	58.477	58.124
	2.2	81.300	84.443		125.211	84.897	96.860	100.542	85.406	84.220
	2.1	92.339	95.815		-787.233	96.967	120.533	132.925	96.819	95.155
	2.02	108.421	112.069		2.473	114.467	159.470	204.895	113.493	111.104
	2.002	113.385	117.035		13.770	119.824	171.886	234.243	118.650	116.033
D4	6	53.076	54.714	54.718	56.572	54.714	56.165	56.356	55.926	55.649
	4	53.105	54.698	54.754	56.555	54.697	56.156	56.342	55.827	55.485
	2	52.515	53.512	53.690	56.260	53.572	55.759	55.972	53.950	53.434
	1.5	56.790	57.572	58.517	62.001	57.644	61.129	61.488	57.782	57.227
	1	67.575	68.213	70.896	76.704	68.255	74.647	75.474	68.156	67.708

who first introduced him to the coupled cluster formalism. His most sincere thanks are also due to Debbie White for her patience and immaculate typing of the manuscript.

REFERENCES

1. W. Meyer, The Third International Congress of Quantum Chemistry, Kyoto, Japan, 1979; see *Horizons of Quantum Cehmistry*, F. Fukui and B. Pullman, Eds. (Reidel, Dordrecht, 1980), p. 3.
2. R. Ahlrichs, Compt. Phys. Commun. *17*, 31 (1979).
3. A.C. Hurley, *Electron Correlation in Small Molecules* (Academic Press, New York, 1976).
4. W. Kutzelnigg, in *Methods of Electronic Structure Theory*, H.F. Schaefer, Ed. (Plenum Press, New York, 1977), p. 129.
5. W. Meyer, Int. J. Quantum Chem. *5*, 341 (1971); Theoret. Chim. Acta *35*, 277 (1974).
6. W. Meyer, in *Methods of Electronic Structure Theory*, H.F. Schaefer, Ed. (Plenum Press, New York, 1977) p. 413.
7. R. Ahlrichs, H. Lischka, V. Staemmler and W. Kutzelnigg, J. Chem. Phys. *62*, 1225 (1975).
8. J. Paldus and J. Čížek, in *Energy, Structure and Reactivity*, D.W. Smith and W.B. McRae, Eds. (J. Wiley and Sons, New York, 1973), p. 198.
9. F.E. Harris, in *Electrons in Finite and Infinite Structures*, P. Phariseau and L. Scheire, Eds. (Plenum Press, New York, 1977), p. 274.
10. H. Kümmel, K.H. Lührmann and J.G. Zabolitzky, Phys. Rep. *36*, 1 (1978).
11. J. Čížek and J. Paldus, Physica Scripta *21*, 251 (1980).
12. R.J. Bartlett and G.D. Purvis, Physica Scripta *21*, 255 (1980).
13. R.J. Bartlett, Annu. Rev. Phys. Chem. *32*, 359 (1981).
14. I. Lindgren and J. Morrison, *Atomic Many-Body Theory* (Springer-Verlag, New York, 1982).
15. P.-O. Löwdin, J. Chem. Phys. *19*, 1396 (1951); *43*, S175 (1965); Phys. Rev. *139*, A357 (1965).
16. P.-O. Löwdin, in *Perturbation Theory and its Applications to Quantum Mechanics*, C.H. Wilcox, Ed. (J. Wiley and Sons, New York, 1966), p. 255.
17. M. Moshinsky, *Group Theory and the Many Body Problem* (Gordon and Breach, New York, 1968).
18. J. Paldus, J. Chem. Phys. *61*, 5321 (1974); in *Theoretical Chemistry: Advances and Perspectives*, H. Eyring and D.J. Henderson, Eds. (Academic Press, New York, 1976), Vol. 2, p. 131.
19. I. Shavitt, Int. J. Quantum Chem. S*11*, 131 (1977), S*12*, 5 (1978).
20. J. Paldus and M.J. Boyle, Physica Scripta *21*, 295 (1980); Phys. Rev. A*22*, 2299 (1980).
21. P.E.M. Siegbahn, J. Chem. Phys. *70*, 5391 (1979); *72*, 1647 (1980).
22. B.R. Brooks and H.F. Schaefer, J. Chem. Phys. *70*, 5092 (1979).
23. C.C.J. Roothaan, Rev. Mod. Phys. *23*, 69 (1951); *32*, 179 (1960).
24. P.-O. Löwdin, Phys. Rev. *97*, 1474, 1490, 1509 (1955).

25. P.-O. Löwdin, Advan. Chem. Phys. *2*, 207 (1959).
26. P.-O. Löwdin, J. Mol. Spectr. *10*, 12 (1963), *13*, 326 (1964); *14*, 112, 119, 131 (1964); J. Math. Phys. *3*, 969, 1171 (1962); 6, 1341 (1965), Int. J. Quantum Chem. *2*, 867 (1968).
27. P.-O. Löwdin and O. Goscinsky, Int. J. Quantum Chem. S*5*, 685 (1971).
28. K.A. Brueckner, Phys. Rev. *100*, 36 (1955).
29. J. Goldstone, Proc. Roy. Soc. A*239*, 267 (1957).
30. J. Hubbard, Proc. Roy. Soc. A*240*, 539 (1957); A*243*, 336 (1958); A*244*, 199 (1958).
31. N.M. Hugenholtz, Physica (Utrecht) *23*, 481 (1957).
32. M. Gell-Mann and K.A. Brueckner, Phys. Rev. *106*, 364 (1957).
33. Cf., e.g., H.A. Bethe, Phys. Rev. *103*, 1353 (1956).
34. C. Bloch, Nucl. Phys. *6*, 329 (1958); *7*, 451 (1958).
35. B.H. Brandow, in Proceedings of the International School of Physics "Enrico Fermi", Course 36, C. Bloch, Ed. (Academic Press, New York, 1966), p. 496; Rev. Mod. Phys. *39*, 771 (1967).
36. T. Morita, Progr. Theoret. Phys. (Japan) *29*, 351 (1963).
37. G. Oberlechner, F. Owono-N'-Guema and J. Richert, Nuovo Cimento B*68*, 23 (1970).
38. M.B. Johnson and M. Baranger, Ann. Phys. (N.Y.) *62*, 172 (1971).
39. H.P. Kelly, Phys. Rev. *131*, 684 (1963), Advan. Theoret. Phys. *14*, 129(1969) and references therein.
40. See, e.g., J.E. Mayer and M.G. Mayer, *Statistical Mechanics* (J. Wiley and Sons, New York, 1940), Ch. 13.
41. B. Kahn and G.E. Uhlenbeck, Physica (Utrecht) *5*, 399 (1938).
42. Cf., e.g., S.T. Epstein, *The Variational Method in Quantum Chemist* (Academic Press, New York, 1974).
43. F. Coester, Nucl. Phys. *7*, 421 (1958); F. Coester and H. Kümmel, Nucl. Phys. *17*, 477 (1960); cf. also W. Brenig, Nucl. Phys. *4*, 363 (1957), J. da Providencia, Nucl. Phys. *61*, 87 (1965).
44. O. Sinanoğlu, J. Chem. Phys. *36*, 706 (1962); *6*, 315 (1964); *14*, 237 (1969); H.J. Silverstone and O. Sinanoğlu, J. Chem. Phys. *44*, 1899, 3608 (1966); Advan. Chem. Phys. *6*, 315 (1964), *14*, 237 (1969) and references therein.
45. R.K. Nesbet, Advan. Chem. Phys. *14*, 1 (1969) and references therein.
46. H. Primas, in *Modern Quantum Chemistry*, O. Sinanoğlu, Ed. (Academic Press, New York, 1965), Vol. 2, p. 45.
47. J. Čížek, J. Chem. Phys. *45*, 4256 (1966); Advan. Chem. Phys. *14*, 35 (1969).
48. H. Kümmel, Nucl. Phys. A*176*, 205 (1971); H. Kümmel and K.H. Lührmann Nucl. Phys. A*191*, 525 (1972); K.H. Lührmann and H. Kümmel, Nucl. Phys. A*194*, 225 (1972); J.G. Zabolitzky, Nucl. Phys. A*228*, 272, 285 (1974); H. Kümmel and J.G. Zabolitzky, Phys. Rev. C*7*, 547 (1973).
49. J. Paldus and J. Čížek, Advan. Quantum Chem. *9*, 105 (1975).
50. R.J. Bartlett and I. Shavitt, Chem. Phys. Lett. *50*, 190 (1977), *57*, 157E (1978).
51. J. Čížek and J. Paldus, Int. J. Quantum Chem. *5*, 359 (1971).
52. J. Čížek, J. Paldus and L. Šroubková, Int. J. Quantum Chem. *3*, 149 (1969).
53. J. Paldus and M.J. Boyle, Int. J. Quantum Chem. (to be published).

54. J. Paldus, J. Čížek and I. Shavitt, Phys. Rev. A5, 50 (1972).
55. M.J. Frisch, R. Krishnan and J.A. Pople, Chem. Phys. Lett. *75*, 66 (1980) and references therein.
56. V. Kvasnička, Phys. Rev. A*25*, 671 (1982).
57. G.D. Purvis and R.J. Bartlett, J. Chem. Phys. (in press).
58. A.P. Jucys, I.B. Levinson and V.V. Vanagas, *Mathematical Apparatus of the Theory of Angular Momentum* (Institute of Physics and Mathematics of the Academy of Sciences of the Lithuanian S.S.R., Mintis, Vilnius, 1960) (in Russian), English translations: Israel Program for Scientific Translations, Jerusalem, 1962, and Gordon and Breach, New York, 1964; A.P. Jucys and A.A. Bandzaitis, *The Theory of Angular Momentum in Quantum Mechanics* (Institute of Physics and Mathematics of the Academy of Sciences of the Lithuanian S.S.R., Mintis, Vilnius, 1964) (in Russian); E. El Baz and B. Castel, *Graphical Methods of Spin Algebras in Atomic, Nuclear and Particle Physics* (M. Dekker, New York, 1972).
59. Cf., Appendix of J. Paldus, B.G. Adams and J. Čížek, Int. J. Quantum Chem. *11*, 813 (1977).
60. J. Paldus, J. Chem. Phys. *67*, 303 (1977).
61. B.G. Adams and J. Paldus, Phys. Rev. A*20*, 1 (1979).
62. B.G. Adams and J. Paldus, Phys. Rev. A*24*, 2302 (1981).
63. B.G. Adams, K. Jankowski and J. Paldus, Phys. Rev. A*24*, 2316 (1981).
64. B.G. Adams, K. Jankowski and J. Paldus, Phys. Rev. A*24*, 2330 (1981).
65. R.A. Chiles and C.E. Dykstra, J. Chem. Phys. *74*, 4544 (1981).
66. H. Nakatsuji and K. Hirao, J. Chem. Phys. *68*, 2053, 4279 (1978); Chem. Phys. Lett. *47*, 569 (1977); K. Hirao and H. Nakatsuji, J. Chem. Phys. *69*, 4535, 4548 (1979); Chem. Phys. Lett. *79*, 292 (1981).
67. H. Nakatsuji, K. Ohta and K. Hirao, J. Chem. Phys. *75*, 2952 (1981).
68. K. Jankowski and J. Paldus, Int. J. Quantum Chem. *18*, 1243 (1980).
69. B.G. Adams, K. Jankowski and J. Paldus, Chem. Phys. Lett. *67*, 144 (1979).
70. R.A. Chiles and C.E. Dykstra, Chem. Phys. Lett. *80*, 69 (1981).
71. S.M. Bachrach, R.A. Chiles and C.E. Dykstra, J. Chem. Phys. *75*, 2270 (1981).
72. H.J. Monkhorst, Int. J. Quantum Chem. Symp. *11*, 421 (1977).
73. J. Paldus, presented at the Propagator Workshop, Sanibel Symposium, 1982.
74. J. Čížek and J. Paldus, Int. J. Quantum Chem., Symp. *6*, 435 (1972); J. Paldus and J. Čížek, J. Chem. Phys. *60*, 149 (1974).
75. J. Paldus, J. Čížek, M. Saute and A. Laforgue, Phys. Rev. A*17*, 805 (1978); M. Saute, J. Paldus and J. Čížek, Int. J. Quantum Chem. *15*, 463 (1979).
76. R.J. Bartlett and G.D. Purvis, Int. J. Quantum Chem. *14*, 561 (1978).
77. J.A. Pople, R. Krishnan, H.B. Schlegel and J.S. Binkley, Int. J. Quantum Chem. *14*, 545 (1978).
78. P.R. Taylor, G.B. Bacskay, N.S. Hush and A.C. Hurley, J. Chem. Phys. *69*, 1971 (1976); Chem. Phys. Lett. *41*, 444 (1976).
79. G.D. Purvis and R.J. Bartlett, J. Chem. Phys. *75*, 1284 (1981).
80. J.A. Pople, R. Krishnan, H.B. Schlegel and J.S. Binkley, Int. J. Quantum Chem., Symp. *13*, 225 (1979).
81. P.E.S. Wormer, F. Visser and J. Paldus, J. Comput. Phys. (in press).

82. Cf., e.g., M. Hestenes, *Conjugate Direction Methods in Optimization* (Springer-Verlag, New York, 1980).
83. J. Paldus, P.E.S. Wormer, F. Visser and A. van der Avoird, J. Chem. Phys. *76*, 2458 (1982).
84. J. Paldus, P.E.S. Wormer, F. Visser and A. van der Avoird, in Proceedings of the Fifth Seminar on Computational Methods in Quantum Chemistry, W.C. Nieuwpoort, Ed. (Max Planck Institute, München, 1982), p. 13.
85. B. Weiner, Int. J. Quantum Chem. *21*, 591 (1982) and references therein.
86. F.A. Matsen, Int. J. Quantum Chem., Symp. *15*, 163 (1981) and references therein.
87. E.R. Davidson, *The World of Quantum Chemistry*, R. Daudel and B. Pullman, Eds. (Reidel, Dordrecht, 1974), p. 17; R. Langhoff and E.R. Davidson, Int. J. Quantum Chem. *8*, 61 (1974).
88. E.R. Davidson and D.W. Silver, Chem. Phys. Lett. *52*, 403 (1977).
89. R.J. Bartlett and I. Shavitt, Int. J. Quantum Chem., Symp. *11*, 165 (1977), *12*, 543(E) (1978).
90. P.E.M. Siegbahn, Chem. Phys. Lett. *55*, 386 (1978).
91. D. Mukherjee, Chem. Phys. Lett. *79*, 559 (1981).
92. O. Goscinski and O. Sinanoğlu, Int. J. Quantum Chem. *2*, 397 (1968).
93. R.J. Bartlett, I. Shavitt and G.D. Purvis, J. Chem. Phys. *71*, 281 (1979).
94. D.J. Rowe, A. Ryman and G. Rosensteel, Phys. Rev. A*22*, 2362 (1980); G. Rosensteel and D.J. Rowe, Phys. Rev. A*24*, 673 (1981).
95. S. Wilson, K. Jankowski and J. Paldus, Int. J. Quantum Chem. (to be published).
96. J. Paldus, J. Čížek and B.W. Cho, unpublished results.
97. B. Jeziorski and H.J. Monkhorst, Phys. Rev. A*24*, 1668 (1981).
98. D. Mukherjee, R.K. Moitra and A. Mukhopadhyay, Pramana *4*, 247 (1975); Mol. Phys. *30*, 1861 (1975), *33*, 955 (1977).
99. A. Banerjee and J. Simons, Int. J. Quantum Chem. *19*, 207 (1981).
100. F. Coester, in *Lectures in Theoretical Physics*, K.T. Mahanthappa and W.E. Brittin, Eds. (Gordon and Breach, New York, 1969), Vol. XI-B, p. 157; R. Offermann, W. Ey and H. Kümmel, Nucl. Phys. A*273*, 349 (1976); R. Offermann, Nucl. Phys. A*273*, 368 (1976); W. Ey, Nucl Phys. A*296*, 189 (1978).
101. I. Lindgren, Int. J. Quantum Chem., Symp. *12*, 33 (1978).
102. J.A.R. Coope, Mol. Phys. *18*, 571 (1970).
103. J. Paldus, K. Jankowski, B.G. Adams and J. Čížek, in *Electron Correlation*: Proceedings of the Daresbury Study Weekend, 1979, M.F. Guest and S. Wilson, Eds. (Science Research Council, Daresbury, 1980), p. 92.
104. N.W. Bazley and D.W. Fox, J. Math. Phys. *3*, 469 (1962).

Chairman's Introduction to Sessions M1 and M2 on June 15, 1982

E. Bright Wilson
Department of Chemistry
Harvard University
Cambridge, Mass 02138, U.S.A.

As Chairman of this session, I feel very privileged. First, it gives me the opportunity to express my tremendous admiration for the willingness of Professor Löwdin, Karin Löwdin, the local committee, and the helpers from far away Florida to undertake the organization of this meeting. It is already a demonstrated success and we are all very grateful.

Secondly, being a Session Chairman allows me to make a few general remarks in the way of personal prejedices which I hope are not totally unconnected with the three plenary lectures of this session.

It is probably acceptable to divide the contributions of quantum mechanics to chemistry into three parts. One thinks perhaps first of the thousands of quantitative calculations of electronic energies from which values of interatomic distances, heats of reaction, barriers to internal rotation, reaction rates, etc. have been obtained. I believe that ultimately this will be the most important type of contribution and I am tremendously impressed by the improvements which are being made in the methods for carrying out such calculations. When I hear talk about CI computations with one million configurations, I stand in some awe!

On the other hand, there are still some difficulties with these numerical methods. Experience shows that variationally better wave functions do not always yield better structures, or other properties of most interest. Molecular properties

P.-O. Löwdin and B. Pullman (eds.), New Horizons of Quantum Chemistry, 61–63.

are often quite sensitive to the degree of balance of the basis set and configurations used. Since there do not yet appear to be any practical methods for estimating error bounds for calculated properties, the sole way of judging the reliability of these calculations is comparison with experiment on the same or at least similar molecules. One can well agree with those who have asked if the label *a priori* is the right description.

However, there is another way in which quantum mechanics has had an immense impact and that is as a tool with which to interpret experimental data and obtain molecular properties. In fact, for most experiments the raw data is almost meaningless, but with the aide of simple models it can often be interpreted quantitatively with quantum mechanics. Thus, microwave spectroscopy yields accurate bond lengths, dipole moments and a long list of additional molecular parameters, but only with the aide of quantum mechanics.

Quantum mechanics has impinged on chemistry in still a third way, which may well be the most important of all three. This is by suggesting broad qualitative ideas that have wide predictive value and considerable importance in interpreting large amounts of observational data. I will just mention two of these concepts - and they are related - the concept of resonance in chemistry and the rules for the effect of orbital symmetry on reaction mechanisms.

Both of these ideas are characterized by their elegant simplicity and by the huge mass of experimental data which supports them. I would like to raise one question about such concepts. It may not be necessary but I think it is esthetically desirable that the effect in question be reasonably well defined and that it not disappear when more accurate wave functions are considered.

The effect of orbital symmetry was first expressed in terms of very approximate wave functions indeed, and that is the most likely way for such ideas to arise. However, the concept can be expressed in a way which holds for much higher levels of approximation. Whatever the level of accuracy of the wavefunction used, it can be employed to give natural orbitals and their occupation numbers. Then instead of correlating reactant and product orbitals in terms of plots of orbital energies, one can plot the occupation numbers of the natural orbitals for reactants and products and correlate the natural orbitals of the same symmetry, as in the conventional diagrams. The thermally forbidden reactions will now be those in which the curves of occupation number as a function of the geometric variable connecting reactant with product show a crossing such that the occupation number for one natural orbital drops strongly on passing from reactant to product while another shows a marked increase.

In the prototype example of resonance involving the two Kekulé structures of benzene, one can use an exactly similar plot of occupation numbers for the natural orbitals versus a geometric parameter which carries the system from one Kekulé structure to the other. In this case there is no strong change in the occupation numbers of the natural orbitals so the process is an allowed one; in fact, the intermediate structure is considerably more stable than either Kekulé form.

My time does not permit me to extend these ideas here but I do believe that very simple and elegant concepts such as these two have had an immense impact on chemistry and I expect there are others waiting to be discovered.

ENERGY TRANSFER ALONG ALPHA-HELICAL PROTEINS

A.S.DAVYDOV

Institute for Theoretical Physics of the Academy of Sciences of the Ukrainian SSR

ABSTRACT

It is shown that the energy of the hydrolysis of ATP molecules can be transported without losses along alpha-helical protein molecules as special collective excited states - SOLITONS. These molecules are also a perfect guide for electron transfer from a donor to an acceptor molecule. A new hypothesis of the mechanism of muscle contraction is suggested.

I. INTRODUCTION

It is known that the energy released in the hydrolysis of adenosine triphosphate (ATP) molecules is the universal energy unit used in a great number of energy transformations in plant and animal cells, ranging from the simplest ones up to a human being.

All these transformations involve protein molecules. The most important property of protein molecules is their very large size (molecular weight is equal to tens and many hundreds of thousands daltons). In this connection the site where the ATP hydrolysis occurs is often separated by a large distance from the site where the energy released is used. Therefore, the question arises concerning the mechanism responsible for the transduction of energy released in ATP hydrolysis along large protein molecules.

In I973 a conference devoted to an investigation of the mechanism of energy transport in biological systems was held in New York [I] .The following three problems were discussed: I) Is there a crisis in bioenergetics? 2) What is the nature of the crisis? 3) How can the crisis be resolved?

Taking into account the small energy ($\sim$ 0,5 ev) released under the ATP hydrolysis some scientists (McClare, D.Green, L.Shohet, K.Staub et al.) held that the energy of characteristic vibrations of carbon monoxide (C = O) of peptide groups contained in all proteins is transferred along protein molecules in biological systems. Weber et al. disputed this opinion asserting that due to the short lifetime (of the order of 10^{-12} sec) of vibrational excitations of individual

P.-O. Löwdin and B. Pullman (eds.), New Horizons of Quantum Chemistry, 65–75.

peptide groups they cannot take part in the excitation transfer. The supporters of the view that the energy is transferred by vibrations refuted this objection by arguing that a special structure of protein molecules considerably enhances the life-time of such excitations. However, they failed to prove convincingly their opinions. Thus, the problem concerning the crisis in bioenergetics remained open. Theoretical investigations conducted at the Institute for Theoretical Physics of the Academy of Sciences of the Ukrainian SSR were aimed to solve this problem. It was shown that the energy of the hydrolysis of ATP molecules can be transferred without losses along alpha-helical protein molecules as special collective excited states - solitons. It was suggested a new hypothesis of one mechanism of muscle contraction. It was also shown that alpha-helical proteins are a perfect guide for the electron transfer from a donor to an acceptor molecule. The results of these investigations are given below.

2. SOLITONS AND EXCITONS IN PROTEIN MOLECULES

All the protein molecules are formed by a linear polymerization of 20 different aminoacids and represent very long polypeptide chains with the periodic repetition of peptide groups (PG) which contain the H, N, C, O atoms. These polypeptide chains can be in different spatial configurations among which the structures called alpha-helices [2] are of special interest.

In alpha-helical proteins the poly-peptide chains are coiled into long helices. The structure is due to an additional hydrogen bonds between the PG, besides the basic chemical link with neighbouring carbon atoms retaining the PG within the primary polypeptide structure.

There are three chains of hydrogen bonds in each helical protein molecule. The first chain is formed by the interactions between the hydrogen atom of the first PG and the oxygen atom of the fourth, between the hydrogen atom of the fourth and the oxygen atom of the seventh, etc. The second chain is due to the hydrogen bond between the hydrogen atom of the second PG and the oxygen atom of the fifth, between the hydrogen atom of the fifth group and the oxygen atom of the eighth, etc. The third chain is formed by the hydrogen bond between the hydrogen atom of the third PG and the oxygen atom of the sixth, between the hydrogen atom of the sixth and oxygen atom of the ninth, etc.

The peptide groups are situated along the three chains of hydrogen bonds at equal distances between each other, forming periodic structures. The energy of hydrogen bond is approximately an order of magnitude less than that of ordinary chemical bonds. Therefore, the displacements of peptide groups from their equilibrium positions along the chains of hydrogen bonds are much easier fulfilled than their displacements along the main peptide chain.

The quasi-periodic structure of an alpha-helical protein molecule provides the collectivization of vibrational excitations of individual PG's. The basic vibrational excitation of the PG corresponds to vibrations C = O. It has an energy of 0,2I ev, and a dipole moment of about 0.3 debaye [3] directed along the chain of hydrogen bonds. Owing to the dipole resonance interactions the vibrations C = O of individual PG's are collectivized. As it was shown by A.S.Davydov et al. [4-8], it is very important to take into account the displacements of equilibrium positions of PG's along hydrogen bonds in a theoretical description of such collective excitations. In this case in the continuum approximation the collective excitations are described by a set of differential equations. If we consider the excitation of each chain of PG's in the protein molecule independently, then the collective excitation corresponding to the excitations C = O in peptide groups spreading along the chain with a constant velocity V will be described by a non-linear Schrödinger equation with a non-linearity coefficient

$$G = 4\chi^2/\varkappa(1-s^2), \quad s = V/V_0, \tag{I}$$

depending on the value of the longitudinal elasticity $\varkappa$ of the peptide chain; om the parameter χ characterizing the connection of PG vibrations with the displacements of their equilibrium positions and on the relation s between the excitation velocity and the longitudinal sound velociyty V_0.
As shown in [4-8], two types of collective excitations of excitons and solitons may appear in each chain of PG's.
The excitons are collective excitations spreading along the chain with a group velocity exceeding the velocity of longitudinal sound waves ($s > 1$). They are described by wave packets smearing with the time. Owing to the rapid motion of an exciton the local deformation cannot follow it. Therefore, the exciton carries along the chain only the energy of an intrapeptide vibration which can be expressed by the formula

$$E_{ex}(V) = E_{ex}(0) + \frac{1}{2} m_{ex} V^2 \tag{2}$$

Here $E_{ex}(0)$ is the internal energy of an exciton which is slightly smaller than the energy of an isolated vibration C = O; the second term in (2) determines the exciton kinetic energy proportional to the squared velocity. The proportionality coefficient m_{ex} is called an effective exciton masa. It is inversely proportional to the energy of resonance interaction, J, between the neighbouring PG's in the chain. While travelling the excitons generate phonons, therefore, they are rapidly decelerated. The solitons are collective excitations formed as a result of the combination of intrapeptide vibrations C = O with a local deformation of the chain. They always move with the velocity V less than that of a longitudinal sound. The energy of a soliton travelling along the chain with velocity is expressed by the formula

$$E_{sol}(V) = E_{sol}(0) + \frac{1}{2} m_{sol} V^2, \quad (s^2 \ll 1). \tag{3}$$

In this case it is very important that the internal soliton energy

$$E_{sol}(0) = E_{ex}(0) - \chi^4/3\varkappa^2 J \tag{4}$$

is less than the internal exciton energy. In other words, when the vibrations are "connected" with a local deformation there appears an excitation with an energy less than the sum of the energies of excitations of its constituents. Since the internal soliton energy is less than the internal exciton energy, the soliton formation is energetically more advantageous than the formation of an exciton. The formula (3) is valid only at small soliton velocities ($s^2 \ll 1$). In the general case, at $s^2 < 1$, the soliton energy is expressed by a more complex formula

$$E_{sol}(V) = E_{sol}(0) + \chi^4 [5s^2 - 1 - (1-s^2)^3] / 3\varkappa^2 (1-s^2)^3 J,$$

from which it follows that at $s \to 1$ the soliton energy would increase up to infinity. Hence, all finite values of the energy $E_{sol}(V)$ correspond to the velocities V which are considerably less velocity V_0 of a longitudinal sound in the chain. Strictly speaking, the continuum approximation used above is not applied in the range of the values $s \approx 1$.

The effective soliton mass considerably exceeds the effective exciton mass

$$m_{sol} = m_{ex}(1 + 8M\chi^4/3\varkappa^3\hbar^2) \tag{5}$$

where M is the mase of the peptide group. The large soliton mass is due to the fact that its motion is accompanied by the motion of a local deformation of the chain. The soliton of a large mass can transfer a great amount of kinetic energy even at small velocities of its motion.

If the position of PG in the chain is characterized by the coordinate $z = na$ where n is an integer and a is the distance between neighbouring PG's, then in the soliton state the probability of distribution of the vibrational excitation in PG's will be expressed by a bell-like function

$$W(z,t) = \frac{aQ}{2\cosh^2[Q(z - z_0 - Vt)]}, \tag{6}$$

called symmetric solitons correspond to a synchronous transfer of a local excitation of the type (6) along three chains of PG's. In this case the helix diameter in the excitation region increases and the distances between neighbouring PG's decrease.
Two others types of solitons are nonsymmetric. They corresponds to the transfer of local excitations along three chains of PG's with phase shifts. In this case in the excitation region the helix bends and slightly increases in its diameter. One of the typce of nonsymmetric solitons corresponds to transfer of local excitation only along two chains. Solitons of this type have the lowest internal energy. Therefore, they are the most stable.
Since the formation of solitons in protein molecules is accompanied by a local deformation, i.e. by a displacement of PG's from their equilibrium positions they can not be excited by light, due to the Frank-Condant principle. By the same principle, the probability of the light emission by solitons is small [IO] . The solitons may be excited by local external effects, for example, by chemical reactions. The probability for an excitation will be the greatest when the edge of a molecule is acted upon. Such an excitation may be realized under the hydrolysis of ATP molecule attached to the end of the protein molecule.
Although the excitons are less stable, they are easily excited by light of a corresponding frequency, since the transition into this excited state is not accompanied by the displacement of equilibrium positions of PG's. As it was shown by Yu.N.Chirgadze et al. [3] ,the study of infrared spectra when the excitons are generated in protein molecules, is an effective means for the analysis of their structure.
The above analytical expressions are obtained in the continuum approximation when discrete chains of PG's are replaced by discontinuous chains of an infinite length. In I979, Hyman, McLaughlin and Scott [II] obtained the numerical solutions for the equations proposed by Davydov et. al. [6] for discrete chains using the computer of Los Alamos Scientific Laboratory (Los Alamos, USA).
They have considered the chains containing 200 PG's. It was assumed that the edge PG's were initially excited and the propagation of the excitation along the chain in the course of time was studied. The numerical calculations showed that if the coupling parameter of vibrational excitations with displacements of PG's exceeds a certain critical value, then an excitation of the soliton type is propagated along the chain. The authors [II] concluded that the numerical calculations support the results in Refs [5-8] obtained analytically.
The numerical calculations made in the Los Alamos Laboratory were repeated by Chris Eilbeck and A.Scott using another program at Alas Computing Division Rutherford Lab. of Heriot-Watt University in Edinburg (England). They made a computer film "Davydov Solitons on the Alpha-Helix". The author thanks Professor Chris Eilbeck for sending the film to the ITP in Kiev. This film is important for two reasons: firstly, it supports the earlier calculation performed at the Los-Alamos Lab. using quite another program and a calculation system; secondly it clearly demonstrates the stability of solitons relative

to acoustic noises appearing in the system in a longitudinal direction. The conditions of soliton excitations in the protein molecule are also elucidated.

The above theoretical investigations showed that the energy of ATP hydrolysis can transfer without the energy loss along alpha-helical protein molecules in the form of solitons. In this case about half the energy released in the ATP hydrolysis transforms into their kinetic energy.

3. THE ELECTRON TRANSFER ALONG PROTEIN MOLECULES BY SOLITONS

The processes connected with the electron transfer along protein molecules play an important role in bioenergetics. It is established that sometimes the electron transfer is realized at distances exceeding 30 Å [I2] .

The problem concerning the possibility of electron transfer from a donor to an acceptor molecule through the protein molecule has often been discussed in the literature. It was suggested that "extra" electrons move along the conduction band of a protein molecule. In particular, such a suggestion was made by the prominent biochemist Szent-Gyorgyi [I3, I4] .

The donor-acceptor model of electron transfer through proteins has recently been developed by Petrov, Ukrainskii and Kharkyanen [I5] et al [I6] . In these works, the motion of an electron in the conduction band of protein was considered without regard for the possible displacements of molecular equilibrium positions of atoms in the molecule.

According to the calculations of Evans and Gergely [I7] and more recent calculations performed by Ito. Pullman, Ladik et al., the forbidden band width of the intrinsic electrons of the protein molecule is of the order of 4-6 eV. Therefore, the proteins are dielectrics rather than semi-conductors. Ukrainskii and Mironov [I8] first pointed ont that in studying the motion of an "excess" electron in the donor-acceptor model the conduction band for an extra electron generated by constant electric dipoles of PG's should be considered rather than the highly placed conduction band of the intrinsic electrons of the molecule.

Each PG in the protein molecule has a comparatively large constant dipole moment (3-4 debye) [I9] . Turner, Anderson and Fox showed [20] that in the field of such a dipole the electron may be in a bound state with a binding energy about 1 eV. Thus, the peptide groups in each of the three chains of hydrogen bonds of the alpha-helical protein molecule form repetitive potential wells for an external electron.

The overlapping of the wave functions of the electron ground states in the neighbouring potential wells causes the collectivization of these states in every chain (the conduction band arises). The electron motion along each chain can be considered independently, since the overlap of the electron wave functions in PG's placed in diffeat

$$Q = \chi^2 / [2a^2 \varkappa J(1-s^2)] \ , \quad s^2 \ll 1 . \tag{7}$$

According to (6), the maximal probability $aQ/2$ corresponds to the excitation of a peptide group with the coordinate

$$z = z_0 + Vt$$

travelling along the chain with velocity V. The region of the chain embraced by an excitation and travelling with velocity V has the approximate sizes

$$\Delta z \approx \frac{2\pi}{Q} = a^2 \varkappa J(1-s^2) / \pi \chi^2 , \tag{8}$$

which do no change with the time. Recall that the wave packets describing the exciton motion smear with the time.
In the region embraced by an excitation the distances between peptide groups decrease by the valus

$$\Delta a(z,t) = \frac{a\chi Q}{\varkappa(1-s^2)\cosh^2[Q(z-z_0-Vt]} . \tag{9}$$

According to (2), (3) and (4) with increasing rigidity $\varkappa$ of the chain and increasing value of the resonance interaction J the difference in the properties of solitons and excitons decreases. In this case according to (8) the soliton length increases. However, in soft chains involving PG's held by hydrogen bonds in protein molecules and a strong connection (Large χ) between the vibrations C = 0 and the chain deformation, the two types of excitations are much different.
The exceptional stability of solitons is caused by two factors:
I. the soliton splitting, i.e. the separation of an exciton excitation and a local deformation, according to (2), (3) and (4), requires an expense of the energy

$$\chi^4 / 3\varkappa^2 J . \tag{IO}$$

2. traveling always with a velocity less than that of a longitudinal sound the solitons do not emit phonons, i.e. do not transform their energy into the energy of heat motion. The great stability of solitons and their motion along the protein molecule involving no energy losses resembles the superconductivity phenomenon in metals.
For simplicity, we have considered above the collective excitations propagating only along one chain of PG's. In real alpha-helical molecules there are three nearly parallel chains of PG's. As shown in Refs. [6,9], in this case three types of solitons can travel along the protein molecule. Solitons of the first type which were

rent chains is very small.
Unfortunately, in a paper by Ukrainskii and Mironov [18] as well as in all the previous papers there were no considerations of local displacements of equilibrium positions of PG's caused by the interaction with an extra electron transferred to a protein molecule from a donor. In this case, as Davydov has shown [21,22], the main feature of its motion, that is a peculiar superconductivity, is lost. Taking into account the displacements of equilibrium positions of PG's the motion of an external electron along the chain is described by a nonlinear Schrödinger equation [21,22]. Its solution shows that the total energy of the electron and the deformation energy of the displacement field (at $V=0$) decrease relative to the bottom energy of conduction band of an external electron by the value

$$\Delta E = \sigma^4 m a^2 / 12 \varkappa^2 \hbar^2, \tag{11}$$

where σ is the parameter of the deformation potential caused by the local displacements of PG's; m is the effective electron mase in the conduction band; $\varkappa$ is the coefficient of the longitudinal elasticity of the chain of PG's.
The probability of electron distribution in the chain in the coordinate system $\zeta = z - z_0 - Vt$ travelling with a constant velocity V is determined by a bell-like function

$$W(\zeta) = aq / 2\cosh^2(q\zeta), \tag{12}$$

where

$$q = am\sigma^2 / 2\varkappa(1-s^2)\hbar^2, \quad s^2 = (V/V_0)^2 \ll 1. \tag{13}$$

In the region where the electron is defined the distance between neighbouring PG's decreases by the value

$$\Delta a(\zeta) = aq\sigma / 2\varkappa(1-s^2)\cosh^2(q\zeta). \tag{14}$$

The electron travelling along the chain together with a local deformation may be called an electrosoliton. The effective electrosoliton mass

$$M_{sol} = m(1 + M\sigma^4 / 6\varkappa^2\hbar^2) \tag{15}$$

considerably exceeds the effective electron mass in the conduction band.
The energy of the electrosoliton travelling with a low velocity is

expressed by the formula

$$E_{sol}(V) = \mathcal{E}_0 - \frac{ma^2\sigma^4}{12\varkappa^2\hbar^2} + \frac{1}{2}M_{sol}V^2, \quad s^2 \ll 1, \qquad (16)$$

where $\mathcal{E}_0$ is the energy of the conduction band bottom.
The electrosoliton can travel along PG's through a helical protein molecule only with a velocity V less than the velocity of a longitudinal sound, V_0 , therefore,it does not lose its energy for emitting phonons.
Consequently, the interaction of an electron with the field of displacements of PG's in a helical protein molecule stabilizes its motion; it goes into a quasi-superconducting state. At small $\varkappa$ and large σ this stabilization is quite important.
Thus, owing to the autolocalization of an electron its motion along the alpha-helical protein molecule is stabilized and assumes a superconducting nature. This is a special type of superconductivity without electron pairing which is characteristic only of one-dimensional systems. This phenomenon seems to be important for elucidating the high efficiency of electron transfer without losing energy in the electron transport chains. Thus, alpha-helical protein molecules may be ideal guides in the electron transfer from a donor to an acceptor molecule.

4. A NEW HYPOTHESIS OF THE MECHANISM OF MUSCLE CONTRACTION

One of the most interesting problems in bioenergetics is to elucidate the molecular basis of converting the chemical energy released in ATP hydrolysis into the mechanical energy of the motion processes.At present the transversely striated muscles of which all the skeleton muscles and cardiac muscles of vertebrates, human being and some invertebrates are composed have received the most study.
The author used the concept of the soliton motion in helical protein molecules [23-26,7] to elucidate the mechanism of cotraction of trans versely striated, muscles at the molecular level.
A muscle fiber is a bundle of several thousands of densely packed parallel myofibrils. Each myofibril consists of repeating segments,or sarcomeres. The sarcomeres are separated by transversal membranes.In both directions from them, thin protein filaments stretch into the sarcomere. Thin filaments are composed of three types of proteins - actin, tropomyosin and tropomin. They penetrate partly into the space between thick filaments placed in the internal region of the sarcomere. Thin and thick filaments are surrounded by the sarcoplasm which involves ATP molecules, Mg^{2+} ions and others.
The contraction of the sarcomere length ahd, consequently, the length of a muscle fibre occurs when under the influence of a nerve impulse the calcium ions come into the sarcoplasm from specific microtubes. The calcium ions stimulate the ATP hydrolysis at the site where thick and thin protein filaments touch one another.

Using electron-microscope investigations it was established that the contraction of the sarcomere length is caused by thin filaments penetrating the space between thick filaments without the contraction of the length of the filaments themselves. The problem has arisen of explaining the mechanism of the sliding of thin filaments relative to thick ones at the molecular levels. What forces cause this sliding? How is the ATP hydrolysis energy transformed into sliding energy? At present there is a widespread opinion that the sliding of thin filaments relative to thick ones is due to an active motion of the heads of myosin molecules of which the thick filaments are formed. It is supposed that under the ATP hydrolysis the head of the myosin molecule lengthens forming a link ("bridge") with a globular actin molecule of the thin fibre, then turns displacing the thin fibre relative to the thick one towards the sarcomere centre and, finally, detaches itself from the actin molecule regaining the previous size and the position in a thick filament.

The idea of the cross-bridges between thick and thin filaments, which get closed pull the thin fibres causing their displacements and then break, does not elucidate the molecular nature of this phenomenon. The questions remain: how is the ATP hydrolysis energy used in the lengthening, the cross-bridge formation, the pull force and the cross-bridge breaking? What is the molecular mechanism of changes at the myosin molecule head that leads to these phenomena? And, finally, why does only the head of the huge myosin molecule (the molecular weight is 500,000) which has a long "tail" (the length is about 1700 A^{o}) in addition to its "head", consisting of alpha-helical protein, take an active part in the sliding mechanism?

Using theoretical investigations of solitons in helix proteins the author proposed a new hypothesis to elucidate the mechanism of shortening in the sarcomere length of muscular fibres. According to this hypothesis, the calcium ions reaching under diffusion the first series of the myosin molecule heads at the ends of thick filaments initiate the hydrolysis of the ATP molecules attached to them. The energy released generates in long helical parts of myosin molecules nonsymmetric solitons which move from the heads of the molecules to their ends placed in the region of the sarcomere centre. The motion of a nonsymmetric soliton is accompanied by a local bending and swelling of a molecule. The heads of myosin molecules placed above the curved parts of helical molecules that run under these myosin molecules attach to thin filaments causing their slight displacement in the direction of the motion of solitons, - that is, to the centre of a sarcomere.

According to this model, the myosin molecule heads attach to thin filaments and detach themselves from thin filaments (as in the model of "cross-bridge" formation and breaking). This movement, however, is due not to the lengthening, turn and contraction of the heads themselves, but to the movement of solitons inside the thick fibre accompanied by the motion of bending parts of helical molecules. In this case the kinetic energy of solitons is converted into the contraction energy or the energy of tension if a load is applied to the muscle. In this model all the parts of the myosin molecule, but not only its head, are active contractile elements.

References

I. Green D.E.: I973, I8I, pp.583-584. See also Ann N.Y. Acad. Sci. I974, p.227 and, particularly, the "General discussion" on pp.I08-II5.
2. Pauling L., Corey R.B., Branson H.R.: I95I, 37, Proc. N.A.S. USA, pp. 729-740.
3. Nevskaya N.A., Chirgadze Yu.N.: I976, Biopolymers, I5, p. 637.
4. Davydov A.S., Kislukha N.I.: I976, Phys. Stat. Sol. (b), 75, pp. 735-742.
5. Davydov A.S.: I977, J. Theor. Biol., 66, pp. 379-387.
6. Davydov A.S., Eremko A.A., Sergienko A.I.: I978, Ukr. Fiz. Zhurnal, 23, pp. 983-993.
7. Davydov A.S.: I979, Intern. J. Quant. Chem., 26, pp. 5-I7.
8. Davydov A.S.: I979, Phys. Scripta, 20, pp. 387-394.
9. Eremko A.A., Sergienko A.I.: I980, Prepr. ITP-80-P, Kiev, pp.I-20
I0. Davydov A.S., Eremko A.A.: I977, Ukr. Fiz. Zhurnal, 22, pp.88I-892.
II. Hyman J.M., McLaughlin D.W., Scott A.C.: I98I, Physica 3 D, pp. 23- 44.
I2. Atanasov B.P., Ostnikova G.B., Sadycov Yu.Ch., Wolkenstein M.V.: I977, Molec. Biology, II, p. 557.
I3. Szent-Gyorgyi A.: I94I, Nature, I48, p. I57.
I4. Szent-Gyorgyi A.: I94I, Science, 93, p. 609.
I5. Petrov E.G., Ukrainskii I.I., Kharkyanen V.N.: I978, J. Theor. Biophys., 73, pp. 29-50.
I6. Davydov A.S.: I978, Phys. Stat. Sol.(b), 90, pp. 457-494.
I7. Evans M.G., Gergely J.A.: I949, Biochem. et Biophys. Acta, 3, pp. I88-I95.
I8. Ukrainskii I.I., Mironov S.L.: I979, Prepr. ITP-79-6IP, Kiev, pp. I-I8.
I9. Bates W.W., Hobbe M.E.: I95I, J. Amer. Chem. Soc., 73, no 5, pp. 2I5I-2I56.
20. Turner I.E., Anderson V.E.: I968, Phys. Rev., I74, pp. 8I-89.
2I. Davydov A.S.: I979, Theor. Math. Phys., 40, no 3, pp. 408-42I.
22. Davydov A.S.: I980, Phys. Stat. Sol.(b), I02, pp. 275-282.
23. Davydov A.S.: I974, Biophysica (Russ.), I9, pp. 670-676.
24. Davydov A.S.: I975, Ukr. Fiz. Zhurnal, 20, no 2, pp. I79-I84.
25. Davydov A.S.: - In.: "Biophysica and Biochemistry of Contraction of Muscles". - M.: "Nauka", I976, pp. 254-257.
26. Davydov A.S.: I974, Studia Biophysica, 47, pp. 22I-228.

COLLECTIVE AND INDEPENDENT-PARTICLE MOTION IN SIMPLE ATOMS AND MOLECULES: A UNIFICATION?

R. Stephen Berry, Gregory S. Ezra and Grigory Natanson
Department of Chemistry and the James Franck Institute
The University of Chicago, Chicago, Illinois 60637, U.S.A.

ABSTRACT

The evidence is reviewed that implies electrons in doubly-excited states of helium behave like the atoms of a linear triatomic molecule. The relation between molecule-like collective motion and independent-particle, solar-system-like motion is explored via the model problem of two interacting particles on concentric spheres. The state of understanding of independent-particle motion for atoms in small polyatomic molecules is surveyed briefly and the proposal is explored that some small molecules, such as dihydrides, might exhibit independent-particle rotational motion in sufficiently highly excited local mode states.

This is a summary and brief review of the accumulating evidence that atomic and molecular few-body problems, that is, the description of electrons in simple atoms and the description of atoms in small molecules and clusters, share characteristics that let us see these problems on a common footing. By exploiting these commonalities in our attacks on these problems, we find powerful new ways to achieve accurate, efficient descriptions of small atoms and molecules, and new insights into the physics of their behavior. There are three components to this presentation: 1) the now-convincing evidence that some atomic states are characterized far better in terms of molecule-like collective vibrations and rotations than of independent-particle, solar-system-like motions; 2) the information that emerges from the set of model problems of the quantum states of particles on concentric spheres (POCS) and from other model problems such as the quantum states of a particle in a Hénon-Heiles potential, and 3) the examples and possibilities that atoms exhibit independent-particle motion in small molecules. One of us reviewed this subject recently for another symposium (Berry, 1982), but even in the brief interval between that presentation and this, new ideas have emerged that move the subject ahead and bear mention here.

P.-O. Löwdin and B. Pullman (eds.), New Horizons of Quantum Chemistry, 77–93.

MOLECULE-LIKE QUANTIZATION IN DOUBLY-EXCITED HELIUM

In 1980, Herrick and Kellman proposed that the "intrashell" states of doubly-excited helium--states in which the two electrons have the same principal quantum number--constitute supermultiplets. (Herrick and Kellman, 1980; Herrick, Kellman and Poliak, 1980). Further, they proposed that the origin of the supermultiplet pattern arises out of a physical situation in which the near-constants of the motion correspond to those of a linear triatomic molecule: (Kellman and Herrick, 1980) the angular momentum of a rigid, linear rotor and the bending motions of an XYX vibrator. They also speculated that stretching vibrations would also emerge. The basis of their proposal is the remarkable fit of the entire set of intrashell states for n = 3,4 and 5 to the patterns, not only of their quantum numbers but of the energy spacings as well, of the rotator-bender supermultiplet. Historically their proposal emerged from their recognition of multiplet rotor series (Kellman and Herrick, 1978). This in turn was stimulated by earlier applications of the model of e-e interaction as the symmetry-breaking interaction of a doubly-excited helium atom described as two hydrogenic electrons, each with its Hamiltonian invariant under O(4) symmetry (Herrick and Sinanoglu, 1975; Sinanoglu and Herrick, 1975; Wulfman, 1973; Wulfman and Kumei, 1973), and also by the spatial distributions exhibited by the electrons even in relatively crude representations, of the lowest members of the rotor series. (Rehmus, Kellman and Berry, 1978).

The pattern of levels for $n_1 = n_2 = 3$ is shown in Figure 1, which is drawn to correspond to molecular quantization. The limited validity of the independent-particle model was well recognized; Herrick and Kellman (1980) had pointed out how poorly the states of the n = 3 shell were fit by a CI expansion, in the sense that no single configuration dominated each and every state.

The molecular model was confirmed by computations from rather good wavefunctions of conditional probability densities for the electrons, in terms of the variables R_1, R_2 and θ_{12}. These densities $\rho(R_2,\theta_{12}; R_1 = \alpha)$ show that the $2s^2\ {}^1S^e$, $2s2p\ {}^3P^o$ and $2p^2\ {}^1D^e$ states of the n = 2 shell do indeed form a rotor series with very similar spatial distributions, as we see in Figure 2. If the independent particle picture were correct, these states would have no particular relationship to one another; in the molecular picture, they should differ only through the relatively small effects of centrifugal distortion. Clearly, the latter is much closer to the behavior of the functions in Figure 2.

Other molecule-like characteristics are a) the distribution for the $2s2p\ {}^1P^o$ state which, in a Hylleraas expansion, appears as a first excited vibrational state strongly coupled by Coriolis interaction to the antisymmetric stretching mode, (Yuh et al., 1981) b) as Figure 3 shows; b) the $2s2p\ {}^1P^o$ and $2p^2\ {}^3P^e$ states, which have no reason to be similar according to an independent-particle picture, do indeed have conditional probability distributions that are quite similar, apart from the mandatory nodes at $\theta_{12} = 0$ and π for the ${}^3P^e$ state (Berry, 1982 ; Ezra and

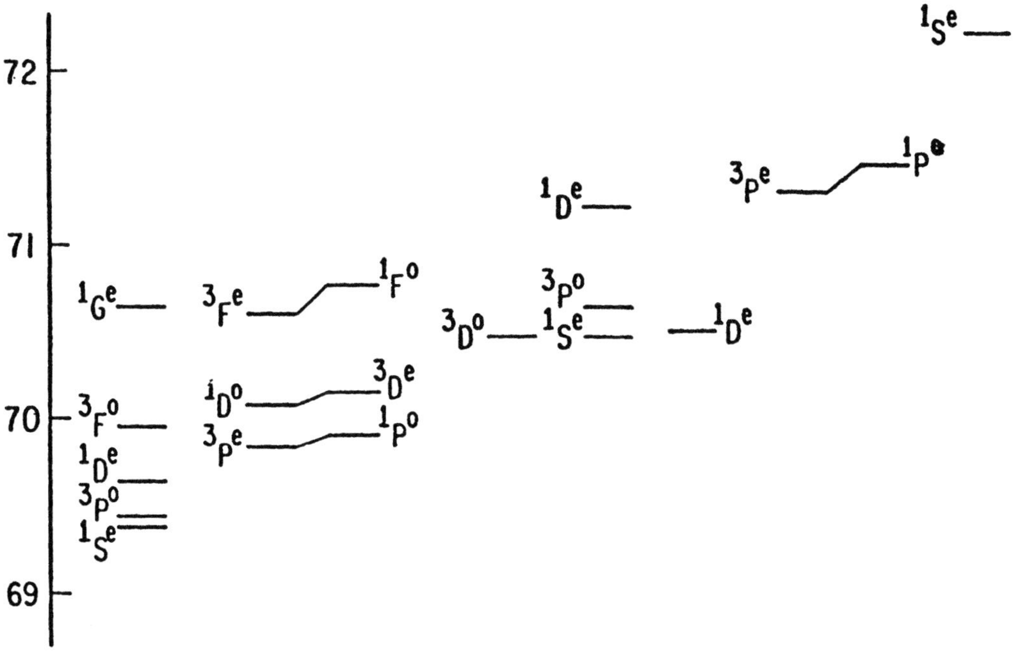

Figure 1. The energy levels of the "intrashell manifold" of states of doubly-excited helium with $n_1 = n_2 = 3$. The energies correspond to states calculated with a Sturmian basis (Ezra and Berry, 1982b) and on the scale of this figure are indistinguishable from energies calculated with Hylleraas functions.

Berry, 1982b); c) the 2s3s $^3S^e$ state has the conditional probability distribution of a first excited state of an antisymmetric stretching mode (Yuh et al., 1981), as Figure 4 shows. Extensive calculations of the states of the n = 2 manifold (Ezra and Berry, 1982b) based on Sturmian configurational functions bear these ideas out. The rotor states are again much alike, the states that should be nearly degenerate partner states of the bending vibration are indeed very similar, and each successive excitation of the bending vibration is associated with

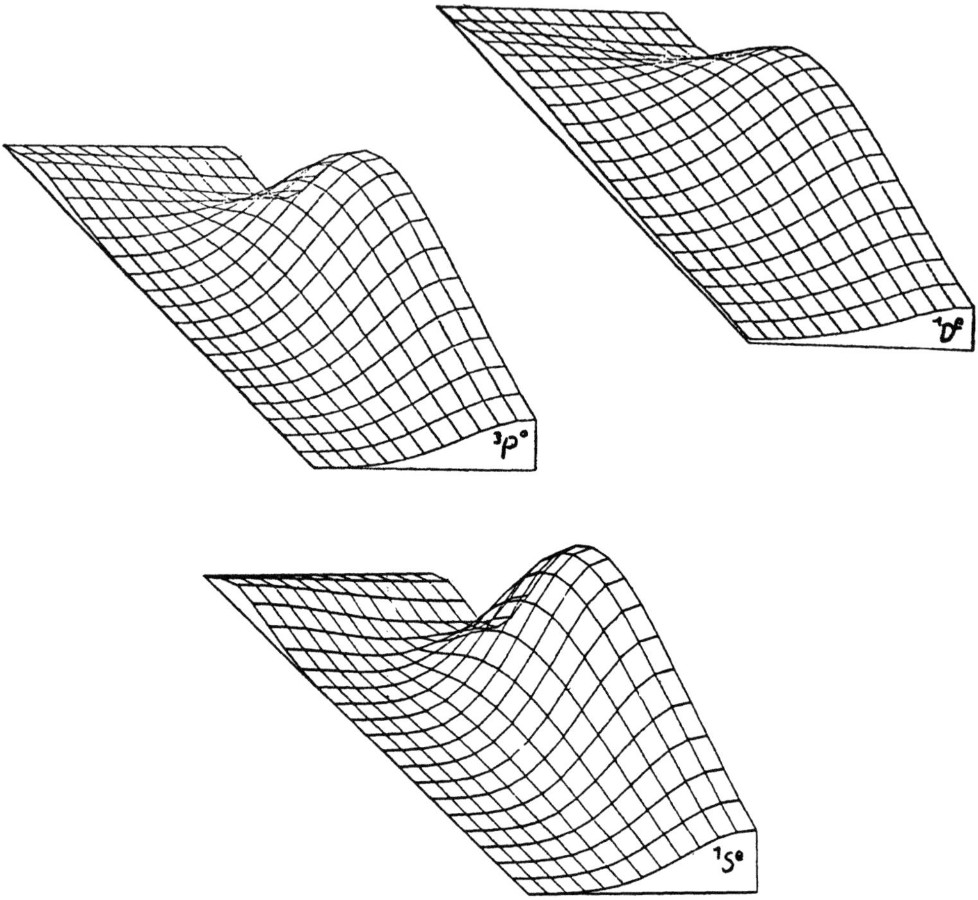

Figure 2. Conditional probability densities for the three pure rotor states of doubly-excited helium with $n_1 = n_2 = 2$. These are the "$2s^2$" $^1S^e$, the "2s2p" $^3P^o$ and the "$2p^2$" $^1D^e$ states. They were calculated with Sturmian functions; the distance R_1 is fixed at its most probable value for all three cases.

a ladder of rotor states.

Two technical points are in order here, for the particular audience this presentation addresses. It is important to recognize the importance of using Sturmian functions or a flexible set of Slater-type functions rather than hydrogenic functions. Because hydrogenic functions admit one-electron states of high angular momentum ℓ only with states of high n, it is impossible to introduce a high degree of angular correlation into the region near the nucleus unless one includes continuum functions into the basis set. There is an open question about the adequacy of Sturmian or other configurational bases for representing states

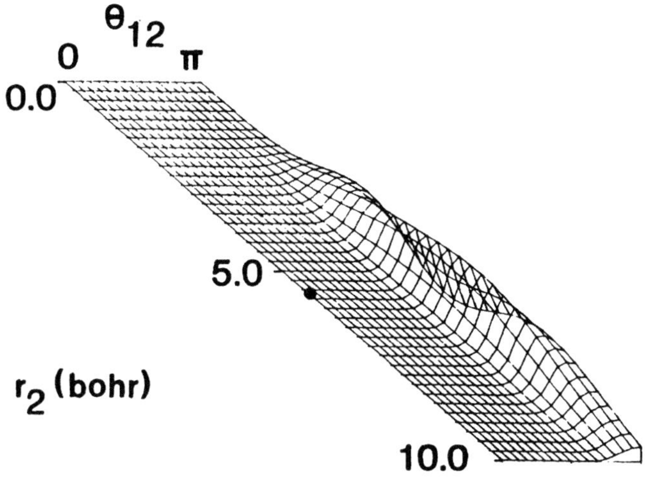

Figure 3. The conditional probability distribution for the "2s2p" $^1P^o$ state of doubly-excited helium, computed from a function expanded in a Hylleraas basis. The value of R_1 is the most probable. This distribution is that of a bending vibration with a node at $R_1 = R_2 = R_{\text{most prob}}$ and $\theta_{12} = \pi$, coupled by Coriolis interaction to the antisymmetric stretching mode.

in which angular and radial correlations are strongly linked, as in the Coriolis-coupled 2s2p $^1P^o$ state and its higher analogues.

The molecular picture is not a panacea for describing doubly-excited states of helium-like atoms and ions. As the kinetic energy of the electrons is raised, e.g., by increase of the nuclear charge, some of the states transform from molecule-like to independent-particle-like. One example is the $2p^2$ $^1D^e$ state which is rotor-like for He and Li^+ but very much like a pure $2p^2$ independent-particle state in Ne^{+8}. The $2s^2$ state remains rotor-like, however, as the nuclear charge is increased because the increase in kinetic energy that tends to make the state become independent-particle-like is compensated by the decrease in separation of the 2s and 2p levels and the corresponding increase in the configuration mixing of $2s^2$ $^1S^e$ and $2p^2$ $^1S^e$ states that lends the lower of these $^1S^e$ states most of its rotor-like distribution.

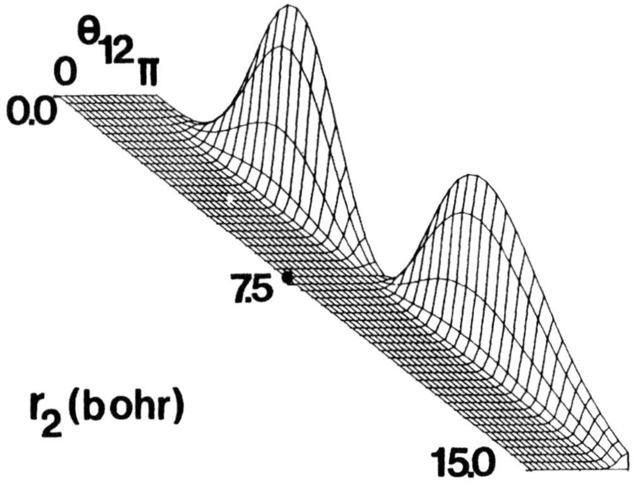

Figure 4. The conditional probability distribution for the "2s3s" $^3S^e$ state, essentially a first excited state of an antisymmetric stretching vibration. Here, R_1 is close to its mean value; if R_1 is made smaller electron 2 moves to larger R along $\theta_{12} = \pi$, and if R_1 is made larger, electron 2 moves to smaller R, also along $\theta_{12} = \pi$.

An obvious question to ask, but a difficult one to answer at this time is "How well does the molecular picture extend to three-electron or many-electron systems?" It may be necessary to have Hylleraas-type representations of excited states of 3-electron systems in order to answer this question. In view of the advances being made for such representations of their ground states, this is no longer an unrealistic expectation (Ho, 1981).

One final caveat must be added. The independent-particle picture does remain the appropriate first approximation for the ground state and singly-excited states of helium and helium-like ions including even H^-. This is seen very clearly in the conditional probability distributions for these states. (Rehmus, Roothaan and Berry, 1978; Rehmus and Berry, 1979). The appearance of molecule-like collective motion in atoms must still be considered the exception, not the rule. However, its appearance at all necessarily opens the range of our expectations to admit varieties of behavior considerably wider than we have previously considered for these systems.

MODEL PROBLEMS: ILLUSTRATIONS OF COLLECTIVE AND INDEPENDENT BEHAVIOR

The problem of two electrons in an atom in three dimensions is richer and more complex than is needed to represent the supermultiplets described by Herrick and Kellman. It is also more elaborate than is necessary to illustrate the transition between collective, molecule-like and independent-particle, solar-system-like behavior. A simpler model containing the essential features of this phenomenon consists of two interacting particles confined to concentric spheres. This system has the advantage that it can be solved by expansion of the wavefunctions in series of spherical harmonics.

The model has one limitation as an analogue of the two-electron problem; there is no natural shell structure in the particles-on-concentric-spheres (POCS) model. Instead, there is an infinite set of solutions for the angular motion, for every choice of the radii. To make the model analogous to the two-electron problem, one must choose radii for the spheres that correspond to n_1 and n_2, the principal quantum numbers of interest, and then truncate the set of solutions to correspond to those contained in the corresponding representation of O(4) x O(4), that is, to the states that can be constructed from the assignments of quantum numbers consistent with n_1 and n_2. The choice of radius can be made by using the relation derived by Kellman and Herrick (1980) between the principal quantum number and the rotational B_e, and then setting $B_e = \frac{1}{4R^2}$

The eigenstates and eigenfunctions for two electrons (Ezra and Berry, 1982a) on the same sphere are remarkably similar to those of two electrons in the helium atom, when the radius of the sphere is chosen to correspond to $n_1 = n_2 = 2$, 3, 4 or 5. Figure 5 shows the level pattern for a value of $R = 13.7$ bohr, corresponding to $n_1 = n_2 = 3$. Figure 6 shows the probability distributions of the corresponding eigenstates. There is, of course, only the one internal variable θ_{12} in this system. As with the intrashell states of doubly-excited helium, we see a rotor series (which, incidentally, exhibits a slight centrifugal contraction with increasing angular momentum), clearly marked states of the bending vibration with almost identical wavefunctions for the nearly-degenerate partner states, and rotor series built in turn on each successive vibrational excitation. The constraint eliminating radial motion has made the system more molecule-like than the helium atom.

A natural extension and comparison with the case of two electrons on a sphere is the parallel case of two particles with electronic masses on a sphere of the same radius but interacting via a repulsive Gaussian potential. (Ezra and Berry, 1982a). The eigenvalues and distributions of this sytem, again for R = 13.7 bohr and for a Gaussian designed to

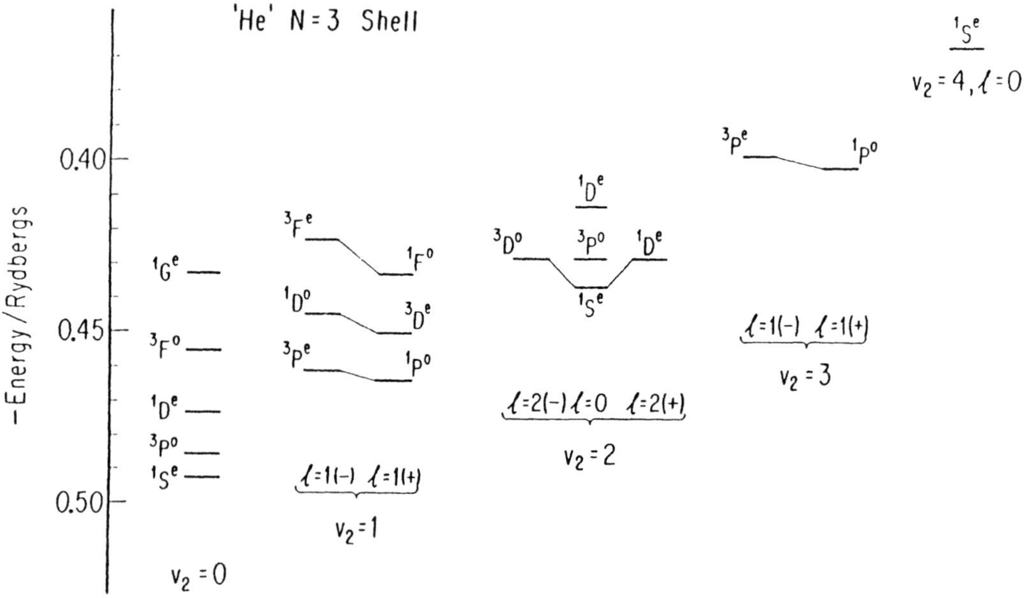

Figure 5. Energy levels for two electrons on a sphere of R = 13.7A, simulating the shell $n_1 = n_2 = 3$ of He.

give a lowest state as similar as possible to that of the corresponding Coulomb problem, are shown in Figure 7 and Figure 8, respectively. The close resemblances to Figures 5 and 6, and to the energy level pattern of Figure 1, are immediately apparent.

The most significant differences between the Coulomb and Gaussian cases are a) the larger splittings of the pairs of vibrational states in the Gaussian model; b) the corresponding larger differences for the Gaussian case between the distributions of some of the vibrational partner states, most notably the $^3D^o$, $^1D^e$ and $^3P^e$, $^1P^o$ states with 2 and 3 quanta, respectively, of bending and no rigid-rotor quanta; and c) the distributions for the highest $^3P^e$ and $^1S^e$ states of the Gaussian system

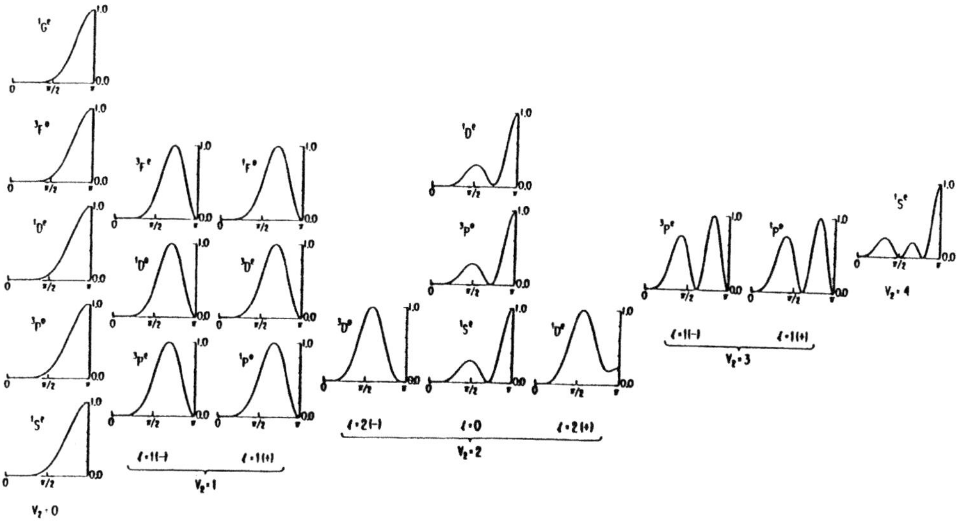

Figure 6. The distributions $|\psi(\theta_{12})|^2$ of the eigenfunctions corresponding to the eigenvalues of Figure 5, for two electrons on a sphere.

that are nearly symmetrical about $\theta_{12} = \pi/2$, a characteristic symptom of independent-particle quantization.

Hence it seems that the Gaussian system, with its short-range force between particles, while still collective in most of the states we are examining, nevertheless exhibits more independent-particle behavior than does the Coulomb system. The reasons for this are clear enough: in the high states of the n = 3 shell, the system has enough energy to surmount the Gaussian barrier around $\theta_{12} = 0$, and the Coulomb potential goes on to infinity. Yet it is striking to see an example in which the Coulomb system, which we traditionally associate so closely with the independent-

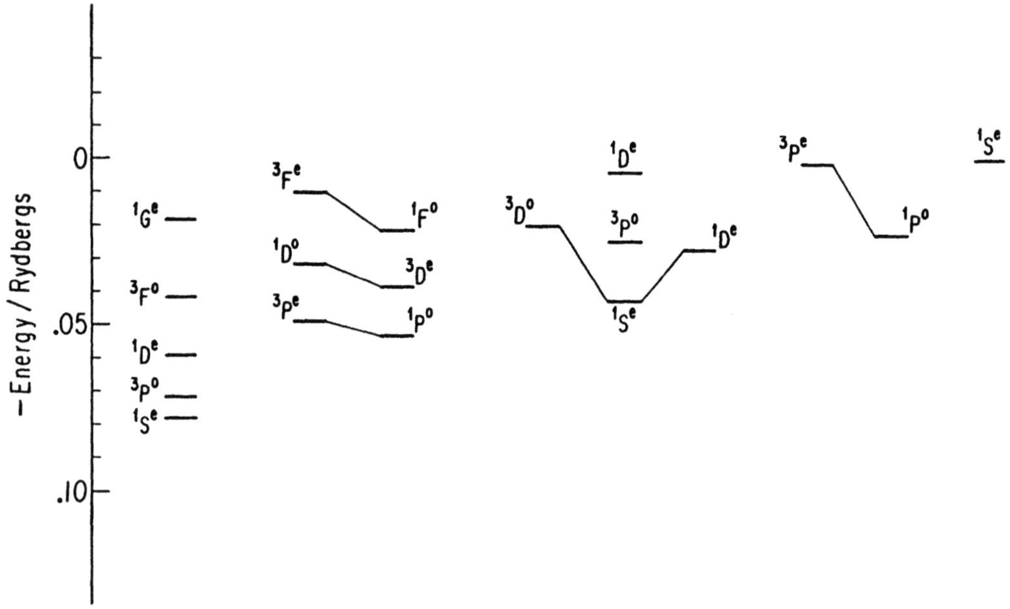

Figure 7. Energy levels for two particles with electron masses interacting via a Gaussian repulsive interaction designed to mimic the lowest states of Figure 6.

particle model, is more molecule-like than the Gaussian system, which is more associated with atom-atom repulsions and collective motions in molecules.

By making the radii of the two spheres different, one can model "intershell" states for which $n_1 \neq n_2$ (Berry, 1982; Ezra and Berry, 1982c). If R_1 is fixed, for example, at 5 bohr and the Schroedinger equation is solved for $R_2/R_1 = 1,2,\ldots$, one can watch the transition from correlated, molecule-like behavior to independent-particle behavior With Coulombic repulsion, the distributions indicate essentially independent-particle behavior for $R_2/R_1 = 10$. If the Coulomb repulsion is

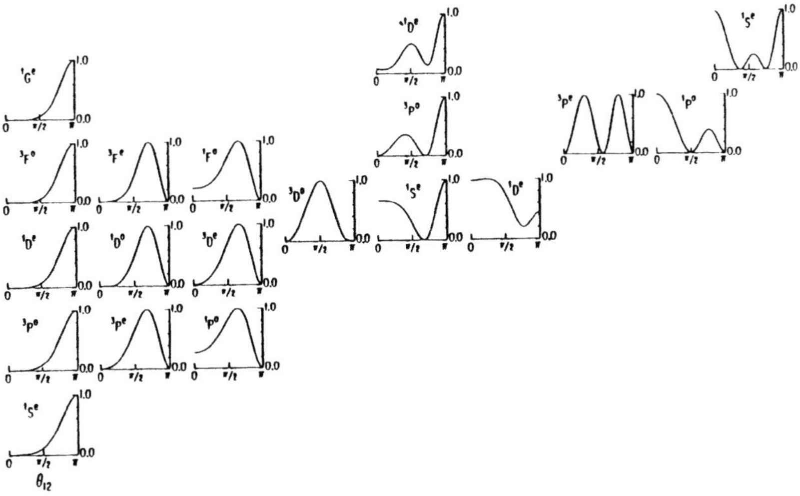

Figure 8. The distributions $|\psi(\theta_{12})|^2$ corresponding to the energy levels of Figure 7.

replaced by a Gaussian $V(R,\theta_{12}) = S \exp[-2(R_1-R_2)^2/W^2]$ (S = 1, w = 5), like that used for Figures 7 and 8, then independent-particle motion sets in when R_2/R_1 is only 3. Recall that the Bohr radius $R \propto n^2$; the relation inferred by Kellman and Herrick (1980) for helium-like atoms is close: $R \propto n^{2.15}$, approximately. These numbers imply that if $n_1 = 3$, the motion is fairly independent when $n_2 \geq 0$ for two electrons in an atom, but is already independent for $n_2 \geq 4$ for the hypothetical Gaussian system.

The ℓ-doubling or Λ-doubling, the splitting of the vibrational pairs deserves a little comment. The order of the splitting in the two-electron atoms is opposite to that of normal triatomic molecules; the order of the splitting in the POCS models is like that of triatomic molecules, with the (+)-partners of the ℓ-doublets lower in energy. The ordering originates in the POCS and linear triatomic molecule, from the coupling of rigid-rotor angular momentum and bending motion: for example the interaction of the two $^1D^e$ states with $V_2 = 2$ (Fig. 5) pushes the energy of the state with $\ell = 2(+)$ down while that of the state with $\ell = 0$ moves up. The order in atoms arises largely from the mandatory nodes in $\rho(\theta_{12})$ and $\psi(\theta_{12})$ at $\theta_{12} = 0$ in the states of lower energy, which keep down the energy of Coulomb repulsion. If an inadequate basis is used for the POCS model, the order is sometimes given incorrectly (Ezra and Berry 1982a).

Another model problem that has attracted much attention is that of the quantum states of a particle in the 2-dimensional Henon-Heiles potential. While most of the attention directed toward this problem has been in the context of the transition between quasiperiodic and chaotic motion, one particular recent study brings the quantum-mechanical Henon-Heiles problem tantalizingly close to the question of independent-vs.-collective motion. Hose and Taylor (1982) making quantitative a proposal of Nordholm and Rice (1974) regarding dominance of single basis functions, have shown that two sorts of approximate constants of the motion arise for this problem, corresponding to two separations of the full Hamiltonian into zero-order and perturbation parts. In both the isotropic, 2-dimensional harmonic separation and the uncoupled harmonic-plus-anharmonic separation, eigenstates of the full Hamiltonian appear that are more than 50% composed of a single eigenfunction of the corresponding zero-order Hamiltonian. Such states generally seem to have all or all but one of the system's quanta in a single mode of that zero-order, separable Hamiltonian. We would like to suggest that these state are analogous to one sort of independent-particle or independent-degree-of-freedom-motion. The many states that do not satisfy the criterion of being over 50% a single basis function presumably correspond to stron mode coupling to which simple mechanical motion or near-constant of the motion corresponds, and which we would have to interpret as a collective motion. This interpretation takes on a broader meaning when it is carried over to the molecular context, particularly in relation to local modes of vibration.

INDEPENDENT-PARTICLE MOTION IN SMALL POLYATOMIC MOLECULES

The description of the eigenstates of nuclear motion in molecules was deeply rooted in the rigid-rotator, near-harmonic-oscillator picture, and led to the picture of nearly separable collective, rigid-body rotations and normal modes of vibration. The limitations of this picture were long recognized in the context of attempts to describe unimolecular dissociation, which generally require energy to flow among degrees of freedom. The normal-mode picture was further confounded by the recognition of local modes, in which, typically, one light atom oscillates nearly independently of all the other motions in the molecule (Henry, 1977; Swafford, Long and Albrecht, 1976; Lawton and Child, 1979,

1980, 1981; Child and Lawton, 1981; Elert, Stannard and Gelbart, 1977; Jaffe and Brumer, 1980). The accessibility of high-overtone states to high-resolution spectroscopy by resonant multiphoton excitation seems to give more and more impetus to forcing the creation of still other alternatives to the normal mode picture (Harter, Patterson and da Paixao, 1978).

There is not yet a clear, unified picture of how to interpret the high-overtone states of small polyatomic molecules. There is not yet even an algorithm for sorting out systematically the quantum states that correspond to any one simplifying model, although the approach of Hose and Taylor (1982) may lead in that direction for some kinds of states. We can only say now that some highly excited states of some polyatomics correspond to nearly-separable simple mechanical motions, and to near-constants of the motion, that have no counterparts in the states near the bottom of the rotation-vibration ladders. That is, a simple near-separability of some states sets in at high energies that is different from the separability appropriate to the low-energy states, and for some of these states, the separability corresponds to independent-particle motion. For other states, such as those characterized by the clustering of levels of rotating, centrifugally distorted octahedra or tetrahedra, the separable motion is still highly collective (Harter, Patterson and da Paixao, 1978).

The known motions characterizable as independent-particle-like in the high-overtone states are stretching motions, radial in direction. We wish to suggest here that there may be states that must also be described by independent-particle excitations, in which the new, near-constant of the motion is the angular momentum of an almost independent atom. This possibility comes to mind when one turns around the conclusion of Kellman-Herrick quantization that says that electrons in atoms may exhibit molecule-like collective oscillations and rotations: Is it possible for atoms in small polyatomic molecules to exhibit independent-particle motion like that of electrons in an atom, that is, motion characterized by nearly constant orbital angular momentum?

The key to the answer seems to lie with the secondary question, "Can local stretching motion of, for example, an O-H bond of H_2O get one hydrogen enough out of the way of the other to permit the bending mode to turn into a hindered internal rotation of an OH in the field of a spectator, Rydberg-like second H-atom?"

We cannot yet provide an unambiguous answer to these questions for any specific molecule, but we can address them by asking whether such states are plausible, in light of what we know about molecular potential surfaces. The problem was attacked in this way (Natanson, Ezra and Berry, 1982): potentials for the H_2O molecule were selected (Sorbie and Murrell, 1978; Murrell et al., 1981). One hydrogen atom was fixed at R_1 = 0.97A, its equilibrium distance from the oxygen atom, and the other hydrogen was fixed at various longer distances R_2 corresponding to the classical turning points of successively higher states of a local stretching mode. For each fixed R_1, R_2 pair, the effective potential $V(\theta_{12}; R_1,R_2)$ for angular motion was evaluated from the Sorbie-Murrell and Murrell-Carter-Mills-Guest (MCMG) potential and for a smoothed version of the Sorbie-Murrell potential. Then the eigenvalues and eigenfunctions for the two interacting hydrogen atoms were obtained for the fixed-radius approximation by using the POCS model described in the preceding section, but with the interaction potential taken as the $V(\theta_{12}; R_1,R_2)$ functions derived from the three-dimensional water potentials.

The results of these calculations would at best be guides to more careful searches for true rotation-vibration eigenstates of H_2O. However the results of the calculations turn out to be ambiguous in their predictions. The Sorbie-Murrell potential, shown for R_2 = 1.9A in Figure 9a, does permit internal rotation. The eigenvalues of the S(+) states are indicated with the potential. The onset of internal rotation can be seen in the transition exhibited by the three distributions shown in Figures 9b, c and d, for the 6s(+), 7s(+) and 8s(+) states: the 6s(+) is localized around the H-H-O configuration but cannot penetrate the barrier at $\theta_{12} = \pi$. The 7s(+) state has maxima at $\theta_{12} = 0$ and at about $\theta_{12} = 2\pi/3$, and tunnels through the barrier at π. The 8s(+) state passes freely through $\theta = 0$ and $\theta = \pi$, and higher states rotate more or less freely. One might expect, in principle, to see a water molecule pass from bending vibrations to hindered internal rotation to nearly free internal rotation as one excites successively higher states of the local stretching mode and examines the ladder of "bending" states built on each. This would correspond to a single molecule exhibiting the three kinds of quantization described by Bratoz and Martin (1965). The ambiguity for H_2O arises because the MCMG potential, as it turns out, will not support states of internal rotation such as those of Figure 9c and d. Motion past the barrier at $\theta_{12} = 0$ in that potential occurs only at energies above the energy of dissociation of H_2O to H + OH. We are left with the questions of a) whether either the Sorbie-Murrell potential or the MCMG potential is accurate enough in the region of R_1 = 1A, $R_2 \cong$ 1.6-2A, $\theta_{12} = 0$ to describe orbital motion reliably, and b) whether, if water is not a good example, we might still expect to see independent orbiting motion of a single light atom in a simple polyhedral molecule.

CONCLUDING REMARKS

We have seen the evidence that some states of simple atoms behave like the rotation-vibration states of linear triatomic molecules, and have examined the possibility that atoms in a simple polyatomic molecule might exhibit independent-particle motion like that of electrons in atoms The two cases have been connected by the model problem of interacting particles on concentric spheres, which displays the transition between the two extreme sorts of behavior and the relative roles of kinematics and dynamics. Together, these bits of information seem to be leading us to a new set of questions about atomic and molecular structure. Some have been mentioned already, but two of the most challenging were not yet made explicit:

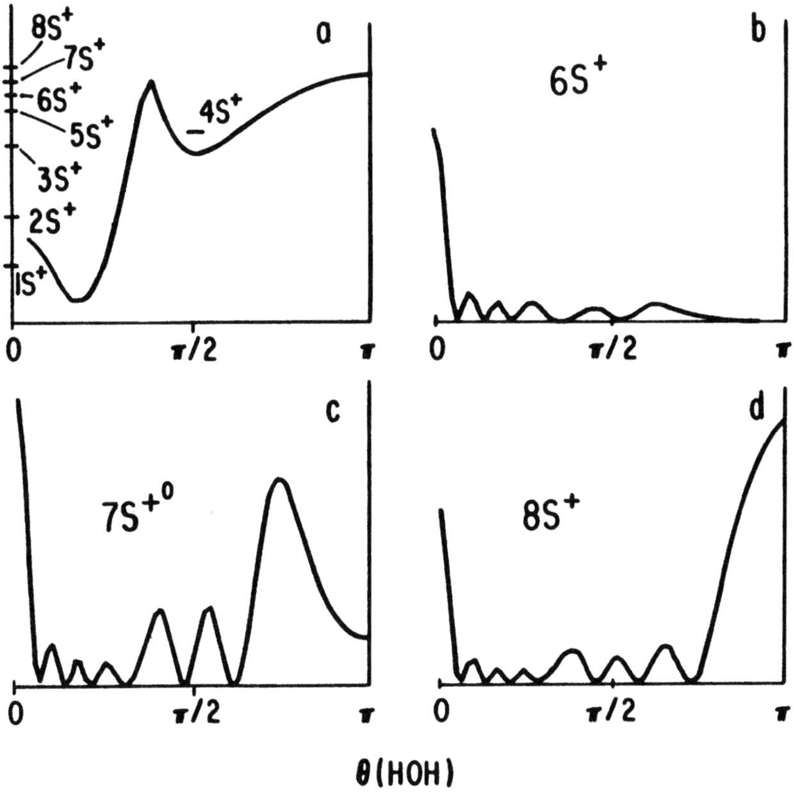

Figure 9. a) The angular potential $V(\theta_{12}; R_1 = 0.97A, R_2 = 1.9A)$, derived from the potential for H_2O given by Sorbie and Murrell (1978); the eigenvalues for the S(+) states through the eighth are indicated;
b) the distribution for the 6s(+) state in the well of a);
c) the distribution for the 7s(+) state of a), showing tunnelling through the barrier at $\theta_{12} = \pi$;
d) the distribution for the 8s(+) state, showing nearly free rotation.

1) What criteria can tell us whether a collective description or an independent-particle description is the most appropriate for states in any given energy range? Do the two descriptions overlap for intermediate cases, or is neither picture an adequate starting point for them?
2) If the molecular, collective picture is so satisfactory for doubly-excited states of helium-like atoms, then how is it that we can still classify groups of these states according to the principal quantum numbers of the two electrons? How can shell structure still persist enough to give useful meaning to terms such as "intrashell states"? Part of the answer to this question seems to be showing itself to us in the forms of the 2s3s $^3S^e$ and $^1S^e$ states, both of which show both shell structure, in their radial behavior, and collective motion, in their angular behavior (Ezra and Berry, 1982b).

When we have answered these questions, we shall have the beginnings of a unification of the atomic and molecular few-body problems in one general framework. At that point, we may well wish to look at the generalization of those ideas into few-body problems on still smaller scales.

ACKNOWLEDGEMENTS

One of us (G.S.E.) would like to thank the UK Science Research Council for the award of a NATO Postodoctoral Fellowship. The research described here by the University of Chicago group was supported by a Grant from the National Science Foundation of the United States.

REFERENCES

Berry, R.S.: 1982, in Proc. Fifteenth Jerusalem Symposium on Quantum Chemistry and Quantum Molecular Biology, J. Jortner and B. Pullman, eds. (Reidel, Dordrecht, Holland, 1982), p. 29.
Child, M.S. and Lawton, R.T.: 1981, Farad, Disc. 71, pp. 1ff.
Elert, M.L., Stannard, P.R. and Gelbart, W.M.: 1977, J. Chem. Phys. 67, pp. 5395ff.
Ezra, G.S. and Berry, R.S.: 1982a, Phys. Rev. A 25, p. 1513-1527.
Ezra, G.S. and Berry, R.S.: 1982b, "Collective and Independent-Particle Motion in Doubly-Excited Two-electron Atoms," (in preparation).
Ezra, G.S. and Berry, R.S.: 1982c, "The Quantum States of Two Particles on Concentric Spheres," (in preparation).
Harter, W. G., Patterson, C.W. and da Paixao, F.J.: 1978, Revs. Mod. Phys. 50, p. 37-83.
Henry, B.R.: 1977, Acc. Chem. Res. 10, pp. 207ff.
Herrick, D.R. and Kellman, M.E.: 1980, Phys. Rev. A 21, pp. 418ff.
Herrick, D.R., Kellman, M.E. and Poliak, R.D.: 1980, Phys. Rev. A 22, pp. 1517 ff.
Herrick, D.R. and Sinanoglu, O.: 1975, Phys. Rev. A 11, pp. 97ff.
Ho, Y.K.: 1981, Int. J. Quantum Chem. 20, pp. 1077ff.

Hose, G. and Taylor, H.S.: 1982, J. Chem. Phys. , xxxx (15 May).
Jaffe, C. and Brumer, P.: 1980, J. Chem. Phys. 73, pp. 5646ff.
Kellman, M.E. and Herrick, D.R.: 1978, J. Phys. B 11, pp. L755ff.
Kellman, M.E. and Herrick, D.R.: 1980, Phys. Rev. A 22, pp. 1536ff.
Lawton, R.T. and Child, M.S.: 1979, Mol. Phys. 37, pp. 1799ff.
Lawton, R. T. and Child, M.S.: 1980, Mol. Phys. 40, pp. 773ff.
Lawton, R.T. and Child, M.S.: 1981, Mol. Phys. 44, pp. 709ff.
Murrell, J.N., Carter, S., Mills, I.M. and Guest, M.F.: 1981, Mol. Phys. 42, pp. 605ff.
Nordholm, K.S.J. and Rice, S.A.: 1974, J. Chem. Phys. 61, pp. 203ff, pp. 768ff.
Rehmus, P. and Berry, R.S.: 1979, Chem. Phys. 38, pp. 257-275.
Rehmus, P., Kellman, M.E. and Berry, R.S.: 1978, Chem. Phys. 31, pp. 239-262.
Rehmus, P., Roothaan, C.C.J. and Berry, R.S.: 1978, Chem. Phys. Lett. 58, pp. 321-325.
Sinanoglu, O. and Herrick, D.R.: 1975, J. Chem. Phys. 62, pp. 886ff.
Sorbie, K.S. and Murrell, J.N.: 1978, Mol. Phys. 29, pp. 1387ff.
Swofford, R.L., Long, M.E. and Albrecht, A.C.: 1976, J. Chem. Phys. 65, pp. 179ff.
Wulfman, C.E.: 1973, Chem. Phys. Lett. 23, pp. 367ff.
Wulfman, C.E. and Kumei, S.: 1973, Chem. Phys. Lett. 23, pp. 367ff.
Yuh, H.-J., Ezra, G., Rehmus, P. and Berry, R.S.: 1981, Phys. Rev. Lett. 47, pp. 497-500.

ARE CLi_6, NLi_5, OLi_4, ETC, HYPERVALENT?

Paul von Ragué Schleyer
Institut für Organische Chemie der Friedrich-Alexander-Universität Erlangen-Nürnberg
Henkestrasse 42, D-8520 Erlangen
Federal Republic of Germany

ABSTRACT

The octet rule fails to predict the stoichiometries of binary lithium compounds of first row elements. There is now gas phase experimental evidence for OLi_3, OLi_4, OLi_5, and CLi_5; many similar perlithiated species are predicted computationally to be highly stable thermodynamically towards all possible dissociation modes. The structures of NLi_4, OLi_4, and FLi_4, for example, are tetrahedral while CLi_5, NLi_5, and OLi_5 favor trigonal bipyramidal (D_{3h}) symmetries. CLi_6^+, and CLi_6 are octahedral. The extra (ninth and tenth) valence electrons occupy totally symmetric (a_1) molecular orbitals, largely lithium-lithium bonding in character. A simple interpretation assigns only an octet of electrons to the central atom, and any additional electrons to the surrounding lithium cage.

1. INTRODUCTION

Efficient quantum mechanical programs facilitate the exploration of fundamentally new combinations of atoms. Our search for novel geometries and bonding arrangements,[1] has now been extended to a reassessment of the octet rule.[2] This work, a collaboration with Ernst-Ulrich Würthwein, Erlangen, and John A. Pople, Carnegie-Mellon University, has employed the Gaussian series of programs written by Professor Pople and his associates,[3] as well as the standard basis sets and Møller-Plesset electron correlation procedures.[4]

Since first row elements possess only four low lying valence atomic orbitals, 2s, $2p_x$, $2p_y$, and $2p_z$, it is widely believed that octet rule violations involving such elements are impossible.[5] Nevertheless, a number of interesting radical anions involving halogenated first row central atoms are known (esr spectroscopy) which involve a 9th electron. However, theoretical analyses of model systems, e.g., CH_3Cl^- and $CFCl_3^-$, show that the "extra" electrons occupy σ^* orbitals, and lead to bond elongation.[6] If "valence" is defined as the number of electron pairs involved in bonding, such radical anions are not truely

P.-O. Löwdin and B. Pullman (eds.), New Horizons of Quantum Chemistry, 95–109.

hypervalent. When bond dissociation is prevented by the imposition of symmetry, molecules like NF_4, NF_5, OF_3, and OF_4 may be calculated, but are indicated to be thermodynamically very unstable.[7] NH_5, at best, is a fleeting species.[8]

Many hypercoordinate first row molecules are known, e.g., CH_5^+, nonclassical carbonium ions, carboranes, and carbide metal carbonyl clusters.[9] While these involve a larger number of nearest neighbours than usual, only an octet of electrons need be assigned to the central atoms. These atoms are not hypervalent. The D_{3h} pentacoordinate S_N2 transition state, e.g., $(FCH_3F)^-$, is particularly well known to organic chemists. In the gas phase, a species with such composition can be observed. However, experimental and calculational investigations agree: The D_{3h} form is a transition structure rather than an energy minimum.[10] The observed C_{3v} dipole-ion complexes, $F^-..H_3CF$, are not hypervalent since one of the halogen atoms is much further away and is not bound directly to carbon.

We have now discovered calculationally a large number of simple molecules, ALi_n, comprised of only first row elements, whose unusual stoichiometries suggest violations of the octet rule. As isolated species, such molecules are indicated to be global energy minima, i.e., they are highly stable thermodynamically towards all possible dissociation modes. A systematic study of all ALi_n possibilities where n=6 or less and A is each of the first row atoms, Be through F, is nearly complete; many of the corresponding cations and anions and some second row systems have been investigated as well. All species are indicated to be highly stable, and can be divided into three categories:

1. Molecules like CLi_4 which violate neither the octet rule nor the normal stoichiometries. Hypovalent species, with fewer than eight valence electrons, are included in this category but will not be discussed here.
2. Effectively hypervalent molecules include those with nine, ten, eleven, or even more valence electrons. These are our principal concern.
3. A final category, exemplified by $BeLi_6$ or $BeLi_5$, have eight valence electrons or less, but do not posess the normal stoichiometries. That is, the number of formal bonds between the central atom and the substituent lithiums exceeds the number of available electron pairs.

2. RESULTS

At the time our work began, there were few experimental or theoretical precedents. Weak lithium atom complexes to ammonia (eq. 1) and to water (eq. 2) had been demonstrated by matrix esr measurements. Calculations by Nicely and Dye[11] and by Kollman and Schaefer,[12] confirmed by our own results, indicate binding energies on the order of 12-14 kcal/mol. Even more interesting was the discovery of Li_3O by Wu, Kudo, and Ihle.[13] By means of high temperature thermochemical measurements, the lithium atom affinity (eq. 3) of Li_2O was deduced to be quite large,

50.7 kcal/mol. The average Li-O bond energy was determined to be 76±3 kcal/mol, and the ionisation potential was 4.54 eV. Despite the rule-breaking stoichiometry and the presence of 9 valence electrons, OLi_3 obviously is a very stable molecule!

NH_3 + Li → $LiNH_3$ -14.5 kcal/mol[12] (1)

OH_2 + Li → $LiOH_2$ -11.7 kcal/mol[12] (2)

OLi_2 + Li → OLi_3 -50.7 ± 10 kcal/mol (expt)[13] (3)

-54.6 kcal/mol (3-21G//3-21G)

-45.9 kcal/mol (6-31G*//3-21G)

-44.8 kcal/mol (MP2/6-31G*//3-21G)

-43.7 kcal/mol (With ZPE)

Why hasn't more attention been paid to this exceptional species? Probably because the structure of OLi_3 is unknown. If the third lithium were attached to another lithium or to both lithiums, but not to oxygen, no violation of the octet rule would be involved. However, OLi_3 is hypervalent. Our calculations indicate all the lithiums to be bound to oxygen. Since the calculated energies are in reasonably good agreement with the experimental values (eq. 3), there is every reason to believe that the calculated structure pertains to the gas phase species. The only minimum on the potential energy surface corresponds to a C_{2v} structure, **1**, only slightly distorted from **2** (the energy difference is only 0.5 kcal/mol at 3-21G). Since the variation of the calculated dissociation energy of OLi_3 (eq. 3) with the size of the basis set is relatively small (electron correlation corrections are insignificant), most of our calculations have been carried out with the efficient 3-21G split valence basis. This permits a systematic evaluation at an uniform level of approximation.

3. ENERGIES

The dissociation energies of a typical collection of perlithiated species are shown in Table 1. These reactions have been chosen to conserve the number of unpaired electrons on both sides of each equation. As can be seen, all of the hypervalent molecules are remarkably stable towards loss of a lithium atom or a Li_2 molecule. However, the most stable form of lithium is the metal, ΔH_f^0 = 0.0 kcal/mol. The heat of formation of atomic lithium, Li, is 38.4 and of Li_2, 50.4 kcal/mol. This helps explain why such perlithiated species have not already been commonly encountered. As isolated entities, e.g., in the gas phase or in a matrix, they may be quite stable, but in condensed phases association and disproportionation may take place.

Nevertheless, it is gratifying that experimentalists, urged on by our calculational results, are now verifying the predictions of theory. Lagow's group in Texas[14] reacts organic substrates, like CCl_4, with

Table 1. Dissociation Energies of Hyperlithiated Neutral Molecules, kcal/mol (3-21G//3-21G). (Number of Valence Electrons)

CLi_5 (9e)	→	$Li + CLi_4$	+ 54.1
	→	$Li_2 + CLi_3$	+ 61.6
NLi_4 (9e)	→	$Li + NLi_3$	+ 55.4
	→	$Li_2 + NLi_2$	+ 57.5
OLi_3 (9e)	→	$Li + OLi_2$	+ 54.6
	→	$Li_2 + OLi$	+106.6
FLi_2 (9e)	→	$Li + FLi$	+ 49.5
	→	$Li_2 + F$	+125.8
CLi_6 (10e)	→	$Li + CLi_5$	+ 14.2[a]
	→	$Li_2 + CLi_4$	+ 65.2
NLi_5 (10e)	→	$Li + NLi_4$	+ 17.0[a]
	→	$Li_2 + NLi_3$	+ 68.6
OLi_4 (10e)	→	$Li + OLi_3$	+ 28.9[a]
	→	$Li_2 + OLi_2$	+ 79.5
FLi_3 (10e)	→	$Li + FLi_2$	+ 9.9[a]
		$Li_2 + FLi$	+ 55.5
OLi_5 (11e)	→	$Li + OLi_4$	+ 15.7
	→	$Li_2 + OLi_3$	+ 40.6
FLi_4 (11e)	→	$Li + FLi_3$	+ 20.4
		$Li_2 + FLi_2$	+ 26.4

[a] Values expected to be larger at correlated levels.

lithium atoms at high temperatures, and subjects the products to flash vaporisation mass spectroscopy. CLi_5, C_2Li_8, C_3Li_{12}, as the corresponding cations, are among the "hyperstoichiometric" species observed. In Jülich, Wu[15] has produced CLi_6 and perhaps CLi_5 by allowing lithium vapor at high temperature to diffuse through a graphite membrane. At my request, Wu has sought and has now found OLi_5 and very probably OLi_4 as well. At the recent ACS meeting, Peterson, Dao, and Castleman[16] reported the experimental observation of ONa_3 and ONa_4 (we have now examined these species calculationally; they are quite like OLi_3 and OLi_4). Finally, "curious suboxides", with known X-ray structures, of the heavier alkali metals, e.g. O_2Rb_9 and O_3Cs_{11}, should be mentioned.[17]

Calculationally, we are finding the phenomenon to be much more widespread than we first imagined. To put the matter in colloquial terms, lithium is a "sticky" element: lithium atoms bind to a variety of molecules (eqs. 1 and 2), but do so particularly well if lithium substituents already are present. Table 2 documents a few miscellaneous examples.

4. STRUCTURES

Hypervalent second row molecules often have structures which reveal the location of lone pair electrons; e.g., in SF_4, these occupy an equatorial position. Our calculated structure of OF_4 is similar, except that C_{4v} instead of C_{2v} symmetry seems to be "preferred". However, this very unstable species can only be examined by imposing C_s symmetry. In contrast, the stable OLi_4 prefers a tetrahedral geometry. Unlike CF_3Cl^-, etc., the O-Li bonds are not lengthened significantly. The average O-Li bond length (1.661 Å) in OLi_3, a nine valence electron species, is actually shorter than the bond length (1.668 Å) in OLi_3^+, an octet molecule. The structures and energies of these hypervalent molecules indicate that the "extra" electrons are neither antibonding nor even nonbonding, but occupy bonding orbitals. As we will see, the fifth valence orbitals of all these species are totally symmetric (a_1) in character.

This results in a peculiar feature. Usually, when electrons are removed from or added to a species, a change in geometry results. For example, CH_4 is tetrahedral, CH_4^{2+} square planar (D_{4h}), but CH_4^+ prefers C_{2v} symmetry. In contrast, NLi_4^+ (eight valence electrons), NLi_4 (9e), and NLi_4^- (10e) all prefer tetrahedral geometries! This is true for all other ALi_4 molecules investigated with eight valence electrons (CLi_4 and OLi_4^{2+}), nine electrons (CLi_4^-, and OLi_4^+), and ten electrons (OLi_4 and FLi_4^+). However, FLi_4, with eleven valence electrons, must singly occupy another valence orbital originally part of the triple-degenerate t_2 set. This leads to Jahn-Teller distortion, and C_{2v} symmetry is preferred.

ALi_5 and ALi_6 molecules with eight, nine, and ten valence electrons are also regular and D_{3h} and O_h symmetries are adopted, respectively. Investigated examples include BLi_5, CLi_5^+, and $BeLi_6$ (8e); CLi_5 and CLi_6^+ (9e); and NLi_5 (10e). With eleven valence electrons, OLi_5 adopts

Table 2. Miscellaneous Association Energies, Kcal/mol (3-21G//3-21G)

LiOH + Li	→	Li_2OH	(C_{2v})	-46.5
				-36.8 (6-31G*//3-21G)
OH + Li_2	→	Li_2OH	(C_{2v})	-106.3
				-90.4 (6-31G*//3-21G)
CH_3Li + Li	→	CH_3Li_2	(C_s)	-21.4
CH_3 + Li_2	→	CH_3Li_2	(C_s)	-35.3
CH_3Li + Li_2	→	CH_3Li_3	(C_s)[a]	-31.9
CH_2Li_2 + Li_2	→	CH_2Li_4	(C_{2v})[b]	-40.6
LiNC + Li	→	Li_2NC[c]	(C_{2v})	-25.7
Na_2O + Na	→	Na_3O[d]	(C_{2v})	-54.9
NaO + Na_2	→	Na_3O[d]	(C_{2v})	-52.7
Na_2O + Na_2	→	Na_4O[d]	(T_d)	-76.6

[a] The $CH_3Li_2^+Li^-$ form is the global minimum.

[b] Nearly "octahedral", with ∡ HCH = 103.7°.

[c] Experimentally observed (C.H. Wu, private communication).

[d] Experimentally observed, see reference 16.

C_{3v} symmetry, and can be regarded, following the Mulliken population, to involve a lithium anion, Li^-, attached to the face of a OLi_4^+ tetrahedron. Likewise, OLi_6, with 12 valence electrons, is not expected to be octahedral. We have not studied all possible distortions, but an apparent D_{3d} minimum has been found. CLi_8, if MNDO calculations are to be believed, should also be a stable species with lower than octahedral symmetry.

The ALi_3 molecules behave somewhat differently. With eight valence electrons, NLi_3 and OLi_3^+ prefer planar, trigonal (D_{3h}) geometries. As already mentioned, the ninth valence electron in OLi_3 induces a small planar distortion to C_{2v} symmetry. FLi_3 with ten valence electrons, prefers a pyramidal (C_{3v}) arrangement, but the D_{3h} form is only 0.2 kcal/mol higher in energy (3-21G//3-21G). FLi_3 was studied some years ago by Allen's group[18] in another context, but they never located the global minimum. These distortions from D_{3h} symmetry as more electrons are added provide the next clue: Li-Li bonding, enhanced by the closer approach of lithiums, is involved. These are not Jahn-Teller effects.

5. BONDING ANALYSIS

Tables 3-6 provide geometries and Mulliken analyses (charges and overlap populations) for the carbon, nitrogen, oxygen, and fluorine based systems. Caveats about Mulliken population analysis notwithstanding, trends in closely related systems are revealed. The most remarkable feature of Tables 3-6 is the invariance of indicated charges on the central atoms as more lithium substituents are added. This is particularly striking for oxygen, where all OLi_n (n = > 3) species, charged or neutral, are indicated to have a negative charge on oxygen of -0.87 $\pm$ .01) (Table 5). A kind of "saturation" effect is involved; the central atom can take up only so many electrons, and no more.

What is the nature of the hypervalent bonding in these perlithiated species? D-orbitals are not involved; none are included in the 3-21G basis set we have commonly employed! In OLi_3 (eq. 3), the 6-31G* basis set (which does include d-orbitals) actually indicates a lower dissociation energy (destabiliziation of OLi_3 relative to the products). Thus, these first row species are quite different from the well known hypervalent second row examples, e.g., PF_5.

The first four valence orbitals in ALi_n molecules are quite normal, e.g., $3a_1$ and $2t_2$ in T_d symmetry. As already mentioned, the key fifth valence molecular orbital also is totally symmetric (a_1). This is the usual situation with AX_n molecules. For example, the fifth valence MO in CF_4 ($4a_1$), has the same character as that in OLi_4. The 2s components on the substituents have the opposite sign than the 2s orbital on the central atom. The antibonding character of the interaction is reduced by outer atom σ 2p contribution, bonding in character. In the ALi_n systems, the fifth valence molecular orbital can be regarded approximately as being nonbonding as far as the A-Li interactions are concerned. However, this MO is Li-Li bonding. This bonding is enhanced by the large lithium coefficients (particularly the diffuse outer functions of the 3-21G Li basis). Since the molecules under consideration generally

Table 3. Geometries and Mulliken Analyses of Perlithiated Carbon Species (3-21G//3-21G)

	C-Li, Å	Charges C	Charges Li	Overlap Populations C-Li	Overlap Populations Li-Li
CLi_4 (T_d)	1.929	-.806	+.202	.569	-.183
CLi_5^+ (D_{3h})	2.022axial 1.999equat	-.902	+.380		
CLi_5 (D_{3h})	2.087axial 2.056equat	-.807	+.151 +.169	.210 .312	.165ea -.120ee
CLi_6 (O_h)	2.058	-.931	+.155	.423	.127

Table 4. Geometries and Mulliken Analyses of Perlithiated Nitrogen Species (3-21G//3-21G)

	N-Li, Å	Charges N	Charges Li	Overlap Populations N-Li	Overlap Populations Li-Li
NLi_3 (D_{3h})	1.739	-.781	+.260	.606	-.098
NLi_4^+ (T_d)	1.800	-.915	+.479	.489	-.132
NLi_4 (T_d)	1.809	-.815	+.204	.416	-.093
NLi_4^- (T_d)	1.789	-.873	-.032	.457	-.052
NLi_5 (D_{3h})[a]	1.844axial 1.856equat	-.934	+.846 -.253	.464 .304	-.340ea .462ee

[a] A C_{3v} ion pair structure, $NLi_4^+Li^-$, is 12.4 kcal/mol less stable.

Table 5. Geometries and Mulliken Analyses of Perlithiated Oxygen Species (3-21G//3-21G) and Sodium Analogs

			Charges		Overlap Populations	
		O-Li, Å	O	Li	O-Li	Li-Li
OLi	$(C_{\infty v})$	1.668	-.547	+.547	.481	
OLi_2	$(D_{\infty h})$	1.583	-.813	+.406	.580	-.011
OLi_3^+	(D_{3h})	1.668	-.876	+.626	.416	-.029
OLi_3	(C_{2v}) 1[a]	1.661av	-.859	+.286av		
	(D_{3h}) 2	1.657	-.856	+.285	.406	.056
OLi_4^{2+}	(T_d)	1.789	-.884	+.721	.329	.002
OLi_4^+	(T_d)	1.745	-.875	+.469	.305	.102
OLi_4	(T_d)	1.728	-.871	+.218	.300	.162
OLi_5	(C_{3v})[b]					
OLi_6	(D_{3d})	1.921	-.877	+.143	.188	.220
ONa	$(C_{\infty v})$	2.015	-.624	+.624	.337	
ONa_2	$(D_{\infty h})$	1.928	-.872	+.436	.439	.079
ONa_3	(C_{2v})	(2.098av)[c]				
ONa_4	(T_d)	2.079	-.895	+.239	.239	.157

[a] $\measuredangle Li_aOLi_a$ = 101.4°, $O\text{-}Li_a$ = 1.676, $O\text{-}Li_b$ = 1.631. [b] Ion pair structure, $OLi_4^+Li^-$, see text. [c] Provisional result.

Table 6. Geometries and Mulliken Analyses of Perlithiated Fluorine Species (3-21G//3-21G)

	Geometries		Charges		Overlap Populations	
	F-Li,Å	∡LiFLi	F	Li	F-Li	Li-Li
FLi ($C_{\infty v}$)	1.520		-.597	+.597	.422	
FLi_2^+($D_{\infty h}$)	1.606	180.0	-.564	+.782	.283	.006
FLi_2 (C_{2v})[a]	1.617	117.9	-.553	+.276	.270	.317
($D_{\infty h}$)[a]	1.594	120.0	-.557	+.278	.275	.185
FLi_3 (C_s)[b]	1.703av	114.0av	-.509	+.169av	.183av	.320av
(D_{3h})[b]	1.594	120.0	-.512	+.171	.320	.183
FLi_4^+ (T_d)	1.846	109.5	-.456	+.364	.118	.239
FLi_4 (C_{2v})[c]	1.843	88.0	-.489	+.324	.152	.670
	1.806	122.3		-.080	.142	-.538
(C_{3v})[c]	1.869x1	96.0	-.491	+.418	.150	.204
	1.806x3	118.7		+.025	.145	-.444

[a] The FLi_2 (C_{2v}) form is 2.60 kcal/mol lower in energy. [b] FLi_3 (C_s) has nearly C_{3v} symmetry. The potential energy surface is very flat; the D_{3h} structure is only 0.2 kcal/mol less stable. [c] These FLi_4 forms differ insignificantly (0.35 kcal/mol) in energy.

have large numbers of Li-Li contacts, the total contribution to the bonding can be quite large. The comparisons in Table 7 are arranged to bring this out clearly. Consider the series of tetrahedral molecules, based on oxygen. Adding a ninth and tenth valence electron to OLi_4^+, to give OLi_4^+ and OLi_4, respectively, results in only a small decrease in the O-Li overlap population (from .33 to .30). In contrast, the Li-Li overlap population increases from 0.00 (OLi_4^{2+}) to 0.10 (OLi_4^+) to 0.16 (OLi_4). This suggests that the decrease in the O-Li distance along the same series (1.79, 1.75, and 1.73 A) is due to more favorable interactions among the lithium atoms rather than to changes in OLi bonding. Note that there are six Li-Li contacts. The nitrogen-based series, NLi_4^+ (8e), NLi_4 (9e), and NLi_4^- (10e), is less regular, but a trend is noted for the Li-Li overlap populations: the antibonding character decreases as electrons are added.

The Mulliken populations suggest the nature of these species in another way. Consider the OLi_n series in Table 5. In OLi, 0.55e has been donated from lithium to oxygen. The next lithium in OLi_2 donates an additional 0.36e. In contrast, the third lithium in OLi_3 contributes very little more, 0.05e. As essentially no further change as additional lithiums are added in OLi_4, OLi_5, and OLi_6. Thus, the extra electrons in these nominally hypervalent species are not associated with oxygen, but rather contribute to Li-Li bonding.

Table 7 also compares the effects of decreasing the electronegativity of the central atom in isoelectronic molecules. This increases the A-Li overlap population and decreases Li-Li bonding. The hypervalent fluorine compounds (also see Table 6) have greater Li-Li than Li-F overlap populations.

We can summarize the bonding situation as follows: the octet of valence electrons in ALi_n molecules is involved in A-Li bonding. The extra electrons in the hypervalent species are not associated with the central atom, but rather contribute to the formation of a metallic lithium cage. The overall structures can be described in terms of a negatively charged center atom, A^-, embedded in a positively charged lithium cluster, Li_n^+.

6. $BeLi_n$ AND BLi_n MOLECULES

The beryllium-lithium and boron-lithium molecules investigated do not violate the octet rule, but do exhibit unusual stoichiometries. BeH_2 and BH_3 are highly electron deficient; the former polymerizes and the latter forms the well known dimer. However, neither species binds H_2; BeH_4 and BH_5 really are van der Waals complexes.[19] With eight valence electrons, BeH_6 can be calculated by imposing octahedral symmetry, but the energy relative to BeH_2 and two hydrogen molecules is very unfavorable. We have also examined BeF_4 but this, too, is only indicated to be a weak $BeF_2.F_2$ complex.

As a substituent, lithium behaves quite differently. The more lithiums are present, the better the energy becomes! 3-21G energies probably are not accurate, but are illustrative. Thus, the addition of Li_2 to Be to give $BeLi_2$ releases only 5.1 kcal/mol of energy.

Table 7. Comparison of Tetrahedral Species (3-21G//3-21G)

	Bond Lengths, Å	Mulliken Populations Charges X	Li	Overlap X-Li	Li-Li
OLi_4^{2+} (8e)	1.789	-.884	+.721	.329	.002
OLi_4^{+} (9e)	1.745	-.875	+.469	.305	.102
OLi_4 (10e)	1.728	-.871	+.218	.300	.162
NLi_4^{+} (8e)	1.800	-.915	+.479	.489	-.132
NLi_4 (9e)	1.809	-.815	+.204	.416	-.093
NLi_4^{-} (10e)	1.791	-.873	-.032	.457	-.052
OLi_4^{2+} (8e)	1.728	-.884	+.721	.329	.002
NLi_4^{+} (8e)	1.800	-.915	+.479	.489	-.132
CLi_4 (8e)	1.929	-.806	+.202	.569	-.183
OLi_4^{+} (9e)	1.745	-.875	+.469	.305	.102
NLi_4 (9e)	1.809	-.815	+.204	.416	-.093
FLi_4^{+} (10e)	1.846	-.456	+.364	.118	.239
OLi_4 (10e)	1.789	-.871	+.218	.305	.162
NLi_4^{-} (10e)	1.789	-.873	-.032	.457	-.052

Reaction of $BeLi_2$ with Li_2 to give $BeLi_4$ is more exothermic, -19.2 kcal/mol. The further reaction of $BeLi_4$, with Li_2 to give $BeLi_6$ is even better, -39.4 kcal/mol. The formation of BLi_5 from BLi_3 + Li_2 is also quite favorable (-45.2 kcal/mol).

These hyperlithiated species can be considered to have metallic cluster structures. With six valence electrons, $BeLi_4$ prefers a square-planar geometry. This is the simplest molecule preferring a D_{4h} global minimum (CH_4^{2+} is also D_{4h}, but is only a local minimum).[20] $BeLi_6$ is octahedral and $BeLi_5$ is trigonal bipyramid (D_{3h}); both have an octet of valence electrons.

7. FAILURES

Although sodium and the other alkali metals are expected to behave in the same way as lithium (see above and Table 5), our search for additional examples involving other substituents has not been successful. Thus, $NeLi_2$ can be calculated by imposing symmetry, but it is unstable thermodynamically; the Ne-Li overlap populations are antibonding. The heavier noble gases may be more favorable in this respect.

Another approach is based on the well known pattern of AX_4 valence molecular orbitals in tetrahedral symmetry, e.g. for CF_4. This pattern, $(a_1)^2(t_2)^6(a_1)^2(t_2)^6(e)^4(t_2)^6(t_1)^6$, suggests that species with 16, 20, and 26 valence electrons might also be stable (as well as those with 8, 10, and 32 electrons). However, we have not found a single instance where this is true. CB_4 (16e) appears to have a tetrahedral minimum, but it is scarcely bound. The C-B overlap population is only .04 and the B-B interactions are antibonding. Tetrahedral C_5 (20e) appears to be the beginning of a three dimensional diamond structure. However, T_d C_5 is not even a local minimum and is over 200 kcal/mol less stable than linear C_5. Since Li and BeH appear to be quite similar (both offer vacant p-orbitals), we had high hopes for $O(BeH)_4$. However, the tetrahedral form could only be calculated by imposing geometrical constraints, the wave function is of lower symmetry and the HOMO was unbound. The calculated dissociation reactions were strongly exothermic. This failure, nevertheless, illustrates a bonding principle we have been emphasizing. The BeH group utilizes sp hybridized orbitals for Be-H bonding, and these are not available to accomodate the hypervalent electrons.

8. CONCLUSIONS

Are CLi_6, NLi_5, OLi_4, etc., hypervalent? The answer depends on what is meant by "hypervalence" but this term, like most chemical terms in common use, has not been carefully defined. The stoichiometries are certainly unusual and violate predictions based on the octet rule. More than eight valence electrons are present and all are involved in bonding.

The opposite viewpoint is equally convincing. Molecules like OLi_4

are electron deficient. Whatever the number of electrons assigned to oxygen, each lithium cannot have a full octet. Suppose we build up OLi_4 by adding valence electrons one by one. The first eight would become involved with the most electronegative atom, oxygen, fulfill its octet requirement and form O-Li bonds. Oxygen having been "satisfied", the next two "hypervalent" electrons would help to alleviate the lithium deficiency. The central atom, oxygen in this case, is not really hypervalent.

Whatever interpretation is preferred, the general phenomenon we call "hypermetallation" is indicated to be extremely general. Further experimental and theoretical attention is warranted.

REFERENCES

1. Kos, A.J.; Jemmis, E.D.; Schleyer, P.v.R.; Gleiter, R.; Fischbach, U. Pople, J.A.: 1981, J. Am. Chem. Soc. 103, p.4996; Apeloig, Y.; Clark, T.; Kos, A.J.; Jemmis, E.D.; Schleyer, P.v.R.: 1980, Israel J. Chem. 20, p. 43. This paper lists earlier references.
2. Lewis, G.N.: 1916, J. Am. Chem. Soc. 38, p. 762; Kossel, W.: 1916, An Phys. 49, p. 229.
3. Gaussian 76: Binkley, J.S.; Whiteside, R.; Hariharan, P.C.; Seeger, R Pople, J.A.; Hehre, W.J.; Newton, M.D.: 1979, QCPE 11, 368; Gaussian 80: Binkley, J.S.; Whiteside, R.; Krishnan, R.; Seeger, R.; DeFrees, D.J.; Schlegel, H.B.; Topiol, S.; Kahn, L.R.; Pople, J.A.: 1981, QCPE 10, 416.
4. Pople, J.A.; Binkley, J.S.; Seeger, R.: 1976, Int. J. Quant. Chem. Symp. 10, p. 1. For the 3-21G basis set, see Binkley, J.S.; Pople, J.A.; Hehre, W.J.: 1980, J. Am. Chem. Soc. 102, p. 939. For absolute energies and fuller descriptions of most of the species in the present paper, see the "Carnegie-Mellon Quantum Chemistry Archive", Whiteside, R.A.; Frisch, M.J.; Binkley, J.S.; DeFrees, D.J.; Schlegel H.B.; Raghavachari, K.; Pople, J.A.: 1981, 2nd Ed., Carnegie-Mellon University.
5. Pauling, L.A.: 1937, J. Am. Chem. Soc. 53, p. 1367.
6. Clark, T.: 1981, J.C.S. Chem. Comm. p. 515 and unpublished calculations; Peyerimhoff, S.D., Buenker, R.J.: 1979, Chem. Phys. Lett. 65, p. 434; Candell, E.; Karafiloglou, P.; Salem, L.; 1980, J. Am. Chem. Soc. 102, p. 855. For recent experimental references, see: Wang, J.T. Williams, F.: 1980, J. Am. Chem. Soc. 102, p. 2860; Hasegawa, A.; Hudson, R.L.; Kikuchi, O.; Nishikida, K.; Williams, F.: 1981, ibid. 103, p. 3436.
7. Allen, L.C.: private communication. Würthwein, E.-U.: unpublished calculations.
8. Olah, G.A.; Donovan,D.J.; Shen, J.; Klopman, G.: 1975, J. Am. Chem. Soc. 97, p. 3559.
9. Brown, H.C. (with comments by Schleyer, P.v.R.), "The Nonclassical Ion Problem", 1977, Plenum Press, New York, pp. 13-19.
10. For leading references, see Wolfe, S.; Mitchell, D.J.; Schlegel, H.B. 1981, J. Am. Chem. Soc. 103, p. 7692 and p. 7694.
11. Nicely, V.A.; Dye, J.L.: 1970, J. Chem. Phys. 52, p. 4795.

12. Trenary, M.; Schaefer III, H.F.; Kollman, P.: 1977, J. Am. Chem. Soc. 99, p. 3885; 1978, J. Chem. Phys. 70, p. 1815.
13. Wu, C.H.; Kudo,H.; Ihle, H.R.: 1979, J. Chem. Phys. 70, p. 1815.
14. For references, see Gurak, J.A.; Chinn, Jr., J.W.; Lagow, R.J.; 1982, J. Am. Chem. Soc. 104, p. 2637. For CLi_5^+, see Jemmis, E.D.; Chandrasekhar, J.; Würthwein, E.-U.; Schleyer, P.v.R.; Chinn, Jr., J.W.; Landro, F.J.; Luke, B.; Pople, J.A.: 1982, ibid. 104, in press.
15. Wu, C.H., private communication. See Wu, C.H.; Ihle, H.R.: 1979, Chem. Phys. Lett. 61, p. 54 and Chandrasekhar, J.; Pople, J.A.; Seeger, R.; Seeger, U.; Schleyer, P.v.R.: 1982, J. Am. Chem. Soc. 105, in press.
16. Peterson, K.I.; Dao, P.D.; Castleman, Jr., A.W.: 1982, 183rd Am. Chem. Soc. National Meeting, Las Vegas, Nevada, Abstract PHYS 173.
17. Simon, A.: 1979, Structure and Bonding 36, p. 81.
18. Kollman, P.; Liebman, J.F.; Allen, L.C.: 1970, J. Am. Chem. Soc. 92, p. 1142. For an earlier study of Li_2F, see Pearson, P.K.; Hunt, W.J.; Bender, C.F.; Schaefer III, H.F.: 1973, J. Chem. Phys. 58, p. 5383.
19. Collins, J.B.; Schleyer, P.v.R.; Binkley, J.S.; Pople, J.A.; Radom, L.: 1976, J. Am. Chem. Soc. 98, p. 3436.
20. Siegbahn, P.E.M.: 1982, Chem. Phys. 66, p. 443; Pople, J.A.; Tidor, B.; Schleyer, P.v.R.: 1982, Chem. Phys. Lett., in press.

INTRODUCTORY REMARKS ON THE TIME-DEPENDENT PHENOMENA AND RELATED TOPICS

Kenichi Fukui
Kyoto University of Industrial Arts and Textile Fibers
Matsugasaki, Sakyo-ku, Kyoto 606 Japan

INTRODUCTION

In general the time-dependent phenomena should be concerned with all of the motions or changes of substance in both classical and quantum mechanical world. In the quantum theory, time is formally incorporated into the time-dependent Schrödinger equation or equivalent alternatives to describe non-stationary systems which have been investigated in quantum mechanics and statistical physics.

The time-dependent problems, however, would mostly remind quantum chemists of those concerning the molecular reaction dynamics. Indeed, the perfect comprehension of the path and the rate of chemical reactions has been one of the biggest targets in experimental and theoretical chemistry. The recent development of molecular beam studies and laser techniques has made detailed analysis of the chemical reaction between the initial and the final states with specific quantum numbers(state-to-state chemistry) almost possible, which, on the other hand, stimulates the progress of theoretical chemistry handling these problems. Other inelastic processes such as excitation, ionization, relaxation, and so on are also important items of the time-dependent problems.

We have four best persons to give us the most recent results covering the significant and fundamental subjects which are of importance in the time-dependent problems. So it seems desirable to refer to complementary topics on the molecular reaction dynamics in the remaining pages.

REACTIVE COLLISION APPROACH AND RELATED AREA

The development of the reactive collision approach is closely related to that of the investigation of the potential energy surface on which chemical reactions are considered to occur as a scattering process. Potential energy surfaces of many-molecular systems have been obtained with the use of the LEPS(London-Eyring-Polanyi-Sato) and the

P.-O. Löwdin and B. Pullman (eds.), New Horizons of Quantum Chemistry, 111–115.

DIM(diatomics-in-molecules) methods. Recently an ab initio Hartree-Fock MO method has been employed to obtain more quantitative energy surfaces thanks to the incessant progress of electronic computers.

In the classical treatment of the reactive scattering, ensembles of trajectories are calculated based on the integration of the equation of motion for nuclei employing an appropriate potential energy surface. A great deal of triatomic reactions, some four-centre reactions, and unimolecular decompositions have been investigated therewith. In order to specify the initial and the final conditions for each trajectory, "quantized" distribution of action variables is sometimes imposed to match the result of the quantum mechanical calculations. Statistical consideration such as the Monte-Carlo procedure is often employed to select the initial condition randomly.

In connection with the calculation of the trajectories, ergodic and stochastic behaviour of molecular system will be of more interest. Hutchinson and Wyatt(1) have discussed the property of quantum mechanical ergodicity and instability based on the analysis of non-stationary wave packets.

Needless to say, one has to be always careful to the limit of the classical approach which will fail in the prediction of the reaction thresholds and the resonance features and the description of the tunneling effect.

Miller(2) has studied the rate constant for unimolecular reactions,

$$H_2CO \longrightarrow H_2 + CO$$

$$\underline{\text{trans}}\text{-HCOH} \longrightarrow H_2CO$$

in the collisionless photochemistry including tunneling corrections, within the RRKM theory. The potential energy surface of formaldehyde employed in the above calculation was that obtained by Goddard and Schaefer(3) on the basis of the SCF-CI computation. The tunneling effects have been found to be quite significant for the rates of $10^9 s^{-1}$ or slower.

Schinke and Lester(4) have performed orthodox classical trajectory calculations for the reaction,

$$O(^1D) + H_2(^1\Sigma_g^+; v, j) \longrightarrow OH(^2\Pi; v', j') + H(^2S)$$

employing analytical fits to an ab initio potential energy surface. They analyzed the rotational and vibrational product distributions as well as the total reaction cross sections, the rate constants, and the angular distribution of the scattered OH product.

The quantum mechanical description of molecular reactive scattering is expected to give an exact result. The quantum mechanical approach is mostly based on the numerical integration of the coupled-channel problem of the Schrödinger equation in an appropriate coordinate system. The S matrix, the reaction probability and the resonance behaviour of the reaction system can be investigated quantitatively in this approach. A complicated computation, however, usually prevents from studying the reaction system involving more than three atoms. Various numerical and approximate methods to reduce the efforts of the calculation have also been devised.

Kuppermann et al.(5) have treated the quantum reaction probability and the rate constant for the collinear reaction

$$F + H_2 \longrightarrow FH + H$$

employing the semiempirical LEPS potential energy surface of Muckerman (6). The results indicate a large degree of population of inversion of the FH product with v'=2 and 3 giving the dominant probability of reaction. It has also been pointed out that resonances play an important role at higher relative translational energies.

Redmon and Wyatt(7) have studied a quantum resonance in the reaction,

$$F + H_2(v=0,\ j=0) \longrightarrow FH(v'=2,\ \text{all}\ j') + H$$

in three dimension. It is shown that resonance structure for the $v=0 \rightarrow v'=2$ process is significantly broadened comparing with the collinear resonance.

Clary and Nesbet(8) have shown that the quantum dynamical population distributions computed from the S matrix elements can be fitted by the surprisal formula(9) in a successful manner for the three dimensional $H + H_2$ exchange reaction. It is of interest since this result suggests that success of the surprisal theory must come from a minute aspect of the collision dynamics, not from the statistical considerations.

The analysis of a metastable state of molecules is of interest in the discussion of activated complexes in complicated reaction systems. In association with the metastable state, Tachibana et al.(10) have studied the theory of resonant states from the standpoint of complex eigenvalue problem. The idea of resonant states will be of importance in high-energy reactions in particular.

Öksüz(11) has discussed the time-stability of an autoionizing state which is encompassed in the solutions of a pseudoeigenvalue equation as well as stationary bound states.

IRC (INTRINSIC REACTION COORDINATE) APPROACH

In general chemical reactions involving polyatomic molecules the IRC approach seems to be practical for only a relatively modest number of the MO calculations are required to obtain the energy surface.

Fukui(12) has defined the IRC as the reaction path of the quasi-static change of nuclear configuration of the reaction system starting from the transition state down to a stable equilibrium point in the steepest descending way in the mass-weighted coordinate system.

The Hamiltonian valid for a narrow region along the IRC is called the IRC Hamiltonian, into which the effect of vibrational nonadiabaticity can be incorporated. Miller et al.(13) have derived the IRC Hamiltonian including (3N-7) normal coordinates (N: number of atoms in the reacting system) that can describe the vibrations orthogonal to the IRC. Gray et al.(14) have studied the tunneling rate of the unimolecular isomerization HNC ⟶ HCN applying the above Hamiltonian.

When the rotational motion exists in a chemical reaction, it is required to discuss the IRC after the separation of the nuclear configuration space from the rotational motion. Fukui et al.(15) have discussed the general classical Hamiltonian in terms of translational, rotational, and internal terms so as to derive the IRC equation in the presence of rotational motion which generally causes a deviation of the IRC.

Tachibana and Fukui(16) have discussed that the internal configuration space associated with a chemical reaction can be divided into cells by each set of the solutions ("orthogonal trajectories") of the differential equations for IRC. Each cell normally contains a stable equilibrium point. The rate of the reaction is then understood wave-mechanically as the motion of a wave packet between the cells.

Kato et al.(17) have examined the rate constant in the case of the proton migration in the enol-form malonaldehyde as a model for this treatment, which will suggest one of interesting directions in chemodynamics.

The investigations hitherto mentioned involve the new theoretical techniques developed in recent years for describing molecular reaction dynamics. This subject merely occupies a partial area of the whole time-dependent problem. Nevertheless, this area seems to attract a good deal of attention of chemists due to the nature of the ambitions of chemistry itself, and will be growing more with the progress and the cooperation of experimental, theoretical, and computational techniques altogether.

REFERENCES

1. J. S. Hutchinson and R. E. Wyatt, Phys. Rev., A23, 1567 (1981).
2. W. H. Miller, J. Am. Chem. Soc., 101, 6810 (1979).
3. J. D. Goddard and H. F. Schaefer, III, J. Chem. Phys., 70, 5117 (1979).
4. R. Schinke and W. A. Lester, Jr., ibid., 72, 3754 (1980).
5. G. C. Schatz, J. M. Bowman, and A. Kuppermann, ibid., 63, 674 (1975).
6. P. A. Whitlock and J. T. Muckerman, ibid., 61, 4618 (1975).
7. M. J. Redmon and R. E. Wyatt, Chem. Phys. Lett., 63, 209 (1979).
8. D. C. Clary and R. K. Nesbet, J. Chem. Phys., 71, 1101 (1979).
9. R. B. Bernstein and R. D. Levine, ibid., 57, 434 (1972); A. Ben-shaul, R. D. Levine, and R. B. Bernstein, ibid., 57, 5427 (1972).
10. A. Tachibana, T. Yamabe, and K. Fukui, J. Phys. B: Atom. Molec. Phys., 10, 3175 (1977); T. Yamabe, A. Tachibana, and K. Fukui, Advances in Quantum Chemistry, 11, 195 (1978).
11. İ. Öksüz, Phys. Rev., A13, 1507 (1976).
12. K. Fukui, J. Phys. Chem., 74, 4161 (1970); in "The World of Quantum Chemistry," R. Daudel and B. Pullmann, eds., D. Reidel, Dordrecht (1974), p.113.
13. W. H. Miller, N. C. Handy, and J. E. Adams, J. Chem. Phys., 72, 99 (1980).
14. S. K. Gray, W. H. Miller, Y. Yamaguchi, and H. F. Schaefer, III, ibid., 73, 2733 (1980).
15. K. Fukui, A. Tachibana, and K. Yamashita, Int. J. Quant. Chem.: Quant. Chem. Symp., 15, 621 (1981).
16. A. Tachibana and K. Fukui, Theoret. Chim. Acta(Berl.) 51, 189 (1979).
17. S. Kato, H. Kato, and K. Fukui, J. Am. Chem. Soc., 99, 684 (1977).

NONADIABATIC EFFECTS IN SLOW ATOMIC COLLISIONS - MANY-TRAJECTORY SEMICLASSICAL APPROXIMATION

E. E. Nikitin
Institute of Chemical Physics, Ac.Sci.
Moscow, USSR

ABSTRACT: The many-trajectory semiclassical approximation for calculation of the classical angle scattering is discussed. It is based on a specific representation of the scattering matrix, following from its piecewise synthesis. Each piece is calculated by solving semiclassical equations for closely spaced molecular adiabatic terms along local common trajectories. To this approximation the differential cross section displays interference effects: generalized Stueckelberg undulations, generalized rainbow undulations, and the slow modulation of rapid undulations due to the relative dephasing of waves scattered on a degenerate molecular term.

1. INTRODUCTION

There are several different versions of the semiclassical approximation in the theory of atomic and molecular collisions. Before the advent of computers these approaches were stimulated mainly by the desire to simplify the scattering problems and thus to avoid lengthy calculations. The technical difficulties have been overcome and direct numerical solution of coupled quantal scattering equations became a routine procedure. Yet the interest to the semiclassical approximation still persists. This approximation provides a transparent picture of scattering events. Moreover, the semiclassical description often permits visualizing the relation between the input (the interaction matrix) and the output (the transition probabilities and cross sections), when the latter arises from the scrambler (the scattering equations).

The existing various semiclassical descriptions stem from two approaches. One, formulated in 1931 by Mott [1,2] for inelastic atomic scattering, takes the relative motion

P.-O. Löwdin and B. Pullman (eds.), New Horizons of Quantum Chemistry, 117–134.

to be classical, while the electronic subsystem is described by quantal nonstationary equations. This is either the impact parameter approximation (if the trajectory is rectilinear), or the common trajectory approximation (if the trajectory is curved). Both are widely used in the collision theory, provided the relative energy is high enough. Limitations on the energy from below depend on the collision type and suffice it to say that very often these prevent using the common trajectory approximation for slow collisions.

Another approach is based on classical mechanics and uses the correspondence principle arguments to construct the scattering amplitude from classical trajectories and the appropriate classical action integrals. A development of this approach known at present as the classical S-matrix approximation is essentially due to Miller [3,4]. This approximation has been discussed in books and reviews [5, 6,7]. In general it is applicable to transitions between states with large quantum numbers.

When the quantum numbers of states (belonging to systems referred to below as quantal subsystems) are not large, and the other degrees of freedom (belonging to a classical subsystem), though not too fast, are of a quasi-classical nature, the formulating of the semiclassical approach is not straightforward. The new ideas proposed in order to widen the applicability of the semiclassical pattern imply either the derivation of self-consistent equations for the time dependent wave function of a quantal subsystem and for trajectories of a classical subsystem [8] or the reduction of the several-state quantal Hamiltonian to an equivalent classical Hamiltonian which is then combined with the Hamiltonian of a classical subsystem [9]. Another approximation, the surface hopping model [10,11,12] is based on the classical motion of atoms over isolated potential surfaces and on the sudden transitions from one surface to another at the crossing or pseudocrossing seams. However, this approach is invalid when two (or more) potential surfaces are close and the nonadiabatic interaction between these is not localized. Besides, the interference effects are neglected here, since the surface hopping model deals with transition probabilities, and not with amplitudes.

These shortcomings of the surface hopping model are very pronounced for atomic collisions where the degeneracy or quasidegeneracy of molecular terms represent a common feature and where there are no extra factors destroying the interference pattern (which are present in the averaging over initial vibrational and rotational phases in molecular collisions).

Still, for slow atomic collisions it is possible to formulate an improved version of the hopping model, free from the above shortcomings. It will be called the many-trajectory semiclassical approximation (MSA) and will be discussed in the present paper.

The list of references contains representative publications only, as a more or less complete list would be too proliferative. However the recently published books cited include all the original references.

2. SOME SPECIFIC FEATURES OF SLOW QUASICLASSICAL ATOMIC COLLISIONS

A successful solution of any scattering problem depends on the completeness of the basis adopted. Though this becomes less essential when resorting to more powerful computers handling hundreds of coupled equations, yet the problem of choosing the optimal basis remains, since the basis showing the fastest convergence usually is that suggesting the simplest physical interpretation of the collision event. For slow collisions the adiabatic basis diagonalizing the electronic Hamiltonian for a fixed molecular axis is known to be the most convenient one. The nuclear motion will then manifest in the nonadiabatic interaction which induces transitions between molecular states. But the use of the adiabatic basis is in general fraught with difficulties arising from the need to account for the additional momenta of electrons occupying the travelling atomic orbital (see e.g [13,14,15]). However for slow collisions these difficulties usually are immaterial.

The nonadiabatic interaction is proportional to the relative velocity of atoms v. This interaction might be low when the global Massey parameter is large

$$\xi_{a'a} = \Delta E_{a'a} R^*/\hbar v = \omega_{a'a}(\infty)\,\tau^* \gg 1 \qquad (1)$$

Here R^* is the characteristic radius of the interaction region, $\Delta E_{a'a}$ is the difference in the electronic energies of the atomic states $|a'\rangle$ and $|a\rangle$. Depending on the process considered $\Delta E_{a'a}$ can vary from hundredths eV for transition between fine structure components to hundreds eV for those between inner atomic states. Note that the Massey parameter in Eq.(1) is written also as the product of electronic frequency $\omega_{a'a}(\infty)$ of free atoms ($R \to \infty$) and the collision time τ^*. The inequality (1) shows that the adiabatic basis set can be optimal.

Assume now that the atomic motion is quasiclassical, i.e.

$$kR^* \gg 1 \tag{2}$$

where k is the wave vector of nuclear motion. Equation (1) implies that the kinetic energy E_{kin} of colliding atoms is restricted

$$E_{kin} \lesssim \Delta E_{a'a}\, kR^* \tag{3}$$

With account for Eq.(2) this means that for slow collisions (in the sense of Eq.(1)) the kinetic energy can exceed the excitation energy, so that many channels will be energetically open. At the same time Eq.(1) implies that the states $|a'\rangle$ and $|a\rangle$ are adiabatically isolated, and transitions between these would have to be improbable.

However that is not the case. The effectivity of nonadiabatic coupling depends not on the global but on the local Massey parameter $\xi_{\alpha'\alpha}(R)$ which is defined for transition between molecular states $|\alpha'\rangle$ and $|\alpha\rangle$, rather than between the atomic states $|a'\rangle$ and $|a\rangle$. For each internuclear distance $\xi_{\alpha'\alpha}$ is

$$\xi_{\alpha'\alpha}(R) = \omega_{\alpha'\alpha}(R)\, \tau_{\alpha'\alpha}(R) \tag{4}$$

where $\omega_{\alpha'\alpha}(R)$ is the frequency of the electronic transition between molecular terms $U_{\alpha'}(R)$ and $U_{\alpha}(R)$; $1/\tau_{\alpha'\alpha}$ stands for the rate of changes in adiabatic functions $|\alpha\rangle$ and $|\alpha'\rangle$. Explicitly

$$\tau_{\alpha'\alpha}^{-1} = \langle \alpha' | \partial/\partial t | \alpha \rangle \tag{5}$$

The derivative in Eq.(5) is calculated along some classical trajectory $R(t)$ which is taken to be independent of the indices of electronic states α' and α. This independence is the crux of MSA. It will be discussed now.

The strong nonadiabatic coupling between states $|\alpha'\rangle$ and $|\alpha\rangle$ occurs when the local Massey parameter is not large, i.e. when $\xi_{\alpha'\alpha}(R) \lesssim 1$. If the extension of the coupling region is large compared to the de-Broglie wavelength for motion along the trajectory adopted, $\xi_{\alpha'\alpha}(R) \lesssim 1$ will be

$$E_{kin}(R) \gg |U_{\alpha'}(R) - U_{\alpha}(R)| \tag{6}$$

where $E_{kin}(R)$ is the local kinetic energy for the interatomic distance R. As the strongly coupled states are close, in the sense of Eq.(6), the local kinetic energy can be expressed via the total energy E and some mean (for two interacting states) potential energy $U(\alpha', \alpha; R)$. Thus, in the region of the essential coupling $U(\alpha', \alpha)$ defines a common local trajectory. The nuclear motion along this trajectory induces nonadiabatic coupling between molecular states $|\alpha'\rangle$ and $|\alpha\rangle$. Clearly, this picture is not restricted to two coupled states: if there are several close terms U_{α}, $U_{\alpha'}$, $U_{\alpha''}$..., one can introduce a mean potential $U(\alpha, \alpha', \alpha'' \ldots)$ which will define the common local trajectory.

Thus, within some interval of the internuclear distance $R_n < R < R_{n+1}$ all molecular terms are binned, and a mean potential is introduced for each bin. Transitions between terms belonging to different bins are neglected, and those between terms of the same bin are described in the semiclassical approximation.

With such a binning many terms will be single, i.e. adiabatically isolated. For these terms a ready solution is given by the JWKB approximation. For coupled terms one has to solve time dependent equations. These will often be comparison equations, the solutions to which have been thoroughly studied for different models (e.g. the Landau-Zener-Stueckelberg model, the Rosen-Zener-Demkov model, the exponential model, etc., see also [14,15]). Note, however, that not only the transition probabilities for these models, but also the so-called dynamical phases which represent nonadiabatic corrections to adiabatic phases are needed here.

When the semiclassical solutions and the accompanying JWKB amplitudes for mean potentials corresponding to each bin are known for every consecutive interval in the whole interaction region, the total solution is constructed by matching at the interval joints. Then a transition to the asymptotics at $R \gg R^*$ will provide the scattering matrix.

It is seen that MSA avoids the restrictions needed for the common trajectory approximation, viz. that the kinetic energy be large compared to each adiabatic potential energy

$$E_{kin} \gg U_{\alpha}(R) \tag{7}$$

It will be emphasized that certain approaches resembling MSA have been used before. An MSA treatment of processes described by decoupled equations (e.g. resonance charge exchange) has been proposed rather long ago [16,17]. For the coupled equations MSA was used implicitly in cases when the solution of a comparison equation was matched to the adiabatic waves. Still, so far main attention was paid to events for which the degeneracy of atomic and molecular states was immaterial.

For degenerate states $\omega_{a'a}$ is zero, so that the adiabatic approximation is usually invalid. However the interatomic interaction partly lifts the degeneracy, and the adiabatic approximation in a restricted range of R becomes again possible. Yet it must account for the remaining degeneracy of some molecular terms.

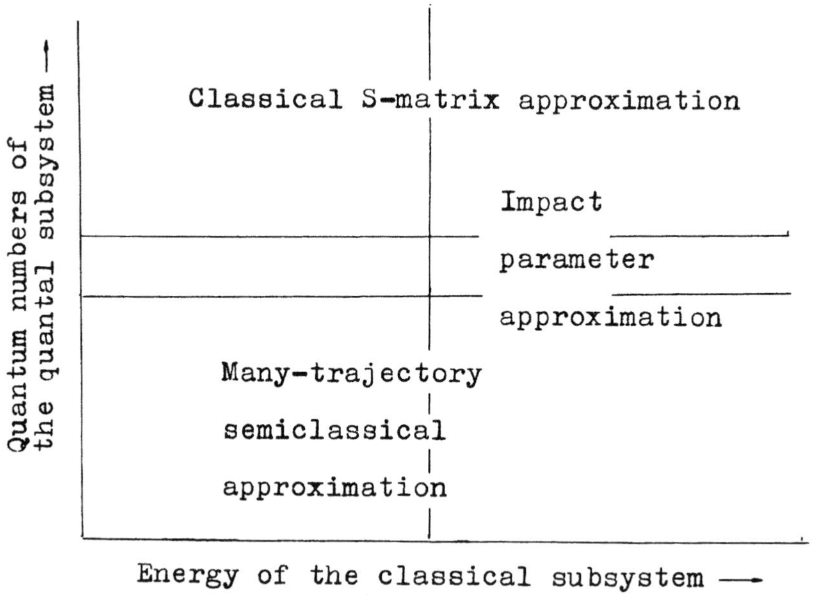

Figure 1. Relation between different approximations in the quasi-classical collision theory

The MSA treatment discussed here refers to cases where the degeneracy is of particular importance. In the limit it describes transitions between the states belonging to one degenerate atomic term, i.e. the transitions responsible for collisional depolarization. In this respect the MSA is a generalization of the sudden matching approximation [18].

The relation between the MSA, the impact parameter approximation, and the classical S-matrix method is presented schematically in Fig.1. Here the appropriate characteristic parameters of colliding atoms are taken to be the kinetic energy of the classical subsystem (relative kinetic energy) and the quantum numbers of the quantal subsystem (electronic quantum numbers).

3. CONSTRUCTION OF A MULTICHANNEL SCATTERING MATRIX BY MATCHING - PIECEWISE SYNTHESIS OF THE S-MATRIX

For definiteness turn here to the scattering equations in the total angular momentum representation. These are the one-dimensional coupled second order differential equations describing motion along the radial coordinate R. Use of the local trajectory approximation means that: (i) quantal scattering equations factorize to independent blocks, and (ii) for each block these equations simplify to first order time dependent equations. The latter is known to be possible only under condition of the JWKB approximation which implies that an incoming wave will not generate outgoing waves until it reaches the region of the turning point for each adiabatic potential [14,15]. If this is not the case, a more sophisticated approach still using the basic idea of MSA [14,15] is needed, but it will not be discussed here.

Assume now that nonadiabatic coupling is restricted to the region $R_1 < R < R_M$. Then introduce the matrix $V(R,R_1)$ which transforms the outgoing adiabatic JWKB waves for $R = R_1$ into outgoing waves for R. If the JWKB conditions are not violated to the right of R_M, then for large $R > R_M$ the matrix V connects outgoing waves within the interaction region with those for free motion at $R \to \infty$. Assume also that to the left of R_1 the adiabatic JWKB phases between R_1 and the turning points R_α^t on each adiabatic term $U_\alpha(R)$ are large. Then for $R = R_1$ the amplitudes of incoming and outgoing adiabatic waves on each term will be simply connected by the phase factor. This directly yields the following expression for the matrix W which transforms the amplitudes of the incoming waves into those of outgoing waves at the same R

$$W(R) = V(R,R_1)A(R_1,R^t)iA(R_1,R^t)V^{tr}(R,R_1) \quad (8)$$

where the superscript tr denotes a transposed matrix. Here factor i takes into account the known correction to the JWKB phase, due to breakdown of the quasiclassical approximation near the isolated turning point, and $A(R_1,R^t)$ are diagonal matrices of adiabatic evolution off the turning points

$$A_{\alpha\alpha'}(R_1,R^t) = \delta_{\alpha'\alpha}\exp\left[i\int_{R_\alpha^t}^{R_1} K_\alpha(R)dR\right]$$

$$K_\alpha(R) = (2\mu)^{1/2}[E - U_\alpha(R) - (J+1/2)^2/2\mu R^2]^{1/2} \quad (9)$$

Here $K_\alpha(R)$ stands for the local wave vectors, J for the total angular momentum, and μ for the reduced mass.

Of course, $V(R,R_1)$ can be represented as the product

$$V(R,R_1) = V(R,R_M)V(R_M,R_{M-1})\ldots V(R_2,R_1) \quad (10)$$

but the essential point here is that with optimal partitioning of the region $R_1 - R_M$ into smaller intervals, each matrix $V(R_n,R_{n-1})$ will be of the block-diagonalized type. Moreover, if within the interval $R_{n-1} < R < R_n$ the coupling is localized in the small region of the extension ΔR_{n^*} centered at R_{n^*}, the matrix V can be represented as

$$V(R_n,R_{n-1}) = A(R_n,R_{n^*})N(R_{n^*})A(R_{n^*},R_{n-1}) \quad (11)$$

Here $A(R_n,R_{n^*})$ are diagonal matrices containing JWKB phase factors, $N(R_{n^*})$ is the matrix of nonadiabatic transition. It is just the N matrix that represents the ultimate result of solving a comparison equation for localized nonadiabatic coupling. The most frequent cases of localized coupling correspond either to crossing, or to pseudocrossing (avoided crossing), or else to steeply diverging terms.

When found, W(R) is used to derive the scattering matrix by removing from W the asymptotic phases of free motion

$$S = -\lim_{R\to\infty} \exp(-ikR-ik'R + i\pi l/2 + i\pi l'/2)W(R) \qquad (12)$$

It is clear from this procedure that each element of the S-matrix will be represented by the sum of exponential terms with the exponents corresponding to JWKB scattering phase shifts for different intervals of R. The pre-exponential factors will depend on the indices of molecular terms and of scattering channels. Thus, the general form of the S-matrix will be set by the piecewise synthesis.

To make representation simple consider an inelastic collision

$$A(a,j,m) + X \longrightarrow A(a',j',m') + X \qquad (13)$$

where X is a structureless particle, and a,j,m are quantum numbers specifying the state of atom A, in particular its electronic angular momentum j and its projection m onto the z-axis of the space-fixed frame. An element of the S-matrix in the J-representation is $S^J_{a'j'l';\,ajl}$ with l',l standing for the relative orbital momentum quantum numbers. Hereafter the subscripts a',a will be omitted bearing in mind that a change in j is generally accompanied by a change in a.

For quasiclassical analysis of the amplitude and the cross section it is important to parametrize the matrix elements $S^J_{j'l';\,jl}$ via functions smoothly depending on large quantum numbers. The latter are J,l',l because the quasiclassical condition requires that $J,l',l \gg 1$. One possibility is to assume that both the exponents, and the pre-exponential factors in the sum expressing $S^J_{j'l';\,jl}$ smoothly depend on J for fixed l' and l. However, analysis of the scattering equations shows that this assumption is invalid. For instance, the equations corresponding to a definite value of J and to parity p contain (for half-integer j) the Coriolis correction proportional to $(-1)^{J-l'+j'}$. The latter factor means that the solution in the fixed exit channel j' does not depend smoothly on J, and neither does $S^J_{j'l';\,jl}$. This and similar arguments suggest that to make the J-dependence smooth one must replace l' and l by other quantum numbers ν' and ν :

$$\nu' = J - l' \;; \qquad \nu = J - l \qquad (14)$$

Accordingly, the matrix elements $S^J_{j'l';jl}$ will be replaced by $S^J_{j'\nu';j\nu}$ using the rule

$$S^J_{j'\nu';j\nu} = \exp[i\frac{\pi}{2}(3\nu - \nu')]\, S^J_{j'l';jl}\Big|_{\substack{l=J-\nu \\ l'=J-\nu'}} \qquad (15)$$

with the phase factor introduced for further simplification.

The basic ansatz about the S-matrix is formulated now as

$$S^J_{j'\nu';j\nu} = \sum_{\Lambda} S^{\Lambda}_{j'\nu';j\nu}(J)\exp[2i\,\delta^{\Lambda}_{j'j}(J)] \qquad (16)$$

where the pre-exponential factors $S^{\Lambda}_{j'\nu';j\nu}(J)$ and phases $\delta^{\Lambda}_{j'j}(J)$ are assumed to be smooth functions of J with other quantum numbers fixed. For large phases $\delta^{\Lambda}_{j'j} \gg 1$, the J-dependence of exponentials is much stronger than that of the pre-exponential factors and this is the crucial point in the quasiclassical analysis.

As for the number of terms in sum (16) it is larger than or equal to the number of coupled equations in the block specified by the total angular momentum J and parity p. Note than any extra integral of motion (e.g. electronic spin in the absence of spin-orbital coupling) ensures additional decoupling of the equations, thus reducing the number of terms. In the extreme, when all equations are decoupled, the number of terms will be equal to that of equations.

Though it is expected that normally $S^{\Lambda}_{j'\nu';j\nu}$ and $\delta^{\Lambda}_{j'j}$ will smoothly depend on J, this is not always the case. A simple example of elastic scattering will illustrate it.

For elastic scattering of an atom A in state $j = 0$ Eq.(16) degenerates to one term, and the scattering matrix is represented by one exponential

$$S^J = \exp[2i\,\delta(J)] \qquad (17)$$

The scattering phase shift is known to smoothly depend on J

only in the absence of resonance scattering. For the latter $\delta(J)$ behaves as a ladder-like function which is obviously not smooth. On the other hand, the S-matrix for elastic resonance scattering can be parametrized via smooth functions, i.e. via the quasiclassical action integrals over classically accessible regions and the transmission coefficient. If this parametrization is represented in the form of Eq.(16), the number of its terms will be infinite. This shows that the assumption about a smooth J-dependence and the final number of terms in representation (16) can be restrictive. Still it enables elucidation of the classical limit of quantal scattering.

4. QUASICLASSICAL LIMITS OF THE AMPLITUDE AND THE CROSS SECTION FOR SCATTERING THROUGH CLASSICAL ANGLES

The region of classical scattering angles is bounded from below by the angle of diffraction scattering $\theta^* \simeq (kR^*)^{-1}$. This region $(\theta > \theta^*)$ comprises small $(\theta \ll 1)$, intermediate, and large $(\theta \gtrsim 1)$ angles. The scattering through small angles including the diffraction region $(\theta \ll 1)$ is described by the familiar eikonal approximation [5,14,15].

Thus for small but classical angles $(\theta^* < \theta < 1)$ MSA overlaps the eikonal approximation. At this limit several classical trajectories of MSA degenerate to one rectilinear trajectory (impact parameter approximation).

The quasiclassical analysis carried out below shows that for scattering through intermediate and large angles the MSA reveals a novel effect that does not exist for small angle scattering. Namely, the transition probability or the differential cross section exhibiting rapid undulations also are slowly modulated. The modulation phase is of the order of θj, and the modulation effect is connected with a change in the relative orbital angular momentum of the colliding atoms. In other words, under the quasiclassical condition $l', l \gg j', j$, when one would expect no marked result of the small change (of the order of j) in the large $(l \gg 1)$ relative angular momentum, there exists an effect which is precisely due to nonconservation of l.

The relation of MSA to the eikonal approximation is illustrated in Fig.2 presenting the validity regions of the two approximations in coordinates θ, ξ for quasiclassical conditions.

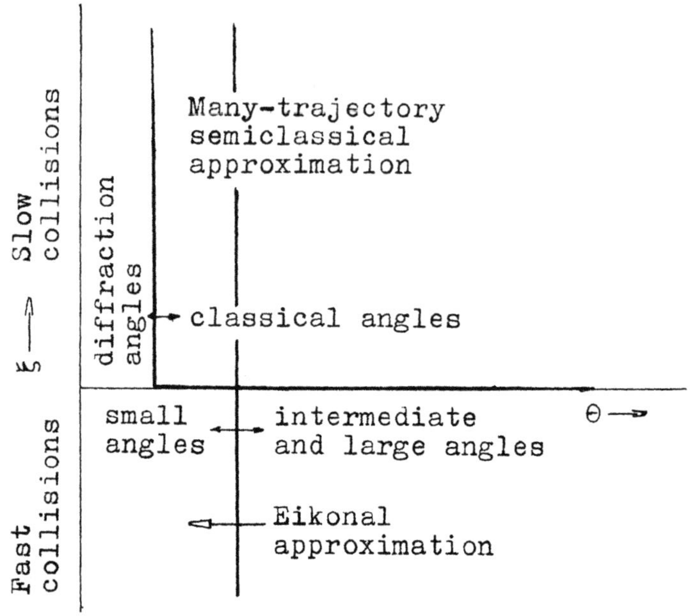

Figure 2. Relation between many-trajectory and eikonal approximations

The scattering amplitude for process (13) in a space-fixed frame is most simply represented in the jmln representation

$$f_{j'm';jm}(\hat{\underline{k}},\hat{\underline{k}}) = \frac{2\pi i}{(k'k)^{1/2}} \sum_{l',n';l,n} i^{l'-l}\, Y^{n'}_{l'}(\hat{\underline{k}}) Y^{n}_{l}(\hat{\underline{k}}) \quad (18)$$

$$[S_{j'm'l'n';jmln} - \delta_{j'j}\,\delta_{m'm}\,\delta_{l'l}\,\delta_{n'n}]$$

The matrix elements $S_{j'm'l'n';jmln}$ are expressed via $S^{J}_{j'l';jl}$ and the Clebsch-Gordan coefficients [14,15].

Rewrite now Eq.(18) with the z-axis directed along the $\underline{k}$ vector, expressing the Clebsch-Gordan coefficients for $\bar{J},l',l \gg 1$ via particular cases of the Wigner D-functions $\Delta^{j}_{m\nu}$ and using the asymptotic representation of spherical functions for classical angles. Replacing the summation over J by integration, we have

$$f_{j'm';jm}(\theta,\varphi) = -\frac{i\exp[i(m-m')\varphi]}{2(\pi k'k \sin\theta)^{1/2}}$$

$$\cdot \int_0^\infty (2J)^{1/2} dJ \; \exp[i(J+1/2)\theta+i(m-m'-1/2)\tfrac{\pi}{2}] A^J_{j'm';jm}(\theta) \qquad (19)$$

$$+ \exp[-i(J+1/2)\theta-i(m-m'-1/2)\tfrac{\pi}{2}] \; A^J_{j'm';jm}(-\theta)$$

where

$$A^J_{j'm';jm}(\theta) = \sum_{\nu'\nu} \exp[-i(\pi+\theta)\nu'] \, \Delta^{j'}_{m'\nu'} \, S^J_{j'\nu';j\nu} \, \Delta^{j}_{m\nu} \qquad (20)$$

Introducing here $S^J_{j'\nu';j\nu}$ from Eq.(16) and taking into account that $\delta^{\Lambda}_{j'j}$ are large (scattering through classical angles), integration over J is performed by the stationary phase method. At this level various kinds of uniform approximation prove to be useful when the stationary phase points are not well separated. When separated, they independently contribute to the scattering amplitude. The stationary phase angular momenta J^* satisfy the equation

$$\pm\,\theta + 2\pi M = 2\,\partial\delta^{\Lambda}_{j'j}/\partial J = \eta^{\Lambda}_{j'j}(J)\Big|_{J=J^*} \qquad (21)$$

where M is an integer. In the extreme case of elastic scattering the stationary values J^* are calculated using an equation of type (21) with $\delta^{\Lambda}_{j'j}(J)$ replaced by the scattering JWKB phase shift $\delta(J)$ and $\eta(J)$ identified with the classical deflection function. Similarly, the function $\eta^{\Lambda}_{j'j}(J)$ in Eq.(21) defined as the derivative $2\,\partial\delta^{\Lambda}_{j'j}/\partial J$ can be called the deflection function for a trajectory with a preset J value indexed by Λ and connecting channels j and j . For j, j' and Λ fixed there can be several solutions to Eq.(21) because of the $\pm$ sign of θ , or for some other reason. These will be identified by the superscript $\varkappa$. In this way $J^* = J^{\Lambda\varkappa}_{j'j}(\theta)$. Thus the trajectories with the preset scattering angle θ connecting channels j' and j will be identified by Λ and $\varkappa$.

For isolated stationary angular momenta J^* the scattering amplitude is a superposition

$$f_{j'm';jm} = \sum_{\Lambda\varkappa} f^{\Lambda\varkappa}_{j'm';jm} \qquad (22)$$

with partial amplitudes

$$f^{\Lambda\varkappa}_{j'm';jm} = -\left(\frac{k}{k'}\right)^{1/2} i \exp[i(m-m')\varphi]\exp\left\{2i\,\delta^{\Lambda}_{j',j}(J) - iw^{\Lambda\varkappa}_{j'j}[(J+1/2)\theta + (m-m'-1/2)\tfrac{\pi}{2}] + iv^{\Lambda\varkappa}_{j'j}\pi/4\right\} \qquad (23)$$

$$\cdot\, B^{\Lambda\varkappa}_{j'm';jm}(J,\theta)[q^{\Lambda\varkappa}_{j'j}(\theta)]^{1/2} \quad ; \quad J = J^{\Lambda\varkappa}_{j'j}(\theta)$$

Consider now each of the three factors explicitly singled out. The first factor corresponds to one of the two exponentials in Eq.(19), $w^{\Lambda\varkappa}_{j'j} = \pm 1$. This sign coincides with that of θ in the left hand side of Eq.(21). The term $v^{\Lambda\varkappa}_{j'j}\pi/4 = \pm\pi/4$, with its sign corresponding to that of the derivative $\partial\eta^{\Lambda}_{j'j}/\partial J$ for $J = J^{\Lambda\varkappa}_{j'j}$ represents an extra phase which arises in integrating over J close to $J^{\Lambda\varkappa}_{j'j}$. The first factor in Eq.(23) varies rapidly with J, and hence with θ.

The second factor $B^{\Lambda\varkappa}_{j'm';jm}$ varies slowly with J and contains only pre-exponential factors $S^{\Lambda}_{j'\nu;j\nu}(J)$ of the scattering matrix in the form (16)

$$B^{\Lambda\varkappa}_{j'm';jm}(J,\theta) = \sum_{\nu'\nu} \exp(-i\pi\nu' + iw^{\Lambda\varkappa}_{j'j}\theta\nu') \cdot \Delta^{j}_{m'\nu'} S^{\Lambda}_{j'\nu';j\nu}(J)\, \Delta^{j}_{m\nu} \qquad (24)$$

Finally, the third factor $q^{\Lambda\varkappa}_{j'j}$ can be identified as the classical differential cross section for the trajectory connecting channels j' and j

$$q^{\Lambda\varkappa}_{j'j}(\theta) = k^{-2}(\sin\theta)^{-1} J \left|\partial\eta^{\Lambda}_{j'j}/\partial J\right|^{-1}\Big|_{J=J^{\Lambda\varkappa}_{j'j}(\theta)} \qquad (25)$$

Equation (23) together with (24) and (25) provide a simple interpretation of the pre-exponential factors $S^{\Lambda}_{j'\nu';j\nu}(J)$

in the representation (16). To this end transform the coordinate system so that the x-axis coincides with vector $\underline{k}$, and the z-axis is normal to the $\underline{k}',\underline{k}$ plane. This transformation is accomplished by rotation through the Euler angles $\alpha = \varphi + \pi/2$, $\beta = \pi/2$, and $\gamma = 0$. Let μ' and μ be projections of j' and j onto the new z-axis. Then

$$f_{j'm';jm} = \exp[i(m-m')(\varphi + \pi/2)] \sum_{\mu'\mu} \Delta^{j'}_{m'\mu'} f_{j'\mu';j\mu} \Delta^{j}_{m\mu} \tag{26}$$

where $f_{j'\mu';j\mu}$ is the scattering amplitude in the new frame. Comparing this expression with Eq-s (23) and (24) it will be found that, provided ν', ν is replaced by μ', μ, $\exp(iw^{\Lambda\varkappa}_{j'j}\theta\nu')S^{\Lambda}_{j'\nu';j\nu}(J)$ coincides (except for the phase factor) with the scattering amplitudes for transition $j,\mu \rightarrow j',\mu'$ induced by classical motion along the trajectory $\Lambda,\varkappa$ connecting channels j' and j and deflected through angle θ.

The representation (22) can be called, in line with the current terminology, a primitive semiclassical approximation. To this approximation the differential cross section $q_{j'm';jm}(\theta) = (k'/k)\,|f_{j'm';jm}(\theta)|^2$ is the sum of differential cross sections for different trajectories $\Lambda,\varkappa$, and the interference terms. Each interference term corresponds to two trajectories $\Lambda,\varkappa$ and $\Lambda',\varkappa'$, and is proportional to $\cos[\delta^{\Lambda}_{j'j}(J^{\Lambda\varkappa}_{j'j}) - \delta^{\Lambda}_{j'j}(J^{\Lambda'\varkappa'}_{j'j}) + \frac{\pi}{4}(\nu^{\Lambda\varkappa}_{j'j} - \nu^{\Lambda'\varkappa'}_{j'j})]$. A sufficient separation of the stationary phase points requires that the argument of cosine is large.

The next simplifying step leading to the classical semiclassical approximation neglects all interference contributions. This is sometimes justified, since this contribution tends to vanish when averaged over small angular or energy uncertainties. To this approximation

$$q_{j'm';jm}(\theta) = \sum_{\Lambda\varkappa} P^{\Lambda\varkappa}_{j'm';jm}(\theta) q^{\Lambda\varkappa}_{j'j}(\theta) \tag{27}$$

where

$$P^{\Lambda\varkappa}_{j'm';jm}(\theta) = \left. |B^{\Lambda\varkappa}_{j'm';jm}(J,\theta)|^2 \right|_{J=J^{\Lambda\varkappa}_{j'j}(\theta)} \tag{28}$$

The factors $P^{\Lambda\varkappa}_{j'm';jm}$ must be interpreted as probabilities of the transition for scattering through the angle θ along trajectory Λ, $\varkappa$.

The integral cross section $\sigma_{j'm';jm}$ calculated to the same approximation is

$$\sigma_{j'm';jm} = \frac{2\pi}{k^2}\int_0^\infty P_{j'm';jm}(J)JdJ \tag{29}$$

where

$$P_{j'm';jm} = \sum_\Lambda P^\Lambda_{j'm';jm}$$

$$P^\Lambda_{j'm';jm}(J) = \left|\sum_{\nu'\nu}\exp[i\,\nu'\eta^\Lambda_{j'j}(J)]\,\Delta^{j'}_{m'\nu'}S^\Lambda_{j'\nu';j\nu}\,\Delta^j_{m\nu}\right|^2 \tag{30}$$

Here again $P^\Lambda_{j'm';jm}(J)$ has the meaning of transition probability ascribed to the trajectory Λ and specified by the angular momentum J. The classical equations (27) to (30) provide the most transparent interpretation of the MSA approach.

Turning now to Eq-s (28) and (29) note that the dependence of the transition probability on θ or J is dual. First, it is the dependence of pre-exponentials $S^\Lambda_{j'\nu';j\nu}$ on J, arising as the result of solving scattering equations in the total angular momentum representation. Second, it manifests in extra modulating factors $\exp(iw^{\Lambda\varkappa}_{j'j}\theta\nu')$ and $\exp[i\,\eta^\Lambda_{j'j}(J)\,\nu']$ which are connected with nonconservation of the relative angular momentum. These modulating factors vary over the range of $\Delta\theta$ of the order $1/j$, or over that of ΔJ corresponding to $\Delta\theta$.

Consider now the common trajectory approximation as a limiting case of MSA. If all the deflection functions $\eta^\Lambda_{j'j}(J)$ are close to some mean deflection function $\bar\eta(J)$, i.e.

$$|\eta^\Lambda_{j'j}(J) - \bar\eta(J)| \ll 1 \tag{31}$$

the stationary momenta J^* are specified by one index $\varkappa$ only, same as for elastic scattering. The number of classical cross sections is also reduced to $q^\varkappa$ only. Notwithstanding the small difference between deflection functions in the sense of Eq.(31), the phase differences $\delta^\Lambda_{j'j}(J^\varkappa) - \delta^{\Lambda'}_{j'j}(J^\varkappa)$ can be arbitrary. In case they are large, the primitive semiclassical approximation or the classical semiclassical approximation of the common trajectory method will be applicable.

The above quasiclassical analysis provides a rigorous definition of the basic concepts of the semiclassical common trajectory approximation, i.e. the space-fixed collision plane, and the impact parameter specifying the classical trajectory. The collision plane is taken to be normal to the classical conserved vector J . This vector serves as the quantization axis for the electronic angular momentum j . The impact parameter b is defined via J by $(J + 1/2) = k_0 b$, where k_0 is the common wave vector before or after collision.

References

1. Mott, N.F.: 1931, Proc.Cambr.Phil.Soc. 27, pp.553-560.
2. Mott, N.F., Massey, H,S.W.: "The Theory of Atomic Collisions", Clarendon Press, Oxford, 1965.
3. Miller, W.H.: 1974, Adv.Chem.Phys. 25, pp.69-177.
4. Miller, W.H.: 1975, Adv.Chem.Phys. 30, pp.77-136.
5. Child, M.S.: "Molecular Collision Theory", Academic Press, London - New York, 1974.
6. "Semiclassical Methods in Molecular Scattering and Spectroscopy", ed.by M.S.Child, Reidel Publ.Co., Dordrecht - Boston - London, 1980.
7. Percival, I.C., Richards, D.: 1975, Adv.Atom.Mol.Phys. 11, pp.1-82.
8. Pechukas, P.: 1969, Phys.Rev. 181, pp.166-185.
9. Mayer, D.H., Miller, W.H.: 1979, J.Chem.Phys. 70, pp.3214-3223.
10. Bjerre, A., Nikitin, E.E.: 1969, Chem.Phys.Lett. 1, pp.179-182.
11. Tully, J.C.: in: "Dynamics of Molecular Collisions", part B, ed.by W.H.Miller, Plenum Press, New York, 1976, pp.217-267.
12. Child, M.S.: in: "Atom-Molecule Collision Theory", ed. by R.B.Bernstein, Plenum Press, New York, 1979, pp. 427-465.
13. Delos, J.B.: 1981, Comm.Atom.Mol.Phys. 10, pp.211-214.
14. Nikitin, E.E., Umanskii,S.Ya.: "Neadiabaticheskie Perekhody pri Medlennykh Atomnykh Stolknoveniyakh" (Nonadiabatic Transitions in Slow Atomic Collisions"), Atomizdat, Moscow, 1979.
15. Nikitin, E.E., Umanskii, S.Ya.: "Theory of Slow Atomic Collisions: Interactions, Interference Effects, Non-adiabatic Coupling Models" - in press.
16. Smith, F.T.: 1963, Bull.Phys.Soc. 9, pp.411-413.

17. Smith, F.J.: 1965, Phys.Lett. 10, pp.290-291.
18. Dashevskaya, E.I., Mokhova, N.A.: 1972, Optika i Spektroskopiya 33, pp.817-825.

TIME DEPENDENT CONSTANTS OF THE MOTION

R.D. Levine
The Fritz Haber Research Center for Molecular Dynamics
The Hebrew University, Jerusalem 91904, Israel

Time dependent constants of the motion enable us to treat time evolving states using techniques and points of view made familiar in stationary state problems. This review introduces the concept and main properties of time dependent constants of the motion and surveys their primary applications with special reference to the maximum entropy formalism. The role of time dependent constants in determining what dynamical information is relevant for a particular experimental situation is emphasized. Particular attention is given to the situation where the true Hamiltonian is unknown and experimental input is used to identify the time dependent constants.

1. INTRODUCTION

Time dependent constants of the motion[1-4] enable us to treat stationary and time evolving states from a unified point of view[4-7]. They are the natural extension of the familiar, time independent, constants of motion and like them can be interpreted in terms of the inherent symmetries of the problem[7]. Notwithstanding their seemingly contradictory character, time dependent constants of motion are not unfamiliar. A simple example is a free particle of mass m in classical mechanics where x_0,

$$x_0 = x - (p/m)t, \tag{1}$$

has a time independent value even though it is an explicit function of time. The constancy of x_0 is, of course, due to that the position x and, in general, also the momentum p do vary with time. x_0 is constant because the implicit time dependence (via the dependence on x and, in general, also on p) is exactly cancelled by the explicit time dependence

$$\begin{aligned} dx_0/dt &= (\partial x_0/\partial x)(dx/dt) + \partial x_0/\partial t \\ &= (p/m) - (p/m) = 0. \end{aligned} \tag{2}$$

P.-O. Löwdin and B. Pullman (eds.), New Horizons of Quantum Chemistry, 135–154.

The extension to the general case is obvious. Let $I(p,x,t)$ be a time dependent constant of motion in classical mechanics. Then, using Hamilton's equations

$$
\begin{aligned}
dI/dt &= (\partial I/\partial x)(dx/dt) + (\partial I/\partial p)(dp/dt) + \partial I/\partial t \\
&= (\partial I/\partial x)(\partial H/\partial p) - (\partial I/\partial p)(\partial H/\partial x) + \partial I/\partial t \\
&= \{I,H\} + \partial I/\partial t \ . \qquad (3)
\end{aligned}
$$

Since, by assumption, $dI/dt = 0$ the implicit time dependence exactly cancels the explicit one

$$\partial I/\partial t = \{H,I\} \ . \qquad (4)$$

Here, as usual, H is the Hamiltonian and the curly brackets denote the Poisson bracket defined by equality between the 2nd and 3rd line of (3). If C is a time independent constant of the motion $\partial C/\partial t = 0$ and equation (4) reduces to the familiar definition of a constant of motion

$$\{H,C\} = 0 \ . \qquad (5)$$

Quantum dynamics replaces the Poisson bracket, $\{\ ,H\}$ by $(i\hbar)^{-1}[\ ,H]$ where the square brackets denote the commutator. Hence a quantum mechanical Hermitian operator I is a time dependent constant of the motion when

$$dI/dt = (i\hbar)^{-1}[I,H] + \partial I/\partial t = 0 \qquad (6)$$

and $I^{\dagger} = I$ or

$$\partial I/\partial t = (i\hbar)^{-1}[H,I]. \qquad (7)$$

To verify that I has indeed a constant mean value consider $Tr(\rho(t)I)$ where $\rho(t)$ is the density operator of the system at the time t. (This includes, of course, the possibility that the system is in a pure state) $\rho(t)$ evolves according to the von Neumann-Liouville equation

$$\frac{\partial\rho}{\partial t} = (i\hbar)^{-1}[H,\rho] \qquad (8)$$

or, cf. (6), $d\rho/dt = 0$. From this we conclude that (a) for any operator I such that $dI/dt = 0$, the expectation value is time independent

$$\frac{d}{dt}Tr(\rho(t)I) = Tr(\frac{d\rho}{dt}I) + Tr(\rho\frac{dI}{dt}) = Tr(\rho\frac{dI}{dt}) = 0 \qquad (9)$$

and (b) $\rho(t)$ itself is an example of a time dependent constant of the motion. In classical mechanics, conclusion (b) is known as Liouville's theorem.

In the example of the free particle the system has two functionally independent constants of motion x_0 and p_0. (p_0 = p for a free particle). Any analytic function f = f(x,p) of x and p can then be converted to a time dependent constant of motion by replacing x and p by x_0 and p_0 respectively

$$df(x_0,p_0)/dt = (\partial f/\partial x_0)(dx_0/dt) + (\partial f/\partial p_0)(dp_0/dt) = 0 \ . \tag{10}$$

In general, a classical system of n degrees of freedom has 2n functionally independent time dependent constants of motion. The proof is based on the observation that subject to mild restrictions on the Hamiltonian, Hamilton's equations generate a unique trajectory for given 2n initial conditions

$$\underset{\sim}{q}(t) = \mathcal{Q}(\underset{\sim}{q}_0,\underset{\sim}{p}_0,t) \tag{11a}$$

$$\underset{\sim}{p}(t) = \mathcal{P}(\underset{\sim}{p}_0,\underset{\sim}{q}_0,t) \ . \tag{11b}$$

Since Hamilton's equations are invariant under time reversal, it is possible to select $\underset{\sim}{q}(t)$ and $\underset{\sim}{p}(t)$ as initial conditions and to integrate backwards in time. At the time -t we should recover the original initial conditions

$$\underset{\sim}{q}_0 = \mathcal{Q}(\underset{\sim}{q}(t),\underset{\sim}{p}(t),-t) \tag{12a}$$

$$\underset{\sim}{p}_0 = \mathcal{P}(\underset{\sim}{p}(t),\underset{\sim}{q}(t),-t) \ . \tag{12b}$$

Indeed, the validity of (12) is a standard check for any computer program which generates classical trajectories. But (12) provides 2n functions of the current values of the coordinates, momenta and time whose numerical values are independent of time. Equation (1) being a special case of (12a). A function of q and p can be converted to a time dependent constant of the motion by replacing q and p by $\mathcal{Q}$ and $\mathcal{P}$ respectively.

Working with the position and momentum operators one can introduce the quantal analogues of $\mathcal{Q}$ and $\mathcal{P}$[3]. For our purpose it will however prove convenient to note an alternative, but equivalent, route for determining the time dependent constant of motion whose value at t=0 is a given operator.

The Heisenberg equation of motion for an operator $A_H(t)$, $A_H(t=0)=A$ where A is time independent, is

$$i\hbar\frac{dA_H}{dt} = [H,A_H(t)] \tag{13}$$

or

$$A_H(t) = U^{\dagger}(t,0)AU(t,0) \tag{14}$$

where U(t,0),

$$i\hbar\frac{dU(t,0)}{dt} = HU(t,0)\,, \tag{15}$$

$$U(0,0) = I$$

is the evolution operator. On comparing (13) and (7) it becomes clear that $A_H(-t)$, where, since $U^\dagger(-t,0) = U(t,0)$

$$A_H(-t) = U(t,0)AU^\dagger(t,0) \tag{16}$$

is a time dependent constant of the motion whose initial value is A. Just as in classical mechanics, quantal constants of motion are generated by propagation of the observables 'backwards' in time. Indeed, since

$$\rho(t) = U(t,0)\rho(0)U^\dagger(t,0) \tag{17}$$

it follows from the unitarity of the evolution operator that

$$\begin{aligned} Tr(A_H(-t)\rho(t)) &= Tr(U(t,0)AU^\dagger(t,0)U(t,0)\rho(0)U^\dagger(t,0)) \\ &= Tr(A\rho(0)) \end{aligned} \tag{18}$$

is independent of time.

Just like their time independent counterparts, the present constants of motion convert solutions of the Schrödinger equation into solutions. Given that $\psi(t)$ is a solution

$$i\hbar\partial\psi/\partial t = H\psi \tag{19}$$

then so is $I\psi$

$$\begin{aligned} i\hbar\partial I\psi/\partial t &= [H,I]\psi + Ii\hbar\partial\psi/\partial t \\ &= HI\psi + I(i\hbar\partial/\partial t - H)\psi \\ &= HI\psi \quad . \end{aligned} \tag{20}$$

Furthermore, non degenerate eigenfunctions of I are solutions of the time dependent Schrödinger equation[6,8]. To see this, let $\phi(t)$ be a non-degenerate eigenstate of I,

$$I(t)\phi(t) = \alpha\phi(t) \ . \tag{21}$$

Then, since

$$I(i\hbar\partial/\partial t-H)\phi(t) = (i\hbar\frac{\partial}{\partial t} - H)I\phi(t) = \alpha(i\hbar\frac{\partial}{\partial t} - H)\phi(t) \tag{22}$$

and since, by assumption $\phi(t)$ is a non degenerate eigenfunction of I it can differ from $(i\hbar\partial/\partial t-H)\phi(t)$ by, at most, a phase factor $f(t)$

$$(i\hbar\partial/\partial t-H)\phi(t) = f(t)\phi(t) \quad . \tag{23}$$

It follows that $\psi(t)$,

$$\psi(t) = \exp(-(i/\hbar)\int^t f(\tau)\,d\tau)\phi(t), \tag{24}$$

is a solution of the Schrödinger equation.

2. DETERMINING TIME DEPENDENT CONSTANTS OF MOTION

The partial differential equations (4) or (7) can be solved by separation of variables[9-11],

$$I_r(t) = \sum_{s=0}^{S} C_s F_{sr}(t) \tag{25}$$

where the C_s are a set of time independent operators and the $F_{sr}(t)$ are time dependent numerical coefficients. The subscript on I serves as a reminder that if the C_s's are linearly independent then one can form as many linear combinations of the type (25) as the number, S+1, of the time independent operators C_s's which are included in the expansion. We shall generate all such linear combinations and distinguish between them in terms of the matrix $\underset{\sim}{F}$ at the initial time. Since we are primarily concerned with collision phenomena we shall often choose the initial conditions at $t \to -\infty$. That is however not necessary and any other choice (e.g. t=0) is equally possible. It is convenient to choose the initial $\underset{\sim}{F}$ as the unit matrix. With this choice I_r is that constant of motion which evolved from the operator C_r.

An equation of motion for the matrix $\underset{\sim}{F}$ is obtained by taking the partial derivative, with respect time, of both sides of (25)

$$\begin{aligned}\partial I_q/\partial t = (i\hbar)^{-1}[H,I] &= \sum_s F_{sq}(t)(i\hbar)^{-1}[H,C_s] \\ &= \sum_r(\partial F_{rq}/\partial t)C_r \quad .\end{aligned} \tag{26}$$

It follows that if the set of operators C_s is closed under the operation of commutation with the Hamiltonian

$$(i\hbar)^{-1}[H,C_s] = \sum_r C_r g_{rs} \tag{27}$$

then, equating coefficients of C_r on both sides of (26)

$$\partial F_{rq}/\partial t = \sum_s g_{rs}F_{sq}(t), \tag{28}$$

and the expansion (25) is an exact representation for the constant of motion.

As already mentioned, a convenient boundary condition for equations (28) is that initially the matrix $\underset{\sim}{F}$ is diagonal. Hence, an alternative way of writing (25) is (cf. (16))

$$UC_rU^{\dagger} = \sum_{s=0}^{S} C_s F_{sr}(t) \quad . \tag{29}$$

Here U is the evolution operator from the initial time to the present.

2.1 The Forced Harmonic Oscillator

As a simple example[10] consider the Hamiltonian for a harmonic oscillator which is perturbed by a time dependent external force f(t)

$$H = \hbar\omega a^{\dagger}a + \hbar f(t)(a + a^{\dagger}) \quad . \tag{30}$$

Here a and $a^{\dagger}$ are the usual annihilation and creation operators so that $a + a^{\dagger}$ is the displacement, from equilibrium, of the oscillator. ω is the frequency.

Consider the four Hermitian operators I (the identity operator), $(a + a^{\dagger})/2$, $(a^{\dagger} - a)/2i$ and $a^{\dagger}a$. These four operators form a closed set under commutation with the Hamiltonian (30) as required by equation (27). Using the basic commutation relation $[a,a^{\dagger}]=1$ the matrix $\underset{\sim}{g}$ defined by equation (27) can readily be constructed

$$\underset{\sim}{g} = \begin{pmatrix} 0 & 0 & -f & 0 \\ 0 & 0 & -\omega & 0 \\ 0 & \omega & 0 & -2f \\ 0 & 0 & 0 & 0 \end{pmatrix} \quad . \tag{31}$$

Four constants of motion can now be computed using (25) and (28). Explicitly, and using the notation U for the evolution operator U(t,0) (cf. (29) and (16)) these are[10]

$$I_1 = UIU^{\dagger} = I \tag{32a}$$

$$I_2 = U[(a + a^{\dagger})/2]U^{\dagger} = (a + a\ )/2 - I\,\mathrm{Re}\gamma(t) \tag{32b}$$

$$I_3 = U[(a^{\dagger} - a)/2i]U^{\dagger} = (a^{\dagger} - a)/2i - I\,\mathrm{Im}\gamma(t) \tag{32c}$$

$$I_4 = Ua^{\dagger}aU^{\dagger} = a^{\dagger}a - a\gamma(t) - a^{\dagger}\gamma^*(t) + I|\gamma(t)|^2 \tag{32d}$$

where

$$\gamma(t) = -i\int_0^t f(\tau)\exp[i\omega(\tau-t)]d\tau \tag{33}$$

and I is the identity operator.

As should be evident from equations (32), we could have picked a smaller set, containing only three operators which is also closed under commutation with H. There are several larger such sets containing e.g. six operators[9] (adding $a^{\dagger 2}$ and a^2) or nine operators (adding further $a^{\dagger}aa^{\dagger}a$, $a^{\dagger}a^{\dagger}a$ and $a^{\dagger}aa$) etc., each set generating a corresponding number of time dependent constants of motion. The same technique has also been applied[12,13] to the forced Morse oscillator.

2.2 The Adiabatic Approximation

Time dependent nearly constants of the motion can be generated in a variety of ways depending on the nature of the Hamiltonian[14-16]. The more obvious ones include truncating the expansion (25) to such operators that are closed under commutation with H_0 where $H = H_0 + V$ and V is 'small'. A less obvious option, explored in section 4 below is to truncate the expansion in such a manner that the expectation value of the remainder is zero, for the particular state of interest. One can also introduce a mean field approximation[14] whereby bilinear products of operators (e.g. $a^{\dagger}a$) are replaced by the corresponding linearized forms (e.g. $\langle a^{\dagger}\rangle a + a^{\dagger}\langle a\rangle$). Another option is the adiabatic approximation[15] which is exact when the rate of change due to the explicit time dependence of the $I_r(t)$'s is slow compared to the natural frequencies of the system. Consider the linear combination

$$A_r = \sum_s C_s W_{sr}(t) \tag{34}$$

where the matrix $\underset{\sim}{W}$ is chosen so as to diagonalize the matrix $\underset{\sim}{g}$

$$\underset{\sim}{g}\underset{\sim}{W} = i\underset{\sim}{W}\underset{\sim}{\Omega} . \tag{35}$$

$\underset{\sim}{\Omega}$ is a diagonal matrix. Then (cf. (27))

$$(i\hbar)^{-1}[H,A_s] = i\Omega_s A_s \tag{36}$$

and the adiabatic constants are given by

$$A_r(t) = \exp(i\int^t \Omega_r dt)\sum_s C_s W_{sr}(t) . \tag{37}$$

Indeed, if the time dependence of $\underset{\sim}{W}$ can be neglected then $\partial A_r/\partial t = i\Omega_r A_r(t)$ and so, using (7) $dA_r(t)/dt = 0$.

For the Hamiltonian (30),

$$\underset{\sim}{W} = \begin{pmatrix} 1 & 0 & f(t) & f(t) \\ 0 & 2f(t) & \omega & \omega \\ 0 & 0 & i\omega & -i\omega \\ 0 & \omega & 0 & 0 \end{pmatrix} \tag{38}$$

with the eigenvalues zero (doubly degenerate) and $\pm\omega$. Corresponding to $I_4(t)$ whose initial value is $a^\dagger a$ (cf. (32d)) we now have the adiabatic constant

$$A(t) = \omega a^\dagger a + 2f(t)(a + a^\dagger)/2 \ . \qquad (39)$$

Integrating (33) by parts we have that

$$\omega\gamma(t) = -f(t) + \int_0^t (\partial f(\tau)/\partial\tau)\exp[i\omega(\tau - t)]d\tau \ . \qquad (40)$$

Hence, when $\partial f(t)/\partial t$ (which is the rate of application of the force on the oscillator) is slowly varying over a period of the oscillator the integrand in(40) averages out to nearly zero. $\gamma(t)$ is well approximated by $-f(t)/\omega$ and exact result (32d) reduces to the adiabatic approximation (39).

3. THE STATE OF A TIME EVOLVING SYSTEM

Determining the state of a time evolving system differs in one essential way from the corresponding stationary problem. It is not sufficient to know the equations of motion. These are given as differential equations and hence their explicit solution requires specifying the boundary conditions. In contrast to the stationary case, here the boundary conditions are neither obvious nor unique. Rather, they are determined by the particular experiment under consideration. Moreover, the initial conditions for real experiments often are not as detailed as those assumed in the traditional computational procedures. This is always the case when classical dynamics is employed since classical initial conditions (all coordinates, all momenta) are twice as many as can be specified in even the most highly resolved quantal system. Furthermore, experiments often do not employ single quantum states. The traditional solution has been to do the dynamics first and then perform the required average (to simulate the initial state of real interest) later. While there is no error of principle in such an approach, both intuition and hard evidence indicate that some of the details generated in the detailed dynamical computations will be averaged out. Time dependent constants of the motion directly identify those dynamical aspects that are relevant to a particular initial state.

Our purpose in this section is then to outline a method whereby the averaging is done first and the time evolution is determined for the initial state of interest. Further discussion of this point of view can be found elsewhere[4,6,11,17].

To illustrate the loss of dynamical detail consider a crossed molecular beam experiment[18] designed to study energy transfer (inelastic collisions. If both beams happen to have a thermal distribution at the same temperature, then the final state will have a thermal distribution at the initial temperature and <u>all</u> dynamical information will average out exactly. The reason, as will be discussed in detail below, is that the time dependent constant of motion which evolved from the unperturbed Hamiltonian, H_0, is known to evolve back into the same H_0 after the collision. H_0 is the sum of H_{01} and H_{02}, the unperturbed Hamiltonians of the two collision partners. For a second experiment consider the case where the two beams both have a thermal distribution but at different temperatures. However, neither H_{01} nor H_{02} share the above mentioned property of their sum. The final state will thus differ from the initial state. Is it now necessary to carry out a full dynamical computation? The two cases mentioned correspond to the initial states whose density operators are

$$\rho_{in} = \exp[-\beta(H_{01} + H_{02})]/\mathrm{Tr}\{\exp[-\beta(H_{01} + H_{02})]\} \tag{41}$$

and

$$\rho_{in} = \exp(-\beta_1 H_{01} - \beta_2 H_{02})/\mathrm{Tr}[\exp(-\beta_1 H_{01} - \beta_2 H_{02})] \tag{42}$$

respectively. Can one not devise a method which will take better advantage of the similarity between (41) and (42)?

3.1 The Stationary Initial State

The common feature of (41) and (42) is that but for the collision they are stationary. This is, of course, a common characteristic of initial states. One wants their time evolution to be solely due to the perturbation. Any state which is stationary under the unperturbed Hamiltonian H_0 can be written as[4,6,9]

$$\rho_{in} = \exp(-\sum_{r=0}^{R} \lambda_r^{in} C_r) \tag{43}$$

where $C_0 = I$,

$$\exp(\lambda_0^{in}) = \mathrm{Tr}[\exp(-\sum_{r=1}^{R} \lambda_r^{in} C_r)] \tag{44}$$

provided the trace converges and the C_r's, r = 0,1,..,R are time independent constants of motion of H_0,

$$[C_r, H_0] = 0, \quad r = 0,1,..,R . \tag{45}$$

That (45) is a sufficient condition follows since $[\rho_{in}, H_0]=0$. In fact, it is also necessary[9], provided the trace in (44) converges.

We are unable to resist pointing out that ρ_{in} as given by (43) is that density operator which is of maximal entropy[9,17] amongst all density operators which have a given set of R+1 expectation values $\langle C_r\rangle^{in}$. The R+1 parameters in λ_r^{in} are equivalent to the R+1 expectation values. One set determines the other.

As an aside it is worth pointing out that, as in all other formulations of scattering theory, we have assumed that all is known about the spectrum of H_0. For molecules with realistic level structure this is by no means the case. Hence we are actively examining the application of algebraic techniques to structural problems[13,19]

3.2 The State of the Interacting System

At this point the formal work is already over. Given the initial state ρ_{in} we can compute the density operator at the time t (cf. (17))

$$\begin{aligned}\rho(t) &= U\rho_{in}U^{\dagger} \\ &= U\exp(-\sum_r \lambda_r^{in} C_r)U^{\dagger} \\ &= \exp(-\sum_r \lambda_r^{in} UC_rU^{\dagger})\end{aligned} \quad (46)$$

where we have used the unitarity of the evolution operator.

The operators $UC_rU^{\dagger}$ which appear in (46) are the time dependent constants of motion whose initial values are the C_r's. Hence the state of the system at time t is given by

$$\rho(t) = \exp(-\sum_{r=0}^{R} \lambda_r^{in} I_r(t)) \quad . \quad (47)$$

Just as a stationary state of a system is a state of maximal entropy subject to constraints which are time independent constants of motion we have shown in (47) that a state of a time evolving system is a state of maximal entropy subject to constraints which are time dependent constants of motion.[4,9-11]

Equation (47) is the central formal result of this review. It shows that only those constants of the motion which are present in the initial state constrain the time evolution of the system. The nature of the interaction is only manifest in determining the explicit form of the constants of the motion (equations (25) (29)), as explicitly demonstrated in equation (49) below.

When the perturbation has a finite duration (e.g. as in a collision) then the $I_r(t)$'s will tend to asymptotic, time independent operators[6]. Letting B_r be the final operator into which $I_r(t)$ will evolve, the final density operator is

$$\rho_{out} = \exp(-\sum_{r=0}^{R} \lambda_r^{in} B_r) \quad . \quad (48)$$

Further discussion of asymptotic constants of motion is provided in an appendix.

Equations (47) and (48) remain a purely formal solution unless an explicit construction of the constants of the motion is possible. When the separation of variables, equation (25), is feasible

$$\rho(t) = \exp\left(-\sum_{s=0}^{S} \lambda_s(t) C_s\right) . \tag{49}$$

Here (cf. (29))

$$\lambda_s(t) = \sum_{r=0}^{R} F_{sr}(t) \lambda_r^{in} . \tag{50}$$

Note that $S \geqslant R$, where S is the number of terms in (25) and hence in the exponent of (49). By taking $\lambda_s^{in} = 0$ for $R < s \leqslant S$, one can make (50) a square matrix equation.

In (47) the (Lagrange) parameters are time independent while the operators carry the time dependence. Conversely in (49). The two, equivalent, expressions are related in the manner made familiar by the Schrödinger and Heisenberg pictures. To see this note that (47) can also be written as

$$\rho(t) = \exp\left(-\sum_{r=0}^{R} \lambda_r^{in} C_{rH}(-t)\right) . \tag{51}$$

3.3 Equations of Motion for the Constraints

Rather than characterize the time dependence of $\rho(t)$ in terms of the (Lagrange) parameters $\lambda_s(t)$ one can use the expectation values of the operators C_s at the time t. These satisfy

$$\begin{aligned} \partial\langle C_r\rangle(t)/\partial t &= \mathrm{Tr}\{C_r \partial\rho(t)/\partial t\} \\ &= (i\hbar)^{-1}\mathrm{Tr}\{C_r[H,\rho(t)]\} \\ &= -(i\hbar)^{-1}\mathrm{Tr}\{\rho(t)[H,C_r]\} \\ &= -\mathrm{Tr}\{\rho(t)\sum_s C_s g_{sr}\} \\ &= -\sum_s \langle C_s\rangle(t) g_{sr} \end{aligned} \tag{52}$$

The solution of the set of coupled linear equations (52) is

$$\langle C_r\rangle(t) = \sum_s \langle C_s\rangle^{in} G_{sr}(t) \tag{53}$$

where $\langle C_s\rangle^{in}$ is the expectation value of C_s in the initial state and the evolution matrix $\underset{\sim}{G}$ is initially diagonal, $\underset{\sim}{G}^{in} = \underset{\sim}{I}$.

Intermediate between the traditional approach where the evolution matrix $\underset{\sim}{U}$ is solved for and hence the dynamics can be considered for any possible initial state and the discussion thus far where the particular initial state (43) is considered, is solving for the evolution matrix $\underset{\sim}{G}$. We note first that, in general, the dimension of $\underset{\sim}{G}$ (i.e. S+1) is far smaller than that of $\underset{\sim}{U}$. ($\underset{\sim}{U}$ is a matrix in quantum state space. Strictly speaking it is denumerably infinite in the example of the forced harmonic oscillator although in practical problems it can be well approximated by a finite matrix. $\underset{\sim}{G}$ is a 'super evolution' matrix as it acts on observables. Its dimension can be quite low, e.g. four in the example above).

Solving the equation of motion for $\underset{\sim}{G}(t)$

$$\partial \underset{\sim}{G}/\partial t = -\underset{\sim}{G}\underset{\sim}{g} \tag{54}$$

$$\underset{\sim}{G}^{in} = \underset{\sim}{I}$$

means that any initial state of the functional form (43), irrespective of the numerical values of the (Lagrange) multipliers, can be solved for. Equivalent to $\underset{\sim}{G}$ is the evolution matrix $\underset{\sim}{F}$, cf. (28)

$$\partial \underset{\sim}{F}/\partial t = \underset{\sim}{g}\underset{\sim}{F} \tag{55}$$

$$\underset{\sim}{F}^{in} = \underset{\sim}{I}$$

or, comparing (55) and (54)

$$\underset{\sim}{F} = \underset{\sim}{G}^{-1} . \tag{56}$$

The reader is invited to verify that the inverse relation between the $\underset{\sim}{F}$ and $\underset{\sim}{G}$ matrices is directly attributed to the unitarity of the evolution operator.

3.4 Sum Rules

Experimental results and models for collision processes can often be stated in the form of a linear transformation between the initial and final expectation values[9,20,21]

$$\langle C_r \rangle^{out} = \sum_s \langle C_s \rangle^{in} G_{sr} . \tag{57}$$

Such relations are often termed 'sum rules'. Comparing (57) with (53) one can identify the coefficients in (57) with the elements of the final (out) $\underset{\sim}{G}$ matrix. Hence, a relation of the type (57) is implied by the presence of time dependent constants of motion. Indeed, the asymptotic values of these constants are directly expressed in terms of the asymptotic $\underset{\sim}{F}$ matrix

$$I_r^{in} = C_r, \quad I_r^{out} = \sum_s C_s F_{sr}^{out} \tag{58}$$

4. UNKNOWN HAMILTONIAN

It was thus far taken for granted that the 'true' Hamiltonian of the system is known. Typically however this is not the case. Indeed the very purpose of collision experiments is to characterize the interaction. Similarly, one often uses known probes to perturb and hence explore the structure of the system.

As is intuitively reasonable, and is demonstrably the case for, say, a thermal initial state, the information that can be extracted from a given experiment depends on the initial state. Can one then define an 'equivalent' Hamiltonian which accounts for all the dynamical information available from the state of the system at a given point in time[22]. (In particular, after the collision), for a given initial state.

4.1 Sufficient Measurements

Let ρ_{in} be given by (43) and $\rho(t)$ be the true but unknown state of the system at the time t. H' is a model Hamiltonian which satisfies the following two conditions: (a) The set of S + 1 operators C_s is closed under commutation with H' (cf. (29))

$$U'C_rU'^{\dagger} = \sum_{s=0}^{S} C_s F'_{sr}(t) \tag{59}$$

and (b) At the time t the matrix $\underset{\sim}{G}'(t)$ which is the inverse of the matrix $\underset{\sim}{F}'(t)$ as determined from (59) correctly predicts the S + 1 expectation values $\langle C_r\rangle(t)$:

$$\mathrm{Tr}[C_r\rho(t)] = \sum_{s=0}^{S} \langle C_s\rangle^{in} G'_{sr}(t) \quad . \tag{60}$$

Subject to these two conditions it is claimed that the model Hamiltonian accounts exactly for the dynamics. Explicitly, the state $\rho'(t)$ which evolved from the true initial state ρ_{in} under the action of the model Hamiltonian H'

$$\rho'(t) = \exp(-\sum_{s=0}^{S} \lambda'_s(t)C_s) \tag{61}$$

$$\lambda'_s(t) = \sum_r F_{sr}(t)\lambda_r^{in} \tag{62}$$

is equivalent to the exact state of the system

$$\rho'(t) \equiv \rho(t) \quad . \tag{63}$$

By imposing the second condition, (b), above it is possible to give up conditions (27) or (29) altogether. H' can be any model Hamiltonian (as long as it is Hermitian) and (60) can be checked by measuring the 'true' expectation values $\langle C_r\rangle(t)$. It is thus not necessary to know $\rho(t)$ to verify (60).

4.2 Proof

For the purpose of the proof of (63) it should be noted that conditions (a) and (b) suffice to insure that

$$\mathrm{Tr}[\rho(t)C_s] = \mathrm{Tr}[\rho'(t)C_s] \ . \quad s = 0,1,..,S \tag{64}$$

To prove (64) note that

$$\begin{aligned}\mathrm{Tr}[\rho'(t)C_r] &= \mathrm{Tr}[U\rho_{in}U^{\dagger}C_r] \\ &= \mathrm{Tr}[\rho_{in} \sum_s C_s G'_{sr}(t)] \\ &= \sum_s \langle C_s\rangle^{in} G'_{sr}(t) \ .\end{aligned} \tag{65}$$

The proof of (63) is based on the logarithmic inequality[23]

$$-\mathrm{Tr}[\rho(t)\ell n\rho(t)] \leqslant -\mathrm{Tr}[\rho(t)\ell n\rho'(t)] \tag{66}$$

where equality obtains if and <u>only if</u> $\rho(t) = \rho'(t)$. Now, using (64)

$$\begin{aligned}-\mathrm{Tr}[\rho(t)\ell n\rho'(t)] &= \sum_{s=0} \lambda'_s(t)\mathrm{Tr}[\rho(t)C_s] \\ &= \sum_{s=0} \lambda'_s(t)\mathrm{Tr}[\rho'(t)C_s] \\ &= -\mathrm{Tr}[\rho'(t)\ell n\rho'(t)] \ .\end{aligned} \tag{67}$$

But $\rho'(t)$ evolved from ρ_{in} under the action of the unitary evolution operator U' so that its entropy is conserved

$$-\mathrm{Tr}[\rho_{in}\ell n\rho_{in}] = -\mathrm{Tr}[\rho'(t)\ell n\rho'(t)] \ . \tag{68}$$

Similarly, $\rho(t)$ evolved from ρ_{in} under the true, if unknown, unitary evolution operator U and hence

$$-\mathrm{Tr}[\rho_{in}\ell n\rho_{in}] = -\mathrm{Tr}[\rho(t)\ell n\rho(t)] \ . \tag{69}$$

Combining (69), (68), and (67) we have equality in (66) and, in view of the 'only if' part of the inequality, we have proved (63).

4.3 How Many Expectation Values Determine the Post Collision State of a System

The model Hamiltonian is required to satisfy (59) and (60). There is always the choice of H' as the true Hamiltonian which shows that the conditions are not empty. In general, one starts with the R + 1 operators C_r required to specify the initial state and measure their expectation values at the time t of interest. Can one then find a model

Hamiltonian H' such that both conditions (59) and (60) are satisfied? If so, the state of the system at the time t is specified and H' fully describes the dynamics, as known at the time t. (It may well however fail to account for the dynamics at other times. If however the time t is, say, after the collision, then other times are, by far, less relevant). If conditions (59) and (60) are not satisfied, the original set of R + 1 operators need to be augmented by additional operators. The more one needs to enlarge the set of operators, the more restrictions are placed on the choice of H' and the better does the experiment characterize the true Hamiltonian. An initial state specified by very few operators will (in general, but not necessarily) not provide as much information about the interaction as a more tightly specified initial state.

4.4 Equivalent Constants of the Motion

In terms of time dependent constants of motion, the equivalent Hamiltonian H' is specified by the condition that its constants I_r'

$$I_r' = U'C_rU'^{\dagger} \tag{70}$$

are constants of motion of the true Hamiltonian for the particular time t,

$$\mathrm{Tr}(\rho_{in}C_r) = \mathrm{Tr}(\rho(t)I_r'(t)) \ . \tag{71}$$

A true constant of motion is required to satisfy (71) for all values of time. The model constants need to satisfy (71) only for the time t of interest, and for the initial state of interest.

4.5 Sum Rules

In applications the result (63) is particularly relevant to the state of the system after the collision. We have previously noted that model Hamiltonians often give rise to linear relations of the type

$$\langle C_r\rangle^{out} = \sum_s \langle C_s\rangle^{in} G'_{sr} \ . \tag{72}$$

It follows that <u>if</u> the values $\langle C_r\rangle^{out}$ as computed from the model agree with the measured ones then all the dynamical information provided by the experiment is summarized by the simplest model which will give rise to (72). Any more detailed model has not really been verified by the experiment.

A thermal initial state is specified by H_0 and I. Conservation of energy and of probability imply that always

$$\langle H_0\rangle^{out} = \langle H_0\rangle^{in} \ , \qquad \langle I\rangle^{out} = \langle I\rangle^{in} \ . \tag{73}$$

Any interaction will satisfy (73) and hence a thermal initial state will provide no characterization whatever of the true interaction.

4.6 The Statistical Limit

As an example of an equivalent evolution consider what is commonly referred to as 'the statistical limit'[25,24]. This is the limit where, after the collision, all final quantum states of the system are equally probable. In other words, for every operator O diagonal in the H_0 representation, the final expectation value is $N^{-1}Tr(O)$ where N is the number of final quantum states. Proof:

$$Tr(O\rho_{out}) \equiv \sum_i O_{ii}(\rho_{out})_{ii}$$

$$\xrightarrow[\text{limit}]{\text{statistical}} N^{-1}\sum_i O_{ii} = N^{-1}Tr(O) \quad . \tag{74}$$

It is usually assumed that the statistical limit can only be valid in some averaged sense[24] and cannot be the result of a unitary time evolution. Here we shall show that as long as the initial state is diagonal in the H_0 representation there exists a unitary evolution matrix which is equivalent to the statistical limit. Hence to probe deeper into the dynamics in the statistical limit it is necessary to employ initial states which are not diagonal in the H_0 representation.

Consider the unitary matrix[26]

$$\begin{aligned} \tilde{U}'_{ij} &= N^{-\frac{1}{2}}\exp(i\alpha_{ij}) \\ \alpha_{ij} &= (2\pi/N)(i-1)(j-1) . \end{aligned} \tag{75}$$

Let $\tilde{U}'$ be an equivalent evolution matrix in the H_0 representation. Then for any diagonal operator O the corresponding constant of motion has the matrix representation

$$\begin{aligned} (\tilde{U}'\tilde{O}\tilde{U}'^{\dagger})_{ij} &= \sum_k U'_{ik}O_{kk}U'^{*}_{jk} \\ &= N^{-1}\exp[(2\pi/N)(i-j)]\sum_k O_{kk} \\ &= N^{-1}\exp[(2\pi/N)(i-j)]Tr(O) \quad . \end{aligned} \tag{76}$$

Apart from an overall numerical factor (i.e. Tr(O)) all these matrices are identical. In particular their diagonal matrix elements are all N^{-1}. Hence, for any initial state ρ_{in}, $Tr(\rho_{in}) = 1$, which is diagonal, $\tilde{U}'$ is equivalent to the statistical limit:

$$(\rho'_{out})_{ii} = N^{-1} \quad . \tag{77}$$

Even in the off diagonal elements of ρ_{out},

$$(\rho_{out})_{ij} = N^{-1}\exp[(2\pi/N)(i-j)] \tag{78}$$

all memory of the initial state has been lost. Only by using a non diagonal ρ_{in} (which is experimentally non trivial) can one hope to extract more details about the dynamics.

The unitary matrix (75) also serves to illustrate the notion of a chaotic time evolution where the asymptotic form of the time dependent constants of the motion is independent of their initial form. Of course, we have not derived (75) from a Hamiltonian but it is possible[15] to trace the origin of this behavior to the eigenfrequencies Ω as defined by (36). When a pair of complex frequencies is possible, the dynamics shows an instability and the asymptotic form of the F or G matrices is dominated by one eigenvector. The asymptotic form of all constants of motion is then the same. The simplest example familiar to us is the SO(2,1) case[15]:

$$\begin{aligned} H &= C_3 + f(t)C_1 \\ [C_1,C_2] &= i\hbar C_3 \\ [C_2,C_3] &= -i\hbar C_1 \\ [C_3,C_1] &= -i\hbar C_2 \quad . \end{aligned} \tag{79}$$

Here, if f(t) is maintained above unity for a sufficient length of time then all the three constants of motion whose initial values are the three C_i's will tend to a common limit.

5. CONCLUDING REMARKS

Time dependent constants of motion enable us to center attention on the evolution of the state of interest

$$\frac{\partial\rho(t)}{\partial t} = \sum_r \frac{\partial\rho(t)}{\partial\langle I_r(t)\rangle(t)} \frac{\partial\langle I_r(t)\rangle(t)}{\partial t} \tag{80}$$

$$\langle I_r(t)\rangle(t) = \mathrm{Tr}[\rho(t)I_r(t)] \quad . \tag{81}$$

All the dynamical information is in the explicit time dependence of the constants of the motion. Different states differ only in the (time independent) weights, $\partial\rho(t)/\partial\langle I_r(t)\rangle(t)$. These weights are determined by the initial state. When a set of S+1 time independent operators C_s, closed under commutation with the Hamiltonian

$$(i\hbar)^{-1}[H,C_s] = \sum_r C_r g_{rs}, \qquad s = 0,1,..,S \tag{27}$$

can be found, one can explicitly solve for the time dependence of the corresponding S+1 time dependent constants of the motion. These S+1 equations of motion (e.g. (28) or (52)) are closed. The S+1 values $\langle C_r \rangle(t)$ at one point in time fully determine the subsequent evolution of the system, including the long time, asymptotic limit.

ACKNOWLEDGMENT

Much of the work reviewed here has been carried out in collaboration with Y. Alhassid, P. Pfeifer, T. Ungar and C.E. Wulfman. The Fritz Haber Research Center is supported by the Minerva Gesellschaft für die Forschung, mbH, München, BRD. This work was supported by the Air Force Office of Scientific Research, Grant AFOSR-81-0030.

APPENDIX: ASYMPTOTIC CONSTANTS OF MOTION

The discussion in the appendix is of relevance to collision problems where the perturbation V vanishes both before and after the collision. Additional details will be found in reference 25.

The in and out wave operators are defined (Sec. 2.7.2 of ref. 24) by

$$\rho(t) = \Omega_+ \rho_{in}(t) \Omega_+^\dagger \tag{A.1}$$

$$\rho(t) = \Omega_- \rho_{out}(t) \Omega_-^\dagger \tag{A.2}$$

or, since $\Omega_\pm^\dagger \Omega_\pm$ is a projector on the continuous spectrum of H_0

$$\Omega_\pm^\dagger \Omega_\pm = P_0 \tag{A.3}$$

we have that

$$\rho_{out}(t) = S \rho_{in}(t) S^\dagger \tag{A.4}$$

with

$$S = \Omega_-^\dagger \; \Omega_+^\dagger \; . \tag{A.5}$$

Here $\rho_{in}(t)$ is the initial state propagated to the time t under H_0 and similarly for $\rho_{out}(t)$. S is the scattering operator.

Using the functional form (43) for ρ_{in} it follows that any constraint C on ρ_{in} will evolve to

$$C_+ = \Omega_+ C \Omega_+^\dagger \; . \tag{A.6}$$

Operators C for which the limit in (A.6) exists have been termed[27] 'asymptotic constants of the motion' and can be shown to commute with H_0. This result provides an additional justification as to why only constants of the motion of H_0 can act as constraints for the initial state.

The operator C_+ as defined by (A.6) can be shown[26] to be a constant of motion of the full Hamiltonian H. Indeed, it can be thought of as the 'diagonal part'[28] of C in the H representation and to be written as an Abelian limit[26,28,29] sometimes known as the 'adiabatic switching' procedure

$$C_+ = \lim_{\varepsilon \to 0} \varepsilon \int_{-\infty}^{0} \exp(\varepsilon t) C_H(-t)\, dt \quad . \qquad (A.7)$$

C_+ can thus be identified as $C_H(t=0)$ while $\lim C_H(t\to -\infty)$ is C.

REFERENCES

1. Sudarshan, E.C.G. and Mukunda, N., Classical Dynamics (Wiley, N.Y., 1974).
2. Anderson, R.L., Kumei, S. and Wulfman, C.E.: 1972, Rev. Fiz. Mexicana 21, 1; Aronson, E.B., Malkin, I.A. and Man'ko, V.I.: 1974, Sov. J. Particles Nucl. 5, 47.
3. Dothan, J.: 1970, Phys. Rev. D2, 2944.
4. Levine, R.D.: 1981, Adv. Chem. Phys. 47, 239.
5. Levine, R.D. and Wulfman, C.E.: 1982, Chem. Phys. Letts. 87, 105.
6. Pfeifer, P. and Levine, R.D.: Phys. Rev. A, in press.
7. Levine, R.D.: 1981, Chem. Phys. Letts. 79, 205.
8. Lewis, H.R. and Riesenfeld, J.: 1969, Math. Phys. 10, 1458; Malkin, I.A., Man'ko, V.I. and Trifonov, D.A.: 1973, J. Math. Phys. 14, 576.
9. Alhassid, Y. and Levine, R.D.: 1977, J. Chem. Phys. 67, 4321.
10. Alhassid, Y. and Levine, R.D.: 1978, Phys. Rev. A 18, 89.
11. Alhassid, Y. and Levine, R.D.: 1979, Phys. Rev. C 20, 1775.
12. Levine, R.D. and Wulfman, C.E.: 1979, Chem. Phys. Letts. 60, 372.
13. Levine, R.D. in Intramolecular Dynamics (Reidel, Dordrecht, 1982) p.17.
14. Levit, S.: 1980, Phys. Rev. C 21, 1594; Alhassid, Y. and Koonin, S.E.: 1981, Phys. Rev. C 23, 1590.
15. Wulfman, C.E. and Levine, R.D.: 1981, Chem. Phys. Letts. 84, 13.
16. Anderson, R.L., Shibuya, T. and Wulfman, C.E.: 1974, Rev. Fiz. Mexicana 23, 257.
17. Levine, R.D. in The Maximum Entropy Formalism (M.I.T. Press, Cambridge, 1979).
18. Levine, R.D. and Bernstein, R.B., Molecular Reaction Dynamics (Clarendon Press, Oxford, 1974).
19. Iachello, F. and Levine, R.D.: Chem. Phys., in press.
20. Levine, R.D.: 1978, Ann. Rev. Phys. Chem. 29, 59.
21. Alhassid, Y., Levine, R.D., Karp, J.S. and Steadman, S.G.: 1979, Phys. Rev. C 20, 1789.

22. Levine, R.D. and Ungar, T.: Phys. Rev. Lett., in press.
23. Ruelle, D., Statistical Mechanics (Benjamin, N.Y., 1969); Falk, H.: 1970, Am. J. Phys. 38, 858.
24. Levine, R.D., Quantum Mechanics of Molecular Rate Processes (Oxford University Press, 1969), Sec. 3.5.1.
25. Pfeifer, P. and Levine, R.D.: J. Chem. Phys., in press.
26. Brenner, J.L., quoted by F.T. Smith in Kinetic Processes in Gases and Plasmas (Academic Press, N.Y., 1969).
27. Thirring, W., A Course in Mathematical Physics (Springer Verlag, Berlin, 1980), V. 3.
28. Primas, H.: 1963, Rev. Mod. Phys. 35, 710.
29. Kubo, R.: 1957, J. Phys. Soc. Japan 12, 570.

ELECTRON MOLECULE SCATTERING AT LOW ENERGIES: RELATIONS BETWEEN DYNAMICS AND STRUCTURE

F.A. Gianturco
Gruppo di Chimica Teorica, Università di Roma
Nuovo Edificio Chimico, Città Universitaria
00100 Rome A D, Italy

ABSTRACT

The collisions between collimated beams of slow electrons ($0.1 \leq k^2 \leq 2$ Ryd) and gaseous molecules provide a very powerful experimental tool for the study of molecular properties, of pathways for energy storage and deposition among the various degrees of freedom and of molecular behaviour during dissociation and/or ionization and attachment of elementary charges.

The corresponding theoretical studies of the last few years have provided great insight into the various possible elementary mechanisms which couple the relative motions of all the involved particles during the above process. They have shown that to combine quantum chemical knowledge of the target structure with the corresponding use of scattering theory allows one to obtain very successfully a detailed picture of events.

Some of the models employed by theory are here reviewed and discussed.

1. INTRODUCTION

The constant aim of theory has always been to develop from first principles treatments of specific ensembles of phenomena, as wide as possible, which would provide at the same time physical insight on the whole event and accurate predictions of its many related processes that could afford experimental detection.

In the case of collision phenomena involving electrons (or positrons) and simple molecules in the gas phase one is faced, at the outstart, with the problem of describing the structures of isolated targets, i.e. the spectral nature of the corresponding hamiltonian for M nuclei and N electrons, together with the continuum nature of the

P.-O. Löwdin and B. Pullman (eds.), New Horizons of Quantum Chemistry, 155–178.

impinging beam of electrons. In other words, with the discrete, and hopefully sparse, distribution of molecular states which play an important role in the event and the (asymptotically) continuum electrons which however need to interact (i.e. change their behaviour within finite regions of space phase) with the above mentioned states in order to cause the observed variety of partial cross sections (integral and differential) that make up the whole set of scattering measurements.

When the range of considered relative energies is increased, one's task is further complicated by the onset of many more processes which involve electronic excitations of targets, dissociative attachments, bound-free electronic transitions, and free-free transitions induced by photon absorption or emission. Moreover, if one limits the study to the lower end of the energy scale, the variety of outgoing channels is accordingly reduced and therefore one might hope to obtain a unified treatment of the phenomenon.

The major interest of theoreticians in recent years has therefore been focussed on elastic processes and rotovibrationally inelastic processes which occur for ground-state molecular targets at collision energies up to $\sim$ 30 - 40 eV. In this more limited range, the dynamical outcomes are particularly interesting in that they provide the testing ground for the various possible theoretical models vis à vis a very detailed and accurate wealth of experimental findings.

It therefore becomes feasible to compare the relative interplay of the two main, different aspects of the collisional process, i.e. the bound-state structures of the molecular targets and their dynamical couplings with the continuum electrons which cause the observed redistribution into final asymptotic channels. It is thus essential at this point, both conceptually and technically, to distinguish between two major spatial regions in the description of the interaction between the molecular core and the additional electron: the long range (LR) region where the various dissociation channels are weakly coupled by multipolar terms of local potentials (and the relative motions of nuclei and electrons then become dynamically comparable) and the short range (SR) region where the strong, non-local many body effects lead to the formation of a compound system and where a unified description of the (N+1) electrons becomes necessary. At the same time, the faster motions of all the electrons may there be effectively decoupled, in a dynamical sense, from the slower moving molecular nuclei.

In the following we are therefore trying to describe how such an interplay could be analyzed, what can be directly learned from the structural properties of the molecular, bound wavefunctions of the core electrons, which are the leading terms that appear in the dynamical coupling and what are the computational approaches most successfully applied to date to simple molecules and which take both aspects into account in an even-handed way. Thus, we will necessarily exclude some of the recent methods that focus either on the structural aspects only or on the treatment of dynamics within purely model potentials: the balance of the two aspects is, in fact, in our opinion a very useful way of carrying into the real world of colliding molecules what has been learned already from the more traditional methods of Quantum Chemistry.

The next Section describes the construction of the interaction between bound and continuum wavefunctions, while Section III reports the more common formulations of the scattering equations and of the wavefunction expansions. Finally, Section IV presents some of the most instructive and recent results for simple targets.

2. THE INTERACTION POTENTIALS

As is well known, the nonrelativistic hamiltonian for an electron incident upon an N-electron molecule with M nuclei may be written, in atomic units, as:

$$H_{tot} = -\frac{1}{2}\nabla^2_{N+1} + \hat{V}_{N+1} + H_N + H_{rot} + H_{vib} = H_{el} + H_{nucl} \tag{1}$$

The first two terms represent the kinetic energy of the incident electron and its interaction with the target electrons and nuclei, H_N is the electronic hamiltonian of the molecule and the vibrational and rotational hamiltonians are simply the nuclear kinetic energy operators in the c. of m. of the system.

The total collision wavefunction for the whole (N+1) electrons system is obviously the solution of the corresponding time-independent Schrödinger equation, subject to the appropriate collision boundary conditions[1]. The essential ingredients for it are usually obtained via an expansion of the form:

$$\Psi = A \sum_{\alpha} (\Phi_i \chi_v \rho_j)_{\alpha} F_{\alpha}(N+1) + \sum_k c_k \phi_k (1...N+1) \tag{2}$$

where Φ_i, χ_v and ρ_j are the electronic, vibrational and rotational eigenfunction of the asymptotic molecular target, which are assumed to be closely coupled by the interaction, whilst the second expansion in eq. (2) involves a set of known L^2-integrable electron correlation functions ϕ_k.

The latter set is included to improve on the necessary slow convergence of the α-summation (especially for situations in which the impinging beam of electrons strongly disturbs the target system) and therefore it compensates for the unavoidable truncation of the first expansion. The k-summation thus includes functions which would otherwise be omitted due to the usual orthogonality constraint between the bound and continuum orbitals of eq. (2). The latter functions obviously depend also on the nuclear coordinates, while the first expansion should include any type of molecular state needed to represent the molecular function to the incoming projectile, i.e. the polarization of the target electrons. Finally, the antisymmetrization operator, $\mathcal{A}$, makes the total wavefunction antisymmetric with respect to exchange of space and spin coordinates of any pair of electrons.

The solution of the Schrödinger equation using the full hamiltonian (1) and the expansion (2) is neither computationally feasible nor necessary to determine most of the physical outcomes of the collision process. In a way similar to the more common considerations of the Born-Oppenheimer (B.O.) approximation, a procedure for separating the dynamical couplings of the relative motions of the various degrees of freedom can also be devised and will be discussed in the next Section. Let us now examine instead what sort of physical interactions are obtained from the operator (1) acting on the full wavefunction of (N+1) electrons within a fixed configuration of M nuclei.

2.1. The electrostatic potential

When the expansion (2) is approximated by a one-term only in the first summation, while the k-summation is altogether suppressed, one is dealing with the lowest level of description for the relevant system, i.e. with the single configuration representation of the unperturbed, usually $^1A_1$ or $^1\Sigma^+$, Hartree-Fock (HF) ground state (GS) of a closed shell molecule:

$$\Psi(1...N+1) = \mathcal{A}[\,\Phi_{GS}(N)\,F\,(N+1)\,] \tag{3}$$

where the electronic variables include both the spatial and spin parts.

The above simplification is equivalent to choosing only the lowest of the diagonal matrix elements generated by applying the hamiltonian (1) to expansion (2) and then integrating. The further approximation of holding the M nuclei fixed in space, as discussed later, disregards the effect of H_{nucl} on the above wavefunction (FN approximation) and therefore yields a set of exact static exchange (ESE) equations for the continuum orbitals F:

$$\{-\frac{1}{2}\nabla_r^2 + V_{GS}(r) - \frac{1}{2}k^2\}\,F(r) = \Sigma_s\{\,(\epsilon_s - \frac{1}{2}k^2)\langle\,\phi_s|F\,\rangle + \langle\phi_s\,|\,\frac{1}{|r-r'|}\,|\,F\rangle\}\,\phi_s(\underline{r}) \tag{4}$$

The s-summation now runs over the occupied, single-particle LCAO-MO's describing the single detor used to represent Φ_{GS}, the spin variable has been integrated out and the ϵ_s are the single-particle energies for the bound orbital ϕ_s's.

The operator $V_{GS}(\underline{r})$ is now the static, direct interaction between the incoming electron and the 'frozen' electronic configuration of the molecular target:

$$V_{GS}(\underline{r}) \equiv V_{GS}(\underline{r}_{N+1}) = < \Phi_{GS}(1 \ldots N) | \hat{V}(r \ldots r_{n+1}) | \Phi_{GS}(1 \ldots N) > \qquad (5)$$

where the integration is carried out over the space and spin coordinates of the N target electrons. In the more general case of polyatomic target, the above potential can be written as a multipolar expansion over symmetry-adapted components[2]:

$$V_{A_1}(\underline{r}) = \sum_{h,l} V^{A_1}_{hl}(r_{N+1}) X^{A}_{hl}(\hat{r}_{N+1}) \qquad (6)$$

where the expansion is performed around the centre of mass of the molecule in question and the radial coefficients can be obtained by numerical quadrature over the charge distribution yielded by SCF-HF calculations of the MO-LCAO type for the target occupied bound states. Here the index h labels different bases of the same irreducible representation A_1 corresponding to the same value of the index ℓ, while the X coefficients are linear combinations of real or imaginary spherical harmonics.

$$X^{p}_{shl}(r_{N+1}) = \sum_{m} b^{p\mu}_{shlm} Y^{m}_{l}(r_{N+1}) \qquad 57)$$

where p denotes the chosen I.R. and μ distinguishes the component of the basis if its dimension is greater than one.

The question of convergence obviously arises immediately and has been extensively examined in the literature by our group[4,5] and by others[6,7]. In the FN approximation, the singularities of the electrostatic interaction lead to an extremely slow convergence of partial waves in eq. (6), especially for diatomic targets where much fewer angular momenta are excluded for symmetry reasons[3]. The existence of nuclei with large Z values and located away from the centre of mass provides obviously one of the hardest situations where to reach acceptable convergence. Thus, the number of partial waves needed to describe the continuum orbital near the excentered nuclei rapidly becomes very large, in order to account for the strong anisotropy and rapid increase of the interaction given by eq. (5). Moreover, in the case of polar species the exceptionally strong and long-range field due to the charge-dipole interaction also causes en extremely slow convergence in the radial integration of eq. (4) where bohr radius values from a few hundreds to a thousand may be needed.

It is therefore clear that even a simple analysis of the one-centre-expanded (OCE) form of the static, direct interaction of eq.(6) can provide some qualitative indication on the corresponding behaviour of the scattering observables and on the effects that the structural properties of each target might have on the relative ease in solving the dynamics according to the frame formulations described in the following Section. As an example of how graphical representation could help physical intuition, a few cases of diatomic molecules [8] are shown in Fig. 1.

The top row shows two mild examples of molecular targets which either do not exhibit permanent dipole moment (H_2) or only have a very small D value (0.044 a.u., CO). In the former case, the target is not much more than a larger atom, with its small equilibrium internuclear distance (1.4 a_0), tiny repulsive quadrupolar bumps that yield barriers of $\sim$ 0.15 eV at 2.85 a_0 perpendicular to the internuclear axis and a splitting of the nuclear charges that only takes place at - 24.5 eV. At a consequence of this behaviour, the broad resonance observed for H_2^- at low energy [9] can only be brought about by extra effects beyond the ESE equations.

For the case of the CO molecule, the similarities with the iso-electronic, homonuclear, N_2 are more striking than the difference due to the presence of a small dipole. The quadrupole bumps near the two atoms (a barrier of 0.50 eV at 3.7 a.u. from the centre-of-mass on the O side, and a barrier of $\sim$0.60 eV at $\sim$4.2 a.u. from the c. of m. on the C side) are in fact very similar to those shown by the N_2 molecule [10], while the tiny polarity can compete only very weakly, and only around the molecule, with the stronger quadrupole (Q = -1.547 a.u.).

In the situation of a target system that is strongly polar, much more marked features can be observed from the spatial behaviour of the static potential (bottom of Fig. 1). Both Li H and Li F, in fact, have supercritical dipoles (D_c = .639 a.u.) and their highly ionic character is apparent from the large anisotropy of the interaction around the nuclei and the huge repulsive barriers on the negative side of the molecules (H and F): 3.12 eV for LiH and 4.46 eV for Li F. The presence of large attractive and angular dependent wells on the Lithium sides, will accordingly force the electron to closely penetrate into the molecule and will require a detailed description of their angular dependence to correctly perform dynamical calculations.

Polyatomic targets are obviously more complex to represent, but also provide a very useful insight into the specific behaviour that is likely to be shown by the impinging electrons through its dynamical observables [4].

As is shown by the above ESE equation, another important feature

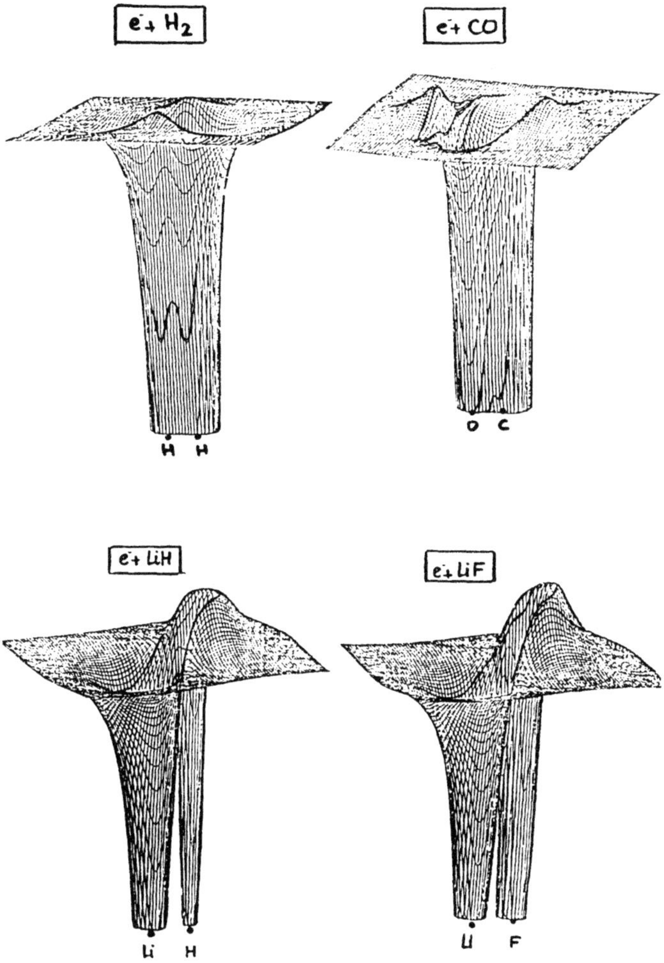

Figure 1. Spatial representation of OCE static potentials for a few diatomic molecules at their equilibrium geometries (from ref.10). See text for comments.

of the interaction is the explicit orthogonalisation of the continuum orbital within the chosen symmetry of each given group of MO's, to the bound orbitals obtained at the SCF or HF level. For a closed shell molecule this is an exact requirement in that it eliminates the one electron exchange terms and makes sure of excluding from closed shells whatever component of $\langle \phi_s | F \rangle$ appears on the bound orbitals. Moreover, it allows to account for part of the Pauli principle requirements and introduces into the unknown continuum orbital F the accurate, short-range nodal structure of the lower-lying bound states given at the level of accuracy of the corresponding SCF or HF description of the target. As previously discussed in the literature[11-14], it is another element of the interaction which very transparently relates to the specific structural properties of the target molecule. It is, however, only part of the antisymmetrisation requirements, as explicitely shown in eq. (4), while the full treatment of exchange requires further analysis and considerations.

2.2. The exchange interaction

The full exchange potential implied by the two summations on the r.h.s. of eq. (4) is much more complex to describe, a fact that can be immediately seen in the case of a diatomic target, only to simplify notation:

$$\sum_{s} \Rightarrow W^{\Lambda}_{\ell\ell'} f^{\Lambda}_{\ell'}(r) = \sum_{\substack{s\ \lambda \\ \ell'\ell'''}} C(\ell,\ell',\ell'',\ell''',\alpha,\lambda)\, Y_{\lambda}(u_{\ell''}, f^{\Lambda}_{\ell'}; r)\, u_{\ell'''}(r) \quad (8)$$

where the C's are Clebsh-Gordan coefficients and both the continuum function F and the bound orbitals ϕ_s have been expanded around the centre of mass of the molecule:

$$F(\underline{r}_{N+1}) = \sum_{\ell} f^{\Lambda}_{\ell}(r_{N+1})\, Y^{\Lambda}_{\ell}(\hat{r}_{N+1})\, r^{-1}_{N+1} \quad (9a)$$

$$\phi_s(r_N) = \sum_{\ell} u^{s}_{\ell}(r_N)\, Y^{m_s}_{\ell}(\hat{r}_N)\, r^{-1}_{N} \quad (9b)$$

The Y_{λ} exchange functions are thus defined by:

$$Y_{\lambda}(u,v;r) = r^{-\lambda-1} \int_{0}^{r} r'^{\lambda} u(r')v(r')dr' + r^{\lambda} \int_{0}^{\infty} r'^{-\lambda-1} u(r')v(r')dr' \quad (10)$$

The main problems in numerically solving eq. (4) are firstly to retain enough terms in the expansions (6), (9a) and (9b) to give convergence, and secondly to accurately represent the integral terms occurring in the exchange operator of eq. (8). In recent years several methods have been suggested and applied to the simpler case of diatomics but it would be outside the scope of this review to analyse them in detail; we refer instead to a very recent examination of their relative merits[15].

In the domain of model exchange potentials, on the other hand, a more transparent relationship could be found between the additional effects of the non-local kernel in the sum on the r.h.s. of eq. (4) and the structural properties of the molecular electronic wavefunctions, thus allowing one to further carry out, albeit often in qualitative terms, the analysis between dynamics and structure that we are here examining.

The simple orthogonality requirement (i.e. only the first integral on the r.h.s. of the ESE equation) is already injecting in the continuum functions, as we said before, the nodal structure of the bound orbitals. On the other hand, it suffers from the restriction that it does not make allowance for the correct two-electron operator and therefore no exchange effects exist between bound and continuum orbitals which belong to different I.R. of the molecular point group. Thus for example, in the ℓ^{-}-N_2 scattering in the resonant $^2\Pi_g$ channel the simple orthogonalisation takes no account of exchange. This is a serious defect since it is known that exchange plays an important role in determining the correct position of the $^2\Pi_g$ resonance.

A number of different model potentials have therefore been proposed to treat the two-electron part of the exchange interaction in the intermediate range of collision energies[16,17,18]. For low energy electrons, however, the first suitable model for an ab initio treatment was proposed by Hara[19], who extended to the continuum the Slater average exchange approximation for bound states[20]. In his treatment the molecule is considered to be made up of electrons as in a free electron gas, while the incident electron is treated in the Born approximation thus allowing to derive a local form of the exchange potential.

The total wavefunction is thus approximated by a single detor of plane waves, with the momentum of those corresponding to the bound electrons determined by the exclusion principle up to the locally defined Fermi level:

$$K_F(\underline{r}) = \{3\pi^2 \rho(\underline{r})\}^{1/3} \tag{11}$$

where $\rho(\underline{r};\ \underline{R})$ is the molecular charge distribution ar $\underline{r}$, for the fixed nuclear geometry $\underline{R}$. The local impulse of the continuum electron is here determined by the same potential as that seen by the bound electrons:

$$E_{kin} = \frac{1}{2} k^2 = \frac{1}{2} K^2(\underline{r}) + V(\underline{r}) \tag{12}$$

$$\text{I.P.} = \frac{1}{2} K_F^2(\underline{r}) + V(\underline{r})$$

where I.P. is the molecular ionization potential. Hence:

$$\frac{1}{2} K^2(\underline{r}) = E_{kin} + \frac{1}{2} K_F^2(\underline{r}) - \text{I.P.} \tag{13}$$

Its substitution into the two-electron exchange kernel of eq.(4) leads then to the HFEG local exchange potential that depends on the position and energy of the colliding electron[19]:

$$W_{HFEG}(\underline{R};\ \underline{r},\ E_{kin}) = -\frac{2}{\pi} K_F^2(\underline{r})\, F(\eta) \tag{14}$$

where:

$$F(\eta) = \frac{1}{2} + \frac{1-\eta^2}{4\eta} \log \left|\frac{1+\eta}{1-\eta}\right| \tag{15a}$$

and:

$$\eta(r,\ E_{kin}) = K(\underline{r}) / K_F(\underline{r}) \tag{15b}$$

The above, local potential therefore introduces in an approximate way the attractive corrections due to the bielectronic repulsion term required by the Pauli exclusion principle and provides the correct complement to the explicit modulation of the continuum functions performed by the orthogonalisation. The latter is essentially making the one-electron exchange terms correctly vanish within the spatial volume of the molecular, closed-shell charge distributions.

The role of the combined procedures to treat exchange, which then lead to the orthogonalised HFEG potential, $V_{OHFEG}(\underline{r}_{N+1})$ in a local form have been recently examined for various diatomics [10,21,22] in a series of studies on its quantitative validity. It was found that the simple orthogonalisation provides a poor representation of the exact results and fails to predict resonances, for example in Π-symmetry e^--CO scattering, even in situations where bound orbitals and continuum functions with the same symmetry exist. The HFEG model, on the other hand, is more successful in producing qualitative agreement while the enforcement of the correct nodal structural for the $F(\underline{r})$ strengthens the potential and greatly improves the agreement with both experiments and more rigorous treatments. The results that have so far been obtained in applications of the V_{HFEG} or the V_{OHFEG} to non-linear polyatomic molecules suggest taht this is also true in the more general situation[3,23].

The direct link of the above potential and the structural properties of the target molecules can be usefully visualized via a graphical representation similar to the one used before for the static interaction[24]. Two illustrative examples are shown in Fig. 2 of this paper for the HCl and LiH molecules, where the polarity of the systems (mild in HCl and very strong in LiH) plays again a very important role. The overall interaction is clearly strongly anisotropic and deeply attractive in the regions of the heavier atoms with a clear dependence on the charge density around the nuclei, which causes stronger attractive behaviour from Cl as opposed to Li.

The ESE equations, however, are not yet correctly describing the full interaction. As the slow electron approaches the molecule, in fact the molecular cloud is responding to the field of the incoming projectile and an additional potential term arises. The induced polarisation interaction, which is a second-order effect caused by the distorsion of the target by the scattering electron, is the most difficult of the three contributions to include properly; the principal difficulty arises from the dynamic (non adiabatic) nature of this interaction, as will be discussed in the final part of this Section.

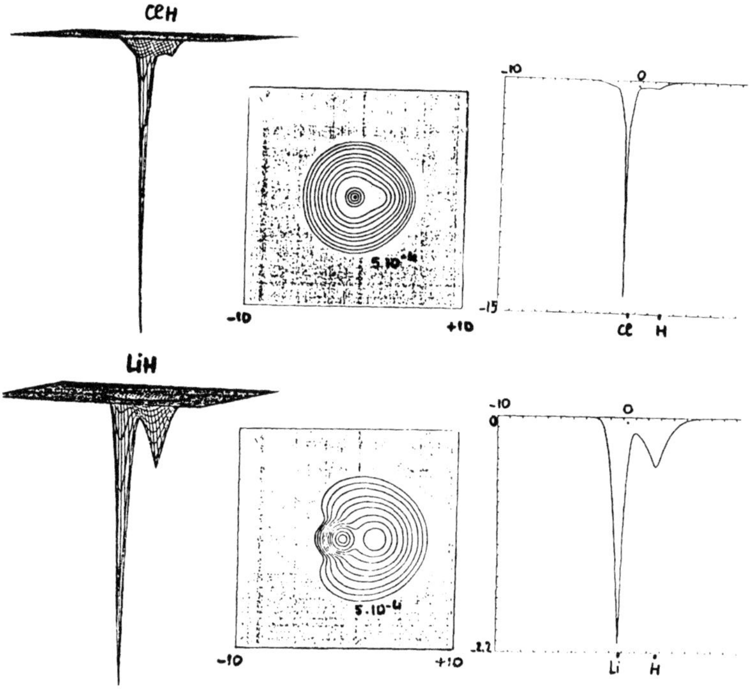

Figure 2. Spatial representation of the local exchange potential discussed in the text, V_{HFEG}, for HCl and LiH targets. The level separations is 5.10^{-4} a.u. and the area covered has a radius of 10 a.u. from the centre of mass. The attractive wells are plotted in eV units (from ref. 24).

2.3. The polarisation potential.

In many low-energy electron-molecule scattering calculations polarisation is included by taking the asymptotic form of an adiabatic potential:

$$V_{Pol}^{asym}(r) \underset{r \to \infty}{\sim} -\frac{\alpha_0(R)}{2r^4} - \frac{\alpha_2(R)}{2r^4} P_2(\hat{r} \cdot \hat{R}) \tag{16}$$

for a diatomic molecule, where α_0 and α_2 are defined with respect to the dipole polarisabilities components parallel and perpendicular to the molecular axis[25]. For non-linear molecules of C_{2v} symmetry, for instance, eq. (16) acquires the following simple form for closed-shell targets of A_1 symmetry:

$$V_{Pol}^{asym}(\underline{r}) \underset{r \to \infty}{\sim} -\frac{\alpha_0(\underline{R})}{2r^4} - \left(\frac{4\pi}{5}\right)^{1/2} \frac{\alpha_2(\underline{R})}{2r^4} X_{2,0}^{A_1}(\hat{r}) \tag{17}$$

$$- \left(\frac{4\pi}{15}\right)^{1/2} \frac{\alpha_2^1(R)}{2r^4} X^{A_1}_{2,2}(\hat{r})$$

where the polarisabilities along the three principal axes of inertia of the molecule, which are the quantities usually quoted in the literature, can be easily expressed[3] in terms of α_0, α_2 and α_2^1. The assumption usually made is then that the above expressions are valid for all values of r greater than the charge cloud of the molecule and are modified at shorter distances by an empirical cut-off factor, implying essentially that V_{Pol} becomes negligible for $r<r_c$, the latter cut-off parameter being determined by fitting some marked features of the dynamical observables such as resonance positions and widths[26,27]. To give a simple indication of the importance of a correct description for the short-range part of the polarisation forces, Figure 3 shows the first two terms of the static potential expansion from eq. (6) for the N_2 molecule[5] and compares them with the spherical part of eq.(16). As one easily sees, in the radial region around the nuclei strong contributions arise from distorting the molecular charge, hence their correct description is bound to play a major role in yielding scattering observables.

To improve upon the above approach, one can then calculate a polarisation potential in the adiabatic approximation by fixing the position of the scattering electron relative to the c. of mass of the molecule and then computing the electronic energy lowering due to this stationary point charge[28,29]. For low energy collisions, this approximation is expected to work well when the scattering electron is far from the target and yields non zero effective potentials at the origin, thus indicating the rather artificial origin of the cutoff procedure, which often compensates for other shortcomings in the overall form of the interaction. On the other hand, near-threshold collisions and those which take place with polar molecular targets are essentially sampling the local, long-range part of the interaction and could therefore be well described by ad hoc procedures which act only within the less important region of the potential.

For higher energy scattering, however, as the electron approaches the molecule it experiences more of the attractive static-exchange potential, thereby locally gaining 'kinetic energy' because of the attractive forces and acquiring a classical velocity comparable to the one of the bound electrons. The approximation that the target relaxes adiabatically in the field of a stationary continuum electron becomes therefore less valid as the latter samples more closely the region of charge around the nuclei.

Non-adiabatic effects were taken into account in elastic electron-

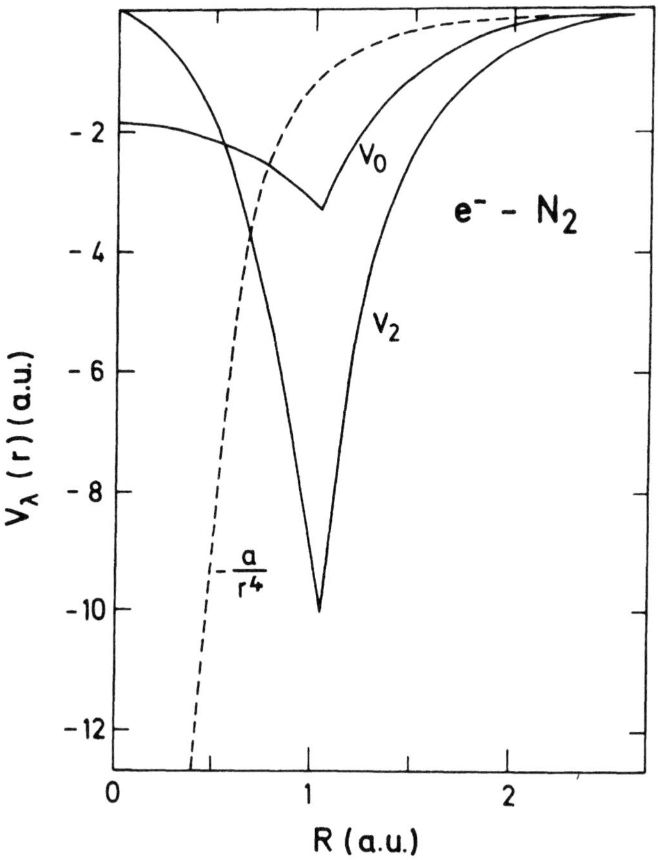

Figure 3. Multipolar expansion of the static potential for ℓ= N_2 interaction (from ref.5), continuous line, and spherical part of the asymptotic polarisation potential (dashed line).

atom processes by Temkin[30] and by Thompson[31], who replaced the ground state wavefunction by another wavefunction which depended on the coordinate of the scattered electron. Thus the target wavefunction for eq.(3), $\Phi_{GS}(1...N)$ is replaced by $\{\Phi_{GS} + \Phi_{Pol}\}$, where Φ_{Pol} depends on N+1. This polarised orbital approach gives rise to a direct potential behaving asymptotically as in eq.(16) as well as an additional exchange contribution. A recent extension to polyatomic molecules has been implemented for ℓ^--CH_4 and ℓ^--H_2O collisions and has yielded a very good, parameter-free, agreement with the experimental findings for integral and differential cross sections[32], as will be discussed below.

Non-adiabatic effects have also been considered in a study of semiclassical exchange potential[33] and in a more recent extension of this work to low-energy ℓ^--H_2 scattering[34], by variationally minimising the second-order Hartree-Fock energy of the perturbed molecule and by introducing a weakening of the potential at short range by dropping monopole terms and by switching off the bound-free interaction whenever the impinging electron happens to be within the charge distribution of the bound electrons. The above procedure is therefore somewhat arbitrary although no free parameters are involved.

From the above discussion it should then be clear how the detailed study of ℓ^--molecule interaction can be very effectively linked to our previous knowledge of target structural properties, and could therefore provide useful ways for assessing the quality of the chosen interaction that is then being used in treating the dynamics. The latter treatment, however, is the ultimate test of theory since it is directly producing the physical observables. It therefore becomes very important to see how far the formulation of the scattering equations depends still on the specific system under study in the sense that the latter ultimately guides our best choice for a realistic treatment.

3. THE SCATTERING EQUATIONS

In the previous Sections we have essentially considered the electronic scattering problem at fixed nuclei (FN) and have analysed in some detail the single center (SC) expansion to treat FN ℓ^--molecule collisions in order to get an efficient and physically transparent way of relating the structural properties of the target molecule to a unified picture of the whole electrostatic interaction. Thus, the SCF description of the (N+1) electron compound provides a simple approach that accounts for exchange and the average effects of the strong, highly anisotropic short-range (and nearly stationary) correlations which dominate the dynamics of a long-lived resonant state. The long-range, permanent multipoles of the target, on the other hand, usually give rise to an additional problem and make the direct scattering event somewhat more difficult to handle, especially when strong permanent dipoles exist and affect near-threshold behaviour of the cross sections[35,36]. The induced polarisation is also a dynamical result of virtual electronic excitations of the target that carries out its effect to rather large distances from the scattering centre, and therefore might cause a nontrivial coupling between electronic and nuclear motions.

The dynamical effect of the nuclei constitutes in fact the second aspect of the electron-molecule collision problem, since one needs to move from an optimal short-range description of a Born-Oppenheimer (N+1) electron compound to an equally efficient, long-range description of a decoupled situation where the competitions between electrostatic interaction and dynamical separation of the nuclear motion becomes important.

An adiabatic approximation in which the nuclei are held fixed during the collision corresponds to the neglect of the H_{nucl} term in eq. (1). This means that when the impinging electron approaches the molecules its angular momentum vector $\underline{\ell}$ couples (just as the bound electrons) to the main molecular axis $\underline{R}$ and therefore $\Lambda = \underline{\ell} \cdot \underline{R}$ becomes a well defined quantum number for the system[1,2]. Whenever long-range forces are not too important, then the spatial region over which Λ remains a good quantum number (a collision constant) includes the asymptotic region where a scattering S-matrix can be defined.

The axial symmetry of the electrostatic interaction, or equivalently the point-group symmetry of the B O molecular electronic wavefunction, leads then to the decoupling of the radial equations originating from (6) either into subspaces of each given projection Λ of the electronic angular momentum on the main axis for linear molecules or into subspaces of each irreducible representation to which the considered molecular wavefunctions belong for the case of polyatomic targets[5].

A breakdown of the adiabatic approximation can occur when the long range forces are important, as in polar molecules. In such cases the definition of a FN scattering matrix leads to divergent total cross sections, since the angular momentum $\underline{\ell}$ no longer couples to $\underline{R}$ in the asymptotic region while still doing so over a large part of the range of action of the interaction potential[37]. To obtain a scattering cross section, it therefore becomes necessary to transform the radial wavefunctions of the FN formulation, $f^{\Lambda}_{\ell\ell'}(r)$, at some suitable value of the radius r, say r_t, to a frame of reference fixed in space (SF) before continuing integration to obtain the correct asymptotic solution[38].

In the new, rotating SF frame the H_{nucl} cannot be neglected and therefore the total hamiltonian for a rigid diatomic target becomes (after integrating out the electronic variables of the bound electrons):

$$H = H^{N+1}_{el} + H_{nucl} = -\frac{1}{2}\nabla^2_{N+1} + \hat{V}(r_{N+1}, \hat{r}_{N+1}\cdot\hat{R}) + B\hat{j}^2 \qquad (18)$$

where B is the rotational constant of the diatomics and $\hat{j}$ its rotational angular momentum. Since now we also have the angular variables for the nuclei, the only good quantum number will be the total angular momentum $\underline{J} = \underline{\ell} + \hat{j}$, and the inversion parity $\eta = (-)^{\ell+\hat{j}}$ of the whole system. The corresponding radial continuum functions will therefore become labelled as $f^{J}_{j\ell,j'\ell'}(r)$ and the set of coupled equations to be solved will break down into a subset for each contributing J value[39].

Different choices can however be made for the set of axis where both nuclei and electrons now move during collisions. Such choices could help in formulating the dynamics in various different ways which in turn could make more or less transparent the introduction of further reductions of the dimensionality for each J- subset needed to be solved[40].

Thus, a fixed frame in the laboratory (LF frame) uses an angular basis of eigenvectors of $\underline{\ell}^2$, $\underline{j}^2$, $\underline{J}^2$, $\underline{J}_z$ and η which are particularly useful to express asymptotic decoupling since the radial potential directly appears on the r.h.s. of the corresponding equation, and vanishes outside the range of action of the electrostatic interaction. A rotating frame held fixed with its main axis along the molecular bond (MF frame), on the other hand, chooses another angular basis of eigenvectors of $\underline{\ell}^2$, $\underline{\ell}_z$, $\underline{J}^2$, $\underline{J}_z$ and $\eta' = \eta(-)^J$, which is particularly useful to express special interactions which $\underline{\ell}_z$ (Λ) is still considered a good quantum number[38,40] during the collision of which the FN formulation is a special example.

Finally, a third choice for the reference frame is one in which the rotating set contains an axis held fixed along the collision coordinate and which always points at the incoming projectile during collisions[41] (BF frame). This choice produces an angular basis of eigenvectors of $\underline{J}^2$, $\underline{J}_z$, $\underline{j}^2$, $\underline{j}_z$ and parity which is particularly good to describe dynamical couplings where the $\underline{j}_z$ can be approximately considered as a good quantum number[42].

For a diatomic molecule, the possible choices of angular basis can then be summarized as follows:

$$| \text{ L F basis} > = Y^{JM}_{\ell j}(\hat{r}, \hat{R}) = \sum_{\substack{m_\ell \\ m_j}} <\ell m_\ell ; j m_j | JM > Y^{m_\ell}_{\ell}(\hat{r})\, Y^{m_j}_{j}(\hat{R}) \tag{19a}$$

$$| \text{ M F basis} > = Y^{JM}_{\ell\Lambda}(\hat{r}, \hat{R}) = Y^{\Lambda}_{\ell}(\hat{r}) \left(\frac{2J+1}{4\pi}\right)^{1/2} D^{J*}_{M\Lambda}(\hat{R}) \tag{19b}$$

$$| \text{ B F basis} > = Y^{JM}_{j\Omega}(\hat{r}, \hat{R}) = Y^{\Omega}_{j}(\hat{R}) \left(\frac{2J+1}{4\pi}\right)^{1/2} D^{J}_{M\Omega}(\hat{r}) \tag{19c}$$

where the various $\hat{r}$, $\hat{R}$ axis have different orientations in each case, as discussed above.

The corresponding coupled equations are given by:

$$\left\{ \frac{d^2}{dr^2} - \frac{\ell(\ell+1)}{r^2} + k_j^2 \right\} f^{J\eta}_{\ell j, \ell' j'}(r) = 2 \sum_{\ell'' j''} \left\{ V^{J\eta}_{\ell j, \ell'' j'} + W^{J\eta}_{\ell j, \ell'' j''} \right\} f^{J\eta}_{\ell j, \ell'' j''}(r) \tag{20}$$

with:

$$\frac{1}{2} k_j^2 = \frac{1}{2} k^2 - Bj(j+1)$$

This is the exact formulation in the LF frame which retains the rotational hamiltonian and couples explicitely the rotational sublevels of the target in a rotational close couples (CC) expansion. It is, however, very hard for it to correctly account for the complete ($j\ell$) couplings till convergence over the whole space because of the high anisotropy and strength of the short range interaction and because of huge number of rotational levels that thus become coupled.

In the rotating MF frame, the equations can be written as:

$$\left\{ \frac{d^2}{dr^2} - \frac{\ell(\ell+1)}{r^2} + k^2 \right\} f^{J\eta}_{\ell\Lambda,\ell'\Lambda'}(r) = 2 \sum_{\ell''} \{ V^{\Lambda\Lambda'}_{\ell,\ell''} + W^{\Lambda\Lambda'}_{\ell,\ell''} \} f^{J\eta}_{\ell\Lambda,\ell''\Lambda'} + 2 \sum_{\Lambda''} R^{\ell,\ell'}_{\Lambda,\Lambda''} f^{J\eta}_{\ell\Lambda,\ell'\Lambda''} \qquad (21)$$

with the coefficients, diagonal in ℓ and r-independent, that contain the B constant:

$$R^{\ell,J\eta}_{\Lambda,\Lambda'} = B \times \sum_{j=|J-\ell}^{J+\ell} (G^{\ell J\eta}_{\Lambda j}) j(j+1) (G^{\ell J\eta}_{\Lambda' j})^{+} \qquad (22)$$

where the matrix elements G's are simply given by a unitary transformation between the two angular representations[10].

One sees in eq. (21) that the specific inclusion of the molecular axial symmetry allows one to separate the (ℓj) couplings but still introduces a constant coupling term which is disregarded instead in the FN formulation of eq. (6). One could therefore employ eq. (21) in the short-range region, where Λ is nearly a good quantum number, but where the rotational splitting can be explicitely included via the Λ-diagonal R coefficients, and then move to the LF equation (20) in the regions where only fewer ℓ contribute and fewer multipoles from the potential expansion (4) are still non-vanishing.

A systematic study of the various dynamical conditions which control the relative ranges of validity of the above equations has been carried out by various authors[7,44,45,46] and the reader is referred to them for further details. Suffice it to say here that the relative interplay of the asymptotic kinetic energy of the continuum alectrons and the range of action of the static multipoles of the molecular target, (vis-à-vis the specific value of its rotational constant B) i.e. the structural properties of the system is again the essential ingredients for choosing the angular basis (19), hence the corresponding CC equations (20) or (21) that will eventually provide collision cross sections.

4. SOME ILLUSTRATIVE EXAMPLES.

It is now of interest to look at some of the very few polyatomic systems which have so far been studied with ab initio methods[3,5,23] to get a general feeling on the accuracy of the computational approaches.

An example of simple triatomic is provided by the H_2S molecule, where various experiments on dissociative attachment[47,48] picture the

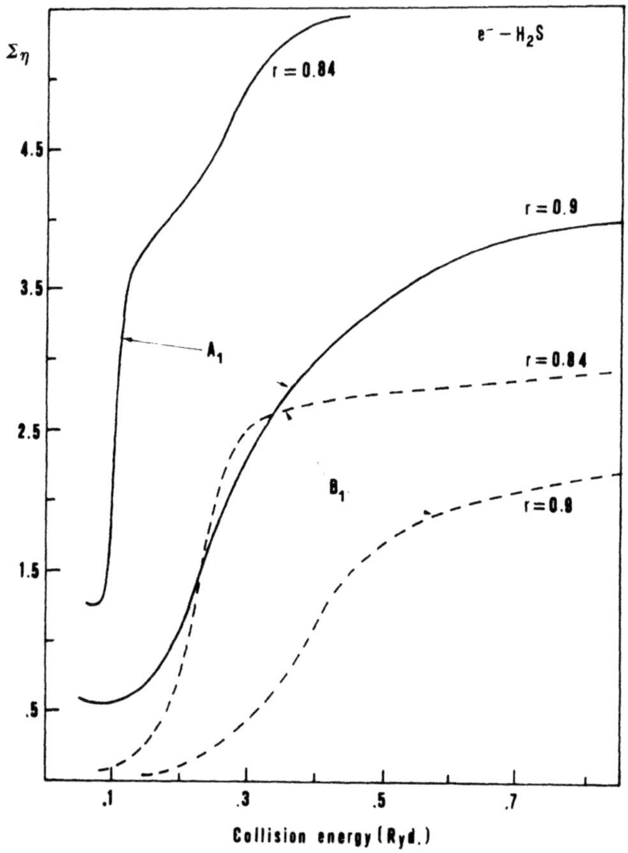

Figure 4. Eigenphase sums for e^--H_2S collisions in the low-energy regime (from ref.3). The continuous lines refe to scattering in the A_1 symmetry, while the dashed lines are for the B_1 symmetry. Two different choice of V_{pol} cut-off forms are shown.

collisional process as going through the formation of an (N+1) electron compound state; correspondingly, a study of the eigenphase sums could provide some theoretical evidence[2] for the existence of a resonant state.

Our results[3] for A_1 and B_1 symmetries are shown in Figure 4 over the range of collision energies examined by the experiments. The resonance parameters can then be obtained by assuming the following energy dependence of the eigenphases[2]:

$$\Sigma \text{ eigenphases} = S(k^2) = \alpha + \beta k^2 + \ldots + \sum_{\ell} \tan^{-1}\left\{\frac{a_i}{b_i - k^2}\right\} \qquad (23)$$

The Figure shows two $S(k^2)$ calculations for two different choices of the cut-off parameter for the polarisation potential, which is clearly a sensitive choice within the model. The stronger polarisation effect with r=0.9 produces too large a background phase, hence makes it harder to identify the two A_1 resonances clearly visible in the top curve. The latter are also in good agreement with the widths and positions suggested by experiments[3].

Further experiments on the angular distribution of the ejected electrons[49] by this molecule suggest that another resonant state exists at 5.2 e V and that it exhibits different behaviour, in the vibrationally inelastic channels, than the lower resonances. The computed dashed line of Fig. 4 (with r = 0.84) corresponds to a B_1 resonance in the energy region close to the second A_1 resonance, again in qualitative agreement with measurements.

Considerable work has also been carried out in recent years on the measurements of cross sections for the scattering of electrons and positrons with methane molecules[50-52], especially in the low-energy region ($E \leq 10$ eV) where interesting Ramsauer effects were detected[53] and predicted by theory[12].

It is therefore interesting to see how the different treatments of the polarisation interaction as discussed before strongly relate to the structural properties of the molecule. In Figure 5, in fact, the spherical part of the local V_{Pol} form as described in eq. (17) is used to also mimick the extra attraction generated by the two-electron exchange terms in eq. (4): the curves marked GT and BC correspond, in fact to a cut-off parameter (0.88 a_0) very close to the centre of mass (GT) and to a cut-off value of about 2.066 a_0, respectively (BC). The latter employs a weaker polarisation since in its calculations the full exchange was properly used. Once ab initio calculations with the polarised orbital approach[30] were performed[32] a much weaker distorsion of the orbitals was also needed, since the presence of the OHFEG interaction is already treating the two-particle exchange and therefore the continuum orbitals are correctly kept out of the spatial region around the C atom, as shown in the Figure.

The corresponding calculations of (elastic + total rotationally inelastic) differential cross sections (DCS) for this system are also shown in Figures 6 and 7, covering an energy range that goes from 3.0 eV to 30 eV hence forcing the impinging electron to sample quite a wide region of ℓ^--molecule interactions. One clearly sees there that the calculations are all in rather good accord with the available data, thus indicating the level of sophistication that present-day treatments of ℓ^--molecule dynamic processes can already reach.

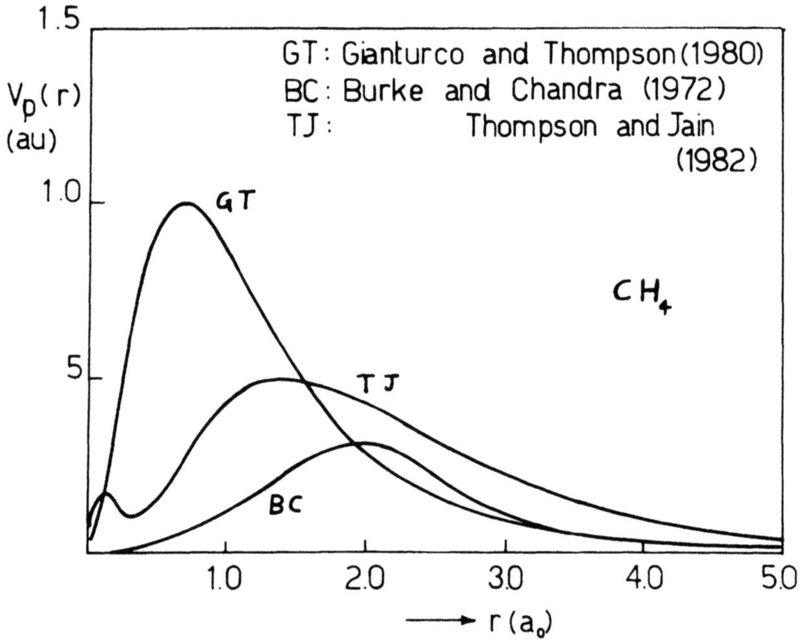

Figure 5. Polarisation potentials for ℓ^--CH_4 interaction, spherical component, as function of C. of m. distance. The curves labelled GT (from ref. 3) and BC (from ref. 11) correspond to different choices of empirical cut-off parameters, while curve TJ (from ref. 32) is computed via the Polarised Orbital method.

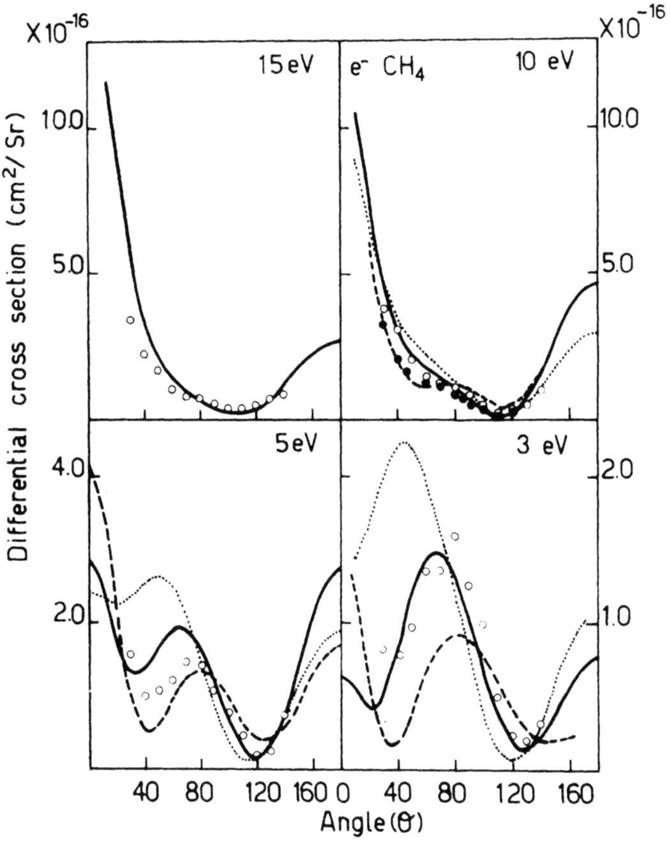

Figure 6. Total differential cross sections (elastic + rotationally inelastic) for ℓ^{-}-CH_4 collisions at various collision energies. The curves refer to GT (...), BC (=--) and TJ (——) calculations of the previous Figure. The experimental points are from ref 52 (o o o) and ref. 54 (. . .).

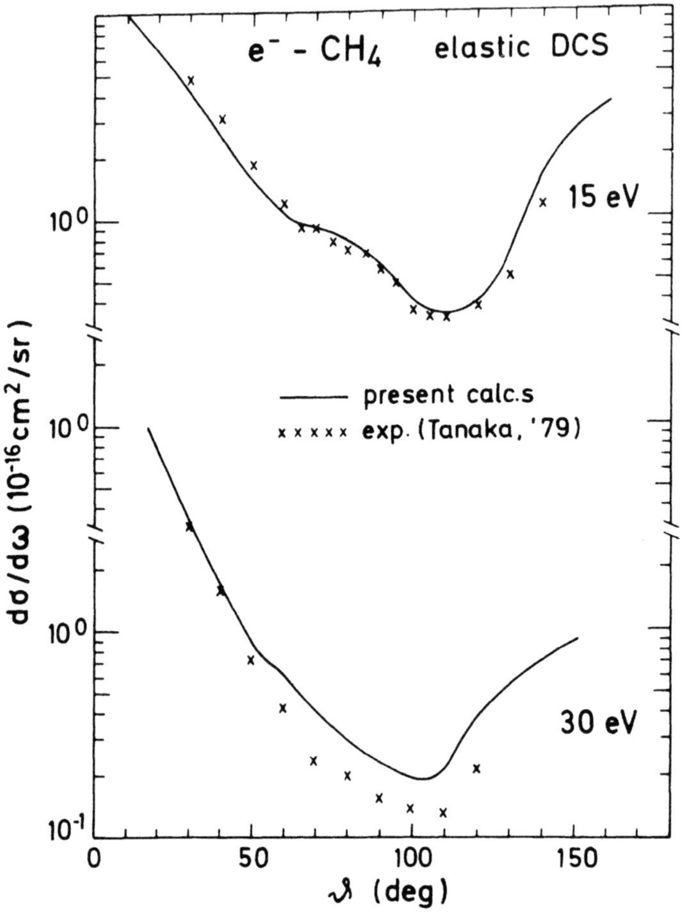

Figure 7. Same as in Fig. 6, for two different collision energies. Only the calculated values from ref. 5 are shown, while the experiments are from ref. 52.

Acknowledgements. I am grateful to Drs A. Jain, D. J. Thompson, H. Tanaka and Vo Ky Lan for sending to me their recent results prior to publication. The financial support of the Italian Nat. Research Council (CNR) is also ackowledged.

REFERENCES

1. For an extensive review on this subject, see the recent article: Lane, N.F.; 1980, Rev. Mad. Phys. 52, p. 29.
2. P.G. Burke, N. Chandra and F.A. Gianturco: 1972, J. Phys. B. 5, p. 2212
3. F.A. Gianturco and D.G. Thompson: 1980, J. Phys. B 13, p. 613.
4. F.A. Gianturco and D.G. Thompson: 1976, Chem. Phys. 14, p. 111.
5. F.A. Gianturco and D.G. Thompson: 1982, in "Electron-Atom and molecule collisions", H. Hinze Ed., ('Plenum Publ. Co., New York).
6. L.A. Collins and D.W. Norcross: 1978, Phys. Rev. A 18, p. 467.
7. M.A. Riley and D.G. Truhlar: 1976, J. Chem. Phys. 65, p. 791.
8. M. Le Dorneuf and Vo Ky Lan, to be published.
9. A. Klonover and U. Kaldor: 1979, J. Phys. B 12, p. 323.
10. Vo Ky Lan, M. Le Dorneuf and J.M. Launay: 1982 in "Electron-Atom and Molecule Collisions", H. Hinze Ed., (Plenum Publ. Co.,New York)
11. P.G. Burke and N. Chandra: 1972, J. Phys. B 5, p. 1696.
12. F.A. Gianturco and D.G. Thompson: 1976, J. Phys. B 9, p. L383.
13. N.F. Lane and S. Geltman: 1967, Phys. Rev. 160, p. 53.
14. M.A. Morrison and L.A. Collins: 1981, Phys. Rev. A 23, p. 127.
15. B.D. Buckley, P.G. Burke and C.J. Noble: 1982, in "Electron-Molecule Collisions", I. Shimamura and K. Takayanagi Ed.s (Plenum Press, New York).
16. J.B. Furness and I.E. McCarthy: 1973, J. Phys. B. 6 p. 2280.
17. D.G. Truhlar and M.A. Brandt: 1976, J. Chem. Phys. 65, p. 3092.
18. D.G. Truhlar, R.E. Poling and M.A. Brandt: 1976, J. Chem. Phys. 64, p. 826.
19. S. Hara: 1967, J. Phys. Soc. Japan, 22, p. 710.
20. J.C. Slater: 1960 "Quantum Theory of Atomic Structures" (McGraw-Hill, New York).
21. M.A. Morrison and L.A. Collins: 1981, Phys. Rev. A 23, p. 127.
22. J.R. Rumble and D.G. Truhlar: 1980, J. Chem. Phys. 72, p. 5223.
23. S. Salvini and D.G. Thompson: 1981, J. Phys. B. 14, p. 3797.
24. Vo Ky Lan and M. Le Darneuf
25. F.A. Gianturco and U.T. Lamanna: 1979, J. Phys. B 12, p. 742.
26. e.g. see: B.D. Buckley and P.G. Burke: 1977, J. Phys. B 10, p. 725.
27. L.A. Collins, R.J.W. Henry and D.W. Norcross: 1980, J. Phys. B 13, p. 2299.
28. M.A. Morrison and P.J. Hay: 1979, Phys. Rev. A 20, p. 740.
29. R.A. Eades, D.G. Truhlar and D.A. Dixon: 1979, Phys. Rev. A 20, p. 867.
30. A. Temkin: 1957, Phys. Rev. 107, p. 1004.
31. D.G. Thompson: 1965, Proc. Roy. Soc. (London) 260, p. 160.

32. D.G. Thompson and A. Jain: unpublished results.
33. T.L. Gibson and M.A. Morrison: 1981, J. Phys. B 14, p. 727.
34. T.L. Gibson and M.A. Morrison: 1982, J. Phys. B, to be published.
35. F.A. Gianturco and N.K. Rahman: 1977, J. Phys. B 10, p. L219.
36. F.A. Gianturco and N.K. Rahman: 1978, J. Phys. B 11, p. 727.
37. D.W. Norcross and L.A. Collins: 1982, Adv. At. Mol. Phys., Vol 18, to be published.
38. E.S. Chang and U. Fano: 1972, Phys. Rev. A 6, p. 173.
39. A.M. Arthurs and A. Dalgarno: 1960, Proc. Roy. Soc. (London) A 256, p. 540.
40. for a detailed discussion see: F.A. Gianturco: 1979, "The transfer of molecular energy by collision" (Springer-Verlag, Berlin, Heidelberg, New York), Chapt. 3.
41. R.T. Pack: 1974, J. Chem. Phys. 60, p. 633.
42. P.M. McGuire and D.J. Kouri: 1974, J. Chem. Phys. 60, p. 2488.
43. B.H. Choi and R.T. Poe: 1977, Phys. Rev. A 16, p. 1821.
44. M. Le Dorneuf and Vo Ky Lan: 1979, in "Electron and photon molecule collisions" B. Schneider and T. Rescigno Eds (Plenum Publ. Co., New York), p. 51.
45. Vo Ky Lan and M. Le Dorneuf: 1979 in "Electronic and Atomic Collisions" K. Takayanagy Ed. (North Holland Publ. Co., Amsterdam), p. 751.
46. N. Chandra: 1977, Phys. Rev. A 16, p. 80.
47. R. Azria, Y. Le Coat and G. Lefevre: 1977, Proc. 10th ICPEAC (Paris Commisariat à l'Energie Atomique), p. 844.
48. R. Azria, Y. Le Coat, G. Lefevre and D. Simon: 1979, J. Phys. B 12, p. 679.
49. M. Tronc: 1981, private communication.
50. K. Rohr: 1980, J. Phys. B 13, p. 4897.
51. T.Okada, M. Kubo, T. Yalamoto, T. Suzuki and H. Tanaka: 1980, Abstracts of VII Int. Conf. At. Phys. (M.I.T. Press), p. 49.
52. H. Tanaka: 1982 (unpublished results).
53. E. Barbarito, M. Basta, M. Calicchio and G. Tessari: 1979, J. Chem. Phys. 71, p. 54.
54. W.R. Newell, D.F.C. Brewer and A.C.H. Smith: 1979, Abstracts in XI ICPEAC Kjoto (Japan), p. 308.

COMPUTATIONAL QUANTUM CHEMISTRY - INTRODUCTORY REMARKS

J.A. Pople
Department of Chemistry
Carnegie-Mellon University
Pittsburgh, Pa 15213
USA

Abstract

Methods currently used in quantitative molecular orbital computations are reviewed. Limitations of some of the simple, widely used techniques are pointed out.

INTRODUCTION

Computational quantum chemistry has now become an investigational technique of major importance. Although the general objective of an approximate solution of the wave equation is common to most work in the field, the variety of techniques used is often bewildering. In attempting to assess the success or failure of techniques in as broad a context as possible, it is important to have a clear understanding of the relation between the various approximations. A full knowledge of the limitations inherent in some of the simpler methods should enable quantum chemists to establish a level of confidence in predictions which are made in advance of experimental work.

A given computational method, if properly defined, constitutes a theoretical model for the whole of chemistry. To be widely applicable and useful, such a model should have the following features:

1. It should be uniquely defined. When applied to the problem of electronic structure in the field of fixed nuclei, this means that the computational technique is unambiguously specified for any number of electrons and any positions. This should permit exploration of any part of a potential surface, including reaction paths connecting local minima representing equilibrium structures.

2. The technique must be computationally efficient, so that widespread testing and application is possible. Much of the advance that has taken place in the last two decades has been through improvements in the efficiency of algorithms.

3. The method should have the property of size-consistency. This means that the errors associated with the approximations should be roughly proportional to the size of the system, so that comparisons can properly be made between large and small molecules. More precisely, size-consistancy can be defined as the property of additivity of energies for completely separated systems:

P.-O. Löwdin and B. Pullman (eds.), New Horizons of Quantum Chemistry, 179–182.

$$E(A + B) = E(A) + E(B) \tag{1}$$

4. Finally, it is desirable that computational results should be accessable to qualitative interpretation. This is often effective in single-configuration (Hartree-Fock) theory but frequently difficult at the multiconfigurational level.

There are three main aspects to theoretical models. In the first place, a basis set has to be chosen, all molecular orbitals being expressed as linear combinations of these basis functions. If a simple determinant or a single electron configuration is used, this leads to single wavefunctions of the Hartree-Fock type which take inadequate account of electron correlation. The second aspect is the correction of Hartree-Fock results for this correlation. There are many ways in which this may be accomplished. Thirdly, approaches are often adopted which avoid Hartree-Fock by starting at a multiconfigurational level. All three of these topics will be touched on in these remarks.

BASIS SETS

The following classes of basis sets are in widespread use:

1. Minimal Sets. These have one function per core and valence atomic orbital of the constituent atoms. They make little allowance for variable atom size and anisotropy; consequently they give poor descriptions of ions and unsaturated molecules. On the other hand, they are economical. A computation on naphthalene ($C_{10}H_8$) requires 58 basis functions.

2. Split Valence or Double-Zeta (DZ) Sets. These double the number of basis functions in the valence shell and give a much improved description of unsaturated molecules. The number of basis functions is almost doubled over a minimal set. For example, a benzene study at this level requires 66 functions.

3. Polarized Basis Sets (DZ + P). In these sets further functions of higher angular quantum number are added for each atom (d-functions for carbon, nitrogen, etc. and possibly p-functions for hydrogen). It is widely recognized that the relative description of different local molecular symmetries cannot be handled adequately without inclusion of polarization functions. Examples are the pyramidal character of nitrogen in amines and the bridging structures of some simple ions such as the ethyl cation. Full polarization basis sets are significantly more costly than split-valence. A relatively small molecule such as propyne already requires 65 basis functions.

4. Large Basis Sets. Further expansion of the basis set to triple-zeta or double-polarization is only possible for very small systems at the present time. A computation on acetylene at This level would already require on the order of 60 basis functions.

At the present time, most computations which have a useful level of reliability are based on DZ or DZ + P basis sets.

ELECTRON CORRELATION METHODS

Once a single-configuration Hartree-Fock (HF) wavefunction has been determined by a variational procedure, the next step is to correct the energy for the effects of electron correlation. There is a wide range of methods for doing this. Each applies within the total space defined by a particular basis set. Some of the most important techniques are:

1. Perturbation Theory. This type of theory, originally due to Møller and Plesset (MP) treats the full Hamiltonian as a small perturbation on the Fock Hamiltonian, for which the Hartree-Fock wavefunction is an exact answer (within the space defined by the basis set). The perturbation series may be terminated at any order. At second order (MP2), the correction is very simple to evaluate and treats all pairs of electrons additively by allowing for double substitutions in the HF electron configuration. Third-order theory (MP3) also treats only electron-pair correlation but has the merit of allowing for pair-pair interactions. Fourth-order theory (MP4) is the first level of theory to allow for mixing of the HF wavefunction with single, triple, and quadruple substitutions. Triple substitutions, which measure three-electron correlation effects, have recently been shown to be important in handling reaction barriers. MP4 is the highest level of theory programmed to date for widespread application.

Møller-Plesset has a number of advantages. It is relatively efficient computationally and can be shown to be size-consistent provided all orders considered are evaluated completely. The principal disadvantage is that the perturbation series may be poorly convergent in systems for which the Hartree-Fock wavefunction is a particularly bad starting point.

2. Limited Configuration Interactions. Variational determination of the best linear combination of the Hartree-Fock determinant and others obtained from it by single and double substitution is computationally feasible for up to about 60 basis functions. It is generally described as configuration interaction with singles and doubles (CISD). Its advantages are that it gives an upper bound to the energy (being variational) and that it can be adapted to derivative algorithms (i.e. calculation of energy derivatives with respect to nuclear coordinates) Its principal disadvantages are (1) it is not size-consistent and (2) it fails to take account of three-electron correlation effects.

3. Coupled Electron Pair Approximation (CEPA). This is an iterative method, closely related to CISD, which avoids the size-consistency difficulty associated with non-interacting pairs (unlinked clusters). It becomes size-consistent but loses the CISD advantage of giving an energy upper bound. When expanded in a Møller-Plesset series, such a method is found to becorrect to third order (MP3) but incorrect in fourth order (MP4). A further disadvantage is that the method may not be invariant under transformation within a degenerate set of molecular orbitals.

4. Coupled Cluster Methods. These methods are closely related to CISD and CEPA. If the CID function is written $(1 + T_2)\Psi_0$ where Ψ_0 is the Hartree-Fock function and T_2 is the double substitution operator (giving a sum of doubly-substituted determinants), then the coupled cluster double function has the form

$$\Psi_{CCD} = \exp\,(\hat{T}_2)\,\Psi_0 \qquad (2)$$

Such a function leads to an energy which is size-consistant but not variational. Like CEPA, however, it is accurate only to third order in the Møller-Plesset series.

MULTICONFIGURATIONAL METHODS

The methods of handling electron correlation listed above are all based on the Hartree-Fock wavefunction as a starting point. In some systems, the single-determinant approximation may be so poor that another, more sophisticated, technique is called for. Thus, it is possible to start with a wavefunction formed as a linear combination of two or more configurational determinants and optimize (using variational methods) both the coefficients of the determinants and the basis expansion coefficients of the orbitals used. Such methods are described as multiconfigurational self consistent field (MCSCF) techniques. Once determined, they may also be used as starting points for fuller treatments involving the remaining unoccupied orbitals.

Although a number of MCSCF algorithmic techniques have been developed in recent years, there remains difficulty in adapting them to the general model approach. Thus, the choice of configurations used in an MCSCF wavefunction should be made in an unambiguous manner for all possible nuclear configurations. The only general way of accomplishing this is to divide the full basis-set space into three orthogonal subspaces. Thus the first is a subspace containing orbitals which are all full (i.e., doubly occupied) in all configurations used. The third is a subspace containing orbitals which are always empty. The second orthogonal subspace is one in which the remaining electrons are assigned to the orbitals in all possible ways. This subspace is often called the active space and the theory is then described as complete active space self consistent field (CASSCF).

If the dimension of the active space is large enough to accomodate all valence atomic orbitals, CASSCF is a well defined procedure and could be the basis of a size-consistent model. However, the principal drawback is that the number of configurations increases very rapidly with the size of molecule. The smallest dimension which could possibly describe all valence configurations of all the atoms in a molecule would be the sum of one for each hydrogen atom and three for each other first row atom. Simple counting of the number of possible configurations (in the absence of symmetry) indicates that this may reach several million for an active space of dimension 12; this is already close to the limit of practicability. It is likely, therefore, that multiconfigurational methods will necessarily involve some degree of configuration selection.

Acknowledgements. Support for this work was provided by the National Science Foundation (Grant CHE-79-01061-01). The aid of the Alexander von Humboldt Foundation is also acknowledged.

AB INITIO CALCULATIONS CLOSE TO THE FULL CI LEVEL OF ACCURACY AND THEIR USE FOR THE INTERPRETATION OF MOLECULAR SPECTRA

Robert J. Buenker
Lehrstuhl für Theoretische Chemie, Universität-Gesamthochschule Wuppertal, Gaußstraße 20, D-5600 Wuppertal 1, West Germany

Sigrid D. Peyerimhoff
Lehrstuhl für Theoretische Chemie, Universität Bonn, Wegelerstraße 12, D-5300 Bonn 1, West Germany

ABSTRACT

A practical method for applying perturbation theory to a completely general multi-reference function is described which is capable of yielding energy results at close to full CI accuracy for all types of electronic states. The advantages of this technique over conventional cluster-type perturbative treatments are pointed out, particularly its self-correcting capability. Various examples of the overall method are presented for different AO basis sets and numbers of electrons, and also for several types of MO or NO transformations. In addition it is discussed how this perturbative CI technique can be generalized to obtain other properties of the associated eigenfunctions. Applications involving current problems of molecular spectroscopy are presented in which the resulting theoretical predictions have been an indispensible aid for the interpretation of the measured data. Furthermore because the method does not rely on a spinless representation of the Hamiltonian it is also possible to extend it in a straight-forward manner to include relativistic terms in the treatment, and thus applications are also reported in which effects of this nature have been computed. Finally, molecular calculations which go beyond the Born-Oppenheimer Approximation are also illustrated, with the coupling of vibrational modes being achieved through the explicit computation of non-adiabatic matrix elements over large-scale CI wavefunctions.

I. INTRODUCTION

Over the past fifteen years a variety of computational methods have been developed which are designed to describe electron correlation effects. The emphasis in these approaches varies from achieving a high degree of elegance in the associated theoretical formulation to producing a program system which is simple enough to be used by non-experts. From the point of view of meeting the needs of experimen-

P.-O. Löwdin and B. Pullman (eds.), New Horizons of Quantum Chemistry, 183–219.

talists for reliable electronic structure calculations, it seems fair to conclude that the main goals for such theoretical techniques should be a guaranteed high level of accuracy combined with a broad range of applicability (all types of electronic states for any nuclear conformation), while at the same time affording a good degree of practicality, particularly from the standpoint of economic feasibility. The present paper describes one such attempt to meet these criteria, beginning with a detailed discussion of the theoretical basis on which the associated calculations are carried out and concluding with a number of specific applications which illustrate to what extent the above cited goals have been attained thereby.

II. COMBINING PERTURBATION WITH CI TECHNIQUES

A. For reasonably large AO basis sets the goal of carrying out a full CI calculation, i.e. one in which all excitation classes are included, by means of the straight-forward solution of the corresponding secular equation is unattainable in practice because of the astronomical dimensions involved. Moreover experience clearly indicates that the results of full CI treatments employing relatively small and inflexible basis sets (1,2) are not sufficiently accurate to allow for truly quantitative predictions of measured data. The alternative approach of relying on a truncated CI treatment which employs a suitably large AO basis also has its disadvantages, although recognition that a space consisting of all single and double excitations relative to a series of the most important configurations present in the expansions of interesting electronic states helps greatly in minimizing inaccuracies in such calculations (3,4). The fact remains, however, that the dimensions of multi-reference, all-singles-and-doubles (MRD-CI) spaces are still too large to be treated efficiently via direct solution of the corresponding secular equations, despite recent advances which have been made in dealing with such problems (5,6).

In view of the difficulties posed by such high dimensionalities on the one hand, but at the same time of the advantages of CI treatments in general for the theoretical description of electronic structure phenomena, it seemed highly desirable in the early 1970's to develop computational methods which can make direct use of CI techniques, but which at a certain point stop solving secular equations explicitly and instead combine the results of such a strictly variational approach with some form of perturbation theory. To accomplish this objective, however, it is inconvenient to employ an ansatz in which the interactions are divided up into a zeroeth-order and a perturbing part (H^o+H'), as for example in conventional Rayleigh-Schrödinger perturbation theory, for the simple reason that the CI method works with the full Hamiltonian from the very beginnging. From this point of view it is far more practical to design the perturbation treatment so that only the wavefunctions are dealt with as series expansions, specifically with the zeroeth-order approximation thereto taken over directly from large-scale (truncated) CI calculations.

The clue as to how best to proceed in this direction can be found by considering the effect of adding a single test species ϕ^t to a large configuration space for which the desired eigenfunctions ψ_n^o are already known explicitly. If one makes the tacit assumption that the strength of the perturbation introduced thereby is not great enough to alter the expansion coefficients in ψ_n^o of the configurations in the original spaces (i.e. all such ratios remain constant) this problem clearly reduces to the solution of a 2x2 secular equation, of which the appropriate Hamiltonian matrix elements shall be referred to as E_n^o, E_n^t and $H_{no,t} = \langle\psi_n^o|H|\phi^t\rangle$ respecitvely. Following Slater (7) the roots of this equation can be expanded in a power series, with the result:

$$ {}^tE_n = E_n^o + \frac{(H_{n^o,t})}{E_n^o - E_t} + \dots \tag{1}$$

and
$$ {}^{C_t}/_{C_n^o} = \frac{H_{n^o,t}}{E_n^o - E_t} + \dots , \tag{2}$$

whereby the index n is used to call attention to the fact that this procedure can be used to treat any of the roots obtained in the original CI space. The above formulae resemble quite closely those derived from first-order Rayleigh-Schrödinger perturbation theory, but there is an important distinction in that the matrix elements are evaluated over the full Hamiltonian operator and not simply over H'.

The energy lowering implied by eq. (1) was first used by Whitten et al. (8) as a criterion in configuration selection techniques, and was also used successfully for this purpose by Shavitt and Davidson and coworkers (9,10) , whereas the authors (3) preferred to solve such 2x2 equations (or their equivalent) exactly in this connection. In the present context, however, the question arises as to whether the individual energy lowerings of the unselected test species can be used to predict their global effect upon inclusion in the overall secular equation. As a first approximation the authors (3,11) therefore suggested that for MRD-CI spaces the actual secular equation eigenvalue $E_n(T)$ at a given selection threshold T could be supplemented by the sum K(T) of the energy lowerings for all unselected configurations to give an estimate for the desired zero-threshold eigenvalue $E_n(0)$. Not surprisingly it was generally found (11) that K(T) overestimates the energy difference $E_n(T)-E_n(0)$, but a simple procedure was quickly found which refines this procedure. In essence it was assumed that a proportionality factor λ exists which is independent of the value of the selection threshold such that the following equation holds to a good approximation:

$$E_n(0) = E_n(T) + \lambda K(T), \tag{3}$$

where $$K(T) = \sum_t^{unsel.} ({}^tE_n - E_n^o), \quad \text{or simply} \tag{4a}$$

$$K(T) = \sum_t^{unsel.} H_{no,t}^{\,2} / (E_n^o - E_t) \tag{4b}$$

with the summations in eqs. 4a,b running only over unselected terms at a given threshold T. By carrying out the required computations at two or more threshold values it is possible to solve for the desired approximation to E_n^o as well as for the optimal λ factor for each root. No new Hamiltonian matrix elements need be computed relative to the truncated CI treatment at the minimal value of T and solutions obtained at high threshold can be used to speed convergence for the analogous results obtained in the larger secular equations.

The above technique has received wide application over the past decade (4,12-13) and experience indicates that multi-reference singles-plus-doubles eigenvalues can be predicted quite generally with it to an accuracy of about 5 % of the difference between the true $E_n(0)$ value and the $E_n(T)$ result obtained at the lowest selection threshold for which explicit diagonalizations are carried out. In practice this capability allows one to achieve 10^{-3}-10^{-4} hartree accuracy in a small fraction of the computer time (0.1 to 10 %) which would be required if the entire MRD-CI space were treated in a purely variational manner. Because of its reliance on statistics to predict a suitable value for the proportionality factor λ in eq. 3 it is important to choose selection thresholds which are low enough to allow for a definite pattern to become established. If one or more of the reference configurations is weakly interacting this observation implies that it may be necessary to employ quite small T values before enough configurations are brought into the secular equations which are solved explicitly to allow for an accurate reading to be taken (see also Ref. 14). Various plotting procedures have been suggested for evaluating λ (11,14-15), perhaps the most intriguing of which involves the use of K(T) itself as abcissa, in which case E(K) takes on a very nearly linear appearance (16) as a consequence of the (approximate) constancy of the proportionality factor.

Finally it should be noted that for potential surface computations the relative errors attendant upon this method can be minimized by employing not only the same set of reference configurations at each point but also identical selected CI spaces at each of the threshold values considered. This procedure insures continuity for the potential surfaces, and gives the same degree of systematicity to these calculations as is enjoyed by simpler theoretical treatments such as SCF procedures.

B. EXTENSION OF THE PERTURBATIVE CI TECHNIQUE TO FULL CI CORRECTIONS

The procedure outlined above combines the CI and perturbation methods in a quite practical manner, but it must be emphasized that the results obtained therewith in the limit correspond only to a zero-threshold MRD-CI space, i.e. with higher than double excitations relative to the reference configurations not included even on a conceptual basis. For a reasonably large number of correlated electrons it is impossible in practice to employ a sufficiently extensive reference set to generate the full CI space by taking all single and double excitations relative to the main configurations, so there is an inherent deficiency if the theoretical treatment is stopped at this point. Consistent with the general discussion in the previous section there is merit in considering how the effect of higher excitations can be accurately taken into account by some further application of perturbation theory. In particular it is of interest to find a term analogous to K in eqs. (3-4) which represents a rough approximation to the energy lowering which occurs in going from a given multi-reference all-singles-and-doubles treatment to the corresponding full CI. There are a number of possible candidates for such a choice, the first of which was given by Davidson and Langhoff (17). On the basis of a rather lengthy argument based on perturbation theory these authors estimated the energy lowering effect of adding triple and quadruple excitation classes to a single-reference all singles and doubles (SD-CI) calculation as $(1-c_o^2) \times (E_{SCF}-E_{SD-CI})$. In end effect this result amounts to simply assuming that the correlation energy missing in a singles-and-doubles CI decreases linearly with c_o^2, the square of the expansion coefficient of the reference configuration. From this point of view it is easy to generalize this result for a multi-reference singles-and-doubles treatment by simply replacing c_o^2 by $\sum^{ref} c_i^2$ for all reference species and by employing the analogous (multi-reference) energy quantities in the second factor in the correction; this generalized full-CI correction was first used by the authors and coworkers to study dissociation processes in which a multi-reference formulation is critical (18).

In a subsequent application for the $^3\Pi$ and $^1\Sigma^+$ states of CN^+ Bruna et al. (19) showed that the full CI energy estimate obtained through application of the above correction term to MRD-CI results was quite different from that indicated from the corresponding corrected SD-CI estimate. Furthermore use of different correction terms (20-21) were found to be similarly dependent on the choice of the reference species. When one compares such findings with those obtained in applying the MRD-CI correction term of eq. (4) it is apparent that a number of clear parallels exist, however, which suggest how best to proceed further. Just as with the sum over energy lowering estimates for unselected configurations, it should come as no surprise that the simple higher-excitation or full-CI correction terms discussed above do not always represent accurate substitutes the highly complex correlation energy quantities being sought. It is possible to assess the magnitude of the errors connected with their use, however, by carrying out the necessary calculations under a variety of conditions; in the present case this procedure involves the use of more than a single set of reference configurations in the MRD-CI treatments.

A typical example showing the change in the various full CI corrections with increasing complexity of the reference set is shown in Fig. 1 for several states of the CH^+ system (22). For a single reference species the two different full CI corrections differ by 0.1 eV for the $^1\Sigma^+$ state and 0.07 eV for $^1\Pi$. For more main configurations the MRD-CI energy decreases in accordance with the variation principle but at the same time the sizes of the various full CI corrections decrease as well. When $\sum^{ref} c_i^2$ increases to 100 % the corrections vanish and the MRD-CI result converges to the full CI value as is clearly seen from the figure in both examples. On this basis it appears possible to predict the correlation energies in both states to within several hundreths of an eV and the transition energy between them to an accuracy of close to 0.01 eV at the full CI level. In the original CN^+ case the variations are much larger, with estimates for the transition energy between $^1\Sigma^+$ and $^3\Pi$ states in this system varying over a range of 1.0 eV depending on the number of reference configurations employed and the type of correction employed. As the $\sum^{ref} c_i^2$ percentage increases to 95 % in the MRD-CI treatments of each state, however, the estimates for this quantity become much more uniform, so that it appears that the $^1\Sigma^+$ state lies 0.1 eV lower than $^3\Pi$ when the full CI limit is approached; the simple SD calculation gives the opposite ordering, and a transition energy of 0.8 eV with the wrong sign. Use of the Davidson-Langhoff correction (17) for a single reference species decreases this error by roughly one half but it still leads to an incorrect sign for the transition energy, while the PSK correction (20) is found to be even less effective in this instance.

As a final example of the full CI correction several internuclear distances for the ground state of N_2 are considered (Fig. 2). Again with single-reference calculations neither the normal CI nor the corrected CI energies give a reliable picture of the true energy difference for this system between the two nuclear arrangements. As the number of reference configurations is gradually increased, however, a much more consistent picture emerges, with plots of the various correlation energy quantities with respect to $\sum^{ref} c_i^2$ approaching well-defined limiting values of 1.0 eV and 6.4 eV respectively for the equilibrium and nearly dissociated structures. By comparison it is much more difficult to estimate the full CI energies on the basis of the various CEPA estimates employing the same AO basis, particularly for the R=4.0 a.u. case. The advantage of having a multi-reference capability is particularly apparent in this example, since it is well known that the Hartree-Fock approximation is quite unreliable for determining the dissociation energy of N_2. More generally it is clear that there is great advantage to being able to test the accuracy of such corrections, by comparing results obtained for a variety of theoretical

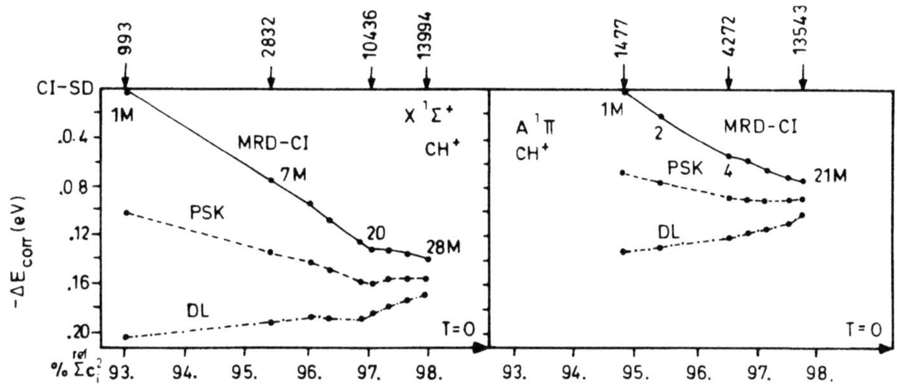

Fig. 1: Estimated full CI and calculated MRD-CI energy for two electronic states of CH^+ (42 AO's) relative to the single-configuration SD-CI energy as a function of the number of reference (main M) states (weight $\sum^{ref} c_i^2$ in percent) in the final CI expansion. The correction is based on the Davidson-Langhoff (17) and Pople-Seeger-Krishnan (20) formula. The orders of secular equations actually solved are given at the top of the figure.

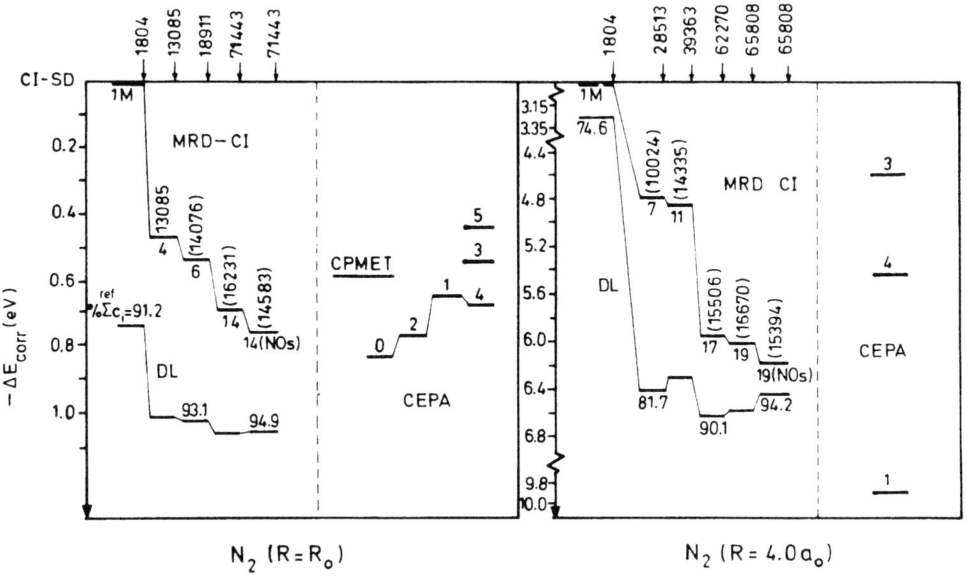

Fig. 2: MRD-CI and estimated full CI energy (generalized Davidson-Langhoff correction) for N_2 at two different internuclear distances (23) and comparison with various CEPA-n results (24); the energies are plotted relative to the single-configuration SD-CI result for various reference sets. The number of configurations and the sizes of generated spaces are indicated, the orders of secular equations actually solved are given in parentheses.

treatments. This feature gives a self-correcting aspect to the method as a whole, similarly as in the T=0 correction first discussed, allowing one to check a given result independent of experience for other systems.

To be able to take advantage of this capability, however, it is necessary to be able to carry out MRD-CI treatments with relatively large numbers of reference configurations, and thus it is important to recognize that the T=0 perturbative correction procedure discussed in Sect. II A can be crucial in affording a practical scheme for arriving at the full CI correction itself. Insistance on diagonalizing all such MRD-CI secular equations explicitly for the T=0 case greatly narrows the class of reference configuration sets which can be treated, and thus ultimately leads to a reduction in the accuracy to which the full CI correction can be computed. As Fig. 2 clearly shows, the errors which occur because of applying such corrections for too simple a reference set can easily fall in the 0.5-1.0 eV range or higher, whereas the T=0 corrections are typically in error by only 0.01 eV. The overall conclusion from this experience is thus that it is best to apply the T=0 and full CI corrections in concert, so that the $\sum^{ref} c_i^2$ percentages attained in the calculations are as high as possible, in particular with uniform values for all types of electronic states and nuclear conformations of interest in a particular study. When this procedure is followed the actual form of the K term used in making the full CI correction (in the sense of eqs. 3-4) is not a critical factor, generally only controlling the slope with which the corresonding energy estimate approaches its limiting value.

Furthermore the choice of one-electron transformation (SCF-MO's vs. MC-SCF MO's or NO's) becomes similarly less critical in such an approach as the size of the reference set increases, raising the pleasant prospect of reducing the number of expensive four-index integral transformations which need to be carried out at a given geometry when a series of electronic states are to be treated. Finally, as discussed earlier, when employing such methods for potential surfaces, it is important to design the theoretical treatment in the same manner for all nuclear arrangements, employing the same (analogous) sets of reference configurations and selected CI spaces at each point so as to guarantee continuity in the ensuing results. Once a pattern has been established at a series of typical geometries, it is generally safe to forego calculations with small numbers of reference configurations and high selection threshold values and simply compute the full CI correction based on the most extensive individual MRD-CI space, thereby minimizing any ambiguities arising in the specific energy value to be taken at a given point in constructing the desired potential surface.

C. FURTHER EXTENSIONS OF THE PERTURBATIVE CI METHOD: AO BASIS CORRECTIONS, PROPERTIES AND SPIN EFFECTS

Before considering specific applications of the computational techniques described above for problems of molecular spectroscopy, it is well to discuss several other areas in which the general perturbative CI approach can be useful. To begin with it is interesting to consider the possibility of arriving at an AO basis energy correction to supplement the T=0 and full CI corrections already discussed. The basic outlines of the computational technique have been given earlier (16) and will be briefly summarized in the present article. Employing the techniques described above a standard AO basis is first treated at close to full CI accuracy. Then additional functions are added to the original basis in small groups. The resulting series of calculations must lead to a lowering of the total energy in each case. It is critical in such an approach to proceed in a quite systematic fashion, so that the orthogonal basis functions of the standard calculation are taken over unchanged in the larger basis sets; this goal can simply be achieved by orthogonalizing each supplemental basis function to the original set, rather than performing a new SCF calculation. In carrying out the configuration selection and arriving at the T=0 correction it is well to keep track of the perturbative energy lowerings separately and to simply supplement the original selected subspaces with the most strongly interacting of the newly introduced configurations. It is also important to use the same sets of reference configurations throughout, thereby impyling that none of the supplemental basis functions should be of qualitative significance for the description of the states desired.

Once this series of calculations is completed a formula similar to eq. (3) is employed to obtain the AO basis correction. The sum of the energy lowerings for each of the basis supplements serves as the K-type term in this case, with the goal being to predict the overall energy lowering when all the new basis functions are added to the standard set simultaneously. To obtain an estimate for the proportionality factor a final calculation presents itself in which several of the supplemental groups are added explicitly and the resulting total energy is compared with what is estimated by simply summing over the appropriate energy lowerings achieved when each supplement group is treated separately. It is imparative both for satisfying practicable computational requirements as well as for attaining sufficient accuracy to avoid calculating a given quantity more than a single time. The big potential advantage of this technique as a whole is the time-saving it offers at the integral transformation stage. If there are M, M' and M" functions in the standard, largest supplemental, and total AO basis sets respectively, the highest-order computational step advances at the rate of $M'(M'')^4$ rather than $(M')^5$ as would normally be required. In effect the method allows a very expensive calculation to be broken up into many smaller parts, the sum of whose computer times is still substantially less than what would be the requirement for a single global treatment in which all interactions are considered explicitly. If the gains in efficiency which have been obtained using analogous methods for the T=0 and full CI energy corrections are any guide, such an approach to the treatment of large AO basis CI calculations should be quite useful in future investigations.

To this point only the calculation of energy quantities has been discussed, but it is clearly advantageous to be able to extend the general perturbative CI method to the determination of other properties as well. It might be thought that such techniques are fundamentally inapplicable in such cases because the variation principle is no longer operative. In eq. (3), for example, one knows that $E_n(T)$ for at least the lowest root of an MRD-CI secular equation monotonically decreases with T while K(T) necessarily changes in the opposite direction, whereas the same cannot be expected for other properties. Nonetheless the success of the perturbative technique implied by eq. (3) is not dependent on the existence of such monotonic characteristics. One simply needs a K-type quantity which always remains roughly proportional to the true difference between the T=0 and selected CI results, something as for the total energy itself. Indeed the monotonicity is strictly speaking only a characteristic of the energy of the lowest root of a secular equation and in actual computations of the T=0 correction it is imperative to add the ΔE_i quantities algebraically in arriving at K(T) for some higher-energy root.

What one can reasonably expect for some other one-electron property is that whenever the addition of new configurations causes a change in the computed (variational) result that the corresponding K(T) quantity will register a change in the opposite direction. The problem resolves itself quite simply to the need for identifying what to use for K-type term when another property than the energy is desired. Especially if a whole series of properties is to be computed the most straightforward mode of attack is to estimate the change in the CI wavefunctions as new configurations are added, taking advantage of the simple first-order relationship of eq. (2). At a given threshold the truncated CI wave functions obtained in explicit diagonalization can be used to make up a density matrix in the usual manner, for example, and then a second density matrix is computed using the coefficients of the unselected configurations obtained via eq. (2) on the one hand and those obtained from the (small) reference secular equation on the other; as usual separate perturbative treatments are done for each root. Appropriate scalar products with the desired property matrices $\langle p \rangle$ are then computed, yielding respectively P(T) and $K_p(T)$ in analogy to eq. 3. The procedure is then made self-correcting in the usual manner by repeating the computations at one or more additional values of T, thereby leading to an estimate for the proportionality factor λ_p in each case. For non-transition properties it is wise to include diagonal contributions for each of the unselected configurations in which case weighting λ_p^2 is called for; it might also be advisable to take more explicit account of the normalization of the final T=0 wavefunction than is implied by eq. 3. A computer program designed to carry out this procedure is currently being constructed.

As a final potential application of such combined CI-plus-perturbation-theory technique, the problem of dealing with spin-containing operators is a worth brief consideration. In principle no additional theoretical considerations are required in this case relative to those met before for spinless operators; eqs. (1-4) are quite general, and for this purpose one need only expand the definition of the Hamiltonian operator contained therein to deal explicitly with terms such as the spin-orbit, spin-spin and Fermi-contact interactions, for example. In practice, however, one is often confronted with the difficulty that the spin contributions to the energy are relatively small and thus severe accuracy problems develop when energy splittings are to be computed.

To alleviate the situation an additional precaution is necessary in applying the overall method, namely to accumulate the energy lowerings effected by weak configurations of different multiplicity (or spatial symmetry) as the main term which is split by the spin-type interaction. The computations are thus divided into parts, one to treat the strong terms in a standard fashion, and one to consider the effect of adding in the weakly interacting configurations responsible for the multiplet splittings. By employing a separate threshold for the weakly interacting species it is possible to compare the change in the sum of such energy lowerings for unselected species with the actual change in the eigenvalues resulting from an appropriate series of explicit diagonalizations; this additional bookkeeping technique thus makes the whole procedure self-correcting for the spin interactions themselves. The alternative method of introducing the spin effects via Rayleigh-Schrödinger perturbation theory (25,26) involves a number of complications which are avoided by such an approach. For one thing it is not necessary to calculate spin-interactions between large-scale CI functions in this case, and in addition one avoids the problem of what to employ in the energy denominator in computing the perturbative interaction; in addition there is less ambiguity about where to safely truncate the sum of perturbating states. In the following section spin-orbit interactions are calculated with a compromise procedure as will be discussed subsequently, so it remains for future work to see how effective such applications of the overall perturbative CI method turn out to be. To terminate the present discussion, it seems warranted to conclude that the basic technique of using high quality CI results as a starting point for a perturbation treatment with a self-correcting capability has a very wide range of application for all types of electronic states and nuclear arrangements, while still retaining a high degree of practicability in actual computations.

III. APPLICATIONS

A. Electronic States

One of the numerous examples in which the perturbative CI calculations worked hand in hand with the experimental measurements is the chlorine molecule. Before this study began only the $X^1\Sigma_g^+$ and the $B^3\Pi_u$ potential energy surfaces were known in the literature; in addition a number of unassigned transitions (27) had been reported, along with the 258 nm "laser line" which was supposed to correspond to an ionic upper state and a lower species of covalent character. High-resolution absorption studies between 73000 and 81000 cm^{-1} combined with fluorescence analysis between 50000 cm^{-1} and 75000 cm^{-1} found an extensive band structure (28,29) which also could not be assigned whereupon MRD-CI calculations (30) were initiated to obtain a clarification. These computations were undertaken for all singlet and some of the triplet states which dissociate to the lowest atomic limits Cl (^2P_u) + Cl (^2P_u), as well as for a large number of higher-lying states with electronically excited or ionic dissociation products as displayed in Fig. 3. Of special interest in this spectrum is the interaction of Rydberg states originating from excitations out of the π_g MO with those corresponding to valence-type transitions into the σ^* orbital; since the Rydberg states exhibit potential energy curves running essentially parallel to the limiting $^2\Pi_g$ ionic curve, while the valence species have potential wells at much larger internuclear distances, the interaction of these states of the same symmetry lead to a variety of avoided crossing and double-well potential surfaces. The two lowest such states which can combine directly

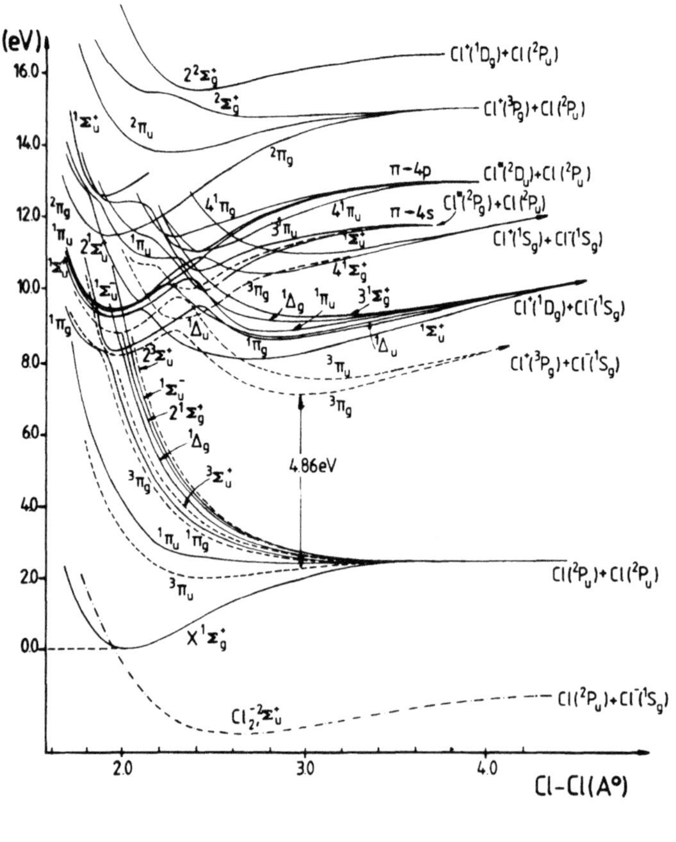

Fig. 3: Calculated potential energy curves for Cl_2 (30).

with the $X^1\Sigma_g^+$ ground state are $2^1\Pi_u$ ($\pi_g \to 4p\sigma$, $\pi_u\sigma_g \to \sigma_u^2$) and $^1\Sigma_u^+$ ($\pi_g \to 4p\pi$, $\sigma_g \to \sigma_u$). The calculations place them in the vertical (Rydberg-like) region at ΔE_e (vertical electronic energy) = 9.16 eV and 9.24 eV respectively. The error limits of roughly 0.15 eV in the standard calculations are such that a definite prediction which of the two states is the lower was not possible, especially in light of the fact that heavy state mixing is also observed for the $^1\Sigma_u^+$ in the vertical range.

An analysis of the measured data on the basis of the calculated potential curves unabiguously assigns the progression between 74000 cm^{-1} and 77000 cm^{-1} to transitions between the $X^1\Sigma_g^+$ and the $2^1\Pi_u$ state, whereby the latter possesses a potential well deep enough to account for at least four vibrational levels. The 0-0 line is thereby assigned at 74443 cm^{-1} (9.23 eV), in excellent agreement with the theoretical ΔE_e predictions of 9.22 eV (MRD-CI value) and 9.16 eV (full CI estimate)*. The concordance between the theoretical predictions for the molecular geometry and the structural data obtained from the measured progression is equally good: r_o=1.90 A° (exptl) vs. r_o=1.94 A° (calc) and ω_e=630 cm^{-1} vs. ω_e=620 cm^{-1} as calculated for the $^2\Pi_g$ ion.

Absorption and emission involving the lowest-lying $^1\Sigma_u^+$ as upper state gives rise to a more complicated pattern according to the theoretical predictions because the barrier between the Rydberg and the valence-shell ionic well is quite small. A vibrational analysis carried out for the $X^1\Sigma_g^+ - {}^1\Sigma_u^+$ transition (29) based on the calculated $^1\Sigma_u^+$ potential energy curves and the pertinent electronic transition moment (which shows a rapid increase with internuclear distance because of the fast change in the character of the $^1\Sigma_u^+$ state in the mixing zone) gives a convincing interpretation of the fluorescence spectrum between 50000 and roughly 72000 cm^{-1}. As expected the spacing of the vibrational levels is very narrow in the neighborhood of the barrier between the two wells; the analysis places v'=35 below the Rydberg well, but indicates that v'=37 is mostly localized inside the same well, while v'=38 is found with higher probability near the second minimum. This prediction is completely consistent with experiment, which shows strong emission to the lowest (v"=0-6) levels of the $X^1\Sigma_g^+$ ground state if v'=37 is populated (ΔE=73450 cm^{-1}, 9.11 eV) while the analogous emission is practically absent if absorption occurs to the v'=38 level (with its higher probability in the outer well). On the basis of the present interpretation the $^1\Sigma_u^+$ (T_o of 9.11 eV) Rydberg well lies below that of $2^1\Pi_u$, i.e. 0.13 eV below the predicted ab initio value for ΔE_e.

* The difference in ground and excited state zero-point energy is 560 cm^{-1} ($X^1\Sigma_g^+$) and 630 cm^{-1} ($2^1\Pi_u$), i.e. only 0.009 eV.

Finally, the strong progression in the 77000 cm^{-1}-81000 cm^{-1} absorption spectrum can also easily be explained on the basis of the CI potential energy curves as $2^1\Sigma_u^+ - X^1\Sigma_g^+$ transitions, whereby the upper state possesses $\sigma_g \to \sigma^*$ character in its inner and $\pi_g \to 4p\pi$ Rydberg character in its outer branches respectively. The calculated energetic difference betweenn both states involved is 78500 cm^{-1} and again agrees extremely well with measured energy difference of 78138 cm^{-1} for the 0-0 transition.

For a better survey the situation is indicated in Fig. 4; spectra are patterned according to the original work of Ref. (29). It is clear that without the knowledge of the ab initio potential energy curves the analysis of the experimental data would have been extremely difficult if not impossible. For the calculations the almost quantitative agreement of a large number of complex states over a large range of internuclear distances was required thereby. For the $2^1\Pi_u$ the portion of experimental interest (Rydberg-like) is relatively easy to represent, but the $2^1\Sigma_u^+$ state definitely requires the correct description of the Rydberg-valence mixing, with this objective only being attainable

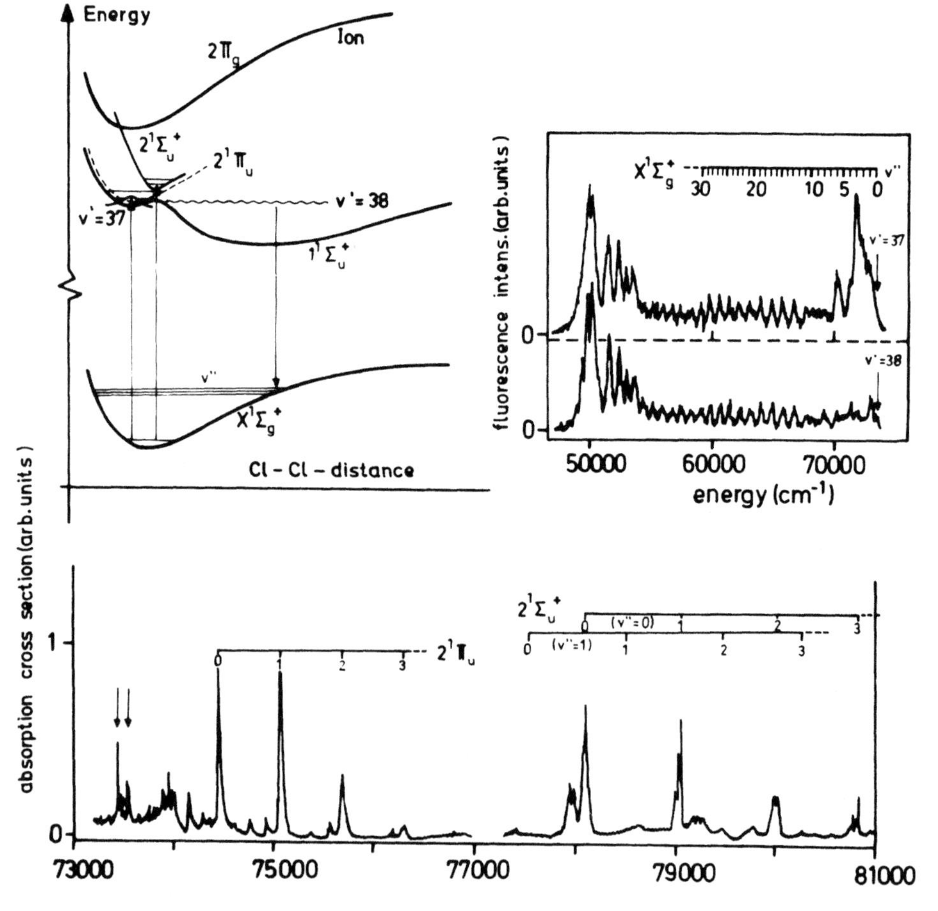

Fig. 4: Experimental Cl_2 spectra and assignment based on the calculated curves.

if a large multi-reference configuration set is employed for the generation of the associated configuration space. The description of the $1^1\Sigma_u^+$ potential curve also contains a Rydberg portion, as well as a Rydberg-valence mixing zone and, at large internuclear separations, the strong ionic Cl^+-Cl^- character for which the T=0 and full CI corrections play an important role in the overall theoretical treatment.

There are various other recent examples in which the calculations have been very useful in unravelling details of the interpretation of complex spectra. The HF and HCl spectra fall in this category, for both of which systems Douglas and coworkers (31) have reported an extensive series of measurements. These authors were able to catalog a large number of measured lines, but state in their original conclusion: "a considerable portion of the HCl spectrum and a small region of the HF spectrum have been analyzed but a theoretical understanding of these spectra is almost completely lacking; furthermore they continue (with respect to HF): "for the present the spectrum must remain an example of the large gap which exists between the simple model of Rydberg states and the complexity of actual Rydberg states". The main cause for this "large gap" can easily be understood on the basis of the calculated potential energy curves for HF displayed in Figs. 5a,b. It is obvious (that similar to the situation discussed in Cl_2) the $B^1\Sigma^+$ ($\sigma\rightarrow\sigma^*$) valence-sehell (ionic) state is repulsive at small internuclear HF-separations and thus leads to numerous avoided crossings with the $\pi\rightarrow np\pi$ and $\pi\rightarrow nd\pi$ Rydberg manifold of the same spatial symmetry. These interactions in turn create double minima, or at least considerable shifts of potential minima relative to the unperturbed Rydberg states (whose potential curves are expected to run roughly parallel to the corresponding $^2\Pi$ ion curve).

In contrast to the $B^1\Sigma^+$ state its triplet counterpart $^3\Sigma^+$ ($\sigma\rightarrow\pi^*$) dissociates into the ground state products and thus shows a much deeper slope at the intersection with the corresponding Rydberg triplet manifold. This crossing occurs for the first such Rydberg state at nearly the minimum of its potential well (Fig. 5b), while the corresponding interaction with higher-lying states occurs primarily at still shorter bond lengths; as a result the perturbation in the triplet manifold is considerably less important than for the singlet states. Unfortunately very little is known about HF triplet states and hence this predicted smaller irregularity in the triplet Rydberg states has not yet been confirmed experimentally. More details about the interactions can be found in the original theoretical reference (32).

The situation is very similar in the electronic spectrum of HCl, whose calculated potential energy curves are displayed in Fig. 6. There is only a quantitative difference relative to the singlet states of HF since the relative position of the $B^1\Sigma^+$ potential curve, and with it the point of intersection with the Rydberg states, is shifted relative to the lighter halide so that less perturbation seems to occur. Apart from the Rydberg-valence interaction a typical Rydberg-like

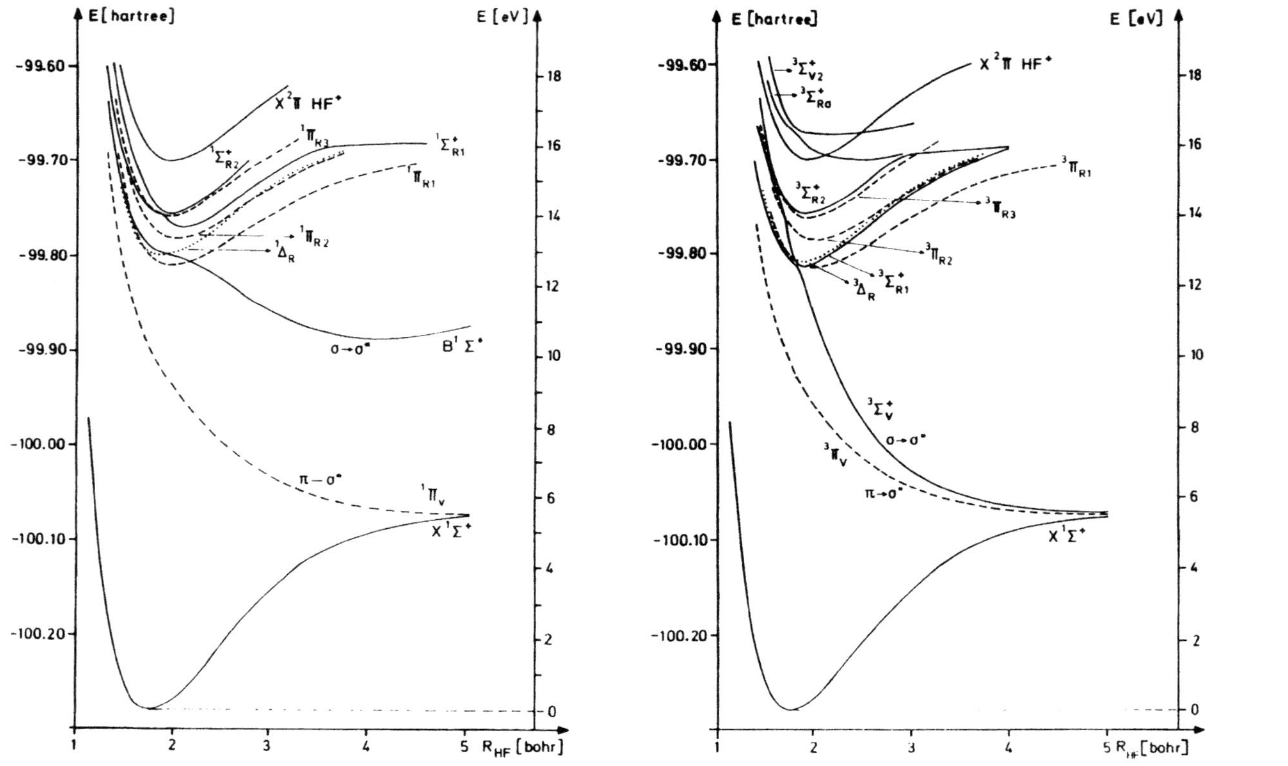

Fig. 5a,b: Calculated potential energy curves (MRD-CI results) for various singlet (a) and triplet (b) states in the HF molecule (32).

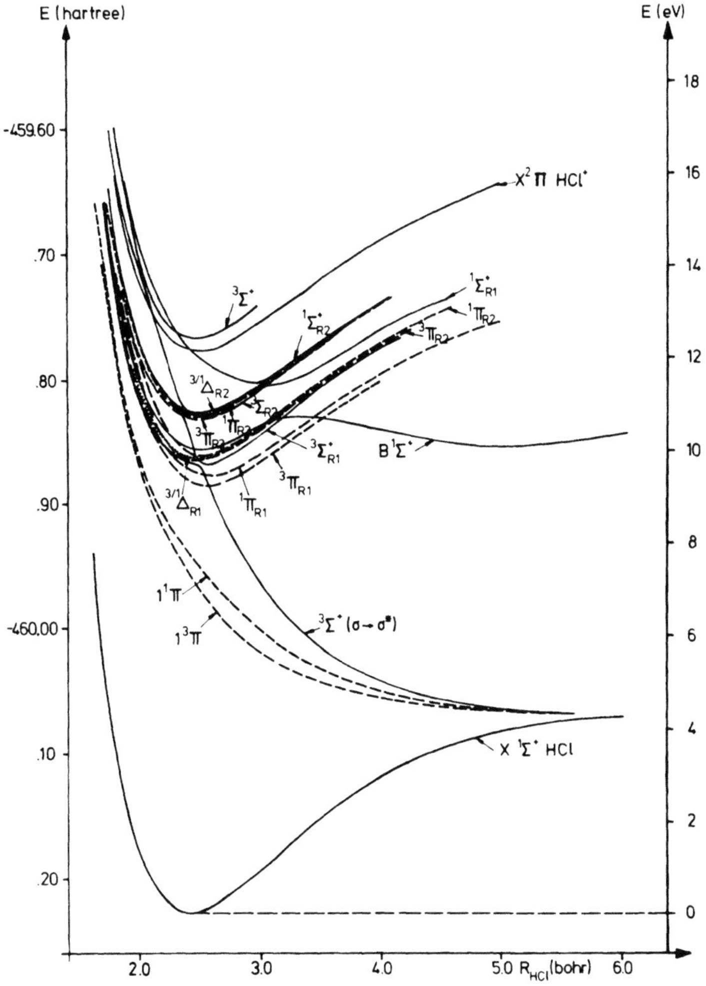

Fig. 6: Calculated potential energy curves (33) for HCl.

pattern is observed in the figure: small singlet-triplet splittings (especially for higher n and l quantum numbers), and practically the same curves for all spatial components of a series (π_{R1} for $\pi\rightarrow$ 3s, Σ^{+}_{R1}, Σ^{-}_{R1}, Δ_{R1} and Π_{R2} for $\pi\rightarrow$ 3p and similar states for $\pi\rightarrow$ 3d). Further details can be found in the original reference (33). These (and many other) examples demonstrate that even in cases in which the interaction of a variety of states plays a key role in the interpretation a sound theoretical assignment can be obtained relatively easily once reliable calculations are available as a guide.

B. PROPERTIES

One of the most important properties in molecular spectroscopy is the intensity of a transition, which is directly related to the square of the electronic transition moment $\langle\psi_{e'}|\sum_i \vec{r}_i|\psi_{e''}\rangle = \vec{R}_{e'e''}$. Typical values obtained for oscillator strengths and radiative lifetimes by employing truncated MRD-CI expansions are displayed in Table 1 for the lowest allowed transitions in NH_3 (34) and for the $^3\Pi_u$ state of N_2 (35). It is obvious that the data predicted by the calculations

Table 1: Calculated oscillator strengths $f_e(r)$ for various transitions in NH_3 and calculated lifetime τ_e for the $C^3\Pi_u$ state in N_2 and comparison with corresponding experimental data.

NH_3	transition	calc. $f_e(r)$	expt. f_e
	X-A ($3a_1\rightarrow$ 3s)	0.089	0.13 (a) 0.079 (b) 0.088 (c) 0.0696(d)
	X-B ($3a_1\rightarrow$ 3px,3py)	0.002	comparable to C←X (e)
	X-C ($3a_1\rightarrow$ 3pz)	0.002	30 times smaller than A← X (e)
N_2	state	calc. τ_e(nsec)	expt. τ_e(nsec)
	$C^3\Pi_u(2\sigma_u\rightarrow\pi_g)$ (calc. $C^3\Pi_u$-$B^3\Pi_g$)	32	36.6±0.5(f) 40.5±1.3(g) 34.8±1.2(h)

$f_e(r) = 2/3|\langle\psi_{e'}|\sum \vec{r}_i|\psi_{e''}\rangle|^2 \cdot \Delta E$; $\tau_e = 1.5/f_e \cdot (\nu_{e'e''})^2$

(a) R.J. Thompson and A.B.F. Duncan, J. Chem. Phys. 14 (1946) 573
(b) E. Tannenbaum, E.M. Coffin, A.J. Harrison. J. Chem. Phys. 21 (1953) 311
(c) K. Watanabe, J. Chem. Phys. 22 (1954) 1564
(d) W.R. Harshbarger, A. Skerbele, E.N. Lassettre, J. Chem. Phys. 54 (1971) 3784
(e) W.R. Harshbarger, J. Chem. Phys. 54 (1971) 2504
(f) K.H. Becker, H. Engels and T. Tatarczyk, Chem. Phys. Letters 51 (1977) 111
(g) J.M. Calo and R.C. Axtmann, J. Chem. Phys. 54 (1971) 1332
(h) A. Pochat, M. Doritch and J. Peresse, J. Chem. Phys. 70 (1973) 936

lie well within the range of corresponding experimental values. Furthermoe it is seen that the spread in experimental data itself is by no means negligible, with this situation found to be the rule rather than the exception as far as absolute intensities are concerned. For this reason a further assessment of the accuracy of the calculated data relative to experiment is not really possible at present. In a more positive sense it can be stated that the determination of absolute intensities with a high degree of reliability can be achieved more easily at present by such calculations than by actual measurements.

One of the few systems for which high-quality intensity measurements are available is the C_2 molecule. In the C_2 Swan ($d^3\Pi_g - a^3\Pi_u$) bands the radiative lifetime has been determined for a number of upper state vibrational levels, and from this information (under the assumption of the r-centroid approximation) the electronic transition moment has been extracted as a function of internuclear CC separation. A comparison of this quantity with the electronic transition moment obtained from MRD-CI calculations is displayed in Fig. 7. The two most recent curves derived from experiment (36,37) are labelled I and II (normalized to $R_e^2(00)=3.57$ at CC=2.44 a.u.), while curve III and the other experimental data indicated originate from earlier measurements as listed elsewhere (37). Curve C stems from an earlier truncated MRD-CI calculation (38) in which the wavefunction expansion contains roughly 1000 terms. It runs fairly parallel to the experimentally determined curves I and II but is too high in absolute magnitude. The two curves A_1 and A_2 originate from a similar calculation (39), but employ CI expansion lengths of roughly 6000 and 9000 SAF's respectively; their curvature is very similar to that of curve C but their absolute magnitude is smaller at every point. The rate of convergence can be estimated from the various calculations at R_{CC}=2.40 a.u. which refer to roughly 1100 (T'=48μh) and 9000 (T_2= 1.4μh) SAF's in the CI expansion. Curve B is the result of another theoretical treatment (40) employing a similarly large CI expansion (orders 6000) in an STO AO basis.

Even though it appears as if the curve labelled A_2 is close to the convergence limit, it is still above curves I and II. On the other hand, the experimental error limits at CC=2.44 a.u. are R_e^2=2.81 a.u. ±50 % (and ± 10 % for the shape of the curve) in curve II; R_e^2=3.57

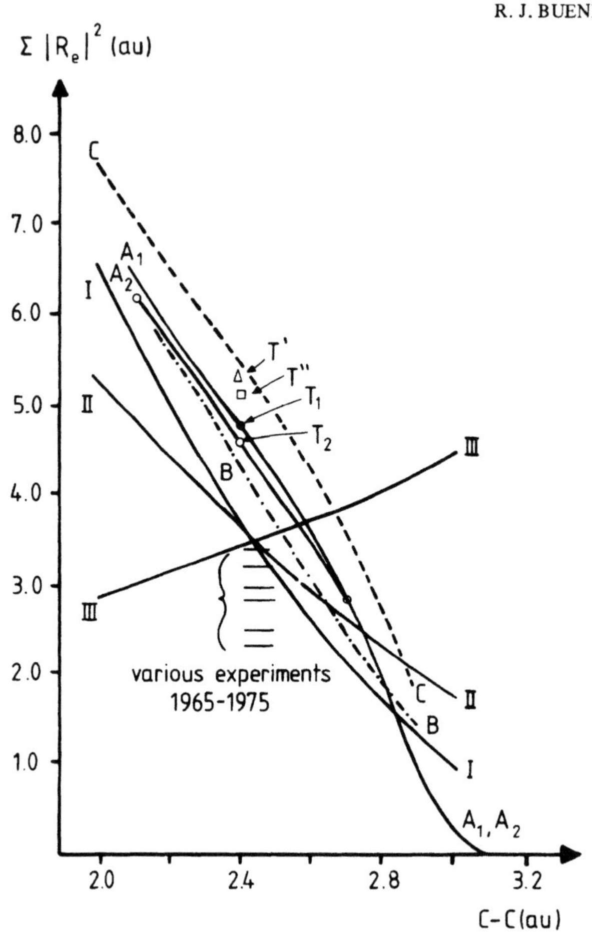

Fig. 7:
Comparison of various curves showing the dependence of the electronic transition moment ΣR_e^2 in the $(d^3\Pi_g - a^3\Pi_u)$ Swan bands in C_2 as a function of the CC separation. Curves derived from experimental labelled I, II, III; curves A_1 and A_2 result from truncated MRD-CI calculations of order 6000 and 9000 respectively; curve C corresponds to a calculation of roughly 1000 CI terms; curve B results from an STO-AO basis. Property convergence is indicated by the various selection threshold values T.

$\pm$ 0.4 a.u. for curve I and $R_e^2 = 3.52 \pm 0.5$ a.u. in another recent measurement (41), so that the true error in the calculations is still difficult to assess. It is hoped that further experience with extrapolation of properties to zero threshold and full CI limit (i.e. via a perturbative treatment which approximately includes the contribution of the non-selected configurations to the properties, as discussed in the first section) will give more information with respect to the absolute error in the property calculations, especially since this knowledge is not likely to be gained in the near future from comparisons with experiment.

Similar calculations for the $e^3\Pi_g - a^3\Pi_u$ Fox-Herzberg bands (39) shows a transition moment curve (Fig. 8) which peaks around CC=2.9 a.u., whereby again the differences in secular equation sizes corresponding to two different thresholds (T_1 leading to approximately 6000 and T_2 to 9000 SAF's) are indicated. It has not been possible to extract the shape of the ΣR_e^2 curve from measurements so far and only upper limits of $\Sigma R_e^2 = 0.45 \pm 0.10$ a.u. and $\Sigma R_e^2 = 0.55$ a.u. have been given for the $\Delta v=0$ series in two independent experimental studies which straddle very nicely the computed curve. The somewhat peculiar shape can be easily understood since it is clear from the potential energy curves plotted in the same figure that the avoided crossing between the $d^3\Pi_g$ and $e^3\Pi_g$ states brings some of the stronger Swan band character into the Fox-Herzberg transition.

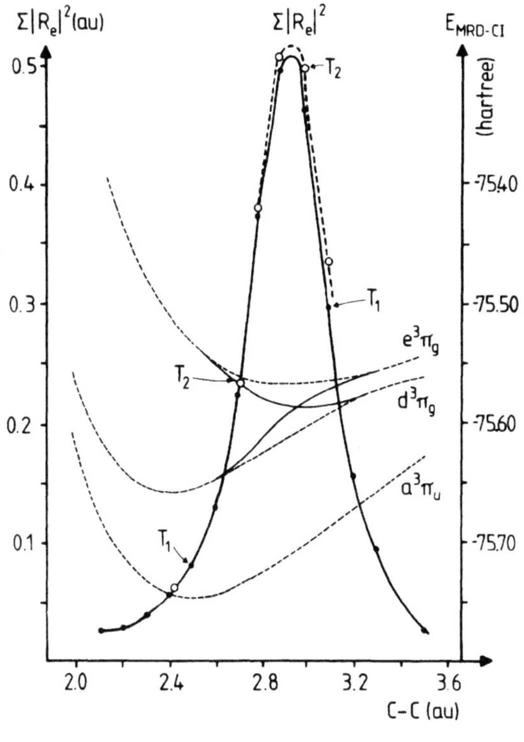

Fig. 8:
Calculated electronic transition moment $\Sigma|R_e|^2$ for the Fox-Herzberg bands in C_2 as a function of the internuclear CC distance. The threshold T_1 corresponds to secular equations of order $\simeq$ 6000, T_2 to orders of roughly 9000. The calculated potential energy curves for the $a^3\Pi_u$ and the $d^3\Pi_g$ and $e^3\Pi_g$ states are also given; the crossing of the latter is indicated.

Further intensities have been calculated in numerous other studies with similarly satisfactory agreement resulting within the error limits of the (generally scarce) experimental data which is available data. A final example shows the good accord for the Einstein coefficient in the $\Delta v=0$ and $\Delta v=1$ infrared transition in the HF molecule (Fig. 9). In summary then it can be stated that the calculations already represent a very powerful tool for obtaining information on absolute intensities and radiative lifetimes, and furthermore that still higher accuracy will probably come as a result systematic theoretical investigations of the convergence behavior such properties rather than by virtue of further developments in the corresponding experimental techniques.

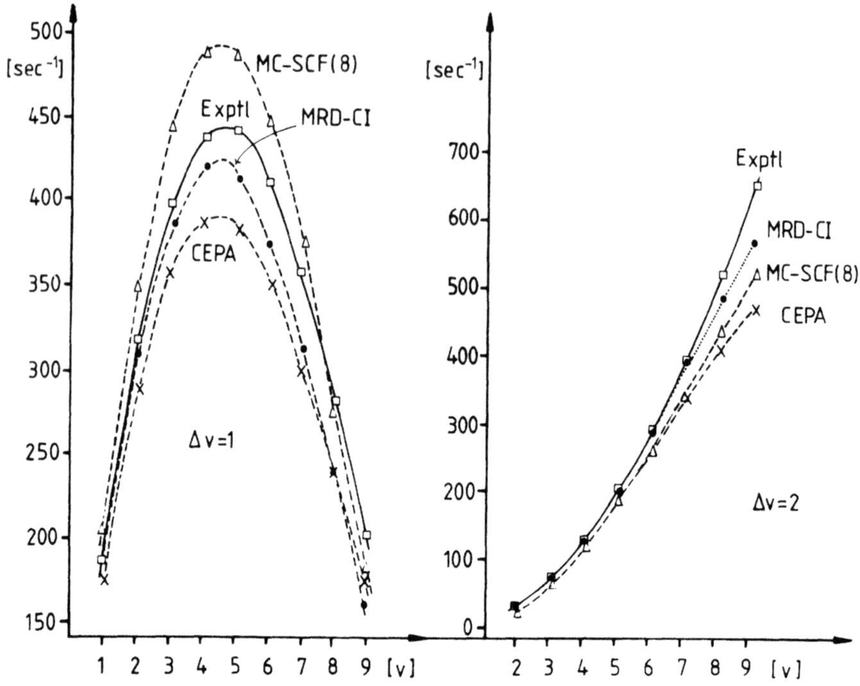

Fig. 9: Calculated Einstein coefficients for Δ v=0 and 1 transitions in the $X^1\Sigma_g^+$ infrared spectrum of the HF molecule obtained from MRD-CI calculations and comparison with corresponding CEPA and MC-SCF results (42) and the data derived from experiments (43). The calculated MRD-CI potential curves are employed for the vibrational treatment in each case.

C. SPIN-DEPENDENT PROPERTIES

One of the goals of electronic structure calculations should be the inclusion of relativistic effects in order to appropriately describe systems containing heavy atoms. Some of the first effects of importance in spectroscopy are spin-orbit and spin-spin interactions, which have been treated for selected molecules such as O_2 and CH_2 in a relatively rigorous (25) manner and for others such as MgH, Xe_2 etc. with various approximations (44). The data to be discussed below result from the use of the full one- and two-particle Breit-Pauli Hamiltonian with large-scale MRD-CI wavefunctions obtained in a gaussian AO basis. Major parts of the respective computer programs have been written by Dr. C. Marian (Bonn), Dr. B. Heß (Bonn/Wuppertal) and Dr. P. Chandra (Wuppertal). Two approaches are taken thereby: first the more conventional Rayleigh-Schrödinger perturbation theory approach in which the first terms of the perturbation sum are evaluated using the MRD-CI wavefunctions, and secondly the procedure in which the additional operators (for example the spin-orbit term) is directly included in the Hamiltonian for which a matrix representation is obtained (hereafter referred to as H_{SO}-CI).

Doublet splittings obtained in this manner by employing essentially double-zeta AO basis are in very good agreement with measured data as seen from Table 2. The zero-field splitting for the $^3\Sigma_g^-$ state

Table 2: Comparison of calculated and observed spin-orbit constant A (cm^{-1}) for a few systems.

		perturbation approach	H_{SO}-CI	Exptl.
H_2S^+	$^2\Pi_u$	-418.4	-418.5	(-432.5 for HS^+)
BO	$A^2\Pi_i$	-120.5	-120.6	-122.3, -116.7
NO	$X^2\Pi_r$	122.0	-	123.13, 122.2
CO^+	$A^2\Pi_i$	-119.0	-	-117.5
CH	$X^2\Pi_r$	26.6	26.6	27.95, 28.0
OH	$X^2\Pi_i$	-138.0	-138.0	-139.2, -139.7
SO^+	$X^2\Pi_r$	+339	-	+340 ± 0.25
SO^+	$A^2\Pi_r$	+145	-	?

For details see the original references (45,46).

of S_2, for which the interacting states of $^1\Sigma_g^+$, $^1\Pi_g$, $^3\Pi_g$ and $^3\Sigma_g^+$ are considered (47), is reproduced essentially quantitatively over a large range of internuclear distances: the calculated formula for the splitting as a function of the vibrational quantum numbers is

$$\lambda_v(\text{calc})\ ^3\Sigma_g^- = 11.82 + 0.05\ (v+\tfrac{1}{2}) + 0.00024\ (v+\tfrac{1}{2})^2$$

which should be compared with the formula derived from measurements

$$\lambda_v(\text{exp})\ ^3\Sigma_g^- = 11.70 + 0.055\ (v+\tfrac{1}{2}) + 0.0011\ (v+\tfrac{1}{2})^2;$$

the spin-spin contribution, not contained in the theoretical formula, is estimated to be smaller than 0.15 cm^{-1} in this case.

In this connection it is also of interest that the spin-orbit matrix elements vary only slightly from a calculation in which all the electrons are correlated to one which employs an core of doubly occupied (inner-shell) MO's, similarly as is done in standard CI calculations. Typical variations are (in cm^{-1}) 406.4/407.8, 195.7/194.7, -196.8/-195.5 for the spin-orbit matrix elements between $^1\Sigma_g^+$, $^1\Pi_g$, $^3\Pi_g$ on the one hand and $^3\Sigma_g^-$ on the other with core=10/core =4 MO's (core=10 corresponds to the five uncorrelated MO's in each sulfur atom, i.e. K and L shells). A detailed analysis (46) of the influence of AO basis on the $^3\Sigma^-$ splitting in SO underlines the importance of affording a good valence-shell description. The spin-orbit matrix element between $X^3\Sigma^-$ and $b^1\Sigma^+$, for example, stays fairly constant (Fig. 10, around 300 cm^{-1}) as long as a double-zeta type valence-shell description is employed, but this value dops to 270 cm^{-1} and 255 cm^{-1} for two smaller (Roos-Siegbahn and STO-4G) AO basis sets. Since the above matrix element makes the largest contribution in the SD results this characteristic is carried over to the total $^3\Sigma^-$ splitting in this system, which falls off from roughly 9 cm^{-1} (DZ) to less than 6 cm^{-1} in the smaller AO basis sets (Fig. 11). The spin-spin contribution to this quantity evaluated in a single-configuration approximation, is around 0.67 cm^{-1} in the relatively flexible basis sets 1-6. Hence the total splitting parameter D in the (representative) AO basis 4 is 9.3 cm^{-1} if the perturbation sum with only one term corresponding to each of the lowest $^1\Sigma^+$, $^1\Pi$, $^3\Pi$ and $^3\Sigma^-$ states is taken, but 10.68 cm^{-1} in the H_{SO}-CI corresponding to the more nearly converged perturbation series. The corresponding measured value is 10.55 cm^{-1} so that the agreement between theoretical and experimental splittings is excellent. This result also is indicative of the advantages of the H_{SO}-CI vis-a-vis the perturbation treatment as soon as the degree of interacting becomes relatively large. This aspect of the calculations will most certainly play an even increasing role as molecular systems with heavier atoms are treated, in which case multiplet splittings reaching magnitudes of an electron volt or more are known.

Table 3: AO basis set description for the results displayed in Figs. 10 and 11.

AO Basis	Description: Oxygen		Description: Sulfur	
1	Dunning[a]	10s6p/5s4p	12s9p/ 6s5p	McLean[b]
2	Dunning[c]	9s5p/4s2p	12s9p/ 6s4p	Veillard[d]
3	Dunning[c]	9s5p/9s2p	12s9p/12s4p	Veillard[d]
4	Dunning[c]	9s5p/3s2p	11s7p/ 6s4p	Dunning[e]
5	Dunning[c]	9s5p/9s2p	11s7p/11s4p	Dunning[e]
6	same as 4, additional s,p Rydberg and s,p bond functions			
7	Roos-Siegbahn[f]	7s3p/5s3p	10s6p/ 6s5p	R-S[f]
8	STO-4G[g]	8s4p/2s1p	12s8p/ 3s2p	STO-4G

[a] T.H. Dunning Jr., J. Chem. Phys. 55 (1971) 716
[b] A.D. McLean and G.S. Chandler, J. Chem. Phys. 72 (1980) 5639
[c] T.H. Dunning Jr., J. Chem. Phys. 53 (1970) 2823
[d] A. Veillard, Theor. Chim. Acta 12 (1968) 405
[e] T.H. Dunning Jr. and P.J. Hay, in: "Methods of Electronic Structure Theory", Vol. 3, Ed. H.F. Schaefer III, Plenum Press (1976)
[f] B. Roos and P. Siegbahn, Theor. Chim. Acta 17 (1970) 209
[g] W.J. Hehre, R.F. Stewart and J.A. Pople, J. Chem. Phys. 51 (1969) 2657

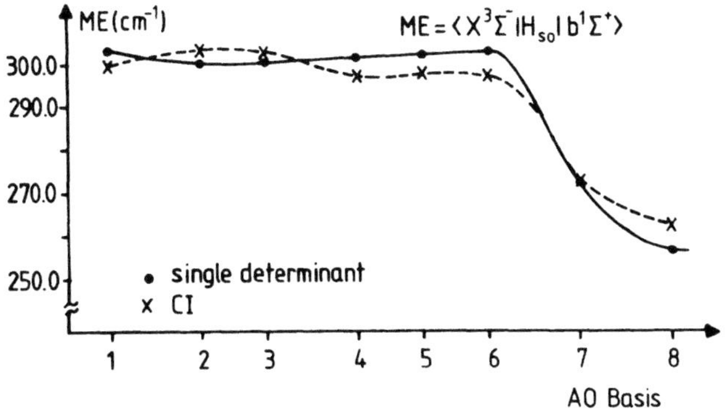

Fig. 10: Spin-orbit coupling matrix element contributing to the $X^3\Sigma^-$ multiplet splitting of the SO molecule evaluated in various AO basis sets described in Table 3.

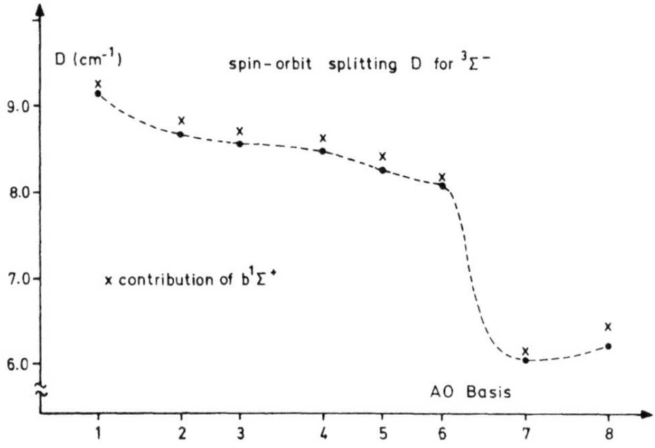

Fig. 11: Spin-orbit splitting of the $X^3\Sigma^-$ state in SO calculated by employing various AO basis sets described in Table 3.

A final example which requires the calculation of reliable potential surfaces, transition intensities and spin-orbit interactions is the study of the predissociation of the $b^4\Sigma_g^-$ state in O_2^+ (48). The calculated potential surfaces in Fig. 12 show that the $^4\Sigma_g^+$ state crosses $b^4\Sigma_g^-$ between v'=5 and 6, which result meshes quite nicely with the fact that predissociation of the $b^4\Sigma_g^-$ is first observed for v'=5 and N=9; the calculated T_e result for $a^4\Pi_u$-$b^4\Sigma_g^-$ is 2.02 eV, while the corresponding measured value is 2.05 eV. Further predissociation is caused by the higher-lying $f^4\Pi_g$ state. Spin-orbit matrix elements connecting the $b^4\Sigma_g^-$ curve with those of $^4\Sigma_g^+$ and $f^4\Pi_g$ are plotted in Fig. 13 and show a very pronounced dependence of these quantities on the internuclear O-O separation in the first case (just as has been found for the $\langle X^2\Pi_{gx}|H_{SO}|X^2\Pi_{gy}\rangle$ splitting) while the spin-orbit interaction between $b^4\Sigma_g^-$ and $f^4\Pi_g$ is much more nearly constant over the range treated. Estimates for the probability of predissociation on the basis of the calculated matrix elements show high values for v'=6 and 7, again consistent with the fact that the line width for v'=6 is observed to be very large. Radiative lifetimes (computed explicitly from $b^4\Sigma_g^-$ (v'=0) to $a^4\Pi_u$ (v"=0-9) to be $\tau = 1.08\cdot10^{-6}$ sec, as compared to the measured value of τ (v'=0) = $1.12\cdot10^{-6}$ sec) are found to be smaller by a factor 10^4 than the value for the non-radiative process, so that predissociation is seen to be far the preferred process as soon as it becomes energetically possible. More details of these calculations may be found in the original reference (48). Even though these examples constitute only the first steps in the treatment of general spin-orbit effects, the results are very encouraging and generalization to heavier atoms seems a realistic possibility in the near future.

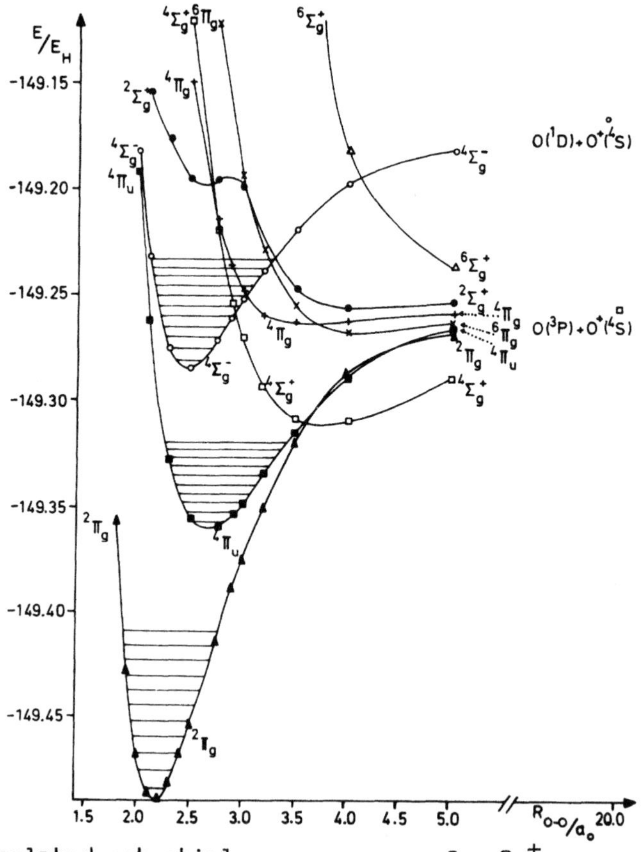

Fig. 12: Calculated potential energy curves for O_2^+.

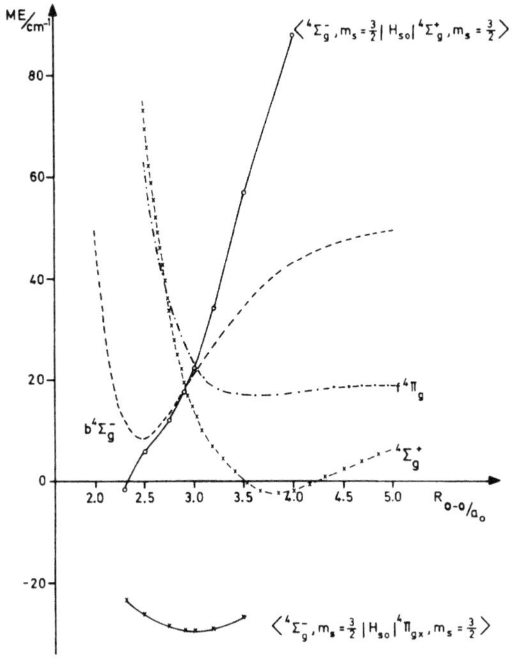

Fig. 13: Spin-orbit coupling matrix elements of $b^4\Sigma_g^-$, $f^4\Pi_g$ and $b^4\Sigma_g^-$, $^4\Sigma_g^+$ states in O_2^+ along with corresponding calculated potential curves.

D. NON-ADIABATIC EFFECTS

If the level of treatment in electronic structure calculations is high, the restrictions imposed by the Born-Oppenheimer approximation become more and more apparent, so that inclusion of non-adiabatic effects in the theoretical description becomes imperative. A typical example thereof is the Renner-Teller interaction in NH_2 for which excellent experimental data are available for comparison (49). The bending potential energy curves for the two lowest X^2B_1 and A^2A_1 electronic states (Fig. 14), which correlate with $^2\Pi_u$ in the NH_2 linear nuclear geometry have been calculated (50) employing an AO basis containing 72 functions (including two f species on nitrogen). The T_{000} transition energy between upper and lower states is predicted to be 11150 cm^{-1} in the CI calculations (full CI estimate) compared to the value of 11126 cm^{-1} derived via an empirical fitting procedure (51) for the measured lines. The non-adiabatic effects producing the Renner-Teller splitting result from the term $(1/2\ \mu r^2)\ \partial^2/\partial\phi^2$ in the kinetic energy expression, (ϕ is the azimuthal angle) which ultimately leads to the non-adiabatic matrix elements $C^{\alpha\beta} = h_{\alpha\beta} = \langle\psi^\alpha | \partial^2/\partial\phi^2 | \psi^\beta\rangle$ and $B^{\alpha\beta} = g_{\alpha\beta} = \langle\psi^\alpha | \partial/\partial\phi | \psi^\beta\rangle$ which are plotted in Fig. 15 as a function of the NH_2 bending coordinate; these results have been calculated by employing a finite difference method (52,53). The off-diagonal term $B^{12} = \langle X^2B_1 | \partial/\partial\phi | A^2A_1\rangle$ is thereby

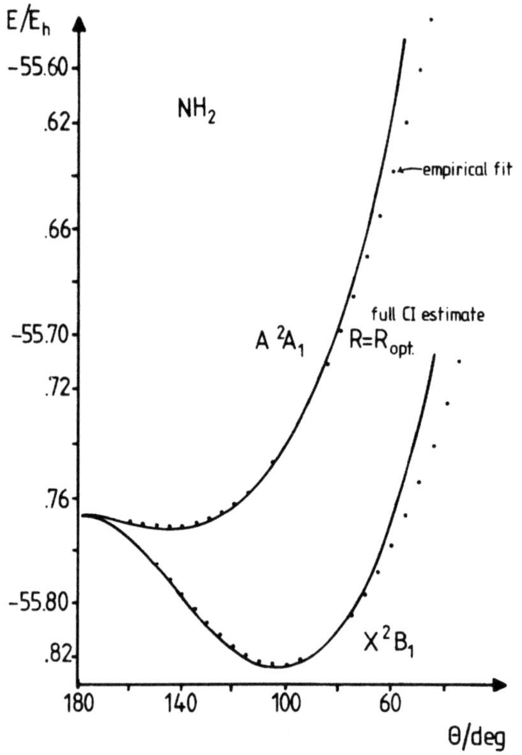

Fig. 14:
Calculated angular potential curves (50) for the two lowest states of NH_2; an AO basis of 72 functions is employed and the NH distance is optimized at each point. The curves obtained from the empirical fitting procedure (51) to the spectral data is also indicated.

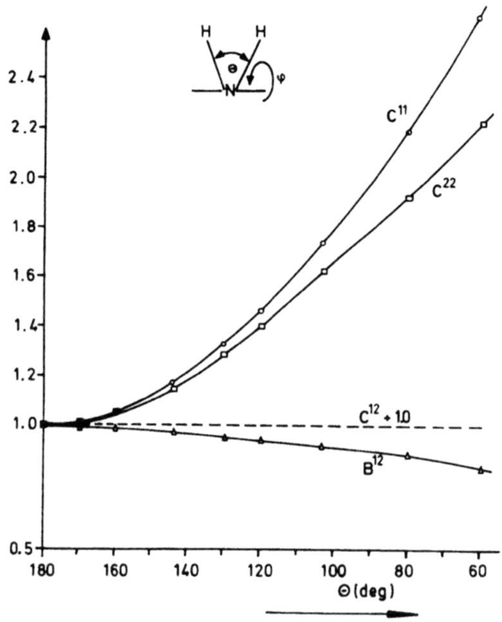

Fig. 15:
Variation of the non-adiabatic matrix elements for the first two electronic states in NH_2 with internuclear bending angle.

responsible for the coupling of bending vibrational levels of the X^2B_1 state with those of the excited $^2A_1$ for rotational (K) quantum numbers greater than 0 (i.e. for levels other than Σ symmetry).

A vibronic CI (54) based on the calculated X^2B_1 and A^2A_1 potential energy curves including the non-adiabatic terms of Fig. 15 leads to the energy levels shown in Fig. 16. It is seen that for $K > 0$ the lowest v_2' excited state levels are mixed strongly with the v_2'' levels in the same energy range, while for higher v' quantum numbers the interaction decreases significantly; this behavior is indeed the pattern observed experimentally (Fig. 16). The most critical approximations in the present theoretical treatment are the assumption of a constant reduced mass in the kinetic energy expression, which in reality changes with both the bending and NH stretching coordinates, and the neglect of vibrational coupling between the symmetric stretching and bending modes. Both effects lead to an overestimation of the ground state ω_2 frequency, as discussed in the original reference (50).

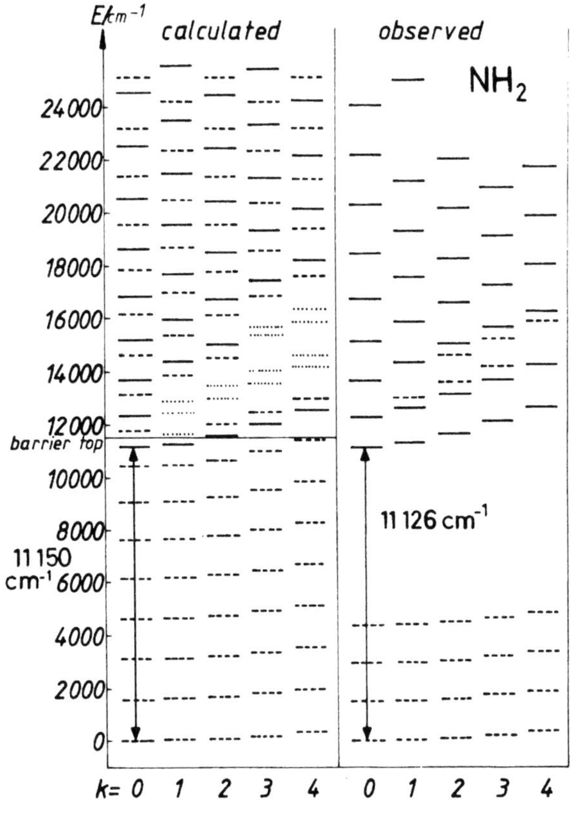

Fig. 16: Comparison between calculated and observed energy levels for the X^2B_1 ground (dashed lines) and A^2A_1 excited (solid lines) states of NH_2. The levels above the barrier which correspond to a heavy mixture of ground and excited state vibrational wavefunctions are represented by dotted lines. The calculated and measured T_0 excitation energies are also indicated.

The empirical fitting procedure (51) takes all $h_{\alpha\beta}$ and $g_{\alpha\beta}$ elements to be constant and chooses a value of 1.0 (±) for each non-zero element, i.e. their limiting value in the linear nuclear arrangement (see Fig. 15). Even though the factor $1/2\ \mu r^2$ decreases the magnitude of the role played by these non-adiabatic matrix elements it is clear that the calculated ab initio Born-Oppenheimer curves and those derived from the empirical fit (with various free parameters in the potential form) employing the above assumption for the $C^{\alpha\beta}$ and $B^{\alpha\beta}$ terms should not completely coincide; this situation underscores the problems encountered in comparing curves resulting from a multi-parameter fit with those derived directly from quantummechnical calculations. When such a comparison is made in the present case (Fig. 14) it is seen that the deviation between the empirically fitted and the ab initio data becomes very large (1900-4500 cm^{-1}) at small internuclear angles, whereby the ab initio Born-Oppenheimer surfaces rise much more steeply, even after optimized NH bond lengths are employed at each angle. The origin for this discrepancy seems to be tied up with the existence of a third (small-angle) state ($B\,^2B_2$), which undergoes a curve crossing (conical intersection) with the $A\,^2A_1$ state in the neighborhood of Θ =70°. For small asymmetric NH stretching distortions it follows that this crossing becomes very strongly avoided, with the result that the lower $^2A_1$ ($^2A'$) state curve takes on a sudden change in slope in this region (following the original $^2B_2$ curve). It seems reasonable to expect that in the empirical fitting procedure (which ignores the existence of this third state) the resulting impact on the measured spectral lines can only be perceived in terms of a (gradual) lowering in the $^2A_1$ potential curve. From this point of view the deviation of the empirical curve in Fig. 14 from its ab inito counterpart which arises at small bonding angles appears to be the consequence of employing an unrealistic (two-state) model on which to base the fitting procedure. This example suggests strongly therefore that the best approach for the interpretation of such a large number of high-resolution lines is a combination of both empirical and ab initio procedures, so that unrealistic models for the potential curve construction can be avoided while still allowing for some empirical adjustments to the ab initio curves in order to obtain the best possible agreement between theory and experiment for the locations of the measured lines.

A second and final example which illustrates a similar point concerns the broad intensity distribution observed in the V-N absorption band system in ethlyene. The major distinction in geometry between ground and $\pi\rightarrow\pi^*$ excited states in this case occurs with respect to the relative orientation of the CH_2 groups, which are coplanar in the N but perpendicular in the V state; as a result it can be expected that the observed spectral progression should result primarily from this torsional motion. The strong irregularity in this band system suggests, however, that either a complex form of the potential is involved and/or that non-adiabatic effects play a significant role. Consistant with this view it is found that the $^1B_{1u}$ ($\pi\rightarrow\pi^*$) and $^1B_{1g}$ (Rydberg → 3py) electronic states undergo an avoided crossing as soon as the torsion

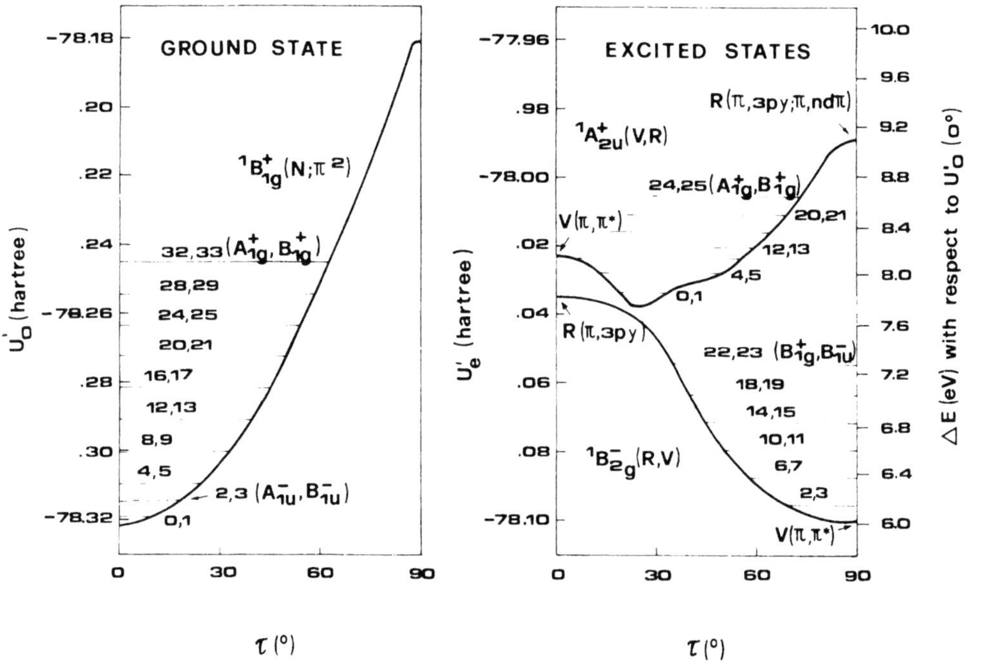

Fig. 17: Estimated full CI potential energy $U_e' = U_e - h_{ee}/2I_r$ (including diagonal non-adiabatic corrections) and some low-lying torsional levels of the ground X^1A_1 and the two excited $^1B_1$ ($\pi \rightarrow \pi^*$) and ($\pi \rightarrow 3py$) states in ethylene.

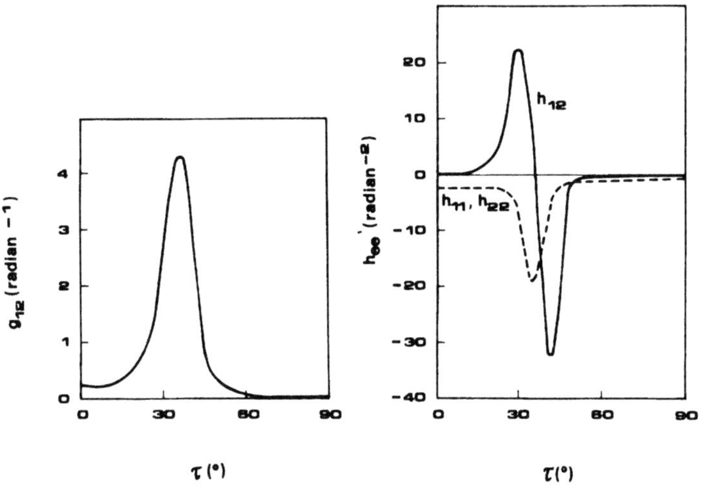

Fig. 18: Calculated nuclear-electronic coupling terms in C_2H_4 for the two $^1B_1$ states as a function of the torsional angle.

progresses out of the planar conformation, thereby opening up the possibility that vibronic mixing occurs between the vibrations level of both upper states for this type of nuclear motion; the corresponding potential energy curves for the ground (N) and the two ($^1B_1$) upper states are given in Fig. 17. The corresponding vibronic CI calculations (55), again carried out on the basis of the ab initio computed potential energy surfaces and the corresponding non-adiabatic matrix elements (Fig. 18), leads to the intensity distribution which is shown in Fig. 19; comparison is made therein with the spectrum which is obtained if transitions to both upper states are calculated separately, i.e. assuming the Born-Oppenheimer approximation. The large distinctions between the results of these two levels of treatment is obvious, especially in so far as the non-adiabatic treatment

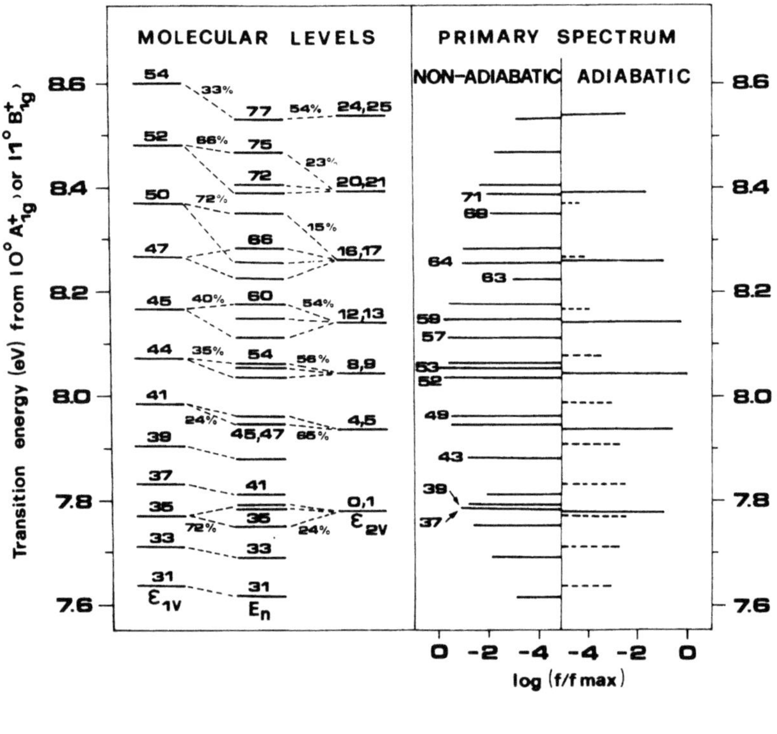

Fig. 19: Several molecular vibrational levels and central portion of the calculated absorption spectrum of C_2H_4 relative to the ground adiabatic states. Molecular energy levels: adiabatic ε_{1v} and ε_{2v} corresponding to the two $^1B_1$ potential curves) on the left and right, and non-adiabatic in the middle; the percentage of mixing of some adiabatic states is also shown. The dashed and full lines in the adiabatic spectrum refer to states $1\,^1B_1$ and $2\,^1B_1$ ($^1B^-_{2g}$ and $^1A^+_{2u}$ in the dynamic point group) respectively.

introduces considerable irregularity and broadening into the theoretical intensity distribution. The two computed maxima fit in well with the observed V-N maximum on the one hand and with at structure at 8.25 eV originally thought to involve pure Rydberg transitions of undetermined assignment. There is still not perfect agreement between the computed and measured intensity results, but it is expected that further broadening and other small modifications of the theoretical spectrum will occur if still other Rydberg states known to undergo similar crossings with the $^1(\pi, \pi^*)$ surface are included explicitly in the vibronic CI treatment. In any event in this case it appears without question that a non-adiabatic treatment is required in order to even qualitatively represent the observed spectral pattern. Furthermore it seems reasonable to expect that such non-Born-Oppenheimer effects become increasingly important for polyatomic systems with many degrees of freedom, especially for high-lying excited states with relatively closely spaced energy levels, in which case the occurrence of conical intersections and avoided crossings become more the rule than the exception in teh overall theoretical treatment.

ACKNOWLEDGMENT

We would like to thank all our associates, in particular Drs. Bruna, Chabalowski, Heß, Marian and Dipl.-Chem. Bettendorff, Hirsch, Klotz and Mr. Quade for considerable help in this research pertinent to the excited states of molecules, and we are grateful to the Deutsche Forschungsgemeinschaft for continued financial support given to our work. The calculations have been carried out at the University of Bonn Computation Center, whose services are gratefully acknowledged, and at the SFB 42 minicomputer PE 3242 at Wuppertal with considerable technical assistance from P. Liebermann and P. Funke.

REFERENCES

1. Tseng, T.J. and Grein, F.: 1973, J. Chem. Phys. 59, 6563

2. Thulstrup, E.W. and Öhrn, Y.: 1972, J. Chem. Phys. 57, 3716

3. Buerker R.J. and Peyerimhoff, S.D.: 1974, Theor. Chim. Acta 35, 33

4. Peyerimhoff, S.D. and Buenker, R.J.: 1975, Adv. in Quantum Chem. 9, 69

5. Brooks, B.R. and Schaefer III, H.F.: 1980, J. Chem. Phys. 48, 157;
Brooks, B.R., Laidig, W.D., Saxe, P., Handy, N.C. and Schaefer III, H.F.: 1980, Physica Scripta 21, 312
Roos, B.O., Taylor, P.R. and Siegbahn, P.E.: 1980, Chem. Phys. 48, 157;
Lischka, H., Shepard, R., Brown, F.B. and Shavitt, I.: 1981, Int. J. Quant. Chem. S15, 91

6. Buenker, R.J.: 1980, in "Quantum Chemistry and Molecular Physics into the 80's", Proc. of the Wollongong Summer Institute, Ed. Burton, P.G., p. 1.5.1

Buenker, R.J.: 1982, "Implementation of the Table CI Method: Configurations Differing by Two in the Number of Open Shells", to appear in the Proceedings of XIIe Congrés dels Quimics Theorics D'Expressio Llatine (Barcelona, September 1981)

7. Slater, J.C.: 1960, "Quantum Theory of Atomic Structure", Vol. I, McGraw Hill Co., p. 119 ff

8. Hackmeyer, M. and Whitten, J.L.: 1969, J. Chem. Phys. 51, 5584; ibid 1971, 54, 3739

9. Shavitt, I. in "Modern Theoretical Chemistry", Vol. 3 (Methods of Electronic Structure Theory), Eds. Schaefer III, H.F., Plenum Press, New York (1977), p. 189

10. Bender, C.F. and Davidson, E.R.: 1969, Phys. Rev. 183, 23

11. Buenker, R.J. and Peyerimhoff S.D.: 1975, Theor. Chim. Acta 39, 217

12. Peyerimhoff, S.D. and Buenker, R.J.: 1980 in "Computational Methods in Chemistry", Ed. Bargon, J., Plenum Press, N.Y., p. 175

13. Bruna, P.J., Hirsch. G., Buenker, R.J. and Peyerimhoff, S.D.: NATO ASI, Kos 1980, "Molecular Ions; Geometric and Electronic Structure"

14. Ha, T.K.: 1979, Chem. Phys. Letters 66, 317; listed as private communication from Davidson, E.R.

15. Sakai, S. and Morokuma, K.: 1980, Chem. Phys. 52, 33

16. Buenker, R.J., Peyerimhoff, S.D. and Bruna, P.J.: 1981, in "Computational Theoretical Organic Chemistry", Eds. Csizmadia, I.G. and Daudel, R.D., Reidel Publ. Co., p. 55

17. Langhoff, S.R. and Davidson, E.R.: 1974, Int. J. Quant. Chem. 8, 61;
Davidson, E.R.: 1974, in "The World of Quantum Chemistry", Ed. Daudel, R. and Pullmann, B., Reidel Co., Dordrecht, Holland, p. 17

18a. Hirsch, G., Bruna, P.J., Peyerimhoff, S.D. and Buenker, R.J.: 1977, Chem. Phys. Letters 52, 442
b. Butscher, W., Shih, S.K., Buenker, R.J. and Peyerimhoff, S.D.: 1977, Chem. Phys. Letters 52, 457

19. Bruna, P.J., Peyerimhoff, S.D. and Buenker, R.J.: 1980, Chem. Phys. Letters 72, 278

20. Pople, J.A., Seeger, R. and Krishnan, R.: 1977, Int. J. Quant. Chem. S11, 149

21. Siegbahn, P.E.M.: 1978, Chem. Phys. Letters 55, 386

22. Results by Bruna, P.J.: 1981, Bonn

23. Results for N_2 by Bruna, P.J.: 1981, Bonn

24. Koch, S. and Kutzelnigg, W.: 1981, Theoret. Chim. Acta 59, 387

25. Langhoff, S.R. and Kern, C.W.: 1977, in "Modern Theoretical Chemistry", Vol. 4, Ed. Schaefer III, H.F., Plenum Press, p. 381

26. Richards, W.G., Triverdi, H.P. and Cooper, D.L.: 1981, "Spin-Orbit Coupling in Molecules", Clarendon Press, Oxford

27. Huber, K.P. and Herzberg, G.: 1979, "Molecular Spectra and Molecular Structure, Constants of Diatomic Molecules", van Nostrand Co.

28. Castex, M.C., Le Calvé, J., Haaks, D., Jordan, B. and Zimmerer, G.: 1980, Chem. Phys. Letters 70, 106

29. Moeller, T., Jordan, B., Gürtler, P., Zimmerer, G., Haaks, D., Le Calvé, J. and Castex, M.-C.: 1982, J. Chem. Phys.

30. Peyerimhoff, S.D. and Buenker, R.J.: 1981, Chem. Phys. 57, 279

31. Douglas, A.E. and Greening, F.R.: 1979, Can. J. Phys. 57, 1650

32. Bettendorff, M., Buenker, R.J., Peyerimhoff, S.D. and Roemelt, J.: 1982, Z. Phys. A 304, 125

33. Bettendorff, M., Peyerimhoff, S.D. and Buenker, R.J.: 1982, Chem. Phys. 66, 261

34. Runau, R., Peyerimhoff, S.D. and Buenker, R.J.: 1977, J. Mol. Spectry 68, 253

35. Shih, S.K., Butscher, W., Buenker, R.J. and Peyerimhoff, S.D.: 1978, Chem. Phys. 29, 241

36. Danylewych, L.L. and Nicholls, R.W.: 1974, Proc. Roy. Soc. A 339, 197

37. Tatarczyk, T., Fink, E.H. and Becker, K.H.: 1976, Chem. Phys. Letters 40, 126;
Tatarczyk, T.: 1976, Ph.D. thesis, University of Bonn

38. Zeitz, M., Peyerimhoff, S.D. and Buenker, R.J.: 1978, Chem. Phys. Letters 58, 487

39. Chabalowski, C.F., Buenker, R.J. and Peyerimhoff, S.D.: 1981, Chem. Phys. Letters 83, 441

40. Arnold, J.O. and Langhoff, S.R.: 1977, J. Quant. Spectry. Radiative Transfer 19, 461

41. Cooper, D.M. and Nicholls, R.W.: 1975, J. Quant. Spectr. Rad. Transf. 15, 139

42. Werner, H.J. and Rosmus, P.: 1980, J. Chem. Phys. 73, 2319

43. Sileo, R.N. and Cool, T.A.: 1976, J. Chem. Phys. 65, 117

44. Ermler, W.C., Lee, Y.S., Pitzer, K.S. and Winter, N.: 1978, J. Chem. Phys. 69, 976;
Wadt, W.R., Hay, P.J. and Kahn, L.R.: 1978, J. Chem. Phys. 68, 1752;
Martin, R.L. and Hay, P.J.: 1981, J. Chem. Phys. 75, 4539

45. Heß, B.A., Marian, C.M., Buenker, R.J. and Peyerimhoff, S.D.: 1982, Chem. Phys. Letters

46. Klotz, R.: 1982, Diplomarbeit Bonn

47. Heß, B.A., Marian, C.M., Buenker, R.J. and Peyerimhoff, S.D.: 1982, Chem. Phys.

48. Marian, C.M., Heß, B.A., Marian, R., Peyerimhoff, S.D. and Buenker R.J.: 1982, Mol. Phys.

49. Johns, J.W.C., Ramsay, D.A. and Ross, S.C.: 1976, Can. J. Phys. 54, 1804

50. Buenker, R.J., Perić, M., Peyerimhoff, S.D. and Marian, R.: 1981, Mol. Phys. 43, 987

51. Jungen, Ch., Hallin, K.-E.J. and Merer, A.J.: 1980, Mol. Phys. 40, 25

52. Galloy, C. and Lorquet, J.C.: 1977, J. Chem. Phys. 67, 4672

53. Hirsch, G., Bruna, P.J., Buenker, R.J. and Peyerimhoff, S.D.: 1980, Chem. Phys. 45, 335

54. Buenker, R.J.: 1978, Gazz. Chim. Ital. 108, 245

55. Petrongolo, C., Buenker, R.J. and Peyerimhoff, S.D.: 1982, J. Chem. Phys. 76, 3655

POTENTIAL ENERGY SURFACES OF CHEMICAL REACTIONS

K. Morokuma, S. Kato, K. Kitaura*, S. Obara, K. Ohta and M. Hanamura†
Institute for Molecular Science, Myodaiji, Okazaki 444, Japan

INTRODUCTION

Understanding mechanisms and rates of chemical reactions is one of the goals of theoretical molecular science. Ab initio molecular orbital studies can contribute substantially to this purpose by providing qualitative and sometimes quantitative information of potential energy surfaces on which chemical reactions take place [1].

For an N atomic system, the potential energy surface $V(\tilde{R})$ is a function of 3N nuclear coordinates or 3N-6 degrees of freedom. When N is large, even for N=4 where 3N-6=6, mapping of the entire potential surface for all degrees of freedom neither is practical nor gives any perspective. It is even unnecessary; what is essential is to clarify critical characteristics of the potential energy surface that control the rate and mechanism of the reaction. The geometry and energy of the transition state is one of such characteristics. Yet the accurate determination of transition state geometries had been limited until recently to triatomic and a few tetraatomic systems.

The latest development of methods and computer programs of the energy derivatives has revolutionalized the search of equilibrium and transition state geometries in ab initio molecular orbital studies. In the usual molecular orbital calculation, the approximate wavefunction $\Psi(\tilde{r};\tilde{R})$ and its energy $E(\tilde{R})$ are calculated for a given nuclear configuration $\tilde{R}$. If the energy derivative method is employed, its first derivatives or gradient $\partial E/\partial \tilde{R}$ with respect to the nuclear coordinates and sometimes its second derivatives or force constants $\partial^2 E/\partial \tilde{R}\partial \tilde{R}$ are calculated analytically at the same time. In the last few years the applicability of the derivative methods has been expanded dramatically. As shown below schematically, the first derivative can be evaluated for almost all commonly used forms of wavefunctions. It can incorporate the effective core potential approximation where core electrons are replaced by an effective potential. Though the second derivatives can be calculated at present only for a Hartree-Fock (HF) wavefunction, in principle

---

* Present address: Department of Chemistry, Osaka City University
† Present address: Department of Chemistry, Tohoku University

P.-O. Löwdin and B. Pullman (eds.), New Horizons of Quantum Chemistry, 221–241.

Method	1st Derivatives	2nd Derivatives
Hartree-Fock	[2][a]	[5]
Multiconfigurational SCF	[3]	√
CI, Møller-Plesset	[4,5]	√
Effective Core Potential	[6]	√

a) Numbers in brackets are references. A check mark means being developed.

they can be extended to other types of wavefunctions, and practical methods and computer programs are being development for this purpose.

Once the energy derivatives are in hand, one can use them for various purposes. The geometry optimization, the full optimization of all the degrees of freedom, can be carried out efficiently to locate stationary points on the potential surface, such as the equilibrium and saddle point (transition state) geometries. The normal coordinate analysis can be performed with ab initio force constants, and the characterization of the equilibrium and the saddle point can be accomplished by finding no and one imaginary normal vibrational frequency, respectively. The normal coordinate with the imaginary frequency is the reaction coordinate at the saddle point. The energy derivatives can be used to define and calculate the intrinsic reaction coordinate that connects the reactant, the saddle point and the product smoothly.

In the present paper, we will show at first an example of geometry optimization: The transition states for the reaction of silaethylene $H_2Si{=}CH_2$ with H_2O and for dimerization of silaethylene. Then we will review the concept of the intrinsic reaction coordinate. The intrinsic reaction coordinate will be used to show that two transition states, found in thermal rearrangement of C_2H_3N, actually connect different reactants and products. We will also use this concept to discuss the E-V energy transfer between $Hg^*(^3P)$ and CO. Because the energy derivative method allows to optimize many nuclear degress of freedom efficiently, one is tempted to challenge chemical reactions involving many atoms. Such an example will be shown for the S_N2 reaction in which the attacking nucleophile is solvated by one or two water molecules. At the end we will propose an approximate method called the effective fragment potential method which is intended to be used for geometry optimization and potential surface study of large systems such as reactions of solvated systems and transition metal complexes, and molecular interactions and reactions of biological systems. The paper will be concluded with a few final remarks.

TRANSITION STATES FOR REACTIONS OF SILAETHYLENE

The double bond containing a silicon atom or atoms has long been proposed, but had until recently eluded isolation and identification. Recent ab initio calculations have indicated for unsubstituted silaethlene $H_2Si{=}CH_2$, the isomeric silylene $HSi{-}CH_3$ is lower or comparable in energy than silaethylene itself [7]. However Nagase, Hanamura and Morokuma have shown that, when alkyl or fluoro substituents are introduced, silaethylene $X_2Si=CH_2$ is more stable than the corresponding silylene and is thermodynamically stable [8]. The latest synthesis of silaethylene with bulky alkyl substituents by Brook et al. comfirms such a prediction [9]. Silaethylenes are also known to be kinetically unstable; they react very easily with polar reagents such as water and also with themselves to dimerize. Bulky substituents prevent reactions and dimerization.

In order to study high reactivities of silaethylene, we have calculated the potential energy surface for reaction of $X_2Si=CH_2$ with H_2O and for dimerization of $H_2Si=CH_2$ [10]. The geometries for the reactants, the reactant complex (an equilibrium geometry), the transition state and the product are optimized with the HF method in the 6-31G basis set. As is shown in Fig.1 for X=H the reactant complex for either reaction is a weak donor-acceptor complex between the reactants. The transition state for either reaction is a four-center transition state in an early stage of reaction. The reaction coordinate at the transition state, shown by arrows in Fig.1, indicates clearly that two new bonds are being formed simultaneously and concertedly while two old bonds are being broken. Silaethylene is strongly polar like $X_2Si^{+\delta}=CH_2^{-\delta}$, and the symmetry rule that prevented a 2+2 cycloaddition of two ethylene molecules does not apply to silaethylene. Relative energies in Table 1 indicate that both reactions should indeed take place with low or no activation energy.

Table 1. Energetics of Reactions of $X_2Si=CH_2$ a)

	Reaction with H_2O				Dimerization	
	X=H		X=F		X=H	
Method	TS	Product	TS	Product	TS	Product
HF(6-31G)	6.7	-70.6	-14.7	-102.6	11.7	-82.1
CI(6-31G) b)	13.5	-53.4	-7.5	-89.1	12.3	-69.1
HF(6-31G**)	11.5	-76.5	-6.5	-102.7	2.7	-89.8

a) The transition state (TS) and the product energies are relative to the reactants, in kcal/mole. The geometries optimized at HF(6-31G) are used.

b) The single reference + single and double excitations + Davidson correction.

II INTRINSIC REACTION COORDINATE

The intrinsic reaction coordinate (IRC) is defined as the non-energy conserving, zero-kinetic-energy classical trajectory that starts at the saddle point to both directions of the normal coordinate having an imaginary frequency [11]. It reaches smoothly the reactants in one direction and the products in the other direction. If one uses the coordinate system $\{q_i\}$ in which the kinetic energy is expressed as $T=\sum_i \dot{q}_i^2$, IRC is

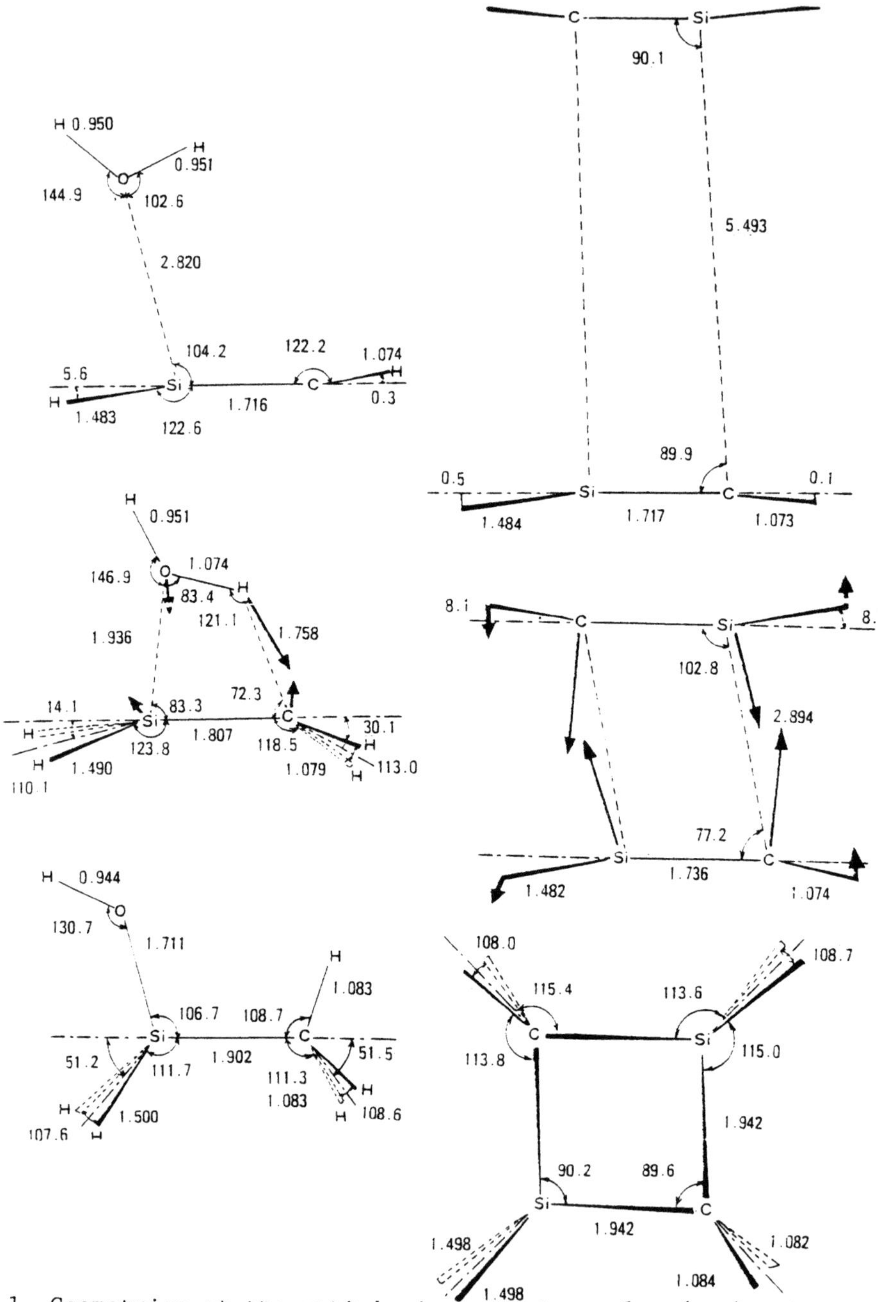

Fig. 1 Geometries of the optimized reactant complex (top), the transition state (middle) and the product (bottom) for the reaction $SiH_2CH_2 + H_2O \rightarrow HOSiH_2CH_3$ (left) and $2SiH_2CH_2 \rightarrow (SiH_2CH_2)_2$ (right). The arrows on the transition states show the vectors of the reaction coordinate. Distances are in Å and angles are in degrees.

equivalent to the steepest descent path which intersects perpendicularly every contour of the potential energy surface. The mass-weighted Cartesian coordinate $q_i = m_i^{1/2} x_i$ is such a coordinate system. Since the kinetic energy depends on the mass, IRC is mass dependent, probing somewhat different part of the potential surface. The word 'minimum energy path' is sometimes used, but it could be misleading without specifying the coordinate system used.

The IRC is a smooth (one-dimensional) curve on a multidimensional surface. Glancing at the trace of IRC gives a good perspective of how the atoms move during the reaction. Fig.2 is the trace of IRC for the unimolecular dissociation reaction: $H_2C{=}CHF \rightarrow HC{\equiv}CH + HF$ [12]. At the beginning of the reaction, the fluorine atom starts to move, which is followed by the motion of H_1. After a substantial move of F and H_1, the two hydrogen atoms H_2 and H_3 in the β-position begin to move, dominating the IRC motion in the neighborhood of the transition state. The motion of H_3 toward F quickly shortens the HF bond length. This trace of IRC suggests qualitatively that the HF molecule formed will have a substantial vibrational excitation.

Another merit of IRC is that it clearly identifies the reactant and the product for a given saddle point. Even if a saddle point is obtained, it is not obvious just from its structure and normal coordinates which reactant and product it leads to. For instance Lohr, Hanamura and Morokuma found two saddle points, shown in Fig.3, on the ground state potential energy surface of rearrangement of C_2H_3N at the 4-31G HF level [13]. The IRC calculation from both of them has re-

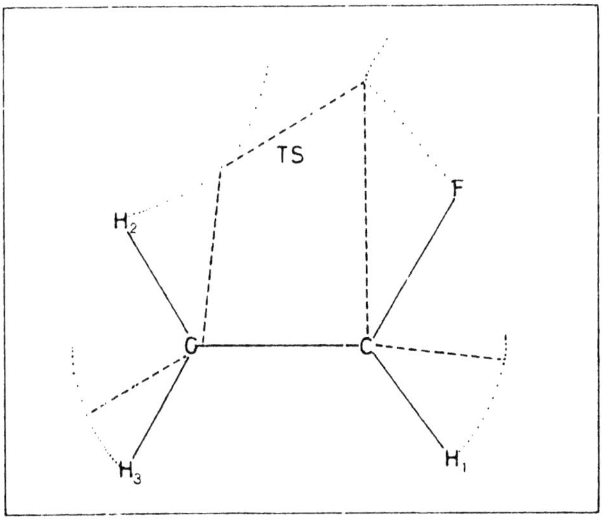

Fig. 2 Change of geometry along IRC for $CH_2{=}CHF \rightarrow CH{\equiv}CH + HF$. The position of the F-substituted C and the direction of the CC axis are fixed. The illustrated positions of atoms are those in the reactant, and each dot denotes the position along IRC after each $\Delta s = 0.4\,amu^{1/2}$ bohr. The positions shown by the broken lines are those at the transition state.

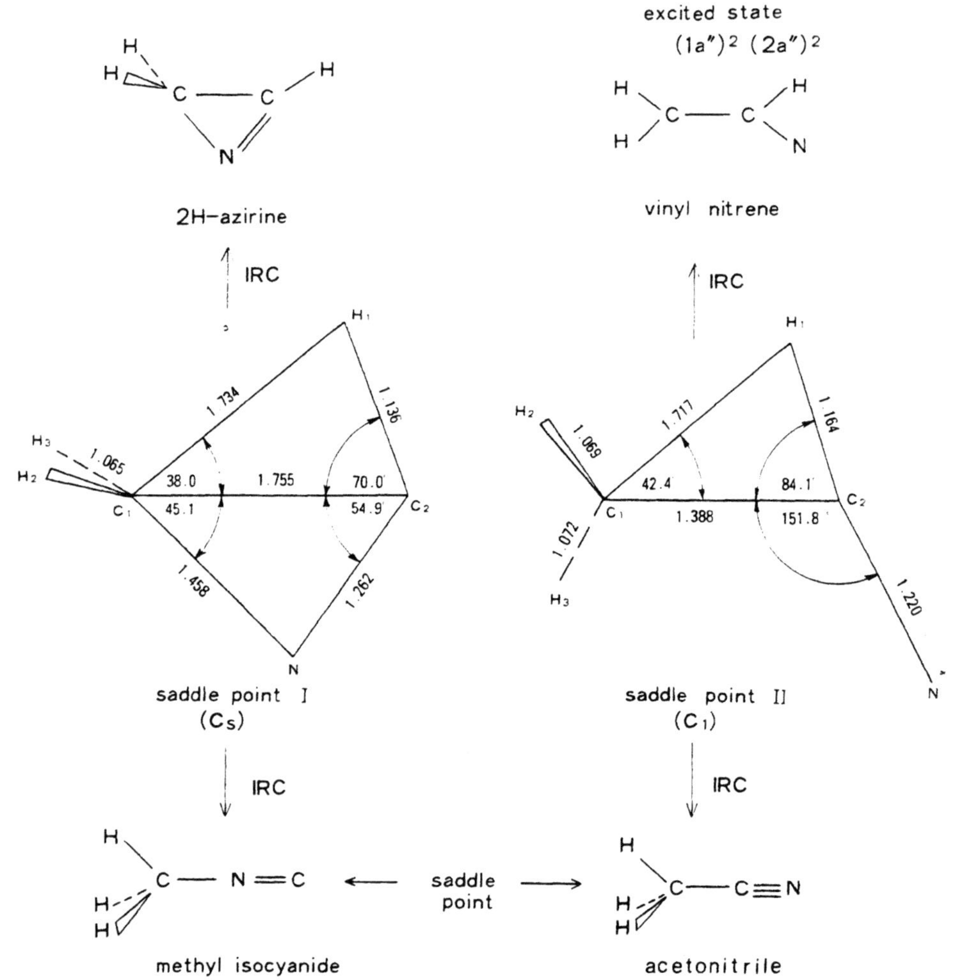

Fig. 3 Some stationary points on the rearrangement potential surface of C_2H_3N. Distances are in Å and angles are in degrees.

solved clearly that the transition state I connects through a path of C_1 symmetry the 2H-azirine and methyl isocyanide, and the transition state II connects through another path of C_s symmetry an excited state $(1a'')^2(2a'')^2$ of vinyl nitrene and acetonitrile. Despite an extensive search there has not been found any saddle point connecting directly 2H-azirine and acetonitrile.

The IRC characteristics, such as the magnitude and direction of its curvature, dictate to a large extent the dynamics of the reaction. If the mass-weighted Cartesian coordinate $\{q_i\}$ is used, the IRC is defined as a solution of the simultaneous differential equation:

$$\frac{dq_i}{ds} = - \frac{\partial W}{\partial q_i} \bigg/ \frac{\partial W}{\partial s} \qquad (i=1,2,\ldots,3N)$$

where $$\frac{\partial W}{\partial s} = [\sum_{i=1}^{3N} (\frac{\partial W}{\partial q_i})^2]^{1/2},$$

W is the potential energy and s is the length along the IRC. For any value of s, one can calculate the 3N-7 normal coordinates, Q_i, which are perpendicular to IRC. Then one can consider s and 3N-7 Q_i's as independent variables, and rewrite the Hamiltonian of the system in the vicinity of the IRC as [14]:

$$H = \frac{1}{[1+\kappa(s)\rho]^2} \frac{Ps^2}{2} + V(s) + \sum_{i=1}^{3N-7} [- \frac{1}{2} \frac{\partial^2}{\partial Qi^2} + \frac{1}{2}\omega_i(s)Q_i^2]$$

where κ is the curvature of IRC, i.e. the magnitude of a vector $\underset{\sim}{\kappa}$ which is parallel to the principal normal of the IRC.

$$\underset{\sim}{\kappa} = (\frac{d^2x_1}{ds^2}, \frac{d^2x_2}{ds^2}, \ldots\ldots, \frac{d^2x_{3N}}{ds^2})$$

ρ is the coordinate in the direction of $\underset{\sim}{\kappa}$ expressed in terms of Q_i's. If $\kappa(s)=0$ throughout the IRC from the transition state to the product, there will be no coupling between the motion along s and the motion along Q_i's. The system is like a boy sliding down a simple straight slide, and his potential energy at the top of the slide will be all converted into the kinetic energy along s. Non-zero values of $\kappa(s)$ cause the coupling of s and Q_i's. The system is like a boy on a curved slide, and as he slides down, the centrifugal force pushes him away from the bottom of the slide width and causes him oscillate along the direction perpendicular to the slide length. The oscillation in the multidimensional case takes place to the direction of $\underset{\sim}{\kappa}$. If $\underset{\sim}{\kappa}$ is projected into normal coordinates or internal coordinates, $\underset{\sim}{\eta}_i$'s,

$$\underset{\sim}{\kappa} = \sum_i \kappa_i(s)\underset{\sim}{\eta}_i$$

the magnitude of $\kappa_i(s)$ indicates into which coordinate the transitional energy is likely to be converted. The energy transfer during a chemical reaction occurs in the region where $\kappa(s)$ is large, and the modes having large $\kappa_i(s)$ are accepting modes.

The analysis of IRC characteristics has been used to discuss semi-quantitatively the energy distribution in translation and internal modes in the following reactions [12,14].

$$CH_3CH_2F \rightarrow CH_2{=}CH_2 + HF$$
$$CH_2{=}CHF \rightarrow CH{\equiv}CH + HF$$
$$CH_2{=}CHF \rightarrow CH_2{=}C{:} + HF$$

In the present paper we will discuss an application of the reaction coordinate concept to a nonreactive electronic → vibrational energy transfer collision.

E-V ENERGY TRANSFER IN Hg^* + CO

The reaction studied is a collision of Hg in excited states with CO to produce the ground state Hg and vibrationally excited CO.

$$Hg(^3P_{0,1}) + CO(0) \rightarrow Hg(^1S_0) + CO(v)$$

The recent experiment by Horiguchi and Tsuchiya has given a non-linear surprisal plot for the product vibrational energy distribution [15], and non-linear surprisal has often been associated with more than one elementary processes [16].

Kato, Jaffe, Komornicki and Morokuma have studied the relevant potential energy surfaces with a double zeta basis set ((9s5p/4s2p) for C and O and (3s3p4d/2s2p2d) for Hg) with the relativistic effective core potential for Hg [17]. A CI calculation with single and double excitations from multiple reference functions, selected and extrapolated with a perturbation method, has been adopted. The spin-orbit interaction was taken into account semiempirically using the experimental atomic value for Hg. The energy derivative method was not used.

The calculated potential energy curve for a collinear approach (γ =0) is shown in Fig.4. Hg($^3P_1$ or $^3P_0$) approaches CO on attractive surfaces,

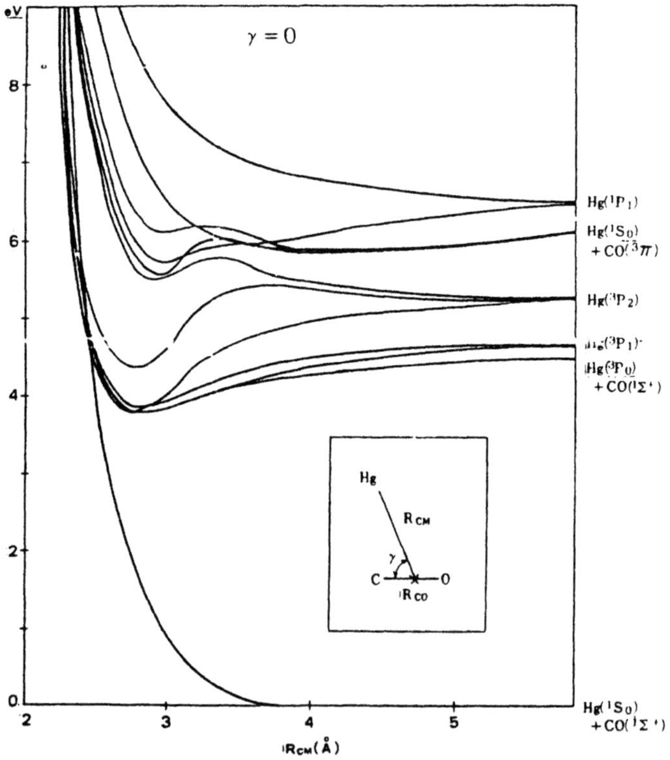

Fig. 4 The potential energy curves for a linear (γ =0) collision of Hg + CO at R_{CO} = 1.129Å.

which at around R_{CM} = 2.4Å cross with the repulsive ground state curve. The effect of the spin-orbit interaction, which is large at large R_{CM}, becomes small for small R_{CM} as an odd electron is transfered from Hg to a CO π^* orbital. Therefore we can and will discuss at small R_{CM} the energy levels of different states without the spin-orbit effect; the spin-orbit effect then only influences the transition probability. The angular dependence indicates that the crossing between the states of interest, $1^3A'$ or $1^3A''$ and $1^1A'$, is more likely to take place at a col-linear geometry (γ =0). The two dimensional potential surfaces at γ =0 for the states of interest, $1^1\Sigma$ and $1^3\Pi$, are shown in Fig.5. With respect to R_{CO} and R_{CM} the upper state is seen to have a flat well, which is also flat with respect to the angle γ. The lowest energy point of the seam of crossing between the two surfaces, marked as A in Fig.5 and occurring at about 4eV in the vicinity of the upper-state well, is the point where the electronic transition from the upper to the lower state is most likely to take place. The seam at this point is found to be parallel to a potential energy contour both in the ground and the excited state. Starting at the point A, we have calculated the zero-energy trajectory. Since the starting point is not the transition state, this is not an IRC in the usual sense but is an effective reaction coordinate after electronic transition. The path, shown also in Fig.5, is straight at first, then bends rather abruptly just before the distance from the point A reaches 1 $(amu)^{1/2}$ bohr, and then is linear again until it reaches

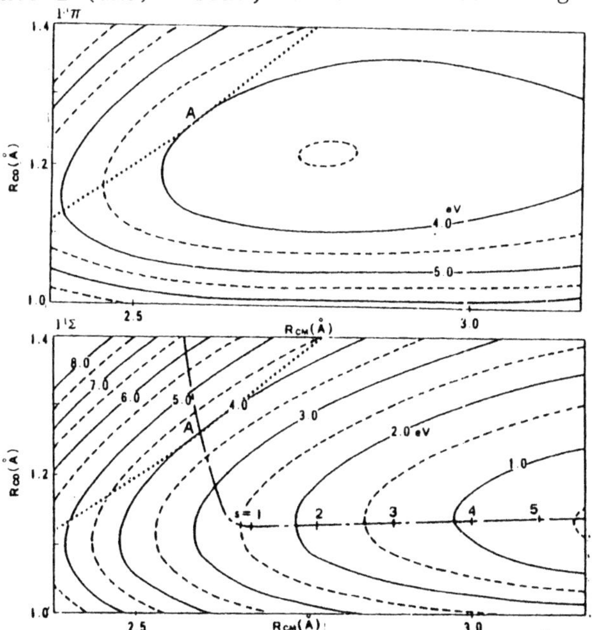

Fig. 5 The potential energy surfaces for the two lowest states, $1^1\Sigma$ and $1^3\Pi$, for a collinear (γ=0) collision of Hg + CO without spin-orbit interaction. The dotted lines are the seam of crossing between the two surfaces, and the point A is the lowest point on the seam. The dot-dash lines on $1^1\Sigma$ are the reaction coordinate starting from A, and the length s along it is shown in $amu^{1/2}$ bohr.

the ground state products. The situation is seen clearly in Fig.6 which shows the curvature $\kappa(s)$ of the reaction coordinate and the normal frequency of the vibrational coordinate perpendicular to the reaction coordinate. The vibrational excitation of CO is expected to take place at the narrow region where $\kappa(s)$ is large, and the potential energy drop of about 1.5eV from the seam to the bend as well as the kinetic and vibrational energy at the seam will be the maximum of energy partitioned into the CO vibrational.

A set of normal (energy-conserving) classical trajectories in two degrees of freedom were calculated on the ground state surface, starting at the seam with an initial kinetic energy E_{KIN} along the direction of s and with an initial Boltzmann distribution of vibrational energy at a temperature T_ω along the direction perpendicular to s. The resultant vibrational energy distribution P(v) in the direction perpendicular to s is shown in Fig.7 as a function of s. One clearly sees that the vibrational excitation actually occurs in the narrow range of $0.5<s<1$. After s=1, there is even a small gradual deexcitation of vibration. The final vibrational distribution thus obtained is shown in Fig.8 for a few initial kinetic energy E_{KIN} and the vibrational temperature T_ω. The experimental distribution is reasonably reproduced at E_{KIN}=0.4eV and T_ω=1000K.

S_N2 REACTION IN HYDRATION CLUSTER

A definite advantage of the energy derivative method is its applicability to a system of many degrees of freedom. We have previously determined a transition state for $(PH_3)_2Pt + H_2 \rightarrow (PH_3)_2Pt(H)_2$, an eleven atom system, where all the degrees of freedom have been optimized [18]. This ability prompted us to look into chemical reactions involving some solvent molecules.

The S_N2 nucleophilic substitution reaction schematically shown as

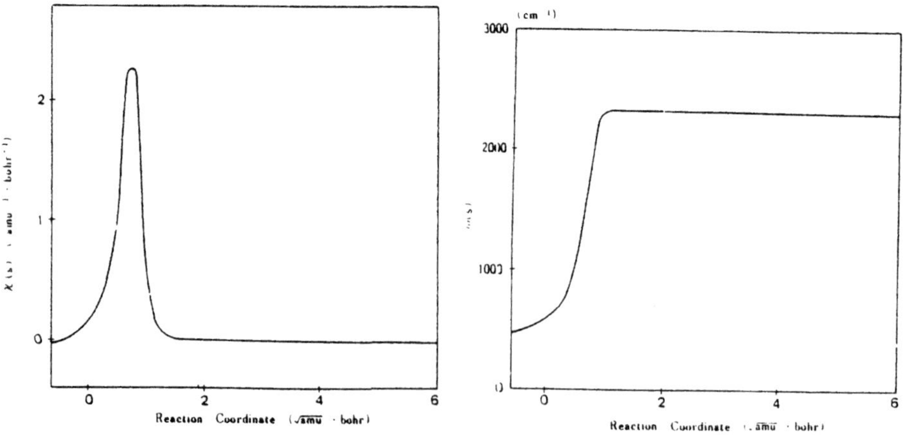

Fig. 6 The curvature $\kappa(s)$ and the vibrational frequency $\omega(s)$ along the reaction coordinate. See Fig.5 for the origin of s.

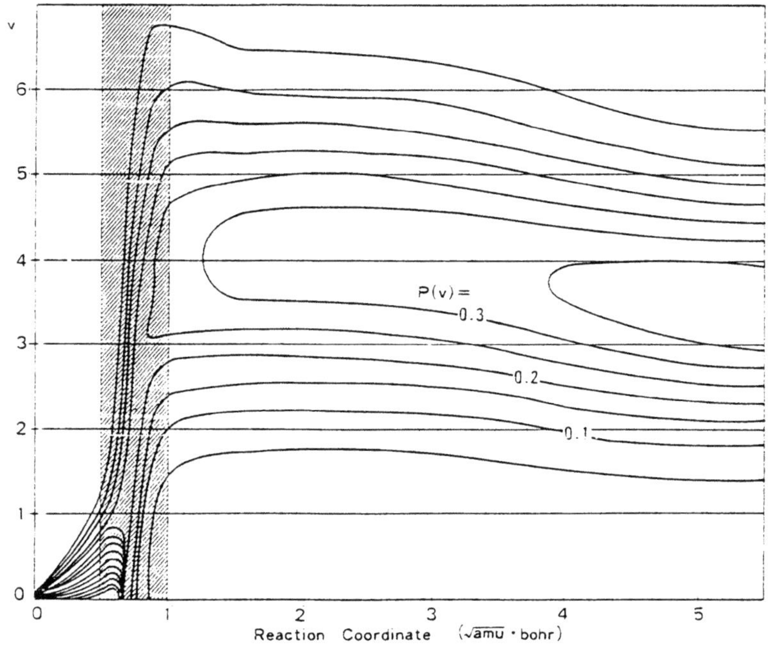

Fig. 7 The probability of finding the system in the state v of the vibration to the direction perpendicular to s at a given reaction coordinate value s. Obtained from trajectories in two degrees of freedom with $E_{KIN}=0$ and $T_{\omega}=300K$. The shaded area, $0<s<1$, shows the region where the vibrational excitation mainly takes place.

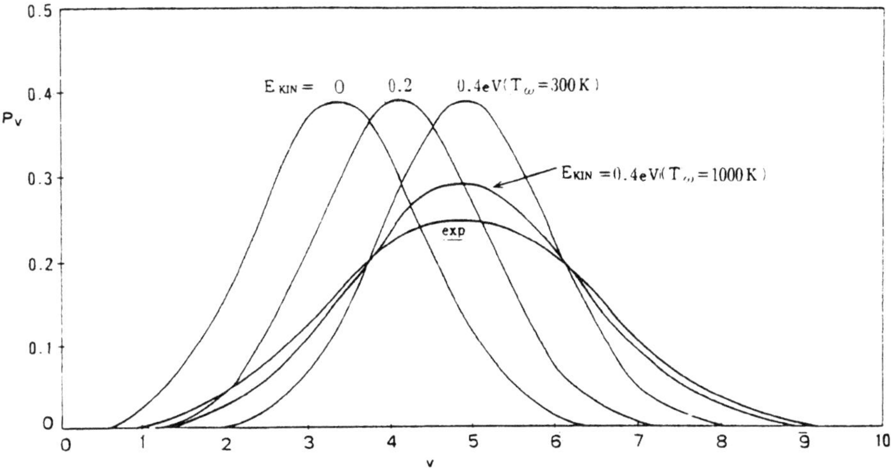

Fig. 8 The product vibrational probability distributions of CO. The calculated distributions obtained from trajectories in two degrees of freedom with the initial kinetic energy E_{KIN} and the vibrational temperature T_{ω}.

$$X^- + H_3CY \rightarrow [\, X{-}\overset{H}{\underset{HH}{C}}{-}Y \,]^- \rightarrow XCH_3 + Y^-$$

is one of the most extensively studied reactions in solution. There are many recent gas phase studies as well. A recent experiment using flowing afterglow technique by Bohme et al. has determined for the first time the rate constants of the following reactions of hydrated clusters [19].

$$OH^-(H_2O)_n + H_3CBr \rightarrow HOCH_3 + Br^-(H_2O)_n, \qquad n=0,1,2 \text{ and } 3$$

The rate constants: $(1.0\pm0.2)\times10^{-9}$ for n=0, $(6.3\pm2.5)\times10^{-10}$ for n=1, $(2\pm1)\times10^{-12}$ for n=2 and $\leq 2\times10^{-13}$ cm^3 $molecule^{-1}$ s^{-1} for n=3, are still many orders of magnitude larger than the solution rate, 2.3×10^{-25}, but decrease progressively with stepwise addition of water molecules. This change has been interpreted in terms of a qualitative model shown in Fig.9. The reactants at first form a reactant complex, which goes over the CH_3 transfer-inversion barrier to reach the product complex which in turn dissociates into products. As the nucleophile is solvated more, the relative stabilization of the reactant complex decreases and the barrier height for CH_3 transfer-inversion increases, until in solution no stable reactant and product complexes are formed and only a net barrier is found.

We have carried out an extensive potential energy search for the following symmetric S_N2 reaction [20]:

$$Cl^-(H_2O)_n + H_3CCl \rightarrow ClCH_3 + Cl^-(H_2O)_n, \qquad n=0,1 \text{ and } 2$$

Though no cluster experiment exists, there are gas phase as well solution studies for this reaction using Cl isotopes. For n=1 and 2, geometry optimization was carried out for equilibrium and transition states with

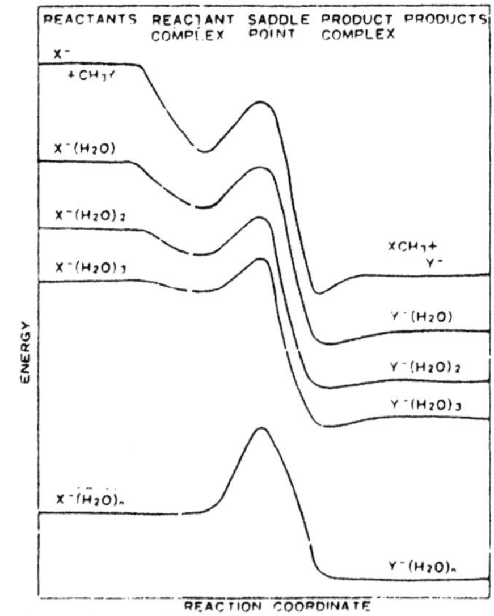

Fig. 9 The quantitative picture of the potential energy surface proposed for S_N2 reaction in clusters and solution.

the 3-21G basis set, for some cases supplemented with a diffuse p function on Cl, at the Hartree-Fock level. CI calculations were also performed for a few cases with a single reference and all single and double excited configurations with frozen core, augmented with a Davidson correction. There exist many ab initio studies for S_N2 reactions without solvent; to be noted in particular is Keil and Ahlrichs' extensive studies with large basis sets and electron correlation [21].

Fig.10. gives for n=0 the energies of the reactant complex,

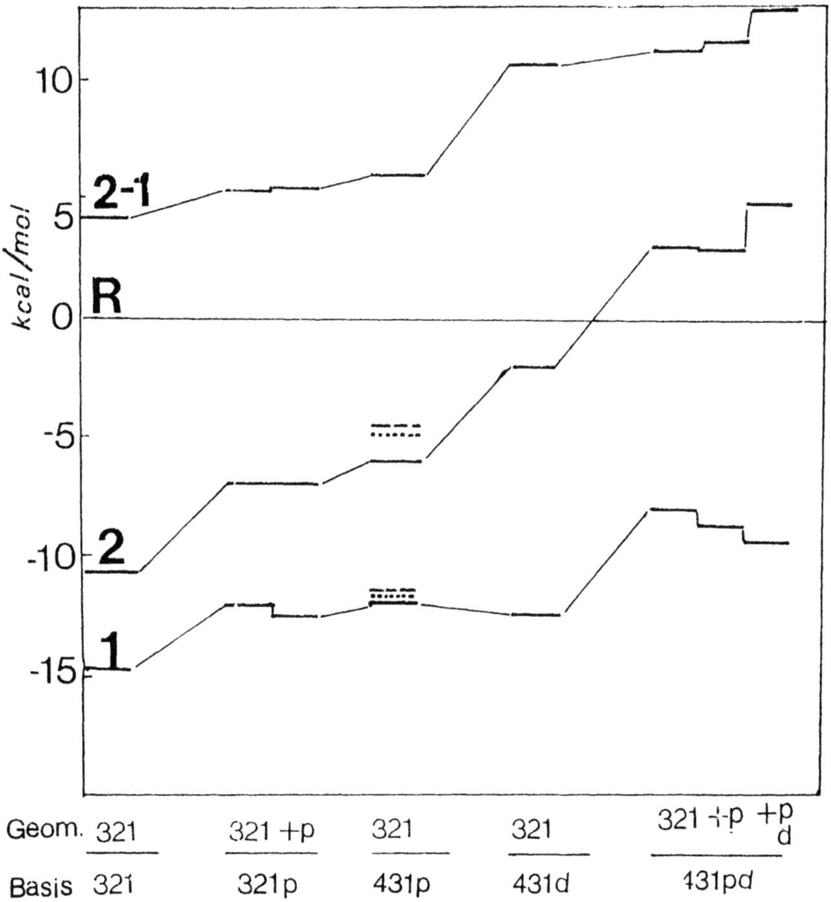

Fig. 10 The basis set and correlation effect on the relative energy of the reactant complex 1 and the transition state 2, relative to the reactants R. Geom denotes the basis set used for geometry optimization, and Basis denotes the one used for energy calculation, all at the Hartree-Fock level. 321 and 431 denote the 3-21G and the 4-31G basis set, and p and d indicate that the basis set was augmented with a set of diffuse p (α=0.06 on Cl) and a set of polarization (d with α=0.6 for Cl and with α=0.65 for C) functions, respectively. The dashed and dotted lines are CI results without and with Davidson correction, respectively.

$Cl^-...H_3CCl$, **1**, and the transition state, $[Cl...CH_3...Cl]^-$, **2**, relative to the reactants Cl^- + H_3CCl for various methods of calculation. The effect of geometry reoptimization for each basis set on the relative energy is small as well as the effect of correlation energy. However, the effect of basis set on the relative energy is extremely large; the relative energy of **2** increases from -11kcal/mol through zero to +5kcal/mol as one increases the size of basis set, and it does not seem to have converged. Experiments suggest that **2** should be lower in energy than the reactants [19], and therefore there appears to be a serious discrepancy between theory and experiment on the absolute barrier height.

Fig.11 shows the calculated potential energy profiles. The results

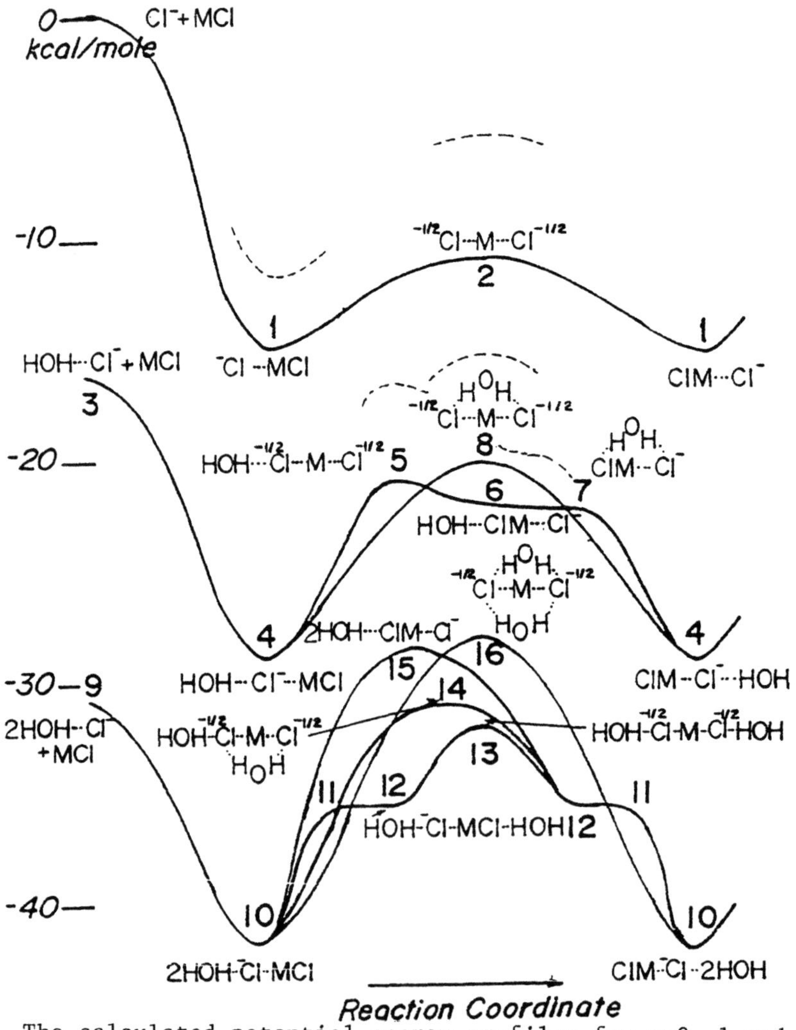

Fig. 11 The calculated potential energy profiles for n=0, 1 and 2, at the 3-21G HF level. The 4-31G + p CI results with Davidson correction at the 3-21G HF geometries are shown by dashed lines.

of CI with a larger basis set are substantially different from 3-21G HF results, but critical energy differences, for instance, between 5 and 8, remain unchanged.

When a hydrated nucleophile is used, in addition to CH_3 transfer-inversion one has to take into account the migration of water molecule(s) from the reactant Cl^- to the product Cl^-. This aspect is not considered in the qualitative picture of Fig.10. For n=1, three possible paths exist. In the simultaneous path, CH_3 transfer-inversion and the H_2O migration takes place simultaneously through the transition state 8. In a sequential path, the CH_3 transfer-inversion takes place first (4→5→6) followed by the H_2O migration (6→7→4). Another sequential path where H_2O migration is followed by CH_3 transfer-inversion is of course energetically equivalent. The simultaneous path is only slightly higher in energy than the sequential path. However, the transition state 8 has a rigid structure where all the reactants are clammed together, while the transition state 5 has a flexible structure where the rotation of H_2O around Cl is relatively unrestricted. Thus, the entropy factor and, as a consequence, the overall rate would be more favorable to the sequential path.

For n=2, several reaction paths have been examined. A more favorable path is a simultaneous CH_3 inversion and H_2O migration through the transition state 14, to form an intermediate 12, which reaches the product complex 10 via an H_2O migration. Slightly lower in energy is the sequential path 10→12→13→12→10, where an H_2O migration takes place first, followed by the CH_3 transfer-inversion and finally by the other H_2O migration. From the entropy argument, the latter path is the most favorable. In this path the energy difference, 12-10, is the energy needed to dehydrate the chloride ion and at the same time to hydrate the Cl end of CH_3Cl, and the energy difference 13-12, is the barrier for CH_3 inversion between monohydrated Cl^- and monohydrated CH_3Cl.

A comparison of potential profiles for n=0, 1 and 2 shows that, as n increases, the stabilization of the reactant complex decreases and the overall barrier relative to the reactant complex as well as relative to the reactants increases, as expected in Fig.10.

The optimized geometries of key species for n=1 and 2 are shown in Figs.12 and 13. Some of the species have low frequency intermolecular vibrational modes, and can move rather freely in the vicinity of the optimized geometries. For instance, in 4, the H_2O molecule keeping the local HOHCl structure more or less unchanged, can swing around in and out of plane of the paper, as if H_2O were connected to H_3CCl through a universal joint Cl.

It is noted in Fig.11 that the H_2O migration processes, 4→7→6 and 10→11→12, take place with little or no activation energy. If one tries to move the H_2O molecule from one end to the other while keeping the Cl...C-Cl axis collinear, the barrier is found to be very high. However, the optimization without restriction leads to a low barrier; in fact, the potential surfaces around 7 and 11 were so flat that it was impossible to locate the transition state. Thus, the path of H_2O migration was followed by fixing a ClCO angle at various values and optimizing all other coordinates. A structure 11' on the path clearly demonstrates the important role of Cl^- played in the process. As the H_2O migration proceeds, the chloride ion accompanies the migrating H_2O molecule by bending away

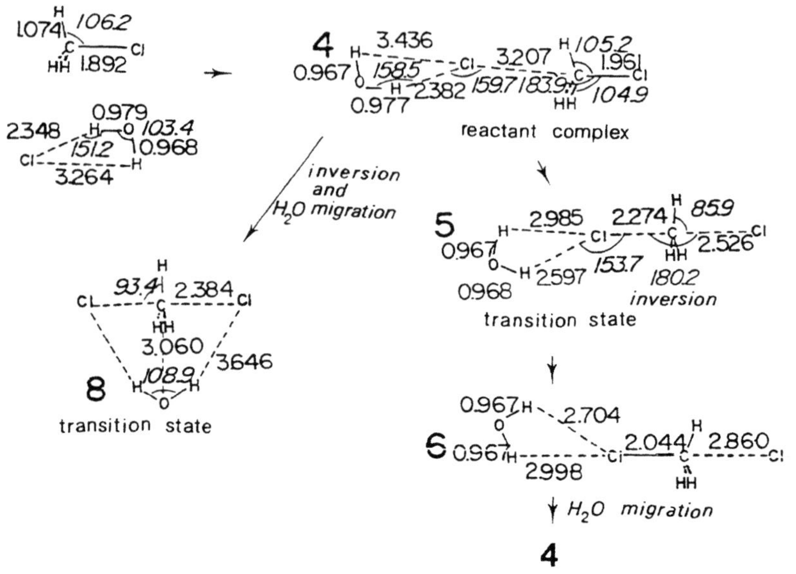

Fig. 12 The 3-21G HF optimized geometries of key species for n=1.

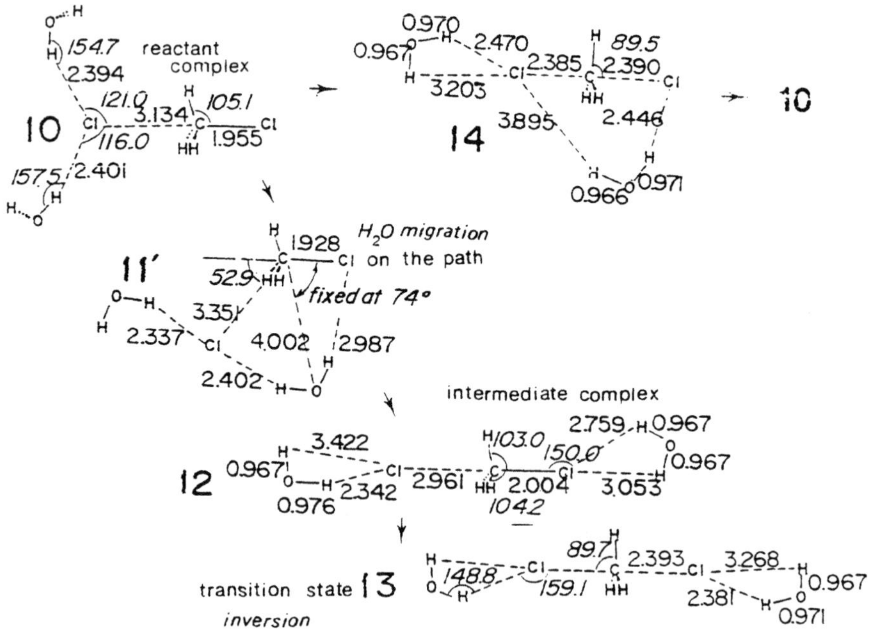

Fig. 13 The 3-21G HF optimized geometries of key species for n=2.

from the C_{3v} axis of CH_3Cl. The energy loss due to bending is obviously more than compensated by the continued hydration energy. Once one end of the H_2O molecule is attracted enough by Cl of CH_3Cl, the chloride ion leaves the H_2O molecule and snaps back to its regular position on the CH_3Cl C_{3v} axis to reach **12**.

From the results of these calculations, one can conjecture the reaction mechanism for larger clusters. If n is not large enough to complete the first solvation shell of X^-, the mechanism would be similar to that for n=2. In the most favorable path (Scheme I below), the reactants $(H_2O)_nX^-$ + MY at first would form a reactant complex $(H_2O)_nX^-$...MY with $\Delta H<0$, and then some of the H_2O molecules would be transfered stepwise to the other end with $\Delta H>0$ but without low or no barrier to form $(H_2O)_{n-m}X^-$...MY$(H_2O)_m$, which then would invert with an activation barrier to form $(H_2O)_{n-m}$XM...$Y^-(H_2O)_m$. The rest of the water molecules would then migrate to the product halide ion to give XM...$Y^-(H_2O)_n$, which would dissociate into the products XM and $Y^-(H_2O)_n$. For a symmetric case X=Y, m would be an integral near n/2. For an asymmetric case X≠Y, m would depends on the interaction energy of X^- and Y^- with H_2O molecules. When the reaction between a hydrated nucleophile $X^-(H_2O)_{n-\ell}$ and a hydrated substrate $MY(H_2O)_\ell$ is investigated, the reactant complex $(H_2O)_{n-\ell}X^-$...$M(H_2O)_\ell$ would go through exothermic H_2O migration processes to join the path described above for $\ell=0$.

If n is equal to or larger than the number f of water molecules required to complete the first hydration shell of X^-, the reaction would have a new feature not present in scheme I. In scheme II, n is assumed, for simplicity, to be equal to f which in turn is assumed to be 4. Since X^- is surrounded by H_2O molecules, the first step of the reactant complex formation involves desolvation, the removal of a first shell H_2O molecule to an outer shell to allow the direct contact of X^- and MY. Therefore this step now would be endothermic. Once the reactant complex is formed, the sequence would be more or less same as in Scheme I. Thus the overall feature of the potential surface beyond this critical hydration number could be substantially different from that of the less hydrated case. A study is in progress along this line.

EFFECTIVE FRAGMENT POTENTIAL METHOD FOR LARGE MOLECULES

In the previous section we described a potential energy surface of reaction including two solvent molecules, a large system at the present ab initio standard. When one is interested in molecular orbital calculations for even larger systems such as solvation clusters, transition metal complexes and enzymic models, one often finds such calculations too expensive to carry out. One recognizes, however, that most of the electrons in large systems are often bystanders, not actively participating in the phenomena of interest but rather only providing some effective field in which the phenomena take place. The fractional charge method by Noell and Morokuma keeps this in mind and replaces secondary solvent molecules by fractional charges, allowing to take their effect on the electronic structure of the solute molecule into account with very little increase in the computational effort[22]. It accounts approximately for the electrostatic contribution to the interaction energy, but does neg-

SCHEME I

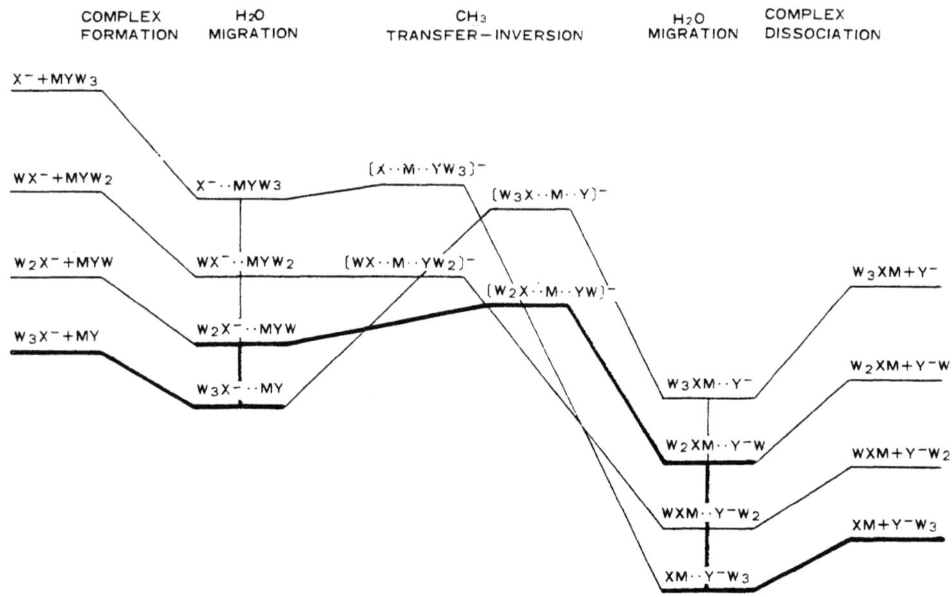

SCHEME II

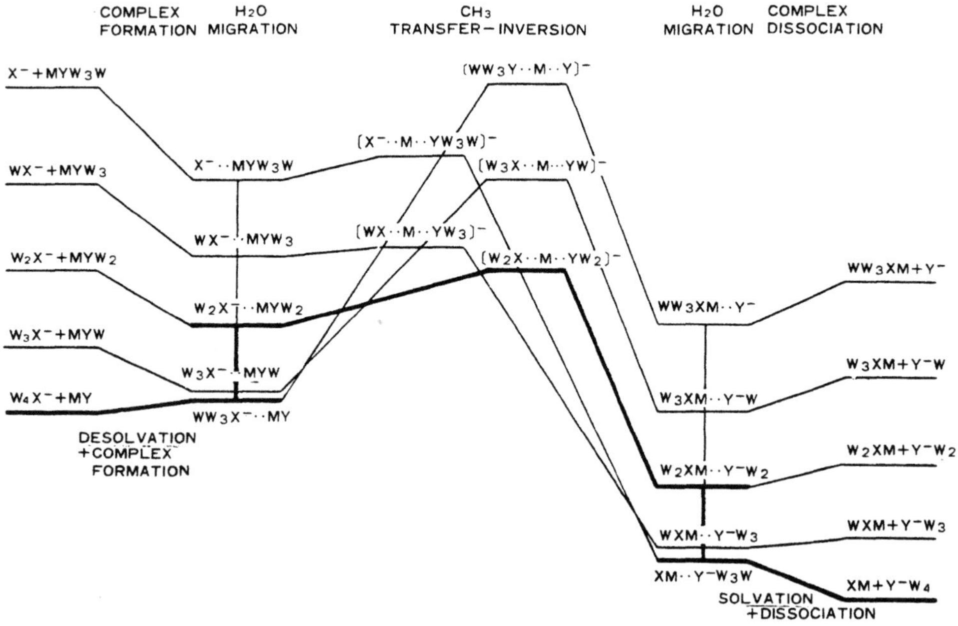

lect completely the exchange repulsion, charge transfer and polarization contributions. The most serious drawback of the method in search for the potential energy surface is the lack of the exchange repulsion; if an geometry optimization is performed, a solvent molecule expressed as fractional charges collapses with solute or other solvent molecules.

We are developing a new approximate method, called the effective fragment potential (EFP) method [23]. The basic idea is, in a sense, similar to that of the atomic effective core potential (ECP) [24], where core electrons of an atom is replaced by a potential consisting of both coulomb attraction and exchange repulsion, and only valence electrons are explicitly considered.

In the EFP approach, for instance, for ammonia only two lone-pair electrons (active electrons) are explicitly included in the calculation, and the rest of the molecule is represented by an effective potential including both coulomb attraction and exchange repulsion. We express the EFP in the following form.

$$V(r_i) = V^C(r_i) + V^P(r_i)$$

$$V^C(r_i) = \sum_I^{center} V_I^C(r_i)$$

$$V_I^C(r_i) = \frac{Z_I^0}{r_{Ii}} - \sum_\xi \frac{1}{r^\xi} \{ \sum_k A_{Ik} (x_i - x_I)^{\ell_{I\xi k}} (y_i - y_I)^{m_{I\xi k}} (z_i - z_I)^{n_{I\xi k}}$$

$$\times \exp(-\alpha_{I\xi k} |r_i - r_I|^2) \}$$

$$V^P(r_i) = \sum_j^{frozen} B_j |\phi_j\rangle\langle\phi_j| \qquad \phi_j\text{: frozen orbitals}$$

The potential parameters are to be determined to reproduce the results of the full ab initio calculation. Any amine, e.g. $((CH_3)_3C)_3N$, would be a two-electron system with a repulsive potential representing the substituents, e.g. bulky tert-butyl groups. For CO, one would explicitly consider, in addition to the carbon σ lone pair, π^* orbitals to accomodate its electron-accepting capability. In calculating intermolecular interaction, the EFP method is expected to take full account of the electrostatic and exchange repulsion interaction, as well as a substantial portion of the charge transfer interaction. One may even device a version where amine is a zero electron system providing just an effective potential; in this version the charge transfer contribution would be ignored.

Since the effective potential contains a correct repulsion, the optimization of the position and orientation of (rigid) solvent molecules as well as the full geometry optimization of the solute molecules can be performed. The energy derivatives can be calculated as easily as in full ab initio and ECP calculations; derivative integrals involving the effective EFP operator can be evaluated without differentiating the complicated operator [6].

An actual parametrization is in progress for NH_3. The 6-31G basis set is used in calculations, and to fit the coulomb potential, two sets of s and p type Gaussian functions are used at each of the nuclei for both $\xi=0$ and 1. Though the final parameters have not been determined, some tests with tentative parameters are encouraging. Fig.14 shows a comparison between a full ab initio and an EFP calculation for $H_3N...H_3N$. In the EFP calculation the electron donor NH_3 is a two-electron EFP system, while all electrons of the electron acceptor NH_3 are explicitly considered.

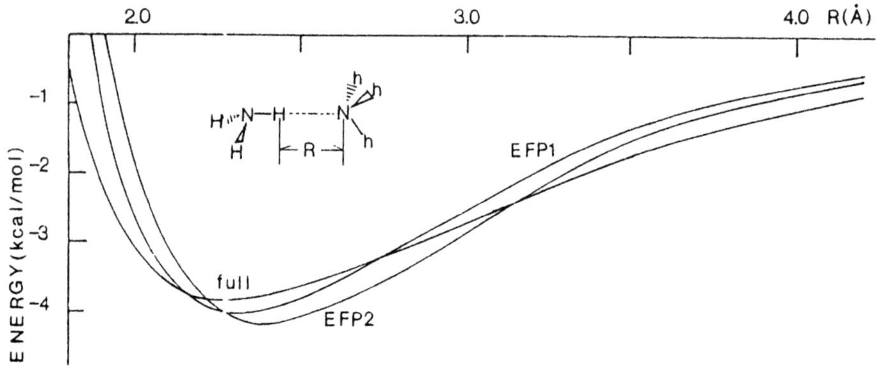

Fig. 14 Preliminary EFP results for $H_2NH...NH_3$. Full is the full ab initio result with the 6-31G basis set. For the electron donor NH_3 EFP1 uses all the AO's, while EFP2 uses only the nitrogen valence (31G) AO's. The geometry of NH_3 is fixed and an NH bond of the electron acceptor is moved along the C_{3v} axis of the electron donor NH_3.

CONCLUDING PEMARKS

In the present paper we have shown that the analytical energy derivative method is an extremely powerful tool for exploring potential energy surfaces of polyatomic reaction systems. The reaction coordinate concept allows, again for polyatomic systems, to probe most important regions of potential energy surfaces, and to extract their characteristics that dictate the dynamical behavior of the systems. With polyatomic systems we do not mean just four or five atom systems but rather even larger systems of several to over ten atoms, where a lot of rich chemistry can be found.

We would like to conclude the paper by pointing out a few areas of chemcal reactions where molecular orbital calculations can make a substantial contributions in the new future: 1) reactions of excited states, 2) reactions of molecular clusters and van der Waals molecules and 3) reactions of transition metal complexes.

ACKNOWLEDGEMENT

The authors are grateful to Prof. L. L. Lohr, Prof. S. Nagase, Drs. R. L. Jaffe, A. Komornicki and Y. Yoshioka for collaboration in some of the work presented in the paper.

REFERENCES

[1] Recent related reviews by the authors: K. Morokuma and S. Kato, in "Potential Energy Surface and Dynamics Calculations", ed. D. G. Truhlar, Plenum, New York (1981), p.243; K. Morokuma, in "Frontiers of Chemistry". ed. K. J. Laidler, Pergamon, Oxford (1982), p.143.
[2] P. Pulay, in "Applications of Electronic Structure Theory", ed. H. F. Schaefer, Plenum, New York (1977), p.153 and references therein; A. Komornicki, K. Ishida, K. Morokuma, R. Ditchfield and M. Conrad, Chem. Phys. Lett. 45, 595 (1977).
[3] S. Kato and K. Morokuma, Chem. Phys. Lett. 65, 19 (1979); J. D. Goddard, N. C. Handy and H. F. Schaefer, J. Chem. Phys. 71, 1525 (1979).
[4] B. R. Brooks, W. D. Laidig, P. Saxe, J. D. Goddard, Y. Yamaguchi and H. F. Schaefer, J. Chem. Phys. 72, 4652 (1980); R. Krishnan, H. B. Schlegel and J. A. Pople, J. Chem. Phys. 72, 4654 (1980).
[5] J. A. Pople, R. Krishnan and H. B. Schlegel, Int. J. Quantum Chem. Symp. 13, 225 (1979).
[6] K. Kitaura, S. Obara and K. Morokuma, Chem. Phys. Lett. 77, 452 (1981).
[7] M. S. Gordon, Chem. Phys. Lett. 54, 9 (1978); J. D. Goddard, Y. Yoshioka and H. F. Schaefer, J. Am. Chem. Soc. 102, 7644 (1980).
[8] M. Hanamura, S. Nagase and K. Morokuma, Tetrahedron Lett. 22, 1813 (1981).
[9] A. G. Brook, F. Abdesaken, B. Gutekunst, G. Gutekunst and R. K. Kallury, Chem. Commun. 191 (1981).
[10] M. Hanamura, S. Nagase and K. Morokuma, to be published; Ann. Rev. IMS 11 (1981).
[11] K. Fukui, S. Kato and H. Fujimoto, J. Am. Chem. Soc. 97, 1 (1975).
[12] S. Kato and K. Morokuma, J. Chem. Phys. 74, 6285 (1981).
[13] L. L. Lohr, M. Hanamura and K. Morokuma, to be published.
[14] S. Kato and K. Morokuma, J. Chem. Phys. 72, 206 (1980).
[15] H. Horiguchi and S. Tsuchiya, J. Chem. Phys. 70, 762 (1979).
[16] R. D. Levine and R. B. Bernstein, Chem. Phys. Lett. 15, 1 (1972).
[17] S. Kato, R. L. Jaffe, A. Komornicki and K. Morokuma, to be published.
[18] K. Kitaura, S. Obara and K. Morokuma, J. Am. Chem. Soc. 103, 2891 (1981).
[19] D. K. Bohme and G. I. Mackay, J. Am. Chem. Soc. 103, 978 (1981).
[20] K. Morokuma, J. Am. Chem. Soc. 104, 3732 (1982); K. Morokuma and S. Obara, to be published.
[21] F. Keil and R. Ahlrichs, J. Am. Chem. Soc. 98, 4787 (1976).
[22] J. O. Noell and K. Morokuma, J. Phys. Chem. 80, 2675 (1976).
[23] K. Kitaura, K. Ohta, Y. Yoshioka and K. Morokuma, to be published.
[24] L. R. Kahn, P. Baybutt and D. G. Truhlar, J. Chem. Phys. 65, 3826 (1976).

AB-INITIO METHODS IN CALCULATIONS OF INTERMOLECULAR INTERACTION ENERGIES

Włodzimierz Kołos

Quantum Chemistry Laboratory, Department of Chemistry
University of Warsaw, 02-093 Warsaw, Poland

A general formulation of the perturbation theory approach to intermolecular interactions is presented and recent progress in computations of various interaction energy components, mostly for model systems, is discussed. In the second part a review is given of the recent work on the nonadditivity of interactions between closed-shell systems.

1. INTRODUCTION

In 1973 at the First International Congress of Quantum Chemistry where, similarly as today, I had the honour to speak on intermolecular interactions (Kołos, 1973), I started by quoting J.O.Hirschfelder (1965) who eight years earlier had complained that this important field of quantum chemistry was in a surprisingly unsatisfactory stage of development. This remark was still valid in 1973. Today, in 1982, one can definitely say that in the nine years that have passed considerable progress has been made in understanding intermolecular interactions and in developing practical formalisms, as well as algorithms, which could be applied to specific systems of physical, chemical or biochemical interest. For obvious reasons I do not intend to give here an exhaustive review of the recent work carried out on the ab initio level in the theory of intermolecular interactions. At the Congress this would be neither feasible nor desirable. Therefore I am going to speak mainly on two selected

P.-O. Löwdin and B. Pullman (eds.), New Horizons of Quantum Chemistry, 243–277.

problems of this broad field of research with which I have been somehow personally connected, i.e. on the perturbation theory approach to intermolecular interactions and on their nonadditivity.

2. PERTURBATION THEORY APPROACH

Intermolecular interactions are weak interactions and they introduce only small changes in the interacting molecules (or closed-shell atoms). Therefore it is natural to develop the theory of intermolecular interactions on the basis of the perturbation theory (PT). This leads, however, to the well known symmetry dilemma. The standard Rayleigh-Schrödinger (RS) PT uses as the unperturbed wavefunction a product of wavefunctions of the interacting molecules. For thi reason RS PT is not able, in a finite order, to yield the exchange energy, and although asymptotically, for large intermolecular separations, it can give reliable results, it fails in the most important region of the van der Waals minimum, where electron exchange is not negligible. In this region a symmetry adapted PT must be used, and a general approach to this problem (Jeziorski and Kołos, 1977; Jeziorski et al., 1978) is presented below.

For simplicity we shall consider the interaction of only two subsystems A and B whose hamiltonians will be denoted H_A and H_B, respectively. In the absence of the interaction the hamiltonian of AB is given by

$$H_0 = H_A + H_B \qquad (1)$$

and the Schrödinger equation for the noninteracting system has the form

$$H_0 \varphi_k = E_k \varphi_k \qquad (2)$$

where

$$\varphi_k = \varphi^A_{n_k} \varphi^B_{m_k} \tag{3}$$

and k = 0 denotes the ground state.

The operator of interaction between A and B is defined as the difference between the complete hamiltonian and H_0

$$V = H - H_0 \tag{4}$$

and the Schrödinger equation for the ground state of the interacting system reads

$$(H_0 + V)\psi = (E_0 + E_{int})\psi \tag{5}$$

where E_{int} denotes the interaction energy, and $E_0 = E_A + E_B$, with E_A, E_B being the lowest eigenvalues of H_A and H_B, respectively.

To solve Eq.(5) a new technique has been proposed (Jeziorski and Kołos, 1977) which has the advantage of a considerable flexibility. It comprises several previously developed perturbational approaches to intermolecular interactions and in a natural way leads to new effective methods.

Let us fix the phase of ψ by imposing the intermediate normalization condition

$$\langle \varphi_0 | \psi \rangle = 1 \tag{6}$$

and introduce the reduced resolvent R_0 of H_0 corresponding to the eigenvalue E_0

$$R_0 = \sum_{k=1} (E_k - E_0)^{-1} |\varphi_k\rangle\langle\varphi_k| \tag{7}$$

It can be easily shown that

$$R_0(H_0 - E_0) = 1 - |\varphi_0\rangle\langle\varphi_0| \tag{8}$$

$$R_0|\varphi_0\rangle\langle\varphi_0| = 0 \tag{9}$$

Eqs. (6-8) enable us to rewrite Eq.(5) in the form

$$\psi = \varphi_0 + R_0(E_{int} - V)\psi \tag{10a}$$

where

$$E_{int} = \langle \varphi_0 | V \psi \rangle \tag{10b}$$

Eqs.(10) can be solved using the iteration procedure

$$\psi^{(n)} = \varphi_0 + R_0(E_{int}^{(n)} - V)\psi^{(n-1)} \tag{11a}$$

$$E_{int}^{(n)} = \langle \varphi_0 | V \psi^{(n-1)} \rangle \tag{11b}$$

When solving Eqs.(11) one has freedom in choosing $\psi^{(0)}$ used to initiate the iteractions. With $\psi^{(0)} = \varphi_0$ one gets the RS PT whereas φ_0 antisymmetrized with respect to permutations of electrons between A and B yields the Murrell-Shaw (1967) - Musher-Amos (1967) (MS-MA) PT. One can, however, generalize Eqs.(11) by introducing two operators F and G

$$\psi^{(n)} = \varphi_0 + R_0(E_{int}^{(n)} - V)F\psi^{(n-1)} \tag{12a}$$

$$E_{int}^{(n)} = \langle \varphi_0 | VG\psi^{(n-1)} \rangle \tag{12b}$$

Eqs.(12) can be shown to be valid if F and G satisfy the relation

$$F\psi = G\psi = \psi \tag{13}$$

for exact ψ. They can be chosen to enforce the proper symmetry on $\psi^{(n-1)}$ thereby speeding up the convergence of the iterative process. Hence the natural choice for F and/or G is the antisymmetrizer A.

Table I. Some possible iteration schemes for solving Eqs. (12)

PT	F	G	$\psi^{(0)}$
RS	1	1	φ_0
MS-MA′	1	A′	φ_0
MS-MA	1	1	$A'\varphi_0$
ISF	A	1	$A'\varphi_0$
EL-HAV	A	A′	$A'\varphi_0$

Table I shows some possible choices of the F and G operators and of $\psi^{(0)}$ (Jeziorski and Kołos, 1977; Jeziorski et al., 1978) which yield various perturbation theories. The operator A' used in this table is related to the antisymmetrizer A by the intermediate normalization (cf. Eq.(6)) which assures the validity of Eq.(10b)

$$A'\psi = \langle \varphi_0 A \psi \rangle^{-1} A\psi \qquad (14)$$

If no symmetry is assumed initially ($\psi^{(0)} = \varphi_0$) and no symmetry is enforced in the iterations (F = G = 1) one gets the RS PT, i.e. $E_{int}^{(n)}$ is equal to the sum of the first n corrections in RS PT plus some terms of order higher than n. The strongest symmetry forcing is seen to produce the Eisenschitz-London (1930) - Hirschfelder (1967) - van der Avoird (1967) (EL-HAV) PT. By releasing one of the symmetry constrains in Eqs.(12) one gets the intermediate symmetry forcing (ISF) PT (Jeziorski and Kołos, 1977; Jeziorski et al., 1978) which is free of a serious drawback of the EL-HAV PT (see below).

It has already been mentioned that by using only an antisymmetrized initial wavefunction one gets the MS-MA PT. One can also show, however, that the same PT results if one uses $\psi^{(0)} = \varphi_0$, F = 1 and G = A' (Jeziorski and Kołos, 1977). This alternative formulation, denoted MS-MA$'$ in Table I, is very convenient in practical applications. In this case there is no antisymmetrizer in Eq.(12a). Thus the same simple algorithm can be used to calculate $\psi^{(n)}$ as in the RS PT. An interesting modification of the MS-MA$'$ method is also possible. In the MS-MA$'$ PT the antisymmetrizer is utilized to calculate $E_{int}^{(n)}$ from Eq.(12b). One can use, however, in Eq.(12a) not $E_{int}^{(n)}$ obtained from Eq.(12b) but the value computed without antisymmetrizing $\psi^{(n-1)}$. This eliminates from Eq.(12a) even the implicitly implied antisymmetrization of the wavefunction. The resulting PT has been shown to have in some cases interesting advantages over the MS-MA PT (Jeziorski et al., 1978).

Table II. Relative errors in the H...H$^+$ interaction energy in the $2p\sigma_u$ state, at R = 12.5 bohr, compute[d] using various perturbation expansions (Jeziorski et al., 1978)

n	MS-MA	ISF	EL-HAV
2	-0.632781	-0.632781	-32.473690
3	0.287968	-0.144906	-7.192288
4	0.187200	-0.034646	-1.617605
5	0.080973	-0.006888	-0.367050
6	0.033580	-0.000826	-0.083704
7	0.012785	0.000229	-0.016139
8	0.003557	0.000272	-0.004381
10	-0.002486	0.000097	-0.000230
12	-0.003749	0.000026	-0.000012
14	-0.004017	0.000007	-0.000001
16	-0.004075	0.000002	

In Table II some results are listed which show convergence of the interaction energy computed using various perturbation expansions for the proton - hydrogen atom system at R = 12.5 bohr, which corresponds to the van der Waals minimum. In this case the reflection in the plane perpendicular to the internuclear axis introduces the same symmetry problem as a permutation of electrons in many-electron systems. The EL-HAV PT is seen to converge rapidly but in low orders, especially in the second order, it gives very poor results. The ISF PT converges equally rapidly and gives excellent results in low orders. The MS-MA PT diverges but, if restricted to low orders, it gives fairly satisfactory interaction energies. In this context, however, an important result obtained recently by Jeziorski et al. (1980) should be mentioned. Convergence of the perturbation expansion requires in practice antisymmetrization of the wavefunction with respect to permutations of the electrons between the interacting systems. This generates all many-electron clusters in $\psi^{(1)}$ which create considerable computational difficulties. The RS PT or MS-MA PT are free of this difficulty but the price one has to pay for it is the pathologically slow convergence (Chałasiński et al., 1977) or divergence, respectively. Jeziorski et al. (1980) have shown, however, that also in these cases by applying a suitable analytic continuation procedure one can get the correct values of the interaction energy.

For the sake of further discussion let us recall the energy partitioning commonly used in perturbational expansions of the intermolecular interaction energy. A symmetry adapted PT gives, at least in principle, the possibility to compute $E_{int}^{(n)}$ for n = 1, 2, ... The interaction energy computed in the n-th iteration can be represented as a sum of the perturbational energies $E^{(m)}$. Thus it is possible to compute $E^{(m)}$ in any order of the perturbation theory. It is also convenient to introduce the concept of the polarization approximation (Hirschfelder, 1967) which consists in using RS PT with perturbation defined in Eq.(4). Thus in each order of the perturbation theory one can define the polarization energy $E_{pol}^{(m)}$, calculated in the polarization approximation, and the exchange energy defined as

$$E_{exch}^{(m)} = E^{(m)} - E_{pol}^{(m)} \qquad (15)$$

Eq.(15) enables us to express the interaction energy in the form

$$E_{int} = E_{pol}^{(1)} + E_{exch}^{(1)} + E_{pol}^{(2)} + E_{exch}^{(2)} + \dots \qquad (16)$$

$E^{(1)}_{pol}$ is usually called the electrostatic energy E_{elst}. $E^{(2)}_{pol}$ has two well known components: the induction and the dispersion energy

$$E^{(2)}_{pol} = E^{(2)}_{ind} + E^{(2)}_{disp} \tag{17}$$

$E^{(2)}_{ind}$ and $E^{(2)}_{disp}$ result when single excitations on A or B, or simultaneous excitations on A and B are included in the first-order correction to the wavefunction, respectively. Analogously one can decompose the second-order exchange energy

$$E^{(2)}_{exch} = E^{(2)}_{exch.ind} + E^{(2)}_{exch.disp} \tag{18}$$

The above partitioning of the interaction energy shows one of the main advantages of the perturbation theory approach to intermolecular interactions. It separates the interaction energy into components which have a well defined meaning. Apart from a clear picture it offers also the possibility to introduce approximations based on physical arguments. This applies in particular to the polarization energies.

The induction energy $E^{(2)}_{ind}$ can still be separated into two parts. For large intermolecular separations, if the interaction hamiltonian is expanded in powers of R^{-1}, one get the long-range induction energy arising from mutual polarization of the electronic charge clouds. For small distances the above approach neglects the induction-penetration energy commonly known as the charge-transfer energy. The separation of the induction energy into long-range and charge-transfer components, although useful in approximate calculations, is however not basis set independent. Other polarization-type contributions to the interaction energy, like E_{elst} or $E^{(2)}_{dis}$ can also be separated into long-range and penetration components.

When solving Eqs.(12) we have tacitly assumed that the exact solutions of the unperturbed Schrödinger equation (2) are known. Practically this is never the case and some approximate wavefunctions and energies of the isolated molecules must be used. A convenient starting point provide solutions of the Hartree-Fock equations

$$H_X^{HF}\varphi_k^{X,HF} = E_k^{X,HF}\varphi_k^{X,HF}, \qquad X = A,\ B \tag{19}$$

This leads to the following partitioning of the complete hamiltonian

$$H = H_0^{HF} + s_A W^A + s_B W^B + tV \tag{20}$$

where

$$H_0^{HF} = H_A^{HF} + H_B^{HF} \tag{21}$$

$$W^X = H_X - H_X^{HF}, \qquad X = A,\ B \tag{22}$$

V is defined in Eq.(4), and s_X, t are formal expansion parameters equal to 1. The screened electron interaction potential W^X is practically identical with a sum of Sinanoğlu's (1964) two-electron fluctuation potentials:

$$W^X = \sum_{i=1}^{N_X} \sum_{j\,i}^{N_X} g^X(i,j) \tag{23}$$

$$g^X(i,j) = r_{ij}^{-1} - (N_X - 1)^{-1}\left[J^X(i)-K^X(i)+J^X(j)-K^X(j)\right] \tag{24}$$

The form of the hamiltonian (20) is suitable for developing a multiple perturbation theory (Jeziorski and Kołos, 1977) which yields for the interaction energy the formal solution

$$E_{int} = \sum_{n=1} \sum_{i=0} \sum_{j=0} t^n s_A^i s_B^j E^{(nij)} \tag{25}$$

Thus

$$E_{int}^{HF} = \sum_{n=1} t^n E^{(n00)} \tag{26}$$

represents the interaction energy of two Hartree-Fock molecules whereas

$$E_{int}^{corr} = \sum_{n=1} \sum_{i=0}' \sum_{j=0}' t^n s_A^i s_B^j E^{(nij)} \qquad (27)$$

is the correlation correction to E_{int}^{HF}, and the "primes" denote that the term i = j = 0 is not included.

For the ground state of two Hartree-Fock molecules in the polarization approximation one has

$$\varphi_{0,pol}^{HF} = \varphi_0^{A,HF} \varphi_0^{B,HF} \qquad (28)$$

and the RS PT based on the zeroth-order wavefunction given by Eq.(28) yields the correlation corrections $E_{pol}^{(nij)}$ to the HF polarization energies $E_{pol}^{(n00)}$. There are no essential difficulties in computation of low-order correlation corrections to $E_{pol}^{(100)}$ and $E_{exch}^{(100)}$. Practically applicable expressions for $E_{pol}^{(nij)}$ and $E_{exch}^{(nij)}$ with arbitrary n have been derived by Szalewicz and Jeziorski (1979).

An alternative formulation of the correlation problem is also possible (Sadlej, 1980). Let us start again with a wavefunction in the form of a product

$$\varphi_{0,pol}^{HF}(t) = \varphi_0^{A,HF}(t)\, \varphi_0^{B,HF}(t) \qquad (29)$$

where $\varphi_0^{X,HF}(t)$ for X = A, B denotes a Slater determinant built of spinorbitals $u_{kX}(t)$ obtained by minimizing the functional

$$E_{pol,AB} = \langle \varphi_0^{A,HF}(t)\varphi_0^{B,HF}(t) | H_0 + tV | \varphi_0^{A,HF}(t)\varphi_0^{B,HF}(t) \rangle \qquad (30$$

under the usual orthonormality conditions for $u_{kX}(t)$. The resulting equations for the spinorbitals are

$$h_A(t)u_{kA}(t) = e_{kA}(t)u_{kA}(t) \qquad (31a)$$

$$h_B(t)u_{kB}(t) = e_{kB}(t)u_{kB}(t) \qquad (31b)$$

where

$$h_A(t) = -(1/2)\Delta - \sum_{\alpha\in A}^{N_A} Z_\alpha/r_\alpha + \sum_{j\in A}^{occ}\left[J_{jA}(t) - K_{jA}(t)\right]$$
$$+ t\left[-\sum_{\beta\in B}^{N_B} Z_\beta/r_\beta + \sum_{k\ B}^{occ} J_{kB}(t)\right] \qquad (32)$$

and an analogous definition holds for $h_B(t)$. For simplicity the dependence of the operators and wavefunctions on the electronic coordinates has been suppressed in Eqs.(29-32).

It is clear from the form of the functional (30) that neither correlation nor exchange effects have been included in the derivation of Eqs.(14). Hence by solving these equations and subtracting from the resulting energy the energies of two isolated Hartree-Fock molecules, one gets the polarization component of the interaction energy of two mutually polarized Hartree-Fock molecules, i.e.

$$E^{HF}_{pol} = E^{HF}_{elst}(t) + E^{HF}_{ind}(t) \qquad (33)$$

The energy functional (30) can also be expanded in powers of t. This results in equations which, in complete analogy with the former approach, yield perturbational corrections (interaction energies) in various orders of t. Equations for $u^{(1)}_{kX}$ obtained in this way determine the best first-order perturbed orbitals for the subsystem X. The same procedure can be performed using $\varphi^{HF}_{0,pol}$ defined as in Eq. (29) but antisymmetrized with respect to permutations of electrons between A and B. If summed through all orders this is equivalent to a Hartree-Fock treatment of the supermolecule AB. Thus it includes in all orders the polarization and exchange energies but no correlation energy. The resulting interaction energy, however, would be different from the value calculated from Eq.(26). One can show (Sadlej, 1980) that the approach sketched above corresponds to the coupled Hartree-Fock (CHF) method (Dalgarno, 1962; Langhoff et al., 1966; Caves and Karplus, 1969) whereas the interaction energy calculated from Eq.(26) corresponds to the uncoupled Hartree-Fock (UCHF) approach (Langhoff et al., 1966). Thus it is clear that E_{int}^{corr} defined in Eq.(27) does not represent pure correlation effects. Using diagramatic technique one can split it into two parts: The first includes terms represented by the ring-containing diagrams. Therefore it is identical with the correlation correction defined with respect to the CHF approach and hence represents the "true" electron correlation effect. The second corresponds to the chain-type diagrams which correct for the selfconsistency effects, i.e., for the difference between the UCHF and CHF interaction energies. Therefore this correction is not due to real electron correlation and is called by Sadlej (1981) "apparent" correlation effect. Thus it follows that the sum $E^{(1)} + E_{ind} + E_{exch.ind}$ computed in the UCHF approach is not equal to the SCF interaction energy resulting from a supermolecule computation. The apparent correlation effect must be included in the former to put on a sound basis the comparison of results obtained in the two approaches.

For $(H_2O)_2$ Jeziorski and van Hemert (1976), neglecting the intramolecular correlation effects, employed the low-order MS-MA PT to compute the various components of the interaction energy. The values of E_{elst}, $E_{exch}^{(100)}$ and $E_{ind}^{(200)}$ obtained for several internuclear distances using fairly extended basis sets, are shown in Table III. In this table their sum, denoted Σ, is compared with the SCF interaction energy computed using the same extended basis sets. The difference between E_{int}^{SCF} and Σ, even at the equilibrium distance R = 5.67 bohr, is seen to be very small. It would be, however, premature to conclude therefrom that in this case the apparent correlation energy is negligible. The difference is due not only to the apparent correlation but also to the higher order induction energy, and to the exchange-induction energy; therefore some cancellation may occur.

Table III. Components of the interaction energy for $(H_2O)_2$ in kcal/mole (Jeziorski and van Hemert, 1976)

R	E_{elst}	$E^{(100)}_{exch}$	$E^{(200)}_{ind}$	Σ	E^{SCF}_{int}
4.80	-17.07	25.11	-9.30	-1.26	0.92
5.20	-11.10	12.03	-4.12	-3.20	-2.77
5.67	-7.12	4.90	-1.63	-3.85	-3.87
7.00	-2.79	0.30	-0.18	-2.68	-2.69
9.00	-1.12	0.00	-0.02	-1.14	-1.14

To get more insight into the structure of intermolecular interaction energy it is instructive to analyse the interaction of beryllium atoms. Their large polarizability enhances the problems encountered in computations of the effects dependent on this property.

Chałasiński and Jeziorski (1977) computed the interaction energy of two beryllium atoms, at several interatomic distances, using the SCF method and the second-order PT corresponding to the UCHF scheme, i.e. from the formula

$$E_{int} = E^{(100)} + E^{(200)}_{ind} + E^{(200)}_{exch.ind} \quad (34)$$

For the same system, and interatomic distance R = 7 bohr, Jaszuński (1980) computed the interaction energy also from Eq.(34) but with the second-order components obtained in a CHF approach. The results are shown in Table IV where, for consistency, we list the UCHF energies computed by the same author.

Three important conclusions follow from Table IV:

(1) $E^{(200)}_{ind}$ and $E^{(200)}_{exch.ind}$ are of a comparable magnitude and therefore the total second-order induction-type contribution

Table IV. Interaction energy of two Be atoms at R = 7 bohr (in millihartree)

SCF	E_{int}	1.808
	$E^{(100)}$	3.222
UCHF	$E^{(200)}_{ind}$	-1.568
	$E^{(200)}_{exch.ind}$	1.161
	E_{int}	2.815
CHF	$E^{(200)}_{ind}$	-2.064
	$E^{(200)}_{exch.ind}$	1.550
	E_{int}	2.707

to the interaction energy is considerably smaller than $E^{(200)}_{ind}$ alone. (2) The apparent electron correlation strongly affects both the induction and exchange-induction energy. Thus it also changes appreciably the polarizability of the Be atom (Sadlej, 1981) but its effec on the interaction energy, due to the opposite signs of $E^{(200)}_{ind}$ and $E^{(200)}_{exch.ind}$, is much smaller. (3) The significant discrepancy between the SCF interaction energy and its value computed in the second-order PT using the UCHF approach persists also if the latter is obtained utilizing the CHF method.

Additional light on the problem of binding in the Be_2 dimer shed new results obtained recently, and reported at this Congress, by Chałasiński (1982b) who studied Be_2 using several approaches and including the dispersion interaction The second-order PT gives a van der Waals minimum in the vicinity of R = 8 bohr, whereas an SCF interaction supplemented with dispersion and exchange-dispersion energies would give a minimum at $R < 6$ bohr. The actual location of the minimum has not been determined since the computations have been carried out for $6 \leqslant R \leqslant 10$ bohr. The results based on the SCF interaction energy seem to be more consistent with those of Liu and McLean (1980) who using an extended STO basis, and a CI approach, obtained $R_e = 4.7$ bohr and $D_e = 2.3$ kcal/mole. No experimental results are available for Be_2. It is, however, very likely that the theoretical results of Liu and McLean (1980) are reliable since their parallel results for Mg_2 and He_2 agree well with the experimental values. Recent computations of Blomberg et al. (1980) and of Chiles and Dykstra (1981) have confirmed the results of Liu and McLean and have spectacularly demonstrated the difficulties encountered in computation of the potential energy curve for Be_2.

Thus an SCF computation for the beryllium dimer seems to give a more reliable interaction energy than a PT (UCHF- or CHF-type) approach. The reason for the failure of the latter, however, is still not clear. From the recent results of Gutowski et al. (1982) one may suppose that the exchange deformation is to be blamed for it. These authors have shown that in the interaction of some singly charged ions, at the distances for which the overlap is not negligible, deformations by purely electrostatic potentials are not sufficient to get agreement between low-order PT and SCF interaction energies. Satisfactory agreement was obtained, however, when interionic exchange effects were included in the potentials responsible for the mutual deformation of the interacting ions. Interaction of neutral atoms is, of course, somewhat different. The results collected in Table IV show, however, that in the particular case of beryllium atoms the deformation and exchange effects are very important. Hence in this case the exchange deformation may be expected to give a considerable contribution to the PT interaction energy.

The problem of interaction of two SCF atoms or molecules, and the structure of the SCF interaction energy, are thus almost completely understood. In the problem of intramolecular electron correlation corrections to the interaction energy considerable progress has also been made. In this case, however, the theory is still rather far away from a stage in which a straightforward implementation would be possible, and would lead to physically meaningful and generalizable results. We shall now briefly indicate some of the problems encountered in computation of various contributions to the interaction energy and in this context some remarks on the intramolecular electron correlation effects will also be made.

The electrostatic interaction energy E_{elst} ($\equiv E^{(100)}_{pol}$) plays an important role in the interaction of polar or ionic systems. Its reliable value, however, can only be obtained if the unperturbed wavefunctions yield correct charge distributions in the interacting molecules, as judged, e.g. from the computed value of the multipole moments. In Table V we list some results obtained for the H_2O molecule using several basis sets in the SCF method. One can clearly see that some even fairly extended basis sets yield dipole moments which considerably deviate from the experimental value, and therefore they cannot give correct E_{elst}. This discrepancy is clearly due to electron correlation and its magnitude gives an indication of the importance of the intramolecular electron correlation effect on the electrostatic interaction energy between polar molecules. On the other hand, one also notices in Table V that properly balanced relatively small basis sets can also give reasonable charge distributions. Semiempirical scaling procedures have been suggested to take care of this defficiency (Kollman and Kuntz, 1976; Umeyama and Morokuma, 1977; Groen and van Duijneveldt, 1978). The value of such a scaling can be judged only from its numerical results. From a pragmatic point of view this may be, within a certain range, a useful procedure. Theoretical objections against it can be raised, however,

Table V. SCF dipole moment of H_2O in a.u. (Clementi, 1976)

basis set[x]	μ
(7,3/3)	0.880
(10,5/4)	1.040
(12,7/7)	1.062
(7,3,1/4,1)	0.772
(10,5,1/4,1)	0.856
(12,7,1/7,1)	0.869
experiment	0.728

[x]In the (k,l,.../mn) symbol k,l,... and m,n denote the number of s, p, ... basis functions on the oxygen and hydrogen, respectively.

since the procedure ignores the penetration (charge overlap) effects and, to be practical, it should be based on the first term in the multipole expansion of the interaction potential.

The first-order exchange energy, $E^{(100)}_{exch}$, represents the main repulsive component of the interaction energy and its computation is considerably more difficult than that of E_{elst}. A convenient formula for $E^{(100)}_{exch}$ has been derived for the case when SCF wavefunctions are used for the monomers (Jeziorski et al., 1976). One can also introduce considerable simplifications if following the original idea of Murrell et al. (1965) only terms resulting from single exchanges, i.e. terms which are of the order of S^2, are retained. Using the many-orbital cluster expansion (Chałasiński and Jeziorski, 1977) one can express $E^{(100)}_{exch}(S^2)$ in a simple form (Bulski et al., 1979)

$$E^{(100)}_{exch}(S^2) = \sum_i \sum_j e(i|j) + \sum_{i>k} \sum_j e(ik|j) + \sum_i \sum_{j>l} e(i|jl) + \sum_{i>k} \sum_{j>l} e(ik|jl) \quad (35)$$

where the indices i and k label the orbitals of atom A, and j, l those of atom B and, e.g., e(ik|j) denotes the corresponding three-orbital contribution. From general considerations and from numerical results for the Ne_2 dimer it follows (Bulski et al., 1979) that the last term in Eq.(35) representing the four-orbital contribution, which is the most expensive one in terms of the computer time, can be neglected. The three-orbital component has been shown to be the dominant one.

Formulae for the terms of the order of S^4 have been derived by van Duijneveldt-van de Rijd and van Duijneveldt (1972). For Ne_2 (Bulski et al., 1979) and for Ar_2 (Chałasiński et al., 1982) their contribution to the interaction energy has been shown, however, to be negligible.

It is also known (Kołos and Leś, 1972) that the accuracy of $E^{(1)}_{exch}$ critically depends on the accuracy of the "tails" of the orbitals. This implies that for large systems methods which provide only locally accurate orbitals may be very useful (Pecul, 1980).

Finally let us mention two recent results which shed light on the effect of the intramolecular correlation correction to $E^{(100)}_{exch}$. Szalewicz and Jeziorski (1979) have shown that for He_2 $E^{(110)}_{exch} + E^{(101)}_{exch}$ amounts to only 2% of $E^{(100)}_{exch}$ at the van der Waals minimum. This optimistic indication fainted, however, when Chałasiński et al. (1982) showed that the total intraatomic correlation correction, i.e. $\sum_{ij\geq 1} E^{(1ij)}_{exch}$, raises $E^{(100)}_{exch}$ by 7%.

$E^{(200)}_{ind}$ and $E^{(200)}_{exch.ind}$ have already been discussed for Be_2 and in this case, similarly as for He_2 (Szalewicz and Jeziorski, 1979), both are of a comparable magnitude and have opposite signs. Hence considerable cancellation occurs Via the London formula the long-range part of $E^{(200)}_{ind}$ is related to the polarizabilities of the interacting systems.

It is known, however, that small standard basis sets fail to give reliable polarizabilities (Werner and Meyer, 1976). Therefore they are also not likely to be effective in computations of $E_{ind}^{(200)}$. Similarly as in the case of E_{elst} one can try to improve $E_{ind}^{(200)}$ with the aid of a scaling procedure utilizing the experimental and theoretical polarizabilities, the latter obtained with the same basis set as the induction energy. With respect to this scaling analogous objections can be raised as in the case of E_{elst}. They may be particularly valid if an important role in the interaction play the charge-transfer effects which are included in $E_{ind}^{(200)}$ but are not directly related to polarizabilities of the interacting systems.

The intraatomic correlation correction $E_{ind}^{(210)} + E_{ind}^{(201)}$ is known only for He_2 where it is small (Szalewicz and Jeziorski, 1979). For larger systems only some partial and indirect information is available if the effect of electron correlation on the polarizabilities is known.

A reliable computation of $E_{disp}^{(200)}$ requires extended basis sets (Jeziorski and van Hemert, 1976) or optimization of virtual orbitals (see, e.g., Chałasiński and Jeziorski, 1976; Chałasiński, 1982b). Otherwise the convergence is poor, and the commonly used small basis sets account for a relatively small part of $E_{disp}^{(200)}$ (see, e.g. Lathan et al., 1976). Also in this case a scaling procedure can be employed and in spite of the formal objections against it Smit et al. (1979) have shown that it leads to fairly realistic values of $E_{disp}^{(200)}$.

Intramolecular electron correlation corrections to $E_{disp}^{(200)}$ present a very subtle problem since the dispersion energy itself is a correlation effect. For the long-range part of $E_{disp}^{(200)}$ the separation of the two correlations is fairly straightforward and correlation corrections to the van der Waals constants can be computed. For shorter distances, however, the separability becomes questionable. For He_2 at the van der Waals minimum the intraatomic correlation correction to $E^{(200)}_{disp}$ has been shown to represent almost 30% of the dispersion energy and nearly 50% of the interaction energy (Chałasiński and Jeziorski, 1976; Szalewicz and Jeziorski, 1979). In addition it converges slowly, i.e. $E^{(210)}_{disp} + E^{(201)}_{disp}$ is considerably smaller than the total correction. It is tempting to suppose that the situation is more optimistic in the case of polar or ionic systems. This may be indeed so if compact and not easily polarizable ions, like Li^+, are involved.

Numerical results for $E^{(200)}_{exch.disp}$ are scarce. For He_2 it represents about 4% of $E^{(200)}_{disp}$ (Szalewicz and Jeziorski, 1979) and similarly for Ne_2 (Chałasiński, 1982b). For $(HF)_2$ however, it becomes relatively more important and amounts to 10% of the dispersion energy (Chałasiński, 1982a).

Let us now return to the SCF computations of the interaction energy which seem to be more popular than those based on the perturbation theory. Using the well known Morokuma (1971) scheme (see also Kitaura and Morokuma, 1976) E^{SCF}_{int} can be decomposed into components which are either identical or closely related to those obtained in the PT approach. It is, however, well established that E^{SCF}_{int}, unless computed using judiciously chosen large basis sets, suffers from the basis set superposition error (BSSE) (see, e.g. Kołos, 1979, and references therein). The counterpoise method of Boys and Bernardi (1970) seems to be very effective in dealing with this problem. BSSE affects, however, in various degrees the components of E^{SCF}_{int}. Sokalski et al. (1982) have recently incorporated the counterpoise method into the Morokuma decomposition of E^{SCF}_{int} thus achieving a milder dependence of various components on the quality of the basis set employed.

When discussing above the various contributions to the interaction energy many difficulties or objections have been mentioned. If collected they would make up a formidable list. In this context we may additionally quote a result of Lischka (1974, 1979) which indicates that for $(HF)_2$ the total intramolecular correlation correction to the interaction energy is repulsive and in absolute value is about as large as the dispersion energy. Thus the outlook for accurate evaluation of intermolecular interaction energies, including all non-negligible effects, is not very optimistic. It is, however, at least somewhat comforting that in many cases complete neglect of intramolecular electron correlation, and a semiempirical evaluation of the dispersion energy from polarizabilities of molecular fragments gives results which are (fortuitously?) reasonable and helpful in understanding the structure and properties of molecular clusters or of condensed phases (see, e.g. Kołos, 1979).

3. NONADDITIVITY OF INTERMOLECULAR INTERACTIONS

In studies of molecular clusters composed of more than two subsystems the problem of additivity of molecular interactions arises. The interaction energy of n subsystems can be represented in the form of the multibody expansion

$$E(n) = \sum_{m=2} E(m,n) \qquad (36)$$

where E(m,n) denotes the m-body contribution to the interaction energy of n subsystems, and the consecutive values of E(m,n) for m = 3, 4, ... can be calculated recurrently.

Sometimes it is convenient to use the relative values of the nonadditive contributions:

$$e(m,n) = E(m,n)/E(2,n) \qquad (37)$$

This transforms Eq.(36) into

$$E(n) = E(2,n)(1 + e(3,n) + e(4,n) + \dots e(n,n)) \quad (38)$$

If the contributions E(m,n) are expressed in the form of perturbation expansions, and in each order the polarization and exchange components are separated, one can easily show that the exchange and induction energies are nonadditive in all orders whereas E_{elst} and $E^{(2)}_{disp}$ are additive. A well known dispersion-type nonadditive contribution to the interaction energy representing the so called triple-dipole effect can be obtained in the third order of the perturbation theory (Axilrod and Teller, 1943). The exchange-type nonadditivity has been shown to be related to the overlap integrals (Kołos and Leś, 1972a,b; Kołos 1973; Magnasco et al., 1976) and therefore it decays exponentially with increasing intermolecular separation. Hence for large intermolecular separations the induction and triple-dipole effects can be expected to give the dominant nonadditive contributions to the interaction energy of polar and non-polar systems, respectively. In the vicinity of the van der Waals minimum, however, the overlap effects, and hence the exchange nonadditivities are not negligible.

Nonadditivity of interaction has been extensively studied for rare gases and for hydrogen atoms with parallel spins (Bader et al., 1971; Novaro and Beltran-Lopéz, 1972; Kołos and Leś, 1972a,b; Musso et al., 1974; González-Tovany et al., 1974; Novaro and Yánez, 1975a,b; Novaro and Nieves, 1976; Magnasco et al., 1976; Bulski and Chałasiński, 1980). In this case, as mentioned above, at large interatomic separations the triple-dipole nonadditivity predominates. At the van der Waals minimum the exchange effects are not negligible and they increase rapidly for shorter distances. The importance of the first-order exchange nonadditivity in determination of the third virial coefficient for rare gases has been demonstrated by Leś (1976).

The nonadditive exchange effects have also been shown to affect the properties of liquids. Helium dimers, in contrast to dimers of the heavier rare gases, are known to be not bound. An interesting question arises, however, whether the liquid helium surrounding could stabilize a He_2 dimer. Sinanoğlu's solvent effect method (Sinanoğlu 1967, 1968, 1974) has been employed to shed light on this problem (Blaisten-Barojas et al. 1976). The triple-dipole effect has been found to have a destabilizing effect on a He_2 dimer. The exchange nonadditivity, however, has been shown to act in the opposite direction and to outweight the former (Blajsten-Barojas et al., 1979).

Of particular interest is the nonadditivity of interaction of beryllium atoms which on the SCF level has been analysed for Be_n, $n \leqslant 5$, clusters (Kołos et al., 1976; Novaro and Kołos, 1977; Daudey et al., 1979). For geometrical configurations occuring in solid beryllium one gets the following numerical values for various terms in expansion (38):

Be_3 - equilateral triangle:

$$E(3) = E(2,3)(1 - 0.764) = 0.236E(2,3) \qquad (39)$$

Be_4 - quasitetrahedron:

$$E(4) = E(2,4)(1 - 1.535 + 0.316) = -0.219E(2,4) \qquad (40)$$

Be_4 - rhombus:

$$E(4) = E(2,4)(1 - 1.666 + 0.866) = 0.200E(2,4) \qquad (41)$$

Be_5 - combination of quasitetrahedron and rhombus:

$$E(5) = E(2,5)(1 - 1.846 + 1.155 - 0.378) = -0.069E(2,5) \qquad (42)$$

Eq.(39) shows that in the equilateral triangle the three-body interaction has a strong bonding effect which decreases the net repulsion to roughly 1/4 of its additive value. It may be also pointed out that the three-body interaction becomes strongest for internuclear distances approximately equal to those in solid beryllium. The latter distances, which are known from experiment, are shorter than any theoretical values for the dimer (see above) and the contraction may be due to the nonadditivity.

In Be_3 there is one three-body attraction against three two-body repulsions. In the quasitetrahedral Be_4, however, there are only six two-body repulsions against four three-body attractions. The net result, as shown in Eq.(40), is attraction in spite of the repulsive four-body contribution. Thus the fourth Be atom enhances the reconciliatory effect of the three-body contribution yielding thereby a stable cluster. In the planar Be_4 on the other hand the strong four-body repulsion prevents the cluster from being stable. The Be_5 cluster (with four Be atoms in a plane forming a rhombus and the fifth atom above the plane forming a quasitetrahedron with three atoms belonging to the rhombus) is again stable with regard to 5Be. It is seen, however, from Eqs.(40) and (42) that the quasitetrahedral Be_4 would not bind the fifth Be atom.

Eqs.(39-42) may give the impression that the multibody expansion converges for Be_n in the sense that $|e(m,n)| > |e(m+1,n)|$. All terms, however, are of the same order of magnitude, and in all cases the last term in the expansion is larger in magnitude than the total interaction energy. Hence for the clusters under consideration a completely wrong interaction energy for Be_n would be obtained by neglecting the n-body contribution.

Results for magnesium clusters (Daudey et al., 1979) are in some respects similar to those for beryllium and in other to rare gas clusters. The three-body attraction is again strongest for geometry which is fairly close to that in solid Mg. The minimum, however, is considerably flatter than for Be and the attraction is weaker. In consequence quasitetrahedral Mg_4 is not bound on the SCF level.

The above SCF results are undoubtedly interesting. The question arises, however, how reliable is the independent particle model in predicting stabilities of Be_n clusters and the answer to it is not clear. The computations by Dykstra et al. (1976) performed using the method of self-consistent electron pairs in the coupled electron pair approximation confirm the SCF results for tetrahedral Be_4 and give roughly the same binding energy. For Be_2 the same authors obtained a very weak binding at interatomic separation of 8.5 bohr i.e. about twice as large as in solid beryllium (4.2 bohr). These results, however, in view of more recent and more accurate computations, discussed above for Be_2, do not seem to be very reliable.

The exceptionally large value of the nonadditive energy in beryllium clusters, especially of the three-body contribution, calls for some rationalization. This can be achieved by decomposing the interaction energy using the Morokuma (1971) method extended to more than two subsystems (Daudey, 1978). The results (Daudey et al., 1979) are shown in Table VI.

The total SCF interaction energy $E_{SCF}(3)$ and its non-additive part $E(3,3)$ are decomposed into the electrostatic, exchange, polarization and delocalization contributions as defined by Morokuma (1971). The first two can be identified with E_{elst} and $E^{(1)}_{exch}$ in the perturbation theory. The polarization and delocalization energies have no counterparts in PT. Their sum, however, can be identified with $\sum_{n=2}(E^{(n)}_{ind} + E^{(n)}_{exch.ind})$. When comparing Be_3 with Mg_3 most striking are the large values of the polarization and de-

Table VI. Decomposition of the three-body contribution to the interaction energy (in kcal/mole) for Be_3 and Mg_3 in equilateral triangle configuration corresponding to the equilibrium distances in the solids

	Be_3		Mg_3	
	E(3)	E(3,3)	E(3)	E(3,3)
elst	-73.86	0	-39.47	0
exch	168.60	-20.94	74.92	-7.16
pol	-81.79	-23.71	-3.75	-1.17
del	-3.23	13.24	-0.37	2.72
SCF	9.72	-31.41	31.33	-5.61

localization energies, as well as their nonadditive components for Be_3. The sum of $E_{pol}(3,3)$ and $E_{del}(3,3)$, i.e. the induction energy including exchange, is also much larger in absolute value for Be_3 than for Mg_3 being negative for the former and positive for the latter. Thus it is evident that the large induction energy for Be_3, in the overlap region, is responsible for the strong three-body attractive energy component. The large induction energy must obviously be related to the 2s-2p near degeneracy in Be atom. The importance of the three-body interactions in elemental clusters is not restricted to the cases considered here in detail. Murrell (1978) has shown that in elemental clusters the three-body interaction is the structure determining factor.

The rapidly growing number of the degrees of freedom in clusters involving molecules makes a study of nonadditivity of interaction a much more difficult problem. In spite of it, however, considerable progress in understanding nonadditivity of molecular interactions has recently been made

The Morokuma (1971) partitioning scheme has been applied to molecular trimers and hydrated ions (Kollman, 1977; Beyer et al., 1979). Some of the results are listed in Table VII.

Table VII. Partitioning of the three-body component of the interaction energy in kcal/mole (Beyer et al., 1979)

	$(HF)_3$		$Li(H_2O)_2^+$		$F(H_2O)_2^-$	
	E(3)	E(3,3)	E(3)	E(3,3)	E(3)	E(3,3)
elst	-10.2	0	-53.9	0	-45.1	0
exch	3.7	-0.0	20.7	-0.3	31.2	-0.0
pol	-1.5	-0.4	-16.6	-2.6	-10.7	-2.2
del	-0.8	-0.1	3.4	0.3	-8.4	-0.3
SCF	-8.8	-0.5	-46.4	-2.6	-33.0	-2.5

For $(HF)_3$ a linear geometry has been assumed. For $Li(H_2O)_2^+$ and $F(H_2O)_2^-$ the water dimer was assumed in its stable geometry and the ion was facing the lone pairs in the proton-donor H_2O and a proton in the proton-acceptor H_2O, respectively, in the two complexes. The results listed in Table VII clearly show that in trimers involving polar molecules or ions, where the induction energy gives a fairly large contribution to the interaction energy, it is just the induction energy which determines the three-body nonadditive interaction energy, similarly as in beryllium clusters.

Extensive SCF studies with large basis sets have been carried out for many geometrical configurations of $(H_2O)_3$ (Clementi et al., 1980b) and $Li(H_2O)_2^+$ (Clementi et al., 1980a). The authors anticipated that the induction energy should be the main factor determining the three-body nonadditive energy, and suggested to approximate it with a

London-type formula which expresses the induction energy in terms of bond polarizabilities of the interacting molecules. Fig. 1 shows the correlations, for 29 clusters of $(H_2O)_3$, between the nonadditivity obtained in the SCF approach, $E_{SCF}(3,3)$, and that calculated from the London formula, $E^{(2)}_{ind}(3,3)$. The agreement is fairly satisfactory and it should be also pointed out that the SCF results, in spite of the large basis sets employed, still suffer from the basis set superposition error.

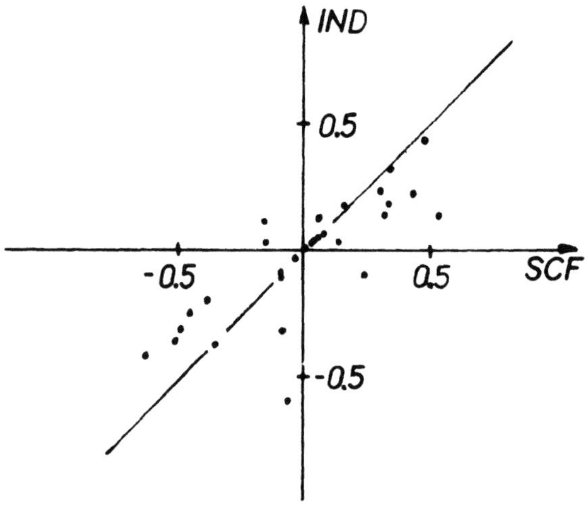

Figure 1. Correlation between the SCF and induction nonadditive components of the interaction energy for 29 configurations of $(H_2O)_3$ in kcal/mole (Clementi et al., 1980b).

It is seen in Fig. 1 that for the water trimer and not very small intermolecular distances the nonadditive component of the interaction energy practically does not exceed 1 kcal/mole. For clusters involving ions it becomes considerably larger (see Table VII). Since, however, the increase is due to stronger induction interaction the relative agreement between the SCF results and those obtained from the

London formula becomes even better, as seen in Fig. 2 where results for $Li(H_2O)_2^+$ are shown. One H_2O molecule has been fixed with respect to Li^+ at the equilibrium distance and E(3,3) is shown as function of the distance between the second H_2O and Li^+. The curves are drawn for several angles $H_2O \cdot Li \cdot H_2O$ as shown in the inset. The solid and broken lines denote the SCF and induction nonadditive energies, respectively (Clementi et al., 1980a).

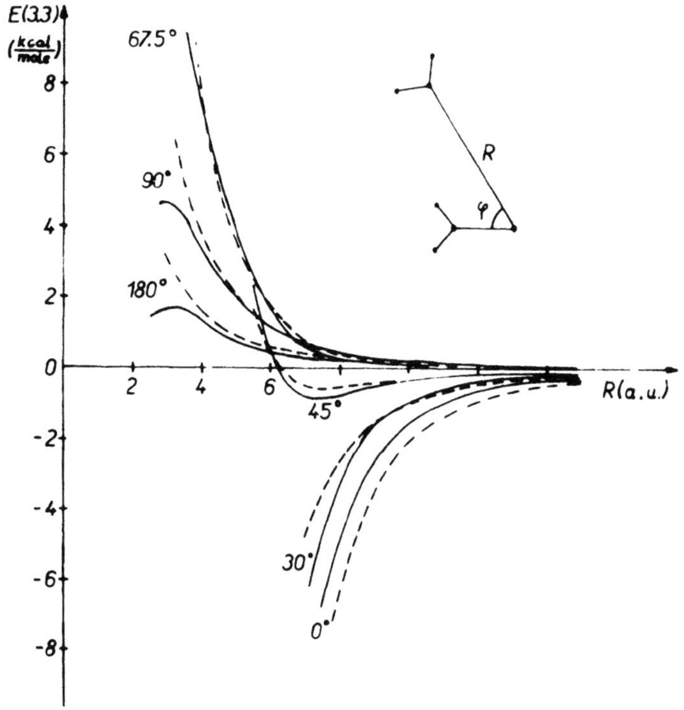

Figure 2. Nonadditive component of the interaction energy for $Li(H_2O)_2^+$.

For clusters involving molecules the nonadditive energy component E(3,3) is a rather complicated functions of the geometry of the system. It is seen from Figs. 1 and 2, however, that it can be reliably represented with a simple analytical London formula. Incorporation of this formula in a Monte Carlo program gives the possibility to determine theoretical stabilities of clusters taking into account the nonadditive three-body interaction. The results for $Li(H_2O)_n^+$ are shown in Table VIII.

Table VIII. Stabilization energies per water molecule of $Li(H_2O)_n^+$ clusters in kcal/mole

n	additive[a]	nonadditive[a]	experimental[b]
1	-33.5	-	-34.0
2	-32.2	-30.1	-29.9
3	-30.3	-25.0	-26.8
4	-28.2	-23.2	-24.2
5	-25.0	-21.0	-22.2
6	-22.7	-19.2	-20.5

[a]Clementi et al., (1980a).
[b]Džidič and Kebarle (1970), Arshadi et al. (1970).

For n = 1 the agreement with experiment is good which indicates that the correlation effects neglected in the SCF approach do not affect appreciably the stabilization energy. For n = 2 the agreement is also good but only if the three-body nonadditive interaction is taken into account. For $3 \leqslant n \leqslant 6$ the agreement is somewhat worse which suggests that the four-body and perhaps even higher multibody interactions are not negligible.

Recently nonadditivity of interaction in $(NH_3)_3$ and $(CH_4)_3$ trimers has been studied on the SCF level using 4-STO basis sets (Novaro et al., 1981a,b). As expected, in $(CH_4)_3$ the nonadditive energy component is similar in sign and magnitude to that in rare gases. The small polarity of the CH bonds gives rise to a negligible induction-type nonadditivity which for the linear configuration of the three methane molecules is attractive in contrast to the repulsive and dominant exchange nonadditivity. The latter, however, decreases exponentially with the CC distance and therefore for larger distances ($R_{CC} \geqslant 3.5$ Å) the former dominates, being of course very small. In $(NH_3)_3$, for $R_{NN} = 3.4$ Å, which is the NN separation in solid ammonia, the signs of the SCF and induction nonadditivities agree, and the exchange-type nonadditivity seems to be roughly of the same order as that due to the induction interaction. When calculated however, for $R_{NN} = 3.0$ Å, which is the SCF equilibrium separation for the most stable, i.e. cyclic trimer (see also Jean, 1980), the nonadditivity becomes large (-2.2 kcal/mole) and is, obviously, predominantly of the exchange type.

Finally one can mention that there has been considerable and growing interest in studies of metal clusters and of interaction between an atom or a molecule and the cluster. This interaction can be regarded as a model of chemisorption on the corresponding metal. The chemisorption energy has been found to depend significantly on the size and geometry of the cluster (see, e.g., Bauschlicher et al., 1975, for Be_nH, Schultz et al., 1979, for Li_nH_2O, and Beckmann and Koutecky, 1982, for Li_nH). Daudey et al. (1979) have shown that in the case of Be_nH the concept of nonadditivity of the interaction energy can be very helpful for analysing and rationalizing the energies of chemisorption on various clusters. It is possible that a similar analysis could be fruitful also in other cases.

Concluding this review one can say that the relations between various symmetry adapted perturbation theories are now well understood, and depending on the problem under consideration a convenient formalism can be chosen for performing numerical computations. There is, however, an urgent need for results of accurate computations, at least for some not very complex model systems. Especially needed are accurate values of those contributions to the interaction energy that are usually neglected, like the intramolecular correlation corrections, exchange corrections of the second order, and all corrections of the order higher than second. Such results could provide a sound basis for introducing approximations unavoidable when dealing with large molecules.

With respect to the nonadditive interaction energy the demand for reliable results for model systems is even greater. The three-body component is known to be important for systems involving ions, and to consist mainly of the induction-type nonadditivity. Little is still known, however, about convergence of the many-body expansion, except for rare gases, about the importance of nonadditivity in understanding, e.g., the structure and stability of metal clusters, the surface phenomena etc. Efficient ways of calculating, as well as approximating the nonadditivities, including the exchange contribution, are needed for studying a host of problems involving condensed phases.

ACKNOWLEDGEMENT

The author is grateful to Drs. G.Chałasiński and B.Jeziorski for interesting and valuable discussions.

REFERENCES

Arshadi, M., Yamdagni, R. and Kebarle, P.: 1970, J.Phys. Chem. 74, 1475.
Axilrod, B.M. and Teller, E.: 1943, J.Chem.Phys. 11, 299.
Bader, R.F.N., Novaro, O. and Beltrán-Lopez, V.: 1971, Chem.Phys.Letters 8, 568.
Bauschlicher Jr., C.W., Liskow, D.H., Bender, C.W. and Schaefer III, H.F.: 1975, J.Chem.Phys. 62, 4815.

Beckmann, H.O. and Koutecky, J.: 1982, to be published.
Beyer, A., Karpfen, A. and Schuster, P.: 1979, Chem.Phys. Letters 67, 369.
Blaisten-Barojas, E., Novaro, O. and Bruch, L.W.: 1979, Mol.Phys. 37, 599.
Blaisten-Barojas, E., Novaro, O. and Sinanoğlu, O.: 1976, Mol.Phys. 31, 1941.
Blomberg, M.R.A., Siegbahn, P.E.M. and Roos, B.O.: 1980, Int.J.Quant.Chem. S14, 229.
Boys, S.F. and Bernardi, F.: 1970, Mol.Phys. 19, 553.
Bulski, M. and Chałasiński, G.: 1980, Theor.Chim.Acta 56, 199.
Bulski, M., Chałasiński, G. and Jeziorski, B.: 1979, Theor. Chim.Acta 52, 93.
Caves, T.C. and Karplus, M.: 1969, J.Chem.Phys. 50, 3649.
Chałasiński, G.: 1982a, Chem.Phys.Letters, 86, 165.
Chałasiński, G.: 1982b, to be published, reported at this Congress
Chałasiński, G. and Jeziorski, B.: 1976, Mol. Phys. 32, 81.
Chałasiński, G. and Jeziorski, B.: 1977, Theor.Chim.Acta 46, 277.
Chałasiński, G., Jeziorski, B. and Szalewicz, K.: 1977, Int.J.Quant.Chem. 11, 247.
Chałasiński, G., van Smaalen, S. and van Duijneveldt, F.B.: 1982, in publication.
Chiles, R.A. and Dykstra, C.E.: 1981, J.Chem.Phys. 74,4544.
Clementi, E.: 1976, "Determination of Liquid Water Structure", Lecture Notes in Chemistry, Vol. 2, Springer, Berlin.
Clementi, E., Kistenmacher, H., Kołos, W. and Romano, S.: 1980a, Theor.Chim.Acta 55, 257.
Clementi, E., Kołos, W., Lie, G.C. and Ranghino, G.: 1980b, Int.J.Quantum Chem. 17, 377.
Dalgarno, A.: 1962, Adv.Phys. 11, 281.
Daudey, J.P.: 1978, unpublished results.
Daudey, J.P., Novaro, O., Kołos, W. and Berrondo, M.: 1979, J.Chem.Phys. 71, 4297.
Dykstra, C.E., Schaefer III, H.F. and Meyer, W.: 1976, J. Chem.Phys. 65, 5141.
Džidič, I. and Kebarle, P.: 1970, J.Phys.Chem. 74, 1466.
Eisenschitz, R. and London, F.: 1930, Z.Phys. 60, 491.
González-Tovany, L., Beltran-Lopéz, V. and Novaro, O.: 1974, J.Chem.Phys. 60, 1694.
Groen, T.P. and van Duijneveldt, F.B.: 1978, unpublished results.
Gutowski, M., Kąkol, M. and Piela, L.: 1982, to be published.
Hirschfelder, J.O.: 1965, J.Chem.Phys. 43, S199.
Hirschfelder, J.O.: 1967, Chem.Phys.Letters 1, 325, 363.
Jaszuński, M.: 1980, Mol.Phys. 39, 777.
Jean, Y.: 1980, Chem.Phys.Letters 69, 216.

Jeziorski, B., Bulski, M. and Piela, L.: 1976, Int.J.Quant. Chem. 10, 281.
Jeziorski, B. and Kołos, W.: 1977, Int.J.Quant.Chem. 12S, 91.
Jeziorski, B., Schwalm, W.A. and Szalewicz, K.: 1980, J. Chem.Phys. 73, 6215.
Jeziorski, B., Szalewicz, K. and Chałasiński, G.: 1978, Int. J.Quant.Chem. 14, 271.
Jeziorski, B. and van Hemert, M.: 1976, Mol.Phys. 31, 713.
Kitaura, K. and Morokuma, K.: 1976, Int.J.Quant.Chem. 10, 325.
Kollman, P.A.: 1977, J.Am.Chem.Soc. 99, 4875.
Kollman, P.A. and Kuntz, I.: 1976, J.Am.Chem.Soc. 98, 6820.
Kołos, W.: 1973, in "The World of Quantum Chemistry" (ed. by R.Daudel and B.Pullman) D.Reidel Publishing Co., Dordrecht.
Kołos, W.: 1979, Theor.Chim.Acta, 51, 219.
Kołos, W. and Leś, A.: 1972a, Chem.Phys.Letters, 14, 167.
Kołos, W. and Leś, A.: 1972b, Int.J.Quant.Chem. 6, 1101.
Kołos, W., Nieves, F. and Novaro, O.: 1976, Chem.Phys.Letters 41, 431.
Langhoff, P.W., Karplus, M. and Hurst, R.P.: 1966, J.Chem. Phys. 44, 505.
Lathan, W.A., Pack, G.R. and Morokuma, K.: 1975, J.Am.Chem. Soc. 97, 6624.
Leś, A.: 1976, Chem.Phys.Letters 37, 399.
Lischka, H.: 1974, J.Am.Chem.Soc. 96, 4761.
Lischka, H.: 1979, Chem.Phys.Letters 66, 108.
Liu, B. and McLean, A.D.: 1980, J.Chem.Phys. 72, 3418.
Magnasco, V., Musso, G.F. and Figari, G.: 1976, J.Chem. Phys. 64, 120.
Morokuma, K.: 1971, J.Chem.Phys. 55, 1236.
Murrell, J.N.: 1978, Chem.Phys.Letters 55, 1.
Murrell, J.N., Randić, M. and Williams, D.R.: 1965, Proc. Roy.Soc. A, 284, 566.
Murrell, J.N. and Shaw, G.: 1967, J.Chem.Phys. 46, 1768.
Musher, I.J. and Amos, A.T.: 1967, Phys.Rev. 164, 31.
Musso, G.F., Magnasco, V. and Giardina, M.P.: 1974, J. Chem.Phys. 60, 3749.
Novaro, O. and Beltrán-Lopéz, V.: 1972, J.Chem.Phys. 56, 815.
Novaro, O., Castillo, S., Kołos, W. and Leś, A.: 1981a, Int.J.Quant.Chem. 19, 637.
Novaro, O., Cruz, S., Castillo, S., Kołos, W. and Leś, A.: 1981b, J.Chem.Phys. 74, 1118.
Novaro, O. and Kołos, W.: 1977, Chem.Phys. 67, 5066.
Novaro, O. and Nieves, F.: 1976, J.Chem.Phys. 65, 1109.
Novaro, O. and Yanez, F.: 1975a, Chem.Phys.Letters 30, 60.
Novaro, O. and Yanez, F.: 1975b, Int.J.Quant.Chem. 9S, 359.
Pecul, K.: 1980, Acta Phys.Polon. A58, 873.
Sadlej, A.: 1980, Mol.Phys. 39, 1249.

Schultz, J.A., McLean, W., Pedersen, L. and Jarnagin, R.C.: 1979, Chem.Phys.Letters 64, 230.
Sinanoğlu, O.: 1964, Adv.Chem.Phys. 6, 315.
Sinanoğlu, O.: 1967, Chem.Phys.Letters 1, 340.
Sinanoğlu, O.: 1968, Adv.Chem.Phys. 12, 283.
Sinanoğlu, O.: 1974, Theor.Chim.Acta 33, 279.
Smit, P.H., Derissen, J.L. and van Duijneveldt, F.B.:1979, Mol.Phys. 37, 501.
Sokalski, W.A., Hariharan, P.C. and Kaufman, J.J.: 1982, to be published.
Szalewicz, K. and Jeziorski, B.: 1979, Mol.Phys. 38, 191.
Umeyama, H. and Morokuma, K.: 1977, J.Am.Chem.Soc. 99,1316.
Van Duijneveldt-van de Rijd, J.G.C.M. and van Duijneveldt, F.B.: 1972, Chem.Phys.Letters, 14, 167.
Van der Avoird, A.: 1967, J.Chem.Phys. 47, 3649.
Werner, H.J. and Meyer, W.: 1976, Mol.Phys. 31, 855.

THE UNITARY GROUP AND THE ELECTRON CORRELATION PROBLEM

Isaiah Shavitt
Department of Chemistry, Ohio State University
Columbus, Ohio 43210, U.S.A.

ABSTRACT

A survey is given of the background and development of the unitary group approach and its graphically-based implementations for electronic structure calculations. The nature of the unitary group and its relationship to the electronic many-body problem are briefly discussed, as well as the reasons, purpose, and methodology of its application to this problem. As much as possible, this discussion is on a mostly intuitive, non-technical level. Finally, recent accomplishments and expected future developments are briefly mentioned.

In this contribution I shall try to address briefly the following questions:

1. What is the unitary group?
2. What has it got to do with the electron correlation problem?
3. How did we get into it?
4. Why did we get into it?
5. How do we use it?
6. What has been accomplished so far?
7. What else do we hope to do with it?

I shall try to address the questions in an intuitive, empirical way, as far as I can, rather than by detailed mathematical exposition. For more rigorous discussions and for the detailed historical background, the reader is referred to the cited literature, and particularly to the elegant exposition by Paldus.[1]

WHAT IS THE UNITARY GROUP?

This is an easy question to answer, at least formally. The unitary group U(n) is the group of all unitary matrices of order n. It is a compact Lie group, which depends analytically on n^2 real parameters. It is a subgroup of the (noncompact) general linear group GL(n) (the

P.-O. Löwdin and B. Pullman (eds.), New Horizons of Quantum Chemistry, 279–293.

group of all regular matrices of order n), but since U(n) contains essentially all of the topological complexity of GL(n) without suffering the burden of being noncompact, it turns out to be entirely adequate for our purpose.[1] This is, in fact, equivalent to saying that nothing of substance is lost if we restrict our orbital basis to be orthonormal.

While U(n) is a continuous group, and thus of infinte order, its irreducible representations (irreps) are all finite dimensional. These irreps can be described most effectively in terms of the chain of subgroups

$$U(n) \supset U(n-1) \supset \ldots \supset U(1). \tag{1}$$

The important property here is that the representation of U(n-1) subduced by an irrep of U(n) is simply reducible, i.e., it contains each irrep of U(n-1) at most once. This allows the construction and unique labeling of a basis for each irrep of U(n) in terms of the contained irreps of the subduction chain (1), and is the key to the Gelfand tableau labeling scheme[2] which, in the application to quantum mechanics, makes possible a systematic organization of the list of configuration state functions (CSF) in a configuration interaction (CI) expansion.[1]

WHAT HAS IT GOT TO DO WITH THE ELECTRON CORRELATION PROBLEM?

The unitary group is not a symmetry group of the Hamiltonian in the same sense as are point groups or the symmetric group S_N (the group of all permutations of N objects). Instead, it reflects a dynamical symmetry,[1] related to the fact that the many-electron Hilbert space on which the Hamiltonian operates is invariant to unitary transformations of the underlying one-electron basis. Stated differently, the second-quantized form of the Hamiltonian is indeed invariant, in a certain sense, to such transformations. Consider the spin-independent nonrelativistic Hamiltonian expressed in second-quantized form in terms of a finite, orthonormal spatial orbital basis $\underset{\sim}{\phi} = (\phi_1, \phi_2, \ldots, \phi_n)$,

$$\hat{H} = \sum_{ij} \sum_{\sigma} \langle i|\hat{h}|j\rangle X^{\dagger}_{i\sigma} X_{j\sigma} + \tfrac{1}{2} \sum_{ijk\ell} \sum_{\sigma\tau} \langle ij|\hat{v}|k\ell\rangle X^{\dagger}_{i\sigma} X^{\dagger}_{j\tau} X_{\ell\tau} X_{k\sigma} \tag{2}$$

where $\hat{h}$ and $\hat{v}$ are the one-electron and two-electron operators in $\hat{H}$, respectively, and where $X^{\dagger}_{i\sigma}$ and $X_{i\sigma}$ are the creation and annihilation operators, respectively, for an electron in orbital i with spin σ (= α or β). Obviously, this Hamiltonian is invariant to all the unitary transformations of the orbital basis described by the elements of U(n) (provided the integrals are also transformed). Furthermore, if we define the spin-averaged "excitation operators"

$$E_{ij} = \sum_{\sigma} X^{\dagger}_{i\sigma} X_{j\sigma} \tag{3}$$

(equivalent to the "basic symmetry operators" discussed by Matsen[3]), and apply the commutation relations of the creation and annihilation operators, we obtain the commutation relation

$$[E_{ij},E_{k\ell}] = \delta_{jk}E_{i\ell} - \delta_{\ell i}E_{kj} , \tag{4}$$

which is exactly the same as the commutation relation for the infinitesimal generators of U(n). We thus identify the E_{ij} with the generators of U(n), and rewrite the Hamiltonian (2) in terms of these generators,

$$\hat{H} = \sum_{ij}\langle i|\hat{h}|j\rangle E_{ij} + \frac{1}{2}\sum_{ijk\ell}\langle ij|\hat{v}|k\ell\rangle e_{ik,j\ell} , \tag{5}$$

where

$$e_{ik,j\ell} = E_{ik}E_{j\ell} - \delta_{jk}E_{i\ell} = N[E_{ik}E_{j\ell}] , \tag{6}$$

the last notation indicating that $e_{ik,j\ell}$ is the normal-ordered product (relative to the true vacuum) of $E_{ik}E_{j\ell}$. This final result, that the nonrelativistic Hamiltonian, projected onto a finite Hilbert space generated by a set of n spatial orbitals, is expressible in terms of the generators of U(n), is the ultimate, and perhaps most important expression of the connection between the unitary group and the many-electron problem.

HOW DID WE GET INTO IT?

The relevance of the unitary group to quantum mechanics was recognized early, particularly by Jordan,[4] but not until the classical studies (see, e.g., Weyl[5]) were augmented by the work of Gelfand,[2] Biedenharn,[6] Moshinsky,[7] and others (for a more complete history and detailed references see Paldus[1,8]) has it become possible to apply the unitary group formalism effectively in quantum mechanical treatments.

The first to take advantage of this formalism was Moshinsky,[9] who applied it to the nuclear shell-model theory. Other applications, mainly to atomic structure, followed (see Paldus[1,8] for more details), but the key development, as far as quantum chemistry is concerned, was the work of Paldus[1,8] who showed how the original formalism (which was general enough to handle both ordinary spin and the isospin of nuclear particles) could be simplified considerably for applications to electronic structure theory. Paldus replaced the relatively cumbersome Gelfand tableau notation by a more compact three-column tableau[8] (now known as a Paldus tableau), and showed how the general formulas for matrix elements of the unitary group generators[2] could be simplified considerably, at least for the elementary generators $E_{i,i+1}$.[8] Many other aspects of the formalism were expounded and developed by him in a clear and elegant manner,[1,8,10-16] and were thus made accessible to other quantum chemists who could apply them in their work.

I have found Paldus's work particularly inspiring, and used it as a starting point for the development of an organizational framework and a plan of attack for large-scale molecular configuration interaction (CI) calculations.[17–20] This plan makes use of a graphical representation of the Gelfand-Tsetlin basis (the basis of configuration state functions labeled by the different Gelfand tableaux or the corresponding simplified Paldus tableaux) to reveal the overall structure of the CI calculation and to devise effective strategies for its organization. This has become known as the "graphical" unitary group approach (GUGA).

The first GUGA-based CI implementation was carried out by Brooks and Schaefer,[21,22] and was followed by an interesting and innovative implementation by Siegbahn.[23–25] We are currently using a new set of computer programs[26] which combines some of the ideas of previous implementations (and particularly some of the ideas of Siegbahn[23–25]) to perform large-scale MCSCF/CI calculations of molecular electronic structure.

WHY DID WE GET INTO IT?

Conceptually, the configuration interaction (CI) method is very simple and straightforward, but if it is to be reasonably efficient, any actual implementation soon becomes exceedingly complex.[27] Because of the slow convergence of the CI expansion, a vast number of terms is needed for satisfactory accuracy in many applications, and efficient organization of the calculations and utilization of spin and other symmetries are very important. In fact, in typical "conventional" CI calculations,[27] a substantial part of the computational effort is devoted to organizational "overhead", formula determination, redundant calculations, and data access.

An important breakthrough which simplified the organizational structure of CI calculations, reduced data access problems, eliminated redundancies, and substantially improved efficiency was the introduction of the "direct CI" approach of Roos and Siegbahn.[28,29] One of its important advantages was that it made CI calculations substantially open ended, i.e., the size of the CI expansion was not limited appreciabl by the memory size of the computer. On the other hand it had a very important limitation: in its early form[28] it was restricted to closed-shell systems and to CI expansions made up of single and double excitations relative to a single closed-shell Slater determinant. While extensions to some other types of systems followed,[29,30] substantial generality was not achieved. A closely related approach, the "vector method", was introduced by Bender and co-workers,[31] and while it was much more general, this was at the expense of expanding in terms of Slater determinants without full spin adaptation. A later version[32] did introduce spin adaptation, but suffered from considerable complexity and other difficulties. Obviously, some new tools were needed, and the unitary group approach provided them.

In his 1976 exposition[1] Paldus wrote about this situation as follows: "...while the former authors [Hausman, Bloom and Bender[31]] work with spin-symmetry nonadapted states (Slater determinants), the latter [spin-adapted]approach of Roos and Siegbahn seems to be difficult to extend beyond the closed-shell biexcited case. It is hoped that the unitary group approach presented here could be very useful in bridging this gap." Time has proved him right, for that is exactly what the new GUGA-based programs do. They allow substantially open-ended, efficient, fully spin-adapted multireference direct CI calculations, i.e., CI expansion which include single and double (and even higher[33]) excitations relative to an arbitrary set of one or more configuration state functions (CSF). They do this by providing:

(a) a procedure for defining a systematically organized, orthonormal, spin-adapted CSF basis,

(b) an efficient procedure for the determination of one-electron and two-electron integral contributions to Hamiltonian matrix elements between the CSF, including easy determination of the position of the relevant CSF in the CSF basis, and

(c) an organizational framework for structuring the calculations so as to eliminate overhead, remove redundancies, and plan the data flow to maximize efficiency.

HOW DO WE USE IT?

Only a very brief and nonrigorous description can be given here of the procedures currently used[20,26] in GUGA-based CI calculations. We begin by considering the graph which describes the expansion set (the set of CSF) for a very small CI expansion (Fig. 1). On this graph each "level" represents one of the n orbitals in the orbital basis. Each "walk" (path) connecting the "head" with the "tail" and passing through one "arc" (line segment) and one "vertex" at each level represents one of the CSF by specifying the successive incorporation and spin coupling of orbitals. The coupling is carried out from the bottom up (the tail of the graph describes the vacuum and the head describes the final spin-adapted N-electron state). The graph is arranged on a lattice in such a way that the slopes of the arcs specify the occupancy and spin coupling of each orbital in the CSF and that the horizontal position of each vertex specifies the intermediate result of the coupling up to that level (in terms of the total occupancy N_i and resultant spin S_i for the first i orbitals - see Fig. 1). Four arc slopes are possible, and are identified by step numbers d_i=0,1,2,3 (as shown at the top level in Fig. 1). These correspond to an empty orbital ($d_i=0$, vertical arc, with $\Delta N_i=N_i-N_{i-1}=0$, $\Delta S_i=S_i-S_{i-1}=0$), a singly occupied orbital coupled to raise the resultant spin ($d_i=1$, $\Delta N_i=1$, $\Delta S_i=\frac{1}{2}$) or lower it ($d_i=2$, $\Delta N_i=1$, $\Delta S_i=-\frac{1}{2}$), and a doubly occupied orbital ($d_i=3$, $\Delta N_i=2$, $\Delta S_i=0$). A set of "weights" are also defined for the arcs and vertices (Fig. 1) which enable the easy identification of the ordinal number of each CSF in the expansion set.[17-20]

The relationship of this scheme with the unitary group U(n) is

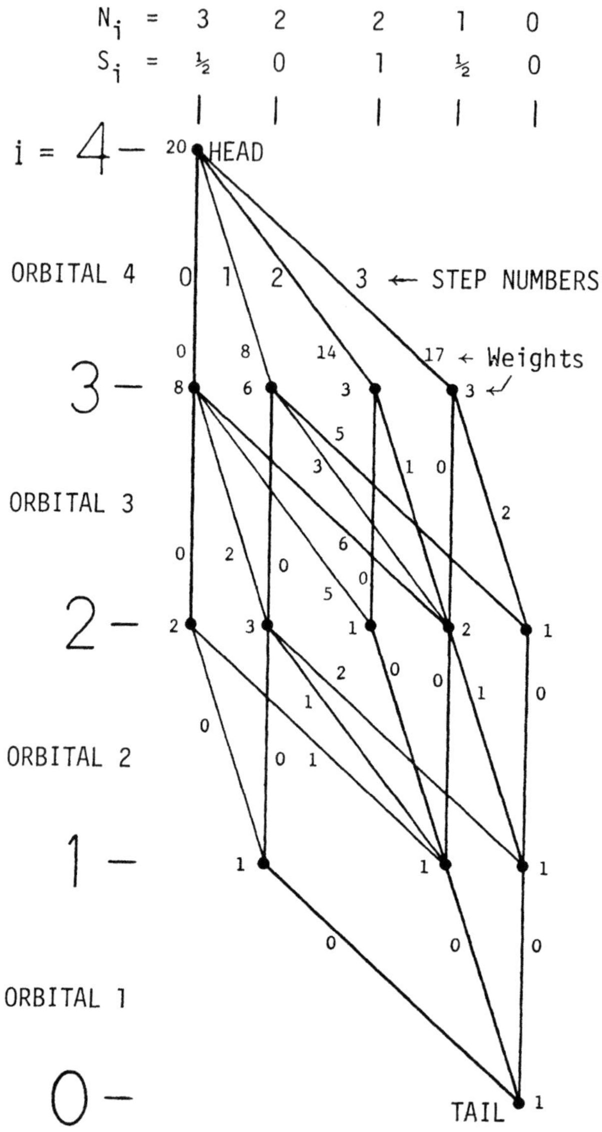

Figure 1. Graph representing the full CI expansion for a three-electron doublet state with four orbitals. The intermediate states represented by the horizontal positions of the vertices are identified above the graph. The step numbers (d_i = 0, 1, 2, 3) are shown for the top level. The vertex and arc weights are also shown (the ordinal number of a walk is equal to the sum of its arc weights, plus one).

that the head of the graph specifies a particular irrep of U(n) (which, in turn, fully specifies the total number of electrons N and the spin states S), each vertex at level i specifies an irrep of the subgroup U(i), and thus each walk specifies one of the members of an orthonormal basis for the irrep of U(n) by identifying the component irrep of U(i) at each level in the subduction chain of Eq. (1). In fact, the full CI expansion set for given N and S provides a complete basis for the corresponding irrep of U(n), and transforms in accordance with that irrep under unitary transformations of the orbital basis. Partial (truncated) CI expansions can also be handled efficiently in this formalism, particularly if the truncated expansion set can be described by simply deleting selected arcs and vertices from the graph (most common CI truncations can be described substantially in this way).

A key element of an effective CI formalism must be an efficient procedure for the determination of Hamiltonian matrix elements. Denoting an arbitrary CSF by the ket symbol $|m\rangle$, we see from Eq. (5) that the matrix elements are given by

$$\langle m'|\hat{H}|m\rangle = \sum_{ij}\langle i|\hat{h}|j\rangle\langle m'|E_{ij}|m\rangle + \tfrac{1}{2}\sum_{ijk\ell}\langle ij|\hat{v}|k\ell\rangle\langle m'|e_{ik,j\ell}|m\rangle. \quad (7)$$

Thus, the matrix elements of the generators E_{ij} and of the generator expressions $e_{ik,j\ell}$, Eq. (6), constitute the "coupling coefficients" which determine the contributions of the one-electron and two-electron integrals, respectively, to the matrix elements of $\hat{H}$. The efficient determination (or retrieval) of these coupling coefficients is essential for any CI treatment, and particularly for direct CI calculations (in which such coefficients need to be obtained again in each iteration in the solution of the matrix eigenvalue equation).

Formulas for matrix elements of the unitary group generators were first derived by Gelfand and Tsetlin,[2] but were too unwieldy for convenient application. More useful formulas were obtained by spin-algebraic techniques by Gouyet et al.,[34] by Drake and Schlesinger,[35] and by Paldus and Boyle.[15] Essentially equivalent results were also obtained purely within the graphical representation approach of GUGA.[18,20,36] The graphical representation of a matrix element of the generator E_{ij} between two CSF is illustrated in Fig. 2. That element vanishes unless the walks representing the two CSF coincide outside the range of levels extending between orbital levels i and j, and is otherwise independent of those coincident portions of the walks. Its value depends on the "loop" formed by the two walks between levels i and j, and can be obtained as a product of "segment values" derived from the individual orbital levels of the loop.[18,20] The procedures for the two-body matrix elements (of $e_{ik,j\ell}$), though somewhat more complicated, are essentially similar.[15,16,20]

Two aspects of this approach to coupling coefficient evaluation are important to our purpose: First, since there is a fairly limited number of segment types which produce nonvanishing factors (segment values) in the evaluation of generator matrix elements, we can reverse the

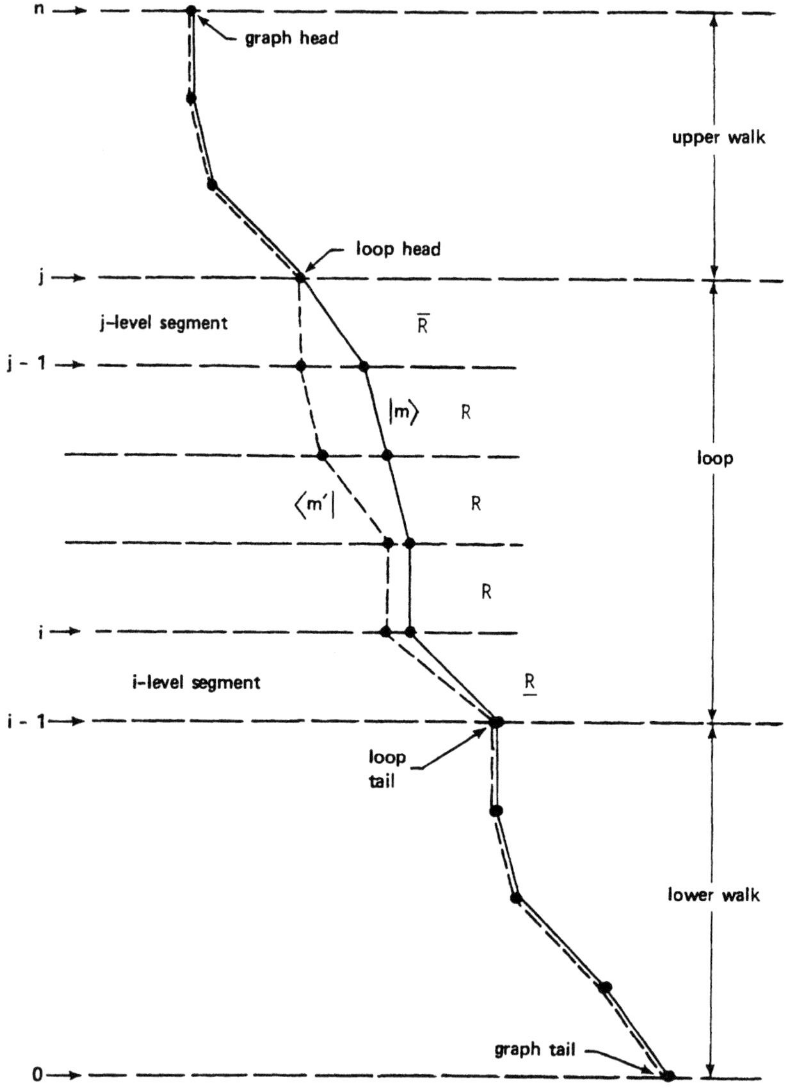

Figure 2. Graphical representation of a generator matrix element $\langle m'|E_{ij}|m\rangle$ ($i<j$). The ket state $|m\rangle$ and the bra state $\langle m'|$ are shown as full and dashed lines, respectively. The matrix element vanishes unless the walks representing the two states coincide outside the loop. The value of the matrix element depends only on the loop, and is obtained as a product of "segment values."

"conventional" approach, in which specific matrix elements of $\hat{H}$ would be considered one by one and their formulas determined by examining the loops formed by their walks on the graph, and instead construct loops by stringing together appropriate segment types with nonvanishing values for any specified generator (or two-body operator), and then connect these loops in all possible ways (without affecting their values) to the head and tail of the graph. This reversal of procedure is of course the key element of the direct CI approach, but is only applicable if there is a convenient way (as there is in GUGA through the arc and vertex weights) for identifying the CSF involved in each generated matrix element. Second, since a segment may be shared by many loops, and since its value is independent of the other segments of the loop, efficient procedures can be organized[18–25] to reduce redundant computation by reducing the number of times any particular segment (or larger section of a loop) is generated and evaluated.

The most efficient organization of GUGA procedures to eliminate redundancies, reduce data handling, and increase efficiency is due to an idea by Siegbahn.[23–25] This approach is designed specifically for multireference CI calculations, which normally include all single and double excitations (or an "interacting" subset of those) relative to a set of reference configurations. It is based on an organization of the orbital levels and the graph as illustrated in Fig. 3.[20,24,26] This graph consists of an "external" part at the bottom, containing the virtual orbital levels (those orbitals which are unoccupied in all the reference CSF), and an "internal" part, containing the closed-shell levels (those which are doubly occupied in all reference CSF) and, at the top, the open-shell levels (singly or unequally occupied in the reference CSF). (This particular vertical ordering of the three sets of levels is preferred for technical reasons[20–22] first pointed out by Brooks and Schaefer.[21])

In a typical CI calculation, the external part of the graph contains by far the most levels, but is also quite simple in structure. As pointed out by Siegbahn,[23–25] it is beneficial to factor the coupling coefficients (and thus the loop values) into internal and external contributions. The former are generated and evaluated by the general procedures described above (and tabulated on a relatively short "formula tape"), while the latter are easily supplied explicitly by the program structure during the diagonalization iterations. This is the approach used in most current GUGA implementations.[24,26,33]

WHAT HAS BEEN ACCOMPLISHED SO FAR?

We shall not enter here into a detailed description of computations and results obtained with unitary group techniques. Instead we shall just mention briefly a few recent achievements to demonstrate the capabilities of the modern implementations of these methods.

Perhaps the most remarkable achievement among GUGA applications

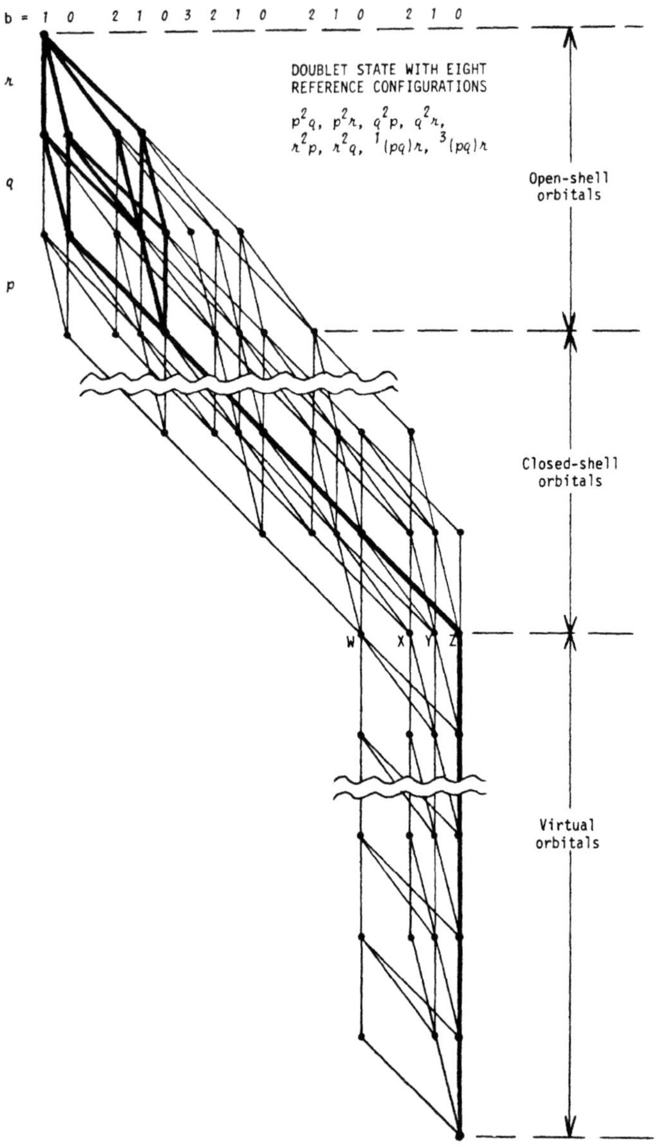

Figure 3. Typical organization of the levels and graph for a multi-reference CI calculation. The heavy lines represent the reference configurations. The complete graph contains all the single and double (and a few higher) excitations relative to the reference functions (various limitations on the expansion set are implemented through an internal-space index vector). The b values shown at the top indicate the resultant spin for the intermediate states at the various vertices ($b_i = 2S_i$). The open-shell and closed-shell levels make up the "internal space," while the "external space" consists of the virtual levels.

has been a very recent calculation on ethylene in a DZP basis including all single, double, and certain classes of triple and quadruple excitations,[33] 1,046,758 CSF in all! This calculation was carried out on a Harris 800 minicomputer, and took 14 hours per iteration. If nothing else, it certainly demonstrates the open-endedness of the approach. A more practical application, one which allows routine calculation of points on a potential curve, was described by Roos,[37] and involved a potential curve for an $^{11}\Sigma_g^+$ state of Cr_2, using about 220,000 CSF. This calculation, on an IBM 3033 computer, took about 20 minutes per iteration. (It was stated, however, that a vectorized version of the program required only 20 seconds per iteration for 200,000 CSF on a Cray 1 vector computer at Daresbury, England.) Our own efforts have been somewhat more modest, and the largest calculation we have carried out so far was on the O_2 molecule, in a (5s4p2d) STO basis set.[38] This was an MCSCF/CI calculation with a full-valence reference set of 14 CSF (in C_{2v} symmetry), and a total of 164,296 CSF, requiring an iteration time of 9.3 minutes on an Amdahl 470/V6 computer (rated at 3.5 mips - million operations per second - about half the speed of a CDC 7600).

In addition to configuration interaction calculations, several other types of applications have been made of GUGA procedures. These include implementation of MCSCF calculations[26,39-41] (including procedures for second-order convergence[40]), energy gradient calculations[41,42] and coupled electron pair approximation (CEPA) treatments.[43] A procedure was also outlined for the extension of GUGA techniques to electron propagator calculations for open-shell states.[44] In this latter application the graph is generalized to have more than one head, in order to allow the representation of (N±1)-electron states on the same graph as the original N-electron system.

WHAT ELSE DO WE HOPE TO DO WITH IT?

In addition to the continued expansion of the capabilities of the current programs (including the implementation of versions for vector computers and array processors), the most important extension required of GUGA techniques is into the realm of nonvariational treatments, particularly for perturbation theory and coupled cluster calculations. In these areas it is hoped that the unitary group approach can provide effective generalizations for open-shell and multiconfigurational zero-order states, particularly in the framework of quasidegenerate perturbation theory (QDPT). A modest start has been made in the implementation of the CEPA[43] and electron propagator[44] methods mentioned previously, and we have started some work on the development of GUGA-based procedures for QDPT, but much remains to be done before this latter effort leads to effective procedures and computer programs.

One of the requirements for an effective many-body treatment is a satisfactory particle-hole formalism. One such approach for the unitary group has been developed by Paldus and Boyle.[45] An alternative approach has also been proposed,[20] based on a "repartitioning" of the Hamiltonian in the form

$$\hat{H} = E_0 + \sum_{ij} \langle i|\hat{f}|j\rangle F_{ij} + \tfrac{1}{2} \sum_{ijk\ell} \langle ij|\hat{v}|k\ell\rangle f_{ik,j\ell} , \tag{8}$$

where E_0 is a reference energy, $\hat{f}$ is a Fock-like operator, and F_{ij}, $f_{ik,j\ell}$ are "modified unitary group operators" defined in terms of a set of reference occupancies μ_i. The μ_i define a spin-averaged reference state with energy E_0, and the modified operators F_{ij} and $f_{ik,j\ell}$ are seen to be normal-ordered forms of E_{ij} and $e_{ik,j\ell}$, respectively, relative to the spin-averaged reference state (some generalization of the normal-ordering concept is implied in the case of an open-shell reference state). A zero-order Hamiltonian can then be defined as

$$\hat{H}_0 = E_0 + \sum_{i} \langle i|\hat{f}|i\rangle F_{ii} , \tag{9}$$

and a general perturbation theory approach can easily be constructed on this basis.

A successful implementation of the nonvariational many-body techniques within the spin-adapted unitary group framework could bring the advantages of the size-consistency and faster convergence characteristics of such methods to a class of problems which have been difficult to treat satisfactorily so far because they require open-shell or multideterminantal reference functions.

Acknowledgment

This work was supported by the National Science Foundation under Grant CHE-7825191 and the National Aeronautics and Space Administration under Grant NSG-2231.

REFERENCES

1. Paldus, J.: 1976, in "Theoretical Chemistry: Advances and Perspectives," Vol. 2, edited by H. Eyring and D. G. Henderson Academic Press, New York), p. 131.

2. Gelfand, I. M., and Tsetlin, M. L.: 1950, Dokl. Akad. Nauk SSSR 71, p. 825, p. 1017.

3. Matsen, F. A.: 1974, Int. J. Quantum Chem. Symp. 8, p. 379.

4. Jordan, P.: 1935, Z. Phys. 94, p. 531.

5. Weyl, H.: 1939, "The Classical Groups, Their Invariants and Representations" (Princeton University Press, Princeton, New Jersey).

6. Biedenharn, L. C.: 1963, J. Math. Phys. 4, p. 436; Baird, G. E., and Beidenharn, L. C.: 1963, J. Math. Phys. 4, p. 1449; Louck, J. D., and Beidenharn, L. C.: 1970, J. Math. Phys. 11, p. 2386.

7. Moshinsky, M.: 1963, J. Math. Phys. 4, p. 1128.

8. Paldus, J.: 1974, J. Chem. Phys. 61, p. 5321.

9. Moshinsky, M.: 1968, "Group Theory and the Many-Body Problem" (Gordon and Breach, New York).

10. Paldus, J.: 1975, Int. J. Quantum Chem. Symp. 9, p. 165.

11. Paldus, J.: 1976, Phys. Rev. A 14, p. 1620.

12. Paldus, J.: 1977, in "Electrons in Finite and Infinite Structures," edited by P. Phariseau and L. Scheire (Plenum, New York), p. 411.

13. Paldus, J.: 1979, in "Group Theoretical Methods in Physics," edited by W. Beiglbock, A. Bohm, and E. Takasugi (Springer, New York), p. 51.

14. Wormer, P. E. S., and Paldus, J.: 1979, Int. J. Quantum Chem. 16, p. 1307; Paldus, J., and Wormer, P. E. S.: 1979, Int. J. Quantum Chem. 16, p. 1321; Paldus, J., and Wormer, P. E. S.: 1981, Int. J. Quantum Chem. 18, p. 841.

15. Paldus, J., and Boyle, M. J.: 1980, Phys. Scripta 21, p. 295.

16. Paldus, J.: 1981, in "The Unitary Group for the Evaluation of Electronic Energy Matrix Elements" (Lecture Notes in Chemistry 22), edited by J. Hinze (Springer, Berlin), p. 1.

17. Shavitt, I.: 1977, Int. J. Quantum Chem. Symp. 11, p. 131.

18. Shavitt, I.: 1978, Int. J. Quantum Chem. Symp. 12, p. 5.

19. Shavitt, I.: 1979, Chem. Phys. Lett. 63, p. 421.

20. Shavitt, I.: 1981, in "The Unitary Group for the Evaluation of Electronic Energy Matrix Elements" (Lecture Notes in Chemistry 22), edited by J. Hinze (Springer, Berlin), p. 51.

21. Brooks, B. R., and Schaefer III, H. F.: 1979, J. Chem. Phys. 70, p. 5092.

22. Brooks, B. R., Laidig, W. D., Saxe, P., Handy, N. C., and Schaefer III, H. F.: 1980, Phys. Scripta 21, p. 312.

23. Siegbahn, P. E. M.: 1979, J. Chem. Phys. 70, p. 5391.

24. Siegbahn, P. E. M.: 1980, J. Chem. Phys. 72, p. 1647.

25. Siegbahn, P. E. M.: 1981, in "The Unitary Group for the Evaluation of Electronic Energy Matrix Elements" (Lecture Notes in Chemistry 22), edited by J. Hinze (Springer, Berlin), p. 119.

26. Lischka, H., Shepard, R., Brown, F. B. , and Shavitt, I.: 1981, Int. J. Quantum Chem. Symp. 15, p. 91.

27. Shavitt, I.: 1977, in "Methods of Electronic Structure Theory" ("Modern Theoretical Chemistry," Vol. 3), edited by H. F. Schaefer III (Plenum, New York), p. 189.

28. Roos, B. O.: 1972, Chem. Phys. Lett. 15, p. 153.

29. Roos, B. O., and Siegbahn, P. E. M.: 1977, in "Methods of Electronic Structure Theory" ("Modern Theoretical Chemistry," Vol. 3), edited by H. F. Schaefer III (Plenum, New York), p. 277.

30. Lucchese, R. R., and Schaefer III, H. F.: 1978, J. Chem. Phys. 68, p. 769.

31. Hausman Jr., R. F., Bloom, S. D., and Bender, C. F.: 1975, Chem. Phys. Lett. 32, p. 483; Hausman Jr., R. F., and Bender, C. F.: 1977, in "Methods of Electronic Structure Theory" ("Modern Theoretical Chemistry," Vol. 3), Edited by H. F. Schaefer III (Plenum, New York), p. 319.

32. Bender, C. F.: 1979, J. Comput. Phys. 30, p. 324.

33. Fox, D., Handy, N., Saxe, P., and Schaefer III, H. F.: 1982, 183rd Amer. Chem. Soc. Natl. Mtg., Las Vegas, Nevada.

34. Gouyet, J.-F., Schranner, R., and Seligman, T. H.: 1975, J. Phys. A 8, p. 285.

35. Drake, G. W. F., and Schlesinger, M.: 1977, Phys. Rev. A 15, p. 1990; Schlesinger, M.: 1981, in "The Unitary Group for the Evaluation of Electronic Energy Matrix Elements" (Lecture Notes in Chemistry 22), edited by J. Hinze (Springer, Berlin), p. 144.

36. Shavitt, I., and Paldus, J.: unpublished.

37. Roos, B. O., Nelin, C., and Ågren, H.: 1982, 183rd Amer. Chem. Soc. Natl. Mtg., Las Vegas, Nevada.

38. Brown, F. B., Shepard, R., Lischka, H., and Shavitt, I.: work in progress.

39. Brooks, B. R., Laidig, W. D., Saxe, P., and Schaefer III, H. F.: 1980, J. Chem. Phys. 72, p. 3837.

40. Shepard, R., and Simons, J.: 1980, Int. J. Quantum Chem. Symp. 14, p. 211; Shepard, R., Shavitt, I., and Simons, J.: 1982, J. Chem. Phys. 76, p. 543.

41. Brooks, B. R., Laidig, W. D., Saxe, P., Goddard, J. D., and Schaefer III, H. F.: 1981, in "The Unitary Group for the Evaluation of Electronic Energy Matrix Elements" (Lecture Notes in Chemistry 22), edited by J. Hinze (Springer, Berlin), p. 158.

42. Brooks, B. R., Laidig, W. D., Saxe, P., Goddard, J. D., Yamaguchi, Y., and Schaefer III, H. F.: 1980, J. Chem. Phys. 72, p. 4652.

43. H. Lischka: 1982, 183rd Amer. Chem. Soc. Natl. Mtg., Las Vegas, Nevada.

44. Born, G., and Shavitt, I.: 1982, J. Chem. Phys. 76, p. 558.

45. Paldus, J., and Boyle, M. J.: 1980, Phys. Rev. A 22, p. 2299; Boyle, M. J., and Paldus, J.: 1980, Phys. Rev. A 22, p. 2316.

FROM HETEROGENEOUS CATALYSIS TO SURFACE CHEMISTRY AND CHEMISORPTION

Massimo Simonetta
Istituto di Chimica Fisica dell'Università e
Centro C.N.R.
Via Golgi 19, 20133 Milano, Italy

The techniques currently used in chemical engineering to design reactors where heterogeneous catalysis is in operation and the necessary kinetic informations are briefly reviewed. A discussion of experimental and theoretical results in the field of chemisorption and surface chemistry is offered with particular emphasis on quantum mechanical and force field calculations for adsorbate-surface bonding.

Many processes of industrial chemistry, especially organic chemistry and petrochemistry, are based on heterogeneous catalysis in the main synthetic step. In the reactor a mixture of gases, usually at high temperature and pressure, flows through the catalyst, a solid in a conveniently dispersed form. The design of a reactor of this kind is one of the most exciting and challenging task for a chemical engineer. Let us consider a special case: the production of methanol from methane or natural gas. The starting hydrocarbons are first transformed to a mixture of CO and H_2. The reaction

$$CO + 2H_2 \rightleftarrows CH_3OH$$

is exothermic so that the equilibrium is shifted to the right at low temperature. However the temperature must be high enough to ensure a convenient rate and to avoid too high temperatures a catalyst must be used. For many years zinc--cromium catalysts were used, at temperatures in the range 300-400°C and pressures up to 350 atm. A small reactor designed years ago for the production of about 100 t/day of methanol, is shown schematically in fig. 1. It is of the conventional quench type. With our catalyst the optimal working temperature was in the range 330-360°C and minimisation of utilities within battery limits suggested a pressure of 280 atm. Many factors had to be considered, for

P.-O. Löwdin and B. Pullman (eds.), New Horizons of Quantum Chemistry, 295–304.

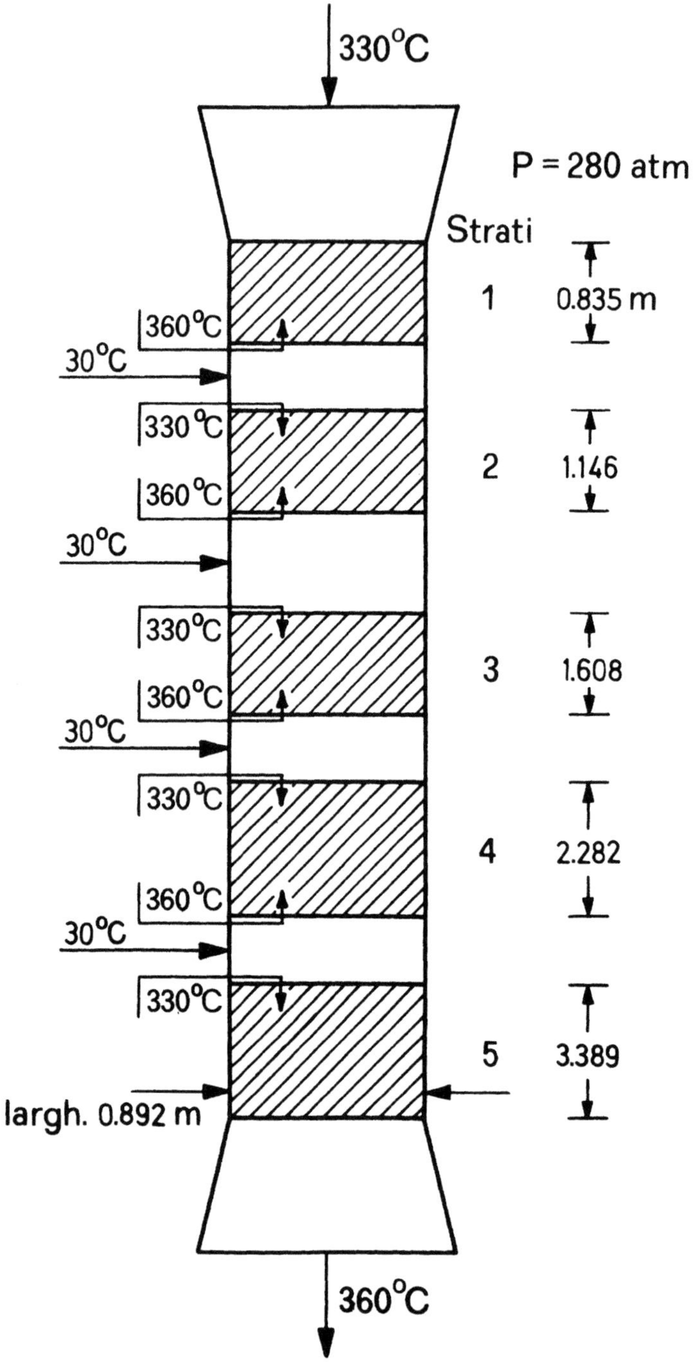

Fig. 1 - Scheme for a methanol synthesis reactor

example corrosion, pressure drop, heat transfer, instrumentation and control and last but not least, the cost. If we concentrate on chemical aspects the central point is the calculation of the amount of catalyst needed for a given production. To this aim one must rely on kinetic experiments. From reaction rates, determined on the same catalyst, at different constant values of temperature and pressure and with different feed gas composition and space velocities it has been possible to build a model consistent with all experimental data, leading to the following phenomenological equation (1):

$$r = \frac{f_{CO}p_{CO}f^2_{H_2}p^2_{H} - \dfrac{f_{CH_3OH}p_{CH_3OH}}{K_{eq}}}{(A + Bf_{CO}p_{CO} + Cf_{H_2}p_{H_2} + Df_{CH_3OH}p_{CH_3OH})^3}$$

where r indicates the moles of CH_3OH formed in unit time on the unit weight of catalyst, p_i and f_i are the partial pressure and fugacity coefficient of component i, K_{eq} is the thermodynamic equilibrium constant for the reaction and A, B, C, D are empirical functions of the temperature. Since

$$\frac{W}{F} = \int_0^x \frac{d_x}{r}$$

where F is the feed in mole/unit time and x is the conversion, the weight of needed catalyst is easily obtained (W). It was found that in order to keep the temperature within the optimal range the catalyst had to be divided in five layers, of different height, as shown in fig. 1. When the reaction was put in operation the expected production was obtained and measured temperatures were all within 2°C from the predicted value. However the whole calculation was based on an empirical equation and empirical values for the A, B, C, D constants. The composition and preparation recipe of the catalyst, to ensure the necessary chemical, thermal and mechanical properties, were the outcome of experimental, in some respect empirical, work. Recently a new catalyst was introduced on the market, containing copper. The new catalyst is more active and allows operation of the reactor at 50 atm. This brings about an enormous saving in power consumption, both for compression and circulation of the gas mixture and allows the use of centrifugal compressors, where there is no waste of energy in the forth and back piston motion. Methanol is now a much cheaper chemical than it used to be, and it is now available as a fuel. Specially when overseas shipping is involved it competes with methane itself. This long preambol wanted to show how important the catalyst can be on the establishment and development of chemical processes and how desirable it would be to be able and "calculate" the catalys that is to predict its optimum composition, efficiency for conversion, selectivity and working conditions. And the enormous effort that chemists and physicists all over the world are producing toward this goal is easily justified. From what it has been said before it is quite clear that understanding heterogeneous catalysis is an awful difficult problem and that an oversemplified approach is needed at the moment. One reasonable model considers the following steps: 1) diffusion of gaseous species from gas phase to surface 2) adsorption of reactants 3) reaction of adsorbed species 4) desorption of products 5) diffusion into the gas phase. Each step can be multistaged. Diffusion takes place first in the gas phase but ultimately within the pores of the catalys where geometrical features become of preminent importance. The structure of the catalyst is sometime very difficult to determine. For example in many cases the catalytic active species is a metal supported on silica, alumina, zeolite. The metallic atoms can form clusters whose dimension structu re and location determine the activity (2). The second step is usually described in connection with the presence of acti sites on the surface, but the physical and chemical nature of these sites may be often rather obscure. The reaction between adsorbed species or with a molecule from the gas phase is the most difficult step. Many studies are available in the literature, where the mechanic aspect of heterogeneou reactions are discussed, mainly on the basis of kinetics, isotope effects, yields, isomer distribution etc. (3), but no interpretation at atomic level is usually given. We can notice here the deep dicotomy between the plentiful of information on the properties of catalysts that offer a safe guide in the practical application and the scanty knowledge that we have in terms of atomic and molecular structure (4). Reaction between adsorbates being so complex it is no surpr se that most studies are devoted to chemisorption. In the last 10-15 years, due to progress in ultra high vacuum techno logy, a number of techniques became available for investiga- tion of clean surfaces and adsorbed species. Let me just mention LEED, AES, UPS, XPS, EELS, SIMS, FEM, FIM, measureme of work functions and others (5, 6). Of course these techni ques are normally applied to systems that are not only much better defined, but also working in quite different conditio from heterogeneous catalysis. The idea is that one should start with structures and possibly chemical reactions that obtain in simple situations before one can master the comple ones. Usually systems under study are kept in high vacuum, often at low temperature, and clean metallic surfaces are involved, or even just the low Miller indexes planes of a single crystal. It is well known that such regular surfaces would perform very poorly as catalysts but again the attitude is to deal with systems that can be clearly defined at the atomic scale. As an example let us mention LEED. By this technique one can determine the crystal structure of a metallic surface, that is one can see if it is reconstructed and how, with respect to the bulk. Then one can observe a mono layer or less of a simple molecule or atom and determine the structure of this layer and its register with respect to the underlying structure. Extracting this information from experimental data is itself a formidable task. From the machine one gets a pattern of diffracted beams before and after absorption and the intensities of each beam as a function of the wave length of the incident electron beam can be measured. The interpretation of the I-V curves is the main problem. It is analogous to the problem in X-ray diffraction but due to the higher scattering amplitude of electrons multiple scattering theory must be applied (7). And of course since the mathematics is rather complex the technique is still at the trial and error stage. This means that a given model is assumed, the I-V curves are calculated by some approximate quantum mechanical treatment and compared with experiment. Then a new model is tried and so on till the best fitting is obtained. In fig. 2 an example is shown: namely two beams produced by C_2H_2 adsorbed on Ni(111) (8). The calculated curves have been obtained by the layer doubling method, using a program that we due to the courtesy of professor Pendry. At given temperature and coverage beams can be collected at different experimental conditions (incidence and azimuthal angles) so that a considerable number of curves is collected and when all can be fitted in a calculation the assumed structure has a good chance to be the correct one. The method is very expensive since it needs a lot of computing time and improvement in theory is badly needed.

Since the experimental approach to chemisorption is so difficult this seems to be an ideal area where theoretical methods should be very welcomed. Of course when complexity is prohibitive for the experimentalist, the problem must be of a hard type also to theoreticians. However in the last few years a considerably effort has been produced toward an understanding of the nature of surface bonds and surface structures. Most treatments are quantum mechanical at different levels of sophistication. Two different models have been mostly used in this context: 1) periodicity on the surface has been recognized and Bloch functions have been introduced, 2) a cluster of atoms was considered sufficient to reproduce the essential features of the system. The first method is to be recommended when regular clean surfaces are investigated, while the second one seems to be the most appropriate in the description of adsorption sites. Most calculations have been restricted to the study of the interaction of a single atom

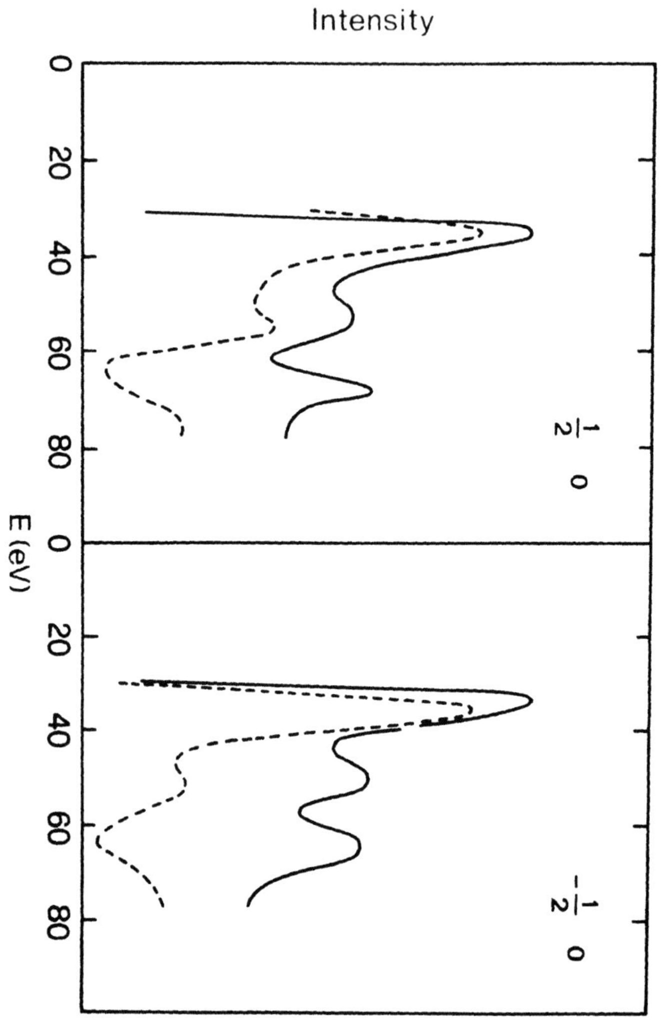

Fig. 2 - Experimental (——) and calculated (---) intensity/energy curves for diffracted beams from C_2H_2 on Ni(111)

or molecule with a metallic surface or cluster. The literature is rich on this subject and I can mention only a few examples. Bloch functions built on large unit cells have been used by Fukui et al. (9, 10) to study the chemisorption of H_2 and ethylene on flat (111) and stepped 3(111)x(100) platinum surfaces. The electronic properties of clean surfaces were investigated first, and the most favourable sites for adsorption have been identified. The smallest activation energy for dissociative adsorption of H_2 was found in a site where bonds are formed with the two Pt atoms at the top and bottom of a step. In the case of ethylene it was found that C-H bond breaking is facile on (110) and stepped surfaces but the effect is absent on (100) and (111) faces.
Messmer has used the SCF X_α method to study metal clusters as models for surface phenomena (11). He obtained very good agreement when comparing calculated density of states curves with the results from photoemission spectra of oxygen chemisorbed on Al(100). The embedding theory has been applied to study chemisorption of H_2 on Ti(0001). The adsorbate and surface adsorption site are embedded in the remainder of the cluster electron distribution. From an approximate SCF cluster model for the lattice a local N-electron subspace is obtained that includes the H_2 adsorbate and it is treated accurately by use of an extended basis set and configuration interaction. Binding energies in good agreement with experiment are obtained (12). Goddard has used his generalized valence bond method to study adsorption of CO on Ni(100). Due to complexity in the calculation the lattice had to be modelled with only 14 Ni atoms. Reasonable agreement with experiment was obtained for the geometry, bond energy, vibrational frequencies and ionisation potentials. This supports the assumption that CO bonds on top of a metal atom. A still different approach is used by Koutecky (13). Clusters made by a small number of alkali metal atoms are investigated by sophisticated molecular orbital methods including extensive multireference CI. Different atomic arrangements are considered and since some of them show biradical character a connection is suggested with catalytic activity of geometric inequalities in the surface.
If besides chemisorption, surface reactivity is the object of analysis, more empirical methods of calculations must be used. We relied an Extended Hückel Theory for chemisorption (14). The adsorption of acetylene on Pt(111) is perhaps the process for which the most extensive investigation has been carried on both experimentally by Somorjai's group (15) and theoretically in our Institute (16). It was found that acetylene adsorbes at low temperature giving a metastable structure where the molecule is bound in a top site parallel to the surface. By heating at room temperature the stable structure obtains, where ethylidyne fragment is perpendiculary bound over a triangular site. The ethylidyne fragment is

formed by hydrogenation of a vinylidene precursor. Calculated and experimental C-Pt and C-C distances are in beautiful agreement, as well as the fragment-metal bond stretching frequency. The same kind of behaviour has been found on Rh(111).

Mono- and difluoroacetylene are expected to show a similar behaviour but since it has been found that the binding energy to the surface increases as the HOMO-LUMO gap in the adsorbate decreases, one would expect the activation barrier for rearrangement from the acetylenic to the vinylidene species to decrease in the order HC≡CH>HC≡CF>FC≡CF. The calculations also show that binding energies are larger for adsorption of acetylene on kinks that on flat surfaces. This finding seems to be in line with the fact that stepped surfaces offer better catalytic efficiency than flat surfaces. It is also worth noticing that E.H. calculations (17) show that the activation barrier for the formation of a kink or a hole or an adatom on a flat surface of metals such as Ni, Rh, Pt is always close to 20 kcal/mole. Since this value is near to values found for the overall reaction in most industrial heterogeneous reactions (18), theory suggests that formation of active sites is one of the steps to be considered when trying to unrevel the mechanism of an heterogeneous process. It is gratifying that recently it has been shown through kinetic studies on single crystal iron catalysts (19) that the catalytic activity for ammonia synthesis is structure sensitive, and varies in the order Fe(111)>Fe(100)>Fe(110). For the Fe(111) face an activation energy of 19.4 kcal/mole has been measured.

Other organic molecules have been adsorbed on Pt(111) such as benzene, naphthalene (20), azulene (21). In the last two examples organic molecules form an ordered layer parallel to the metal surface, and it was suggested that adsorbate-adsorbate interactions are predominant over adsorbate-substrate interactions in determining the structure. This seems to be an ideal situation for a force field calculation, which has been indeed carried on, with standard parametrisation (22). As a result it has been possible to confirm for naphthalene the existence of a glide plane in the pgg surface group and determine the most favourable orientation of the organic molecule. In the case of azulene the structures with and without glide plane show about the same packing energy and no choice can be made on this basis.

Another different approach has been used in the dynamical study of dissociation of H_2 on a Pt(111) surface. A molecular beam experiment has shown that the probability of dissociation is independent of the azimuthal angle and decreases with increasing the incidental angle. Also a barrier of about 2.5 kcal/mole was found (23). The same process has been studied by a Monte Carlo quasi classical trajectory method within the framework of a simple LEPS potential, which describes the interaction of a diatomic molecule with a rigid solid surface (24). The results are very encouraging. When the rigid surface has been replaced by an Einstein solid the results stood almost unchanged, and the assumption that only a small fraction of the energy was exchanged was completely confirmed (25).

I started describing an industrial reactor and the kind of calculations that are necessary to the chemical engineer to correctly design the vessel. At the end empirical, semiempirical,ab initio and dynamical calculations have been presented for a number of simple systems where chemisorption or surface reactions occur. There is still a tremendous gap between the two areas, involving complexity, accuracy, reliability. We are far away from the possibility of replacing kinetic experiments with ab initio computations of reaction rates for heterogeneous systems of chemical interest. Neverthless we are marching in the right direction, altogether theoreticians and experimentalists; and we have marvellous equipments and excellent scientists: I feel the goal can't be missed.

REFERENCES

(1) G. Natta, G. Mazzanti, and I. Pasquon: La Chim. e l'Ind. 1955, 37, p. 1015.
(2) P. Gallezot, A.I. Bienenstock, and M. Boudart: Nouv. J. Chimie 1978, 2, p. 263.
(3) P. Parayre, V. Amir-Ebrahimi, and F.G. Gault: J. C. S. Faraday I 1980, 76, p. 1723.
(4) G. Somorjai: Chemistry in Two Dimensions: Surfaces, Cornell Univ. Press, Ithaca, 1981.
(5) G. Ertl, J. Küppens: Energy Electrons and Surface Chemistry, Verlag Chemie, Weinheim, 1974.
(6) M.W. Roberts, C.S. Mc Kee: Chemistry of the Metal-Gas Interface, Clarendon Press, Oxford, 1978.
(7) J. Pendry: Low Energy Electron Diffraction, Academic Press, New York 1974.
(8) G. Casalone, M.G. Cattania, F. Merati, and M. Simonetta: Surf. Sci. in the press.
(9) H. Kobayashi, S. Yoshida, H. Kato, K. Fukui, and K. Tarama: Surface Sci. 1979, 79, p. 189.
(10) H. Kobayashi, S.Yoshida, H. Kato, K. Fukui, K. Tarama, and M. Yamaguchi: Surface Sci. 1980, 97, p. 329.
(11) P.P. Messmer, and D.R. Salahab: in Computer in Chemical Education and Research, Ed. E. Ludena, N. Sabelli, A. Whal, Plenum Press, New York 1977.
(12) P. Cremaschi, and J.C. Whitten: Surface Sci. 1981, 112, p. 343.
(13) G. Pacchioni, H. Beckmann, and J. Koutecky: Chem. Phys. Letters 1982, 87, p. 151.

(14) A.B. Anderson, and R. Hoffmann: J. Chem. Phys. 1974, 61, p. 4545.
(15) L.L. Kesmodel, L.H. Dubois, and G.A. Somorjai: J. Chem. Phys. 1979, 70, p. 2180.
(16) A. Gavezzotti, and M. Simonetta: Surface Sci. 1980, 99, p. 453.
(17) M. Simonetta: Nouveau J. de Chimie 1982, 6, p. 65.
(18) A. Nielsen: An investigation on Promoted Iron Catalysts for the Synthesis of Ammonia, Jul Gjellerups Forlag, Copenhagen 1956.
(19) N.D. Spencer, R.C. Schomaker, and G.A. Somorjai: J. of Catalysis 1982, 74, p. 178.
(20) D. Dahlgren, and J.C. Hemminger: Surface Sci. Letters 1981, 109, p. L513.
(21) D. Dahlgren, and J.C. Hemminger: Surface Sci. 1982, 114, p. 459.
(22) A. Gavezzotti, and M. Simonetta: Surface Sci. Letters 1982, 116, p. L207.
(23) M. Salmeron, R.G. Gale, and G.A. Somorjai: J. Chem. Phys. 1979, 70, p. 2807.
(24) G.F. Tantardini, and M. Simonetta: Surface Sci. 1981, 105, p. 517.
(25) G.F. Tantardini, and M. Simonetta: Chem. Phys. Letters 1982, 87, p. 420.

MOLECULAR INGREDIENTS OF HETEROGENEOUS CATALYSIS

Gabor A. Somorjai
Department of Chemistry, University of California, Berkeley

1. PREAMBLE

The purpose of this paper is to present a review and status report to my colleagues in theoretical chemistry of the rapidly developing surface science of heterogeneous catalysis. The art of catalysis is developing into science. This profound change provides us with opportunities not only to understand the molecular ingredients of important catalytic systems but also to develop new and improved catalyst. The participation of theorists to find answers to important questions is sorely needed for the sound development of the field. It is my hope that some of the attendees of this conference may be persuaded to tackle some of the outstanding problems of heterogeneous catalysis that I attempt to identify in this paper.

For this purpose the paper is divided into several sections. The brief Introduction to the methodology and recent results of the surface science of heterogeneous catalysis is followed by a review of the concepts of heterogeneous catalysis. Then, we describe the experimental results that identified the three molecular ingredients of catalysis. The a) structure, b) carbonaceous deposit and c) the oxidation state of surface atoms. Each section is closed with a summary and a list of problems that, in my view, require theoretical and experimental scrutiny. Finally we describe attempts to build new catalyst systems and the theoretical and experimental problems that appeared in the course of this research.

2. INTRODUCTION

Modern surface science during the past 15 years developed a large number of techniques that can be utilized to uncover the molecular ingredients of heterogeneous catalysis. The atomic surface structure, the chemical bonding of adsorbates, the composition and oxidation states of surface atoms can all be determined within the sensitivity of less than 1% of a monolayer (about 10^{13} atoms per cm^2) (1). A partial list-

P.-O. Löwdin and B. Pullman (eds.), New Horizons of Quantum Chemistry, 305–337.

ing of the most frequently employed tecniques for surface analysis is displayed in Table 1.

Catalysis is a kinetic phenomenon; reactant molecules adsorb, rearrange and desorb from the surface continuously. It is generally carried out at high pressures (atmospheres) and the kinetic parameters, rate, activation energy and the product distribution reflect the activity and the selectivity of the catalysts to obtain the desired products. In order to unravel the molecular ingredients of heterogeneous catalysis, we had to combine catalytic reaction rate studies that were carried out at high pressures (several atmospheres) with surface analysis of the catalyst surface that usually required ultra high vacuum conditions. We therefore developed techniques (2) that permited us to do both in the same instrument. In the low-pressure-high-pressure apparatus (Figure 1) a small area sample was placed in the center of an ultrahigh vacuum chamber. For the high pressure phase of the study, a small cup with a total volume of 50 cm^3 could be placed over the crystal sample to isolate it from the rest of the chamber. This isolation cell could be pressurized to over 100 atmospheres, as required, using a mixture of gaseous reactants, thus providing a small, high-pressure reaction chamber. This was connected to a gas chromatograph that monitored both rate and product distributions. The structure and composition of the surface can be determined by electron scattering before and after the high pressure experiments by removal of the cup. The crystal sample could be heated during both low and high pressure experiments. Our samples are single crystals or polycrystalline foils of about one cm^2 surface area. Such small area catalysts are entirely adequate for surface analysis as well as for detecting the product distributions and the rates obtained during catalytic reaction studies.

The small area single crystals are model catalysts. Models are used frequently in physical sciences to uncover the workings of a complex system. Of course, it must be ascertained that the model system behaves similarly to that of the more complex chemical system that is to be ultimately utilized. The model small area catalysts were compared with dispersed high surface area catalysts in the same reaction under identical experimental conditions. The same turnover rates, product distributions and activation energies were obtained for the ring opening of cyclopropane (3), and the hydrogenation of benzene over platinum (4), and for carbon monoxide hydrogenation over rhodium (5). It appears that for catalyzed reactions that do not exhibit much structure sensitivity, small area crystal and large area dispersed particles of the same metal yield identical catalytic reaction parameters.

We have concentrated our studies of molecular scale catalysis on transition metals which catalyze hydrocarbon conversion reactions and on the hydrogenation of carbon monoxide. Platinum, an excellent catalyst for dehydrocyclization reactions that produce aromatic molecules from alkanes, or for isomerization, was one of the metals we studied intensively. The other was rhodium, which was found to produce oxygenated organic molecules from carbon monoxide and hydrogen, selectively.

TABLE 1

Table of some of the frequently utilized surface characterization techniques to determine the structure and composition of solid surfaces. Adsorbed species present at concentrations of 1% of a monolayer can be readily detected.

SURFACE ANALYSIS METHOD	ACRONYM	PHYSICAL BASIS	TYPE OF INFORMATION OBTAINED
Low energy electron diffraction	LEED	Elastic backscattering of low energy electrons	Atomic surface structure of surfaces and of adsorbed gases
Auger electron spectroscopy	AES	Electron emission from surface atoms excited by electron, x-ray or ion bombardment	Surface composition
High resolution electron energy loss spectroscopy	HREELS	Vibrational excitation of surface atoms by inelastic reflection of low energy electrons	Structure and bonding of surface atoms and adsorbed species
Infrared spectroscopy	IRS	Vibrational excitation of surface atoms by absorption of infrared radiation	Structure and bonding of adsorbed gases
X-ray and ultraviolet photoelectron spectroscopy	XPS UPS	Electron emission from atoms	Electronic structure and oxidation state of surface atoms and adsorbed species
Ion scattering spectroscopy	ISS	Elastic reflection of inert gas ions	Atomic structure and composition of solid surfaces
Secondary ion mass spectroscopy	SIMS	Ion beam induced ejection of surface atoms as positive & negative ions	Surface composition
Extended X-ray absorption fine structure analysis	EXAFS	Interference effects in photoemitted electron wave function in x-ray absorption	Atomic structure of surfaces and adsorbed gases
Thermal desorption spectroscopy	TDS	Thermally induced desorption or decomposition of adsorbed species	Adsorption energetics and composition of adsorbed species

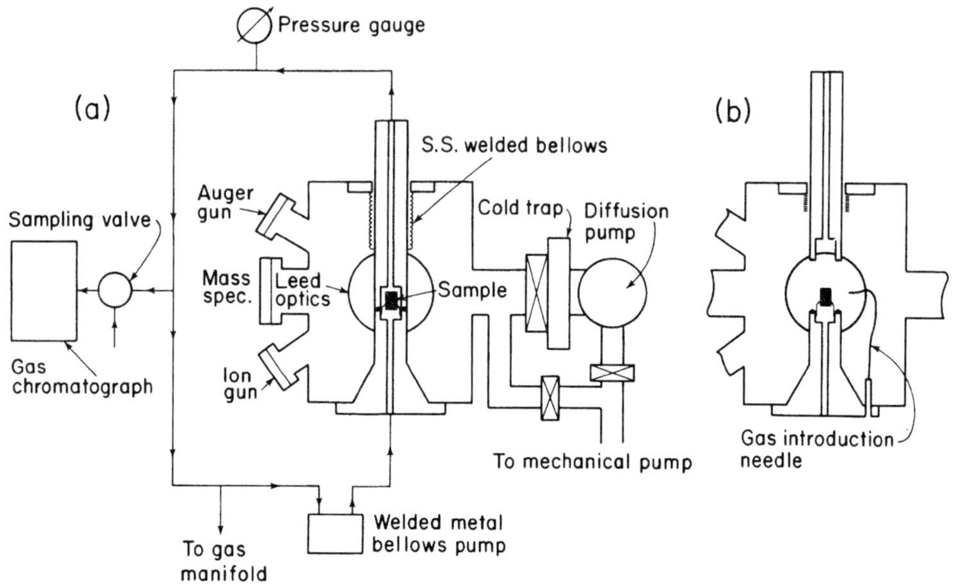

Fig. 1: Schematic diagram of the low pressure-high pressure apparatus for combined surface analysis and catalysis studies.

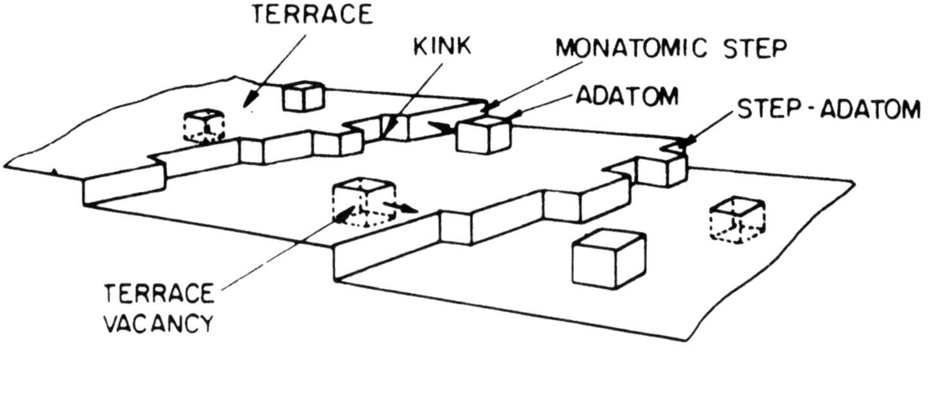

Fig. 2: Model of a heterogeneous solid surface, depicting different surface sites. These sites are distinguishable by their number of nearest neighbors.

Iron that catalyzes ammonia synthesis from nitrogen and hydrogen as well as the CO/H_2 reaction and was also a focus of our investigations. We investigated in detail the photodissociation of water over oxide surfaces and the hydrogenation of graphite to methane; both reactions are catalyzed by alkali metal ions. Out of these studies came the identification of three necessary ingredients of selective molecular scale catalysis.

1) Atomic surface structure.
2) An active carbonaceous deposit.
3) The proper oxidation state of surface atoms.

We shall review each of these reaction parameters in order to understand how the catalytic reactions occur after describing some of the important concepts of heterogeneous catalysis.

3. CONCEPTS OF HETEROGENEOUS CATALYSIS

The word "catalyst" was introduced by Berzelius (1836) to describe the property of certain substances which facilitate chemical change without being consumed during the reaction. The original definition has stood the test of time; catalysts aid the attainment of chemical equilibrium by reducing the potential energy barriers in the reaction path. For example, hydrogen and oxygen will remain as an unreacted mixture indefinitely, although in the presence of a platinum or nickel gauze, they react instantaneously to produce water (1).

Catalysis is a kinetic phenomenon. The speed of a catalysed reaction is often described in terms of a "turnover rate", which reveals how many product molecules form per second at a given temperature and pressure.

Since the catalytic action occurs at specific sites on solid surfaces, often called "active sites", the rate can be increased phenomenally by using very high surface area catalysts. Catalysts with surface areas of $1\text{-}10^2$ m^2/gm are quite common. These are usually prepared by finely dispersing the active metal or compound catalyst on high surface area "supports", which usually consist of refractory oxides or oxyhydroxides (alumina, zirconia, silica, etc.) The catalyst dispersion, defined as the number of surface atoms per total number of atoms, may range from unity (all of the atoms are on the surface) to ~ 0.01 and the catalyst particles may range in size between 10-200Å. This dispersed system must maintain structural and chemical stability for thousands of hours under the conditions of high temperature (400-900K) and high pressure ($1\text{-}10^2$ atmospheres) frequently encountered in the chemical technology. The design of new, more stable, catalysts which resist chemical attrition and sintering of the small particles is a constant concern of the catalytic scientist.

The specific rate, the number of product molecules per catalyst unit area per second, (or the number of product molecules per surface atom per second) at a given temperature and pressure provides a figure of merit to compare the "activity" of different catalysts for the same reaction under similar experimental circumstances. Although only a fraction of the catalyst surface is active during the reaction this fraction is usually unknown; thus the specific rate provides a lower limit of the catalyst's activity.

Specific rates must be in the range of 10^{-4}–10 to be part of a viable process in chemical technology. As a consequence, most catalytic reactions that involve the breaking or rearrangements of C-H, C-C, H-H, C=O and N≡N bonds are carried out in the temperature range of 400–900K since activation energies range from a low of 10–15 kcal/mole (for hydrogenation of unsaturated hydrocarbons) to a high of 40–50 kcal/mole (for dehydrocyclisation or hydrogenolysis of hydrocarbons). There are, of course, several exceptions; the catalytic isotope exchange of hydrogen ($H_2 + D_2 \longrightarrow 2HD$) occurs at temperatures as low as 77K on most transition metal surfaces and the hydrogenation of ethylene to ethane is readily performed at 300K. The oxidation of ammonia to nitric oxide, a highly exothermic reaction, is usually carried out in the range 1100–1200K.

By multiplying the specific rate by the reaction time used in a given study, the turnover number, T.N., (number of product molecules per catalyst surface site) can be obtained. The T.N. must be greater than unity to assure that the reaction is catalytic (i.e. that there is more than one "turnover".) If the T.N. is less than unity we may be dealing with a stoichiometric reaction and our solid surface may be acting as a reactant rather than a catalyst. Thus, to demonstrate its catalytic character, the reaction must be carried out for long enough to obtain a turnover number that is greater than unity.

Many chemical reactions of interest lead to the formation of several different, but all thermodynamically feasible, products. It is often of greater interest to produce one of the many different molecules selectively, rather than increasing the overall rate A selective catalyst will facilitate the formation of one product molecule while inhibiting the production of other, different molecules, even though the formation of all of the species is thermodynamically feasible. Thus, the blocking of reaction pathways which lead to the formation of unwanted molecules is as important an attribute of a good catalyst as the lowering of the potential energy barrier along the reaction pathway to the desired product molecule.

The reaction probability, R.P., is defined as the number of product molecules formed per number of reactant molecules incident on the catalyst surface. It is readily obtained by dividing the specific rate of catalysis by the incident reactant flux. R.P. is generally very low, and varies in the range 10^{-12}–10^{-6} for most catalytic reactions in the pressure and temperature ranges generally employed in the chemical industry. The reason for this is the long surface residence time for the adsorbed molecules and reaction intermediates, (usually in the range of 10^{-1}–10^{3} sec.) which leads to a surface which is covered and inaccessible to the impinging molecules. The desorption of the product molecules from the catalyst surface frees surface sites, onto which the reactant molecules may adsorb. The desorption is always an endothermic reaction and appears to be the slow step in the catalytic process for most reactions running at high pressures.

The long surface residence times of adsorbed intermediates have another interesting consequence. The molecules may diffuse over long distances (10-10^4Å) within their surface lifetime, visiting many catalytic sites where they can undergo consecutive reactions and molecular rearrangements. Thus, a catalyst surface may possess many different active sites, all of which are likely to be accessible to the adsorbed reactants.

The coverage of adsorbed molecules, σ, [molecules/cm^2] is usually determined by multiplying the incident flux, F, [molecules/cm^2 sec] by the surface residence time, t, [sec].

$$\sigma = F \times t$$

The flux can be estimated from the reactant pressure, $F = P(2\pi MRT)^{-1/2}$, and the residence time $t = t_o \exp[\Delta H_{ads}(\theta)/RT]$ depends on the heat of adsorption, ΔH_{ads}, the temperature, T, and the preexponential factor, t_o, (which varies in the range 10^{-17}-10^{-13} (6) depending on the degree of mobility of the adsorbed molecules). The heat of adsorption is a measure of the strength of the chemical bond between the adsorbed molecule and the surface. It depends on both the coverage and the structure of the catalyst surface. As the coverage is increased the adsorbed molecules are packed closer together in the surface monolayer. If the interaction among the adsorbates is repulsive, the heat of adsorption per molecule decreases as the coverage in the monolayer is increased. This is the case for molecular CO adsorption on most metal surfaces (7). If the adsorbate-adsorbate interaction is attractive, the heat of adsorption per molecule may increase with coverage. This occurs, for example, during the growth of oxide monolayers on metal surfaces, (8).

The surface structure sensitivity of bonding of adsorbed molecules and as a result, the structure sensitivity of catalytic reactions is well documented, (9,10). The surface structure is heterogeneous on the atomic scale (Fig. 2) and there are many surface sites that are distinguishable by their number of nearest neighbours: there are surface atoms that can form terraces, steps and kinks, and these structures may, in turn, contain point defects, adatoms and vacancies. The binding of adsorbed atoms or molecules at each site may be different, giving rise to heats of adsorption that vary with the surface structure. This effect is particularly noticeable in adsorption and desorption studies carried out on single crystal surfaces. These surfaces can be prepared with different relative concentrations of terrace, step and kink sites. The structure sensitivity of bonding can lead to structure sensitivity in many catalytic reactions. These have been described in the literature (11).

It is often thought that the most active catalysts consist of surfaces which can form bonds of intermediate strength with adsorbates. If the surface chemical bonds are too strong, the reaction intermediates either have surface lifetimes which are much too long, or they may simply never come off the surface, forming stable surface compounds such as oxides, carbides or nitrides instead. On the other hand, if the surface bonds to adsorbates are too weak, the needed chemical bond-breaking or rearrangements might not occur before desorption. Chemical experience is in agreement with the contention that intermediate bond strength leads to superior catalyst activity. The structure of the catalyst surface, however, is one of the key factors which control chemical selectivity.

The catalytic process may be subdivided into several elementary reaction steps. These are a) adsorption, b) surface diffusion, c) molecular rearrangements at active surface sites and d) desorption. There are several techniques capable of studying the elementary reaction step. Predominant among them are (1) molecular beam surface scattering and (2) thermal desorption. The reader is referred to recent review papers describing these methods of investigation (12,13)

The catalyst, unfortunately, does not have an indefinite life, but becomes "poisoned" or deactivated after a certain period of time. It is thought to reflect good catalyst performance if seven pounds of products are produced per pound of catalyst before the catalyst must be regenerated. Catalyst deactivation and regeneration are important and major fields of catalysis science. There are many possible reasons for deactivation. For example, the catalyst may restructure as a consequence of selective adsorption of impurities from the reactant stream, such as sulphur or nitrogen. Alternatively, carbon may be deposited, blocking the active surface, due to side reactions leading to complete dehydrogenation and decomposition of organic molecules during hydrocarbon conversions. The use of additives, (introduced either during catalyst preparation or with the reactants) to prevent restructuring of the catalyst or the blocking of active sites is being explored intensively by many research laboratories.

3.1. Summary

Single crystal surfaces of small area (~ 1 cm^2) can be used as model catalysts in studies of the surface science of heterogeneous catalysis. The structure, composition and oxidation states of surface atoms can readily be determined by the various techniques of surface science that were developed in recent years. A new technology that permits _in situ_ determination of catalytic reaction parameters (rates, activation energies, product distributions) at high pressures (atmospheres) and the surface properties at low pressures (~ 10^{-9} torr) has been employed. Molecular information about the elementary surface reaction's steps, adsorption, surface diffusion, chemical rearrangement and desorption becomes increasingly available. The reaction probabilities of most heterogeneous catalyzed reactions under industrial conditions used at present are very low (10^{-12} - 10^{-6}) and the surface residence times of reacting molecules are long (seconds). The causes of the catalyst activity (specific rates), selectivity (desired product distribution) and resistance to poisoning (to assure long lifetimes) are the subjects of most research studies. The atomic structure, the carbonaceous deposit that forms during hydrocarbon reactions and the oxidation states of surface atoms are the molecular ingredients of heterogeneous catalysts that have been identified.

3.2. Questions that need theoretical and experimental scrutiny

- Are the desorbing molecules in energy equilibrium with the surface (translational and internal energy content)? What is the nature and efficiency of energy transfer among the reactants, the surface and the products?

- Improvements in the theories that underlie many surface science techniques are needed. Ion scattering from surfaces and high resolution electron energy loss spectroscopy are foremost among them!

- How does bonding to the solid surface change when adsorption occurs at the solid-liquid instead of the solid-gas interface?

4. STRUCTURE SENSITIVITY OF CATALYTIC REACTIONS

4.1. The structure sensitivity of ammonia synthesis over iron catalysts

Single crystal catalysts can be prepared with a variety of surface structures by cutting the solid along different crystallographic orientations. We have studied ammonia synthesis over the (111), (100) and (110) crystal faces of iron at operating pressures approaching those used industrially (20 atm) (9). The structure of these different crystal faces are shown in Figure 3. The Fe(111) face was found to be the most active. It catalyzes the synthesis of ammonia from hydrogen and nitrogen at a rate which is at least 420 times faster than that of the ordered (110) surface. The rate of ammonia synthesis is 32 times faster over the Fe(100) face than that over the ordered (110) surface. These results are consistent with the low presure chemisorption studies of Ertl and his co-workers (14) who reported that the rate of dissociative chemisorption of nitrogen is greatest on the (111) face of iron and least on the (110) face and with previous suggestions that this step is rate determining in the ammonia synthesis reaction. The (111) plane of iron contains both 7-coordinated and 4-coordinated atoms, whereas the (100) and (110) planes contain only 6-coordinated and 4-coordinated atoms, respectively. The large variation in the rates of ammonia synthesis on the various iron crystal planes lend considerable support to the proposals by Boudart and others (15) regarding the importance of iron clusters and 7-coordinated iron atoms as components of the active site on the catalyst surface.

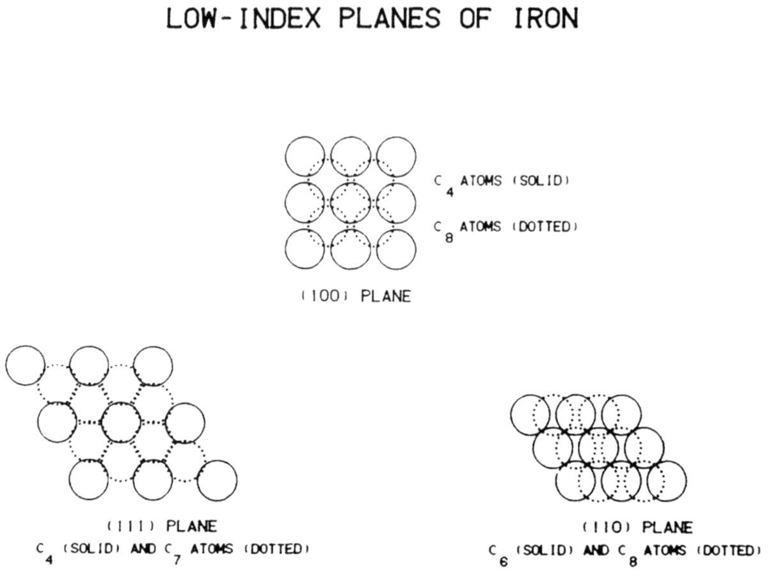

Fig. 3: Idealized atomic surface structure for the low-index planes of Iron Fe(100), Fe(111) and Fe(110).

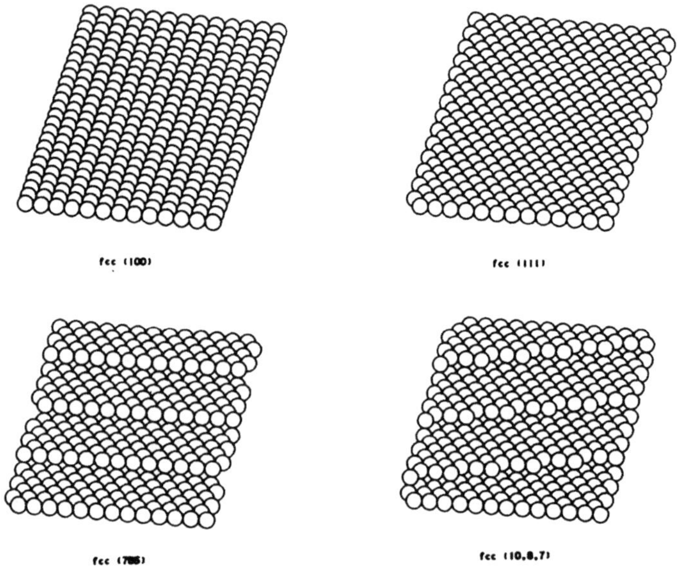

Fig. 4: Idealized atomic surface structures for the flat Pt(111) and Pt(100). the stepped Pt(755) and the kinked Pt(10,8,7) surfaces.

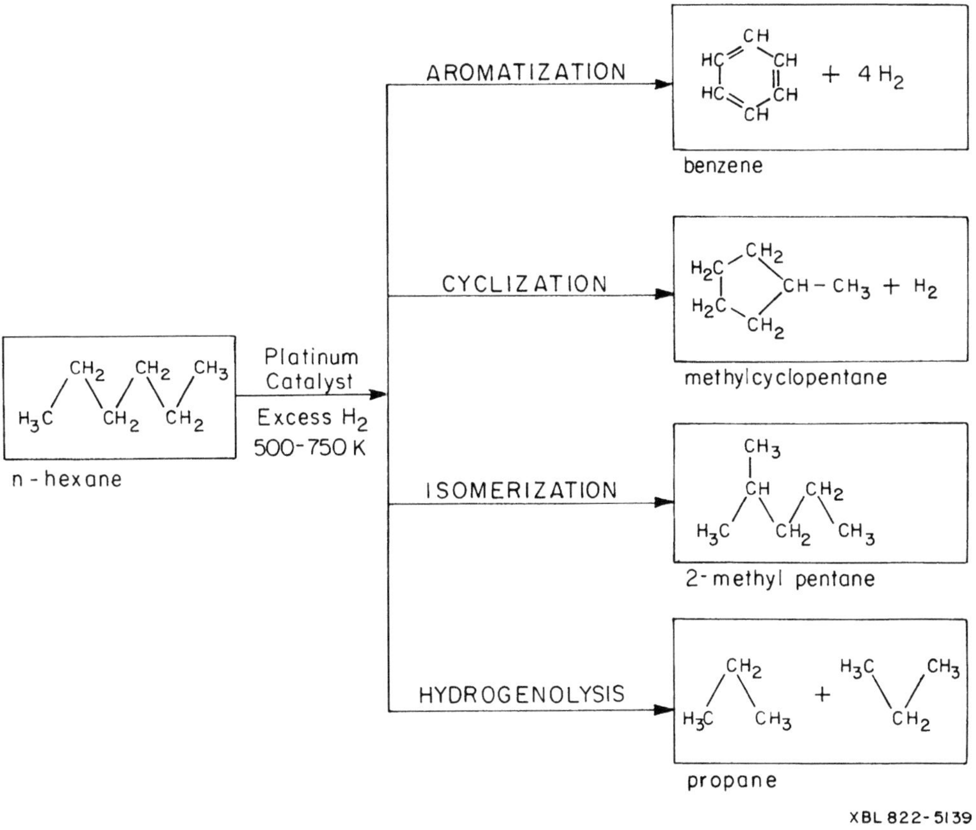

Fig. 5: Skeletal rearrangement reactions of hydrocarbons are catalyzed by platinum with high activity and unique selectivity. Depicted here are the several reaction pathways which occur simultaneously during the catalyzed conversion of n-hexane, C_6H_{14}. The isomerization, cyclizationa nd aromatization reactions that produced branched or cyclic products are important in the production of high octane gasoline from petroleum naptha. The hydrogenolysis reaction that involves a C-C bond yields undesirable gaseous products.

4.2. The structure sensitivity of hydrocarbon conversion reactions on platinum surfaces.

The surface structures of platinum crystals that display different catalytic behavior are shown in Figure 4. The two flat surfaces, the (111) and (100) crystal faces, have hexagonal and square atomic arrangements respectively. Each surface atom is surrounded by six or four nearest neighbors in the surface plane. There (lower two sketches in Figure 4) are also surfaces with periodic arrays of steps of one atom in height. The structure of the terraces and steps can have the same six-fold of four-fold symmetry as the flat surfaces. The widths of the terraces are variable depending on the angle of cuts of the single crystal surface which determines the crystallographic orientation of the surface. There may be kinks or ledges in the steps that can be present in large concentrations if the surface is suitably prepared (1). The step and ledge sites have fewer nearest neighbors (lower coordination number) and lower symmetry when compared to atoms in the flat surfaces.

The reactions of n-hexane or n-heptane over platinum surfaces are excellent probes of structure sensitivity (16). The various reactions of n-hexane are displayed in Figure 5. The reactions leading to branched isomers (isomerization) and cyclic molecules (dehydrocyclization) are especially desirable for producing high octane gasoline from petroleum naphta. The third type, hydrogenolysis reactions that involve the scission of C-C bonds yields low molecular weight gaseous products that are undesirable when producing gasoline. The reactivity of the flat (111) and (100) sufaces for the dehydrocyclization of n-hexane and n-heptane are compared in Figure 6. The hexagonal (111) surface is 3 to 7 times more active than the platinum surface with the square unit cell (16). In Figure 7 the selectivity of stepped platinum surfaces are compared with the flat surfaces under identical conditions for the same reaction. Maximum aromatization activity is obtained on stepped surfaces with terraces about 5 atoms wide with hexagonal orientations as indicated by reaction rate studies over more than 15 different crystal surfaces with varied terrace orientations and step and kink orientations.

The reactivity patterns displayed for alkane isomerization reactions is completely different to that for aromatization (17). Our studies revealed that maximum rates and selectivity (rate of desired reactions/total rate) for butane isomerization reactions are obtained on flat crystal faces with square (100) unit cell. Isomerization rates for this surface are 4 to 7 times higher then those for hexagonal surfaces as shown in Figure 8.

For the undesirable hydrogenolysis reactions that require C-C bond scission we found that the two flat surfaces exhibit very similar reaction rates. However, the distribution of hdyrogenolysis products varies sharply over these two surfaces. The hexagonal surface displays high selectivity for scission of the terminal C-C bond whereas the (100) square surface always prefers cleavage of C-C bonds located in the

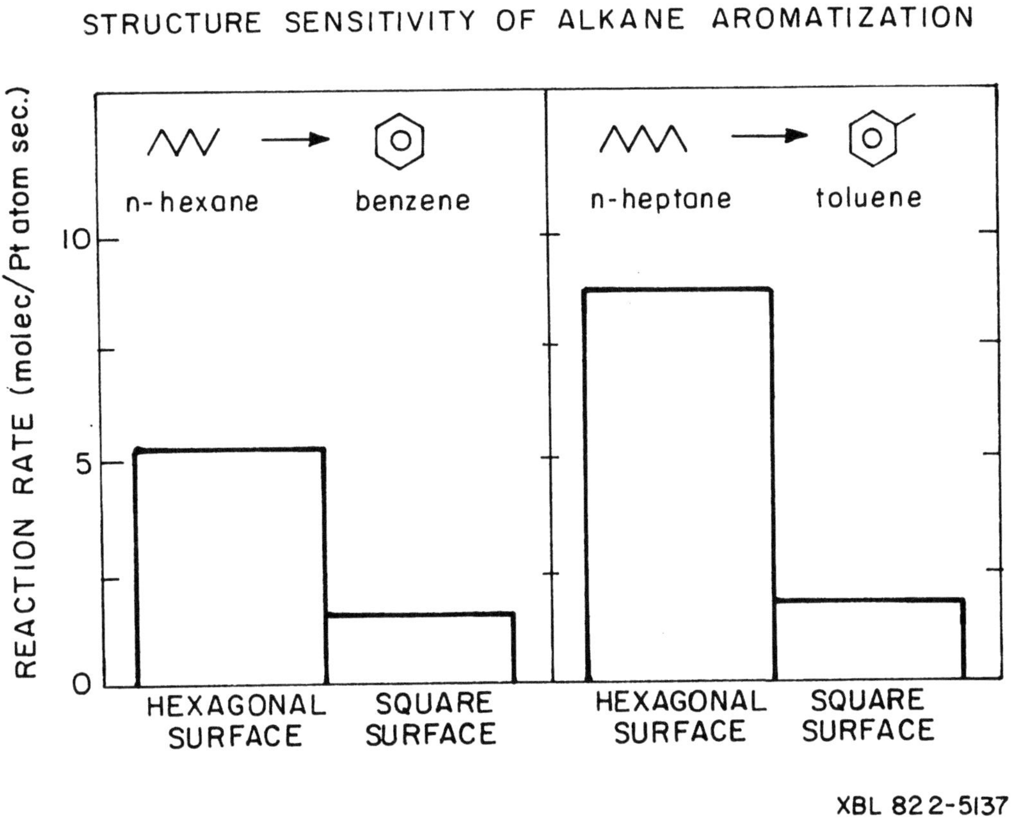

Fig. 6: Dehydrocyclization of alkanes to aromatic hydrocarbons is one of the most important petroleum reforming reactions. The bar graphs shown here compare reaction rates for n-hexane and n-heptane aromatization catalyzed at 573K and atmospheric pressure over the two flat platinum single crystal faces with different atomic structure. The platinum surface with the hexagonal atomic arrangement is several times more active than the surface with a square unit cell over a wide range of reaction conditions.

center of the reactant molecule (18). The hydrogenolysis rates increase markedly (3-5 fold) when kinks are present in high concentrations on the platinum surfaces (Figure 8).

Since different reactions are sensitive to different structural features of the catalyst surface we must prepare the catalyst with appropriate structure to obtain maximum activity and selectivity. The terrace structure, the step or kink concentration or a combination of these structural features is needed to achieve optimum rates for a give reaction. Our studies indicate that H-H and C-H bond breaking processes are more facile on stepped surface then on flat crystal faces, while C-C bond breaking is aided by ledge or kink sites that appear to be most active for breaking any of the chemical bonds during hydrocarbon conversion reactions. Since molecular rearrangements must also occur in addition to bond breaking it is not surprising that the terrace structure exerts such an important influence on the reaction path that the adsorbed molecules are likely to take.

The differences in the chemical behavior of terrace, step and ledge atoms arises not only from their different structural environment but also from their different electronic charge densities that result from variation of the local coordination. Electron spectroscopy studies reveal altered density of electronic states at the surface irregularities; there is a higher probability of electron emission into vacuum at these sites (lower work function) indicating the redistribution of electrons.

While many reactions of organic molecules are structure sensitive there are some that show no structure sensitivity: the ring opening of cyclopropane and the hydrogenation of cyclohexene to cyclohexane are two structure insensitive reactions when carried out on various metal surfaces.

It should be noted that the metal surface structures that are stable in the presence of an adsorbed layer of hdyrocarbons may be unstable in an oxidizing environment (19). In the presence of chemisorbed oxygen other crystal faces may be more stable. As a result, changing the reducing chemical environment into oxidizing or the other way around could cause a massive reorganization of the surface structure. One should not expect all the catalyst surfaces to remain stable under the varied conditions encountered during catalytic reaction studies. One of the reasons for the application of certain additives or the use of appropriate oxide supports is to stabilize the surface structure of the catalyst as more extreme conditions for the catalytic chemistry and for its regeneration are used.

4.3. Summary

The ammonia synthesis over iron, and most complex hydrocarbon reactions over transition metal catalysts are surface structure sensitive. The selectivity to obtain the desired products can be controlled

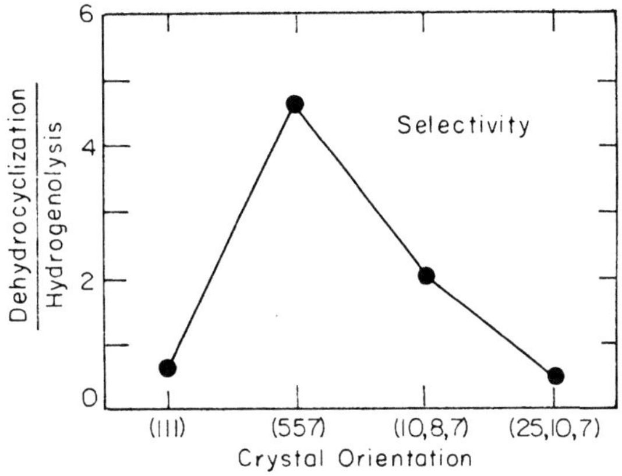

Fig. 7: The selectivity of various platinum surfaces for the dehydrocyclization reaction. The stepped (557) surface with five atom wide terraces of (111) orientation appears to be the most active.

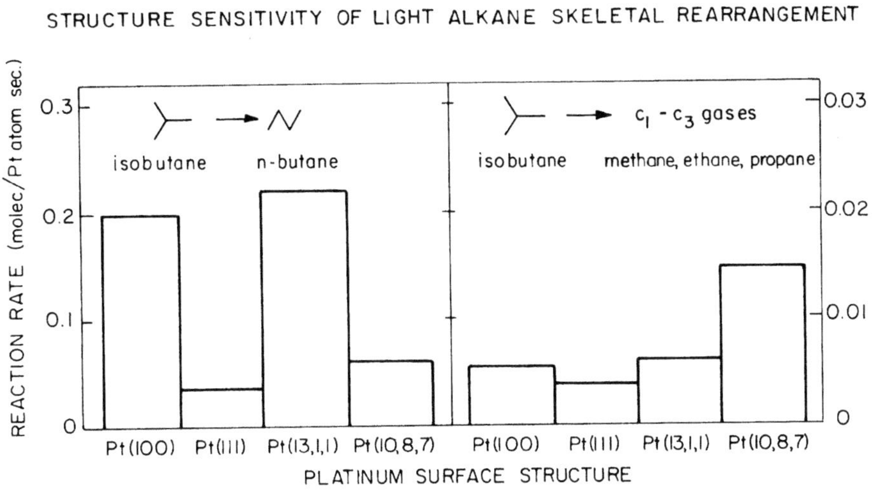

Fig. 8: Reaction rates are shown as a function fo surface structure for isobutane isomerizatriona nd hydrogenolysis catalyzed at 570K and atmospheric pressure over the four platinum surfaces. The rates for both reaction pathways are very sensitive to structural features of the model single crystal catalyst surfaces. Isomerization is favored on the platinum surfaces that have a square (100) atomic arrangement. Hydrogenolysis rates are maximized when kink sites are present in high concentrations as in the Pt(10,8,7) platinum crystal surface.

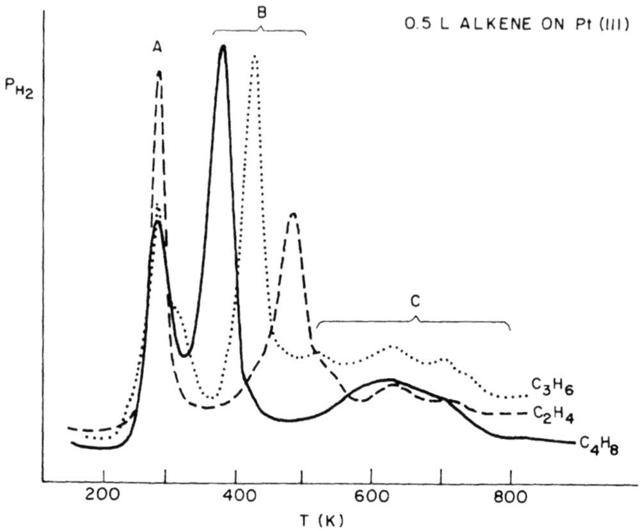

Fig. 9: Hydrogen thermal desorption spectra illustrating the sequential dehydrogenation of ethylene, propylene, and cis-2-butene chemisorbed on Pt(111) at about 120 K (β = 12 K/sec).

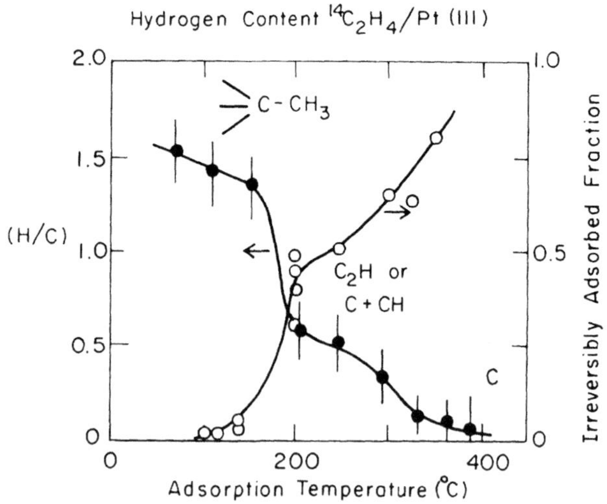

Fig. 10: Composition and reactivity of ^{14}C-ethylene chemisorbed on Pt(111) at temperatures between 20 and 370°C; the irreversibly adsorbed fraction determined by radiotracer analysis displays an excellent correlation with the average hydrogen content (H/C) of the strongly adsorbed species.

by tailoring of the catalyst structure. Surface irregularities, steps and kinks can be present in large surface concentrations and are especially active in breaking large binding energy chemical bonds (N≡N, C=O, C-C, C-H) with near zero activation energy. Additives are often used to stabilize the desired surface structure of catalysts.

4.4. Questions that need theoretical and experimental scrutiny

- How does the chemical bonding of small molecules to the surface vary with the atomic structure of sites?

- What are the electronic ingredients (charge density and angular charge distribution) that make such a site active for bond breaking?

- How charge donor or acceptor species influence the catalytic activity at these sites? How selective adsorption of atoms (O, S, K, C) aid or poison bonding at the active surface sites?

- Are surface atom mobilities great enough under catalytic conditions to generate new active sites continuously?

5. CARBONACEOUS OVERLAYERS

Within milliseconds after the start of a hydrocarbon reaction at atmospheric pressures an equivalent of a carbonaceous layer is deposited on a catalyst surface and the reaction then occurs in the presence of this deposit (20). Indeed a clean platinum or other clean transition metal surface can not readily catalyze hydrocarbon conversion reactions. Most organic molecules adsorb irreversibly and form strong metal-carbon bonds. Upon heating, and in the absence of hydrogen, sequential dehydrogenation occurs until a graphitic overlayer remains (21). Figure 9 shows the sequential dehydrogenation of several alkenes that were adsorbed on platinum at low temperatures, and then heated.

Our studies using C-14 labeled organic molecules (16), thermal desorption and hydrogen-deuterium exchange (17) indicate a strong temperature dependence for the residence time and the composition of the organic deposit on platinum surfaces. Below about 120°C the layer can be rehydrogenated and removed from the metal surface as rapidly as the turnover time for organic reactants. Thus, at these low temperatures the metal surface remains clean during facile reactions, like the hydrogenation of ethylene that takes place with a low activation energy (less than 15 kcal/mole) at these low temperatures. The temperature programmed thermal desorption (22) of hydrogen and AES studies indicate that the stoichiometry of this mobile layer is C_2H_3. With increasing temperature above 120°C an increasingly larger fraction of the surface carbon becomes irreversibly adsorbed as it looses more hydrogen. This layer resides on the surface much longer than the turnover time of the organic reactants that may undergo isomerization, dehydrocyclization or hydrogenolysis. These reactions have high activation energy (20-45 kcal/mole) and are, therefore, carried out at higher temperatures, in the range of 150-400°C, to obtain reasonably high turnover numbers. Figure 10 shows the increase of the concentration of an irreversibly adsorbed carbonaceous layer and the simultaneous decline of the reversibly adsorbed mobile organic layer as a function of temperature (23). The stoichiometry of this tenacious layer on the catalytically active metal is C_2H. Our detailed investigations exploring the catalytic effect of preparing the carbon deposit from a variety of preadsorbed organic molecules show relatively minor alterations of the selectivity of hydrocarbon conversion reactions. The activity of the catalyzed reaction was inversely proportional to the concentration of the irreversibly adsorbed carbon layer.

The structure of the adsorbed hydrocarbon monolayers was submitted to detailed studies by LEED and HREELS (25). In the temperature range of 300-400 K the adsorbed alkenes form alkylidyne molecules that are shown in Figure 11. The C-C bond closest to the metal is perpendicular to the surface plane and its 1.5Å length corresponds to a single bond. The carbon atom that bonds the molecule to the metal is located in a three-fold site equidistant 2.0Å from the nearest metallic neighbors (24). This bond is appreciably shorter than the covalent metal-carbon bond (2.2Å) and is indicative of multiple metal-carbon bonds of the carbene or carbyne type. While this layer is ordered, on heating to about 100°C it disorders and hydrogen evolution is detectable by a mass spectrometer that is attached to the system. As the molecules dehydrogenate the disordered layer is composed of CH_2, C_2H and CH type fragments that can be identified by HREELS (25). Only after heating to about 400°C do the fragments loose all their hydrogen and the graphite overlayer forms. These sequential bond breaking processes which occur as a function of temperature are perhaps the most important and unique characteristics of the surface chemical bond. While the surface remains active in the presence of organic fragments of C_2H stoichiometry it looses all activity when the graphitic monolayer forms.

Thus, we have two seemingly contradictory experimental observations On the one hand many catalyzed reactions are strongly dependent on the structure of the surface. On the other hand, the surface that is catalytically active is covered with about a monolayer of carbonaceous deposit that seem to mask the structure of the solid catalyst. In order to maintain the observed structure sensitivity not all the metal surface sites are likely to be covered by the organic layer, but some of them must be available to perform the catalytic function. We have been able to titrate the remaining clean platinum surface sites by CO adsorption (20) which adsorbs readily on the uncovered metal at low pressures, about 10^{-6} torr and 25°C, but does not adsorb on the organic overlayer under these conditions. Figure 12 shows a fraction of the clean platinum surface as a function of the carbon surface concentration. (θ/θ_{clean}) is the fraction of bare platinum surface as compared to the site concentration on the initially clean metal surface. This fraction decreases rapidly with increasing temperature as seen in Figure 11. At about 400°C the carbonaceous overlayer that is irreversibly adsorbed

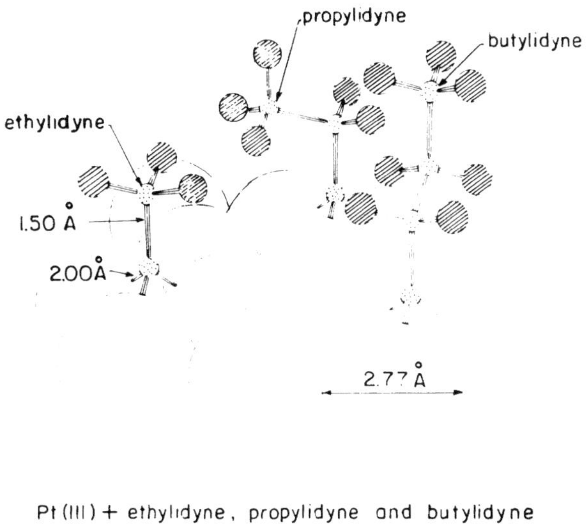

Fig. 11: Atomic surface structure for alkylidyne species chemisorbed on Pt(111).

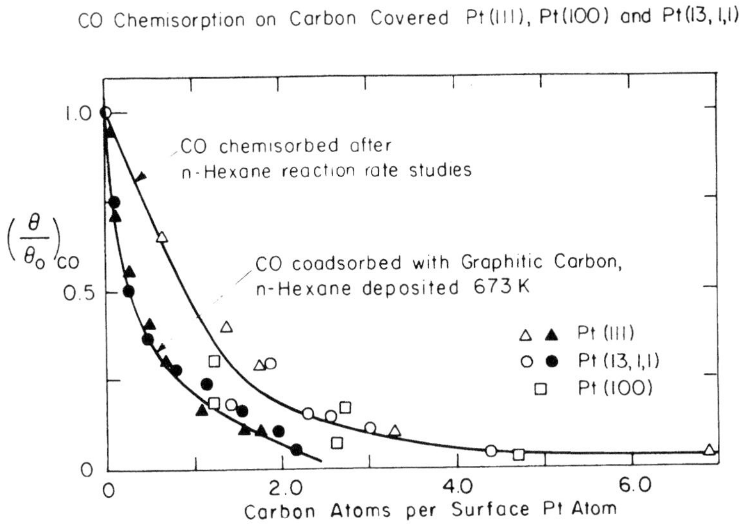

Fig. 12: Fractional concentrations of uncovered platinum surface sites determined by CO adsorption-desorption as a function of surface carbon coverage on the (100). (111) and (13,1,1) platinum surfaces. A comparison is made between the CO uptake determined following n-hexane reaction studies and CO uptake determined when CO was coadsorbed with "graphite" surface carbon.

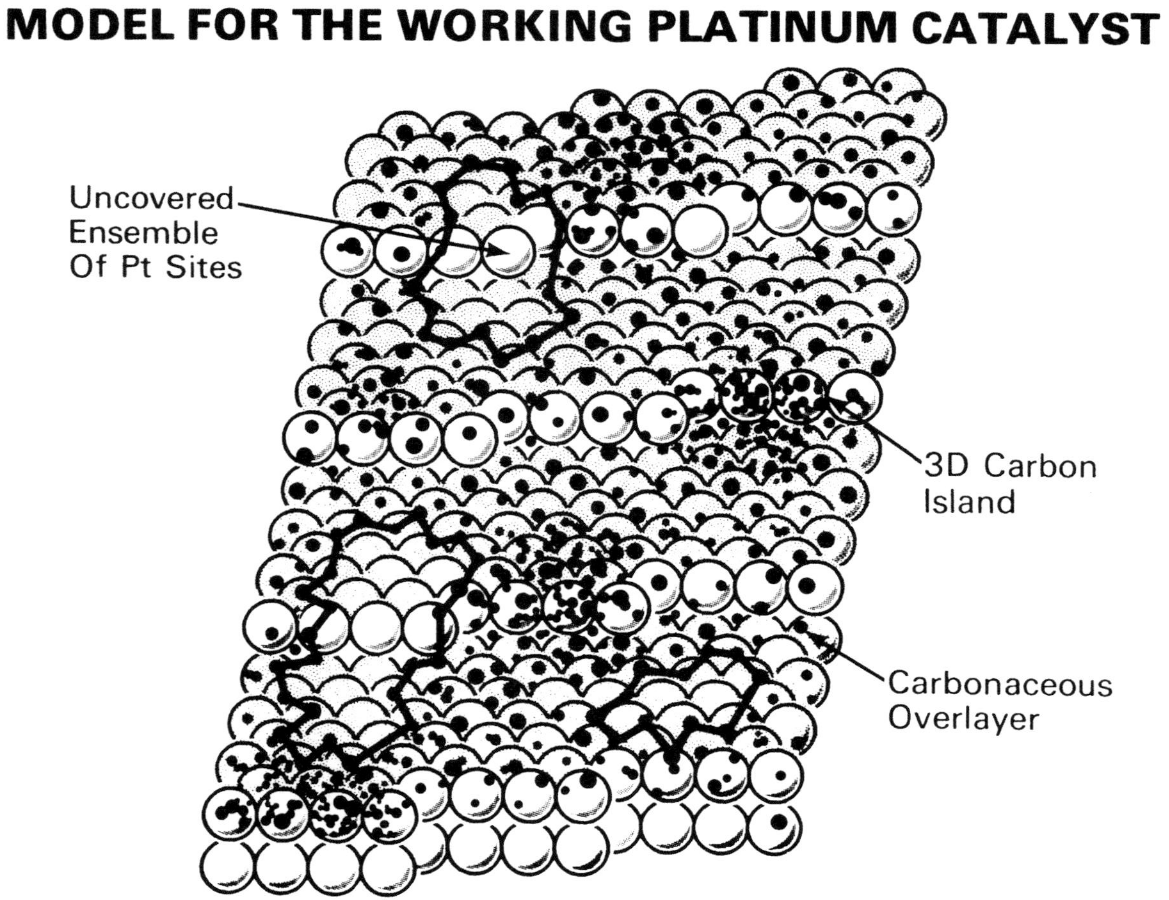

Fig. 13: Model for the working surface composition of platinum reforming catalysts.

gradually looses its remaining hydrogen and becomes graphitized. The graphite layer is a catalyst poison and the catalytic activity is irretrievably lost when the surface is covered with graphite.

Hydrocarbon reactions are always carried out in the presence of excess hydrogen. Increased hydrogen concentration in the reactant feed helps to rehydrogenate a larger fraction of the irreversibly adsorbed carbonaceous layer and to slow the rate of graphitization. In the absence of excess hydrogen in the reactant mixture a complete monolayer graphitization is fast.

As a result of these studies a molecular model (20) of the working platinum catalyst can be proposed and is shown in Figure 13. Most catalyzed reactions could not occur on either the bare metal surface or on the surface that is completely covered by the irreversibly deposited organic layer. Hydrocarbon conversion must occur on the bare metal islands that are kept clean by the presence of hydrogen in the reactant mixture. However, the intermediates or the products that form may not readily desorb from the bare metal sites because they are too strongly bound. It is likely then that these species diffuse over onto the carbonaceous overlayer from which desorption commences perhaps with the additional transfer of hydrogen. By providing sites at which the reaction products are bound only weakly the carbonaceous overlayer facilitates the desorption of the organic molecules. It should be noted that the active carbonaceous deposit that contains CH and CH_2 fragments stores about ten times more hydrogen than the clean metal surface which could be made available to the desorbing organic molecules.

The fractional concentration of the bare metal islands and of the carbonaceous overlayer can be manipulated not only by hydrogen pressure but by additives such as alkali promoters or by other transition metals that are used as alloying agents. Gold and tin were found to beneficially affect the activity, and change the selectivity and surface residence time of adsorbed reaction intermediates and product molecules by: a) alteration of the structure and active site concentration of bare metal islands, b) aiding the rehydrogenation of the carbonaceous layer on the active catalyst and by c) slowing the rate of its dehydrogenation to the graphitic form that poisons the catalyst surface.

5.1. Summary

The metal surface is largely covered with a carbonaceous deposit during hydrocarbon reactions. The stoichiometry of this layer changes as a function of temperature ($C_2H_3 \longrightarrow C_2H \longrightarrow C$) as the layer dehydrogenates with increasing temperature. The residence times of the dehydrogenated species (C_2H, CH, C) are greater than the reaction turnover time. Hydrogen transfer and hydrogen storage appears to be the main functions of this deposit. Under catalytic reaction conditions only 5-10% of the metal surface is uncovered and responsible for much of the catalytic chemistry. The structure of adsorbed organic molecules change with increasing temperature. Each structural intermediate is stable in a finite temperature range while the thermodynamically most stable configuration is attained at the highest temperatures. A molecular model of the working metal catalyst is proposed, consisting of bare metal islands in the carbonaceous deposit.

5.2. Questions that need theoretical and experimental scrutiny

- Why do adsorbed molecules undergo sequential bond breaking and structural rearrangements as the temperature is increased instead of forming their more stable configuration at lower temperatures? Why is there a small activation energy (on the order of RT) to form the more stable molecular structure or change location on the surface?

- Explain the bonding trends of molecules on transition metals across the periodic table and their relation to atomic and band structure properties of the metals.

6. THE OXIDATION STATE OF SURFACE ATOMS

Several experimental studies have been published in recent years which clearly indicate the importance of oxidation states other than the zero valent metallic state for catalyzed reactions. We shall show two examples of the importance of the presence of higher oxidation states of transition metal ions at the catalyst surface.

6.1. Carbon monoxide hydrogenation over rhodium

Rhodium was reported to yield predominantly C_2 oxygenated products, acetaldehyde and acetic acid (26) when prepared under appropriate experimental conditions. Our studies using unsupported polycrystalline rhodium foils (27) have detected mostly methane along with small amounts of ethylene and propylene under very similar experimental conditions. This product distribution was identical to that obtained by Vannice (28) over supported rhodium catalyst as was the activation energy for methanation (about 24 kcal/mole). It appears that most of the organic molecules form following the dissociation of CO by the rehydrogenation of CH_x units in the manner similar to alkane and alkene production from CO/H_2 mixtures over other more active transition metal catalysts (iron, ruthenium and nickel).

However, when rhodium oxide, Rh_2O_3, was utilized as a catalyst, large concentrations of oxygenated C_2 or C_3 hydrocarbons were produced, including ethanol, acetaldehyde and propionaldehyde (29). Furthermore, the addition of C_2H_4 to the CO/H_2 mixture yielded propionaldehyde, indicating the carbonylation ability of Rh_2O_3. Under similar experimental conditions over rhodium metal C_2H_4 was quantitatively hydrogenated to ethane and carbonylation activity was totally absent. Clearly, higher oxidation state rhodium ions are necessary to produce the oxygenated organic molecules. Unfortunately, Rh_2O_3 reduced rapidly in the CO/H_2 mixture to the metallic state with drastic alteration of the product distribution from oxygenated hydrocarbons to methane. In order to stabilize the rhodium ion, lanthanum rhodate, $LaRhO_3$, was prepared by incorporating Rh_2O_3 into La_2O_3 at high temperatures (30). Over this stable catalyst the formation of oxygenated products from CO/H_2 predominated.

The reason for the change of selectivity in CO/H_2 reactions upon alteration of the oxidation state of the transition metal is due largely to the change of heats of absorption of CO and H_2 as the oxidation state of the transition metal ion is varied (30). This is demonstrated in Figure 14. The CO adsorption energy is decreased upon oxidation while the heat of adsorption of H_2 is increased, presumably due to the formation of hydroxyl groups. This, in turn, changes the relative surface concentrations of CO and H_2. In addition, the metal is primarily active for hydrogenation and CO dissociation while the oxide can perform carbonylation and has reduced hydrogenation activity. As shown in Figure 11, the active $LaRhO_3$ catalyst seems to have both rhodium metal and rhodium ion sites, as indicated by the presence of the thermal desorption peaks of CO and H_2, to provide both optimum carbonylation as well as hydrogenation activity so necessary to obtain C_2 or C_3 oxygenated hydrocarbon molecules.

We have also found that the product distribution that results from the CO/H_2 reaction at high pressures over $LaRhO_3$ is highly temperature dependent (30). At low enough temperatures, below 250°C, the formation of methanol predominates. This can occur by the direct hydrogenation of the undissociated carbon monoxide molecule. As the temperature is increased from 250 to 300°C, the formation of C_2 or C_3 oxygenated hydrocarbon molecules ethanol, propanol, acetaldehyde and propionaldehyde predominates. These molecules need the presence of both CH_x units that form as a result of the dissociation of carbon monoxide followed by the partial rehydrogenation of the surface carbon and the insertion of CO into these fragments to form the oxygenated product. In a finite temperature range between 250 and 350°C the CO dissociation and hydrogenation ability and the carbonylation ability of the catalyst are about identical. As the temperature is increased to above 350°C, the formation of methane predominates. The CO concentration is too low on the surface at these temperatures to produce CO insertion and the rehydrogenation characteristics of the material control the product distribution (29,30).

One of the difficulties in preparing selective catalysts for hydrocarbon conversion is the poor thermodynamic stability of higher oxidation states of transition metal ions in the reducing environment. It appears that a strong metal support interaction that permits the incorporation of high oxidation state transition metal ions into the supporting refractory oxide or sulfide crystal lattice often promotes the kinetic stabilization of the desired oxidation state, as long as the catalytic reaction temperatures are appreciably below the decomposition temperature of the binary oxide.

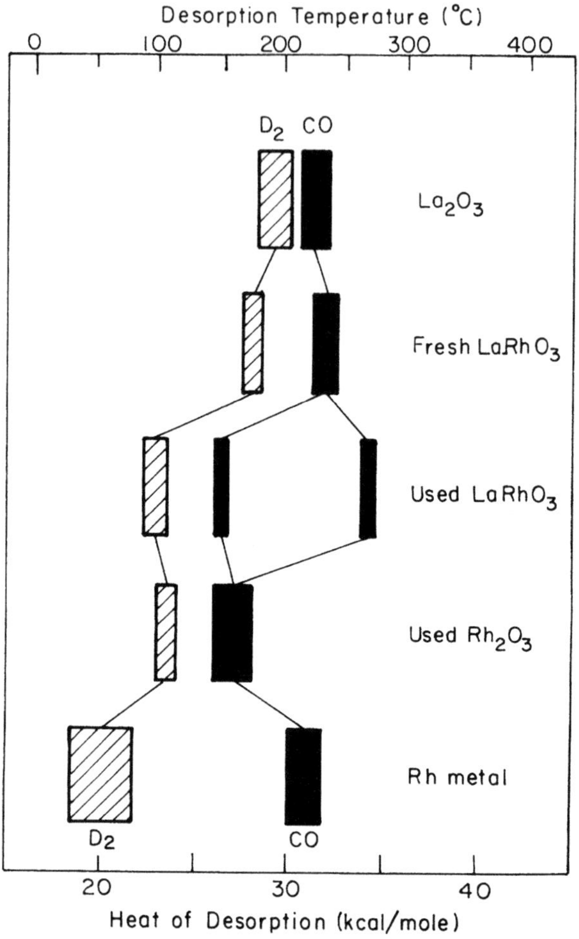

Fig. 14: Heat of desorption (kcal/mole) of CO and D_2 from La_2O_3 fresh and used $LaRhO_3$, used rhodium oxide and rhodium metal. The spread of each value represents the variation with surface coverage rather than experimental uncertainty.

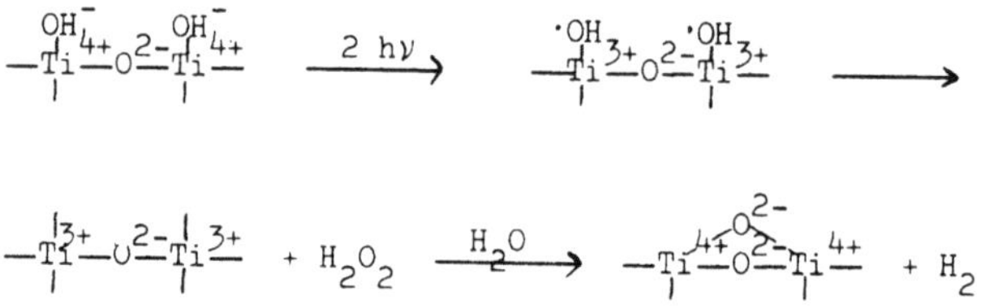

Van Damme and Hall, JACS 101 4373

Fig. 15: The suggested mechanism for the catalyzed photodissociation of water on Titanium oxide surfaces.

6.2. Photodissociation of water over reduced $SrTiO_3$ surfaces

When strontium titanate is illuminated with band gap radiation, about 3.1 electron volts, in the presence of water vapor or aqueous alkali hydroxide solutions hydrogen and oxygen are detected (31). The presence of Ti^{3+} ions at the surface plays a crucial role in dissociating water and is a necessary ingredient of the photocatalytic process (32). Figure 15 shows the catalytic cycle that explains most of the experimental results found so far (33). The reduced $SrTiO_3$ surface, which has a high concentration of oxygen vacancies and Ti^{3+} ions, readily adsorbs water and dissociates it in the dark. As a result the Ti^{3+} species are oxidized to Ti^{4+} and the oxygen vacancies are filled with the fragmented adsorbates. Upon illumination, the photoelectron generated reduces H^+ to H and the electron vacancy traps the negative charge of the hydroxyl ion to convert it to OH radical which in turn produces oxygen through, presumably, a peroxide intermediate. As both hydrogen and oxygen desorb after atom recombinations the Ti^{3+} surface sites and the oxygen vacancy are regenerated. There is evidence for the rapid exchange of oxygen from the water molecules with the oxygen ions at the surface and for the reduction and oxidation of Ti^{3+} sites during the photochemical reaction (34) Thus the importance of the proper oxidation state of transition metal ions to carry out catalyzed reactions is again demonstrated.

6.3. Summary

The catalytic reaction path changes markedly with the oxidation state of the surface atoms. While many transition metals dissociate CO and readily hydrogenate organic molecules, higher oxidation state metal ions insert molecular CO into alkene double bonds or into organic fragments and show reduced hydrogenation ability. Redox catalytic reactions at surfaces often require changes of oxidation states of transition metal ions at the catalyst surface. Often surface atoms must be present in several different oxidation states (metal and metal ion) at the catalyst surface to be able to carry out complex catalytic reactions. A major challange of catalyst preparation is to find ways to stabilize surface atoms in the desired oxidation states.

6.4. Questions that need theoretical and experimental scrutiny

- How do carbon monoxide and dinitrogen dissociate at the metal surface?

- How does chemical bonding change with the oxidation state of the metal ions? How do electron donors (alkali metals) and electron acceptors (halogens, oxygen, sulfur) adsorbed on the surface alter the oxidation states of surface atoms?

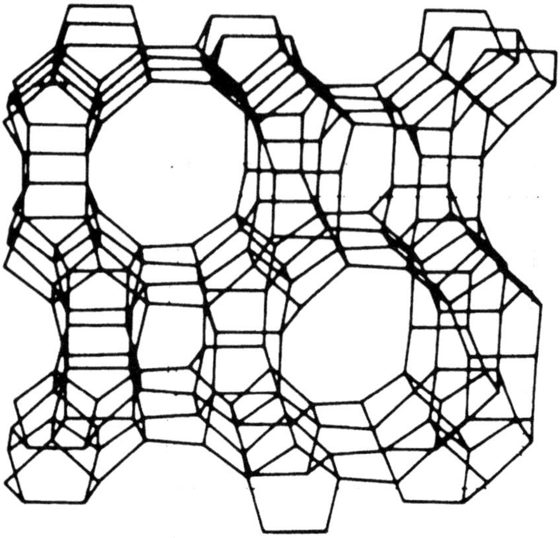

Fig. 16: One of the important zeolites, mordenite. $Na_8\ Al_8\ Si_{40}\ O_{96} \cdot 24\ H_2O$, viewed along (001) axis.

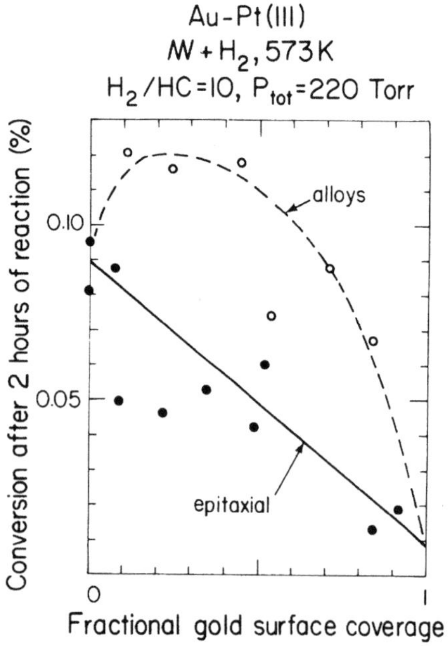

Fig. 17: Activity and selectivity changes for n-hexane conversion as a function of gold coverage in the gold-Pt(111) alloy system.

7. THE BUILDING OF NEW HIGH TECHNOLOGY HETEROGENEOUS CATALYSTS

As a result of combined surface science and catalytic studies on well defined model catalyst systems many of the molecular ingredients of heterogenous catalysis have been uncovered. The model of the working platinum catalyst reveals not only how selective hydrocarbon catalysis occurs over platinum surfaces but also how we might modify or certainly optimize these working catalyst systems. Once we understand the molecular details of the catalyst operation we can design new high technology catalysts by using the scientific understanding that was obtained on the molecular scale. Indeed, the field of heterogeneous catalysis is ready to be transformed from an art to catalysis science that permits the fabrication of high technology catalyst systems.

There exists, already, a group of high technology catalysts that has been in use in the chemical industry over the past 12 years. These are the zeolites (35), one of which is shown in Figure 16. Zeolites are aluminum silicates that can be prepared with variable pore sizes which have molecular dimensions. This gives rise to size and shape selectivity in hydrocarbon conversion reactions since molecules that are small enough to enter the pores can undergo catalyzed reactions. In addition, by controlling the silicon to aluminum ratio of the zeolite crystal the catalytic behavior of these aluminum silicates can be drastically altered. Zeolites, at present, are the largest volume heterogeneous catalysts in use.

Using the molecular ingredients of heterogeneous catalysis that were described above we may control the structure, the concentration of the carbonaceous deposit and the oxidation states of atoms on the catalyst surfaces. The structure of a catalyst may be controlled by epitaxial deposition of one metal on top of another metal or on an oxide or a sulfide that serve as suitably structured supports. The concentration and hydrogen content of the carbonaceous deposits in hydrocarbon conversion reactions or of oxide layers under oxidizing conditions can be altered by the addition of other transition or alkali metals. And finally, the oxidation state of surface ions may be stabilized by incorporation of the transition metal ion into the crystal lattice of refractory oxides or by the addition of electron acceptors, halogens, etc. Below we shall give brief descriptions of several attempts to build a new catalyst.

7.1. The catalytic activity of the gold-platinum system

Bimetallic alloy systems are noted for their higher activity and selectivity for many hydrocarbon reactions as compared to single metal component catalysts. In order to explore the reasons for this we studied the gold-platinum system (36,37). Gold was epitaxially deposited on the (100) and (111) crystal faces of platinum and the structure of the monolayer and the activity for various hydrocarbon reactions as a function of gold surface concentration was monitored. Then by heating the system gold was diffused into platinum and the catalytic activity of the freshly formed alloy was again determined for the same chemical reaction as that used in the presence of the gold overlayer on platinum. Gold forms an ordered overlayer on the platinum surfaces (28). Figure 17 shows the activity for dehydrogenation, hydrogenolysis and isomerization of n-hexane as a function of gold coverage in the epitaxial case. It appears that the activity of platinum is linearly decreased by the presence of a gold overlayer at high reactant pressures. This result, of course, is not surprising as gold is inactive for all these catalytic reactions (37).

Very different results are obtained when gold was difused into platinum at elevated temperatures. The gold-platinum alloy that forms showed a much higher activity for isomerization than clean platinum (29).

Simultaneously the activity for dehydrocyclization and hydrogenolysis decreased exponentially with gold coverage. As a result, the selectivity for isomerization became very large. Gold tends to break up the large coordination number sites by substitution more than the lower coordination number sites; thus reactions that require three-fold sites are more likely to be affected by the presence of gold in the crystal lattice than those reactions involving one or two neighboring platinum atoms. This simple argument can explain the observed selectivity (29). However, this cannot explain the observed increase in chemical activity; the gold alloy is a more active catalyst than clean platinum for isomerization. The presence of gold may decrease the concentration of the carbonaceous deposit and the residence times of the reaction intermediates. The presence of gold may inhibit the formation of carbonaceous polymers, which allows more facile rehydrogenation of the carbon. This can certainly account for the increased rates obtained on the gold alloy surface for isomerization. It is also known that bimetallic alloys poison more slowly than pure metallic catalysts in many hydrocarbon conversion reactions. Perhaps the presence of a second component metal also decreases the rate of conversion of the active carbonaceous fragments to the inactive graphitic phase.

Atomic scale studies of several bimetallic systems including platinum-iridium, iridium-gold and platinum-rhenium systems are in progress and no doubt will help to elucidate the important effects of bimetallic catalysts; the increased activity, higher selectivity and greater resistance to poisoning.

7.2. The effect of alkali metals on the chemisorption of CO and on the reactivity of surface carbon

Alkali metals are frequently used as promoters in many catalyzed surface reactions. These include the ammonia synthesis on iron, the catalyzed hydrogenation of carbon monoxide on various transition metal surfaces and coal gasification in the presence of water. Alkali metals when present on transition metal surfaces affect the catalytic activity in several ways. One important effect is their influence on the heats of absorption of adsorbates such as carbon monoxide. HREELS and TDS

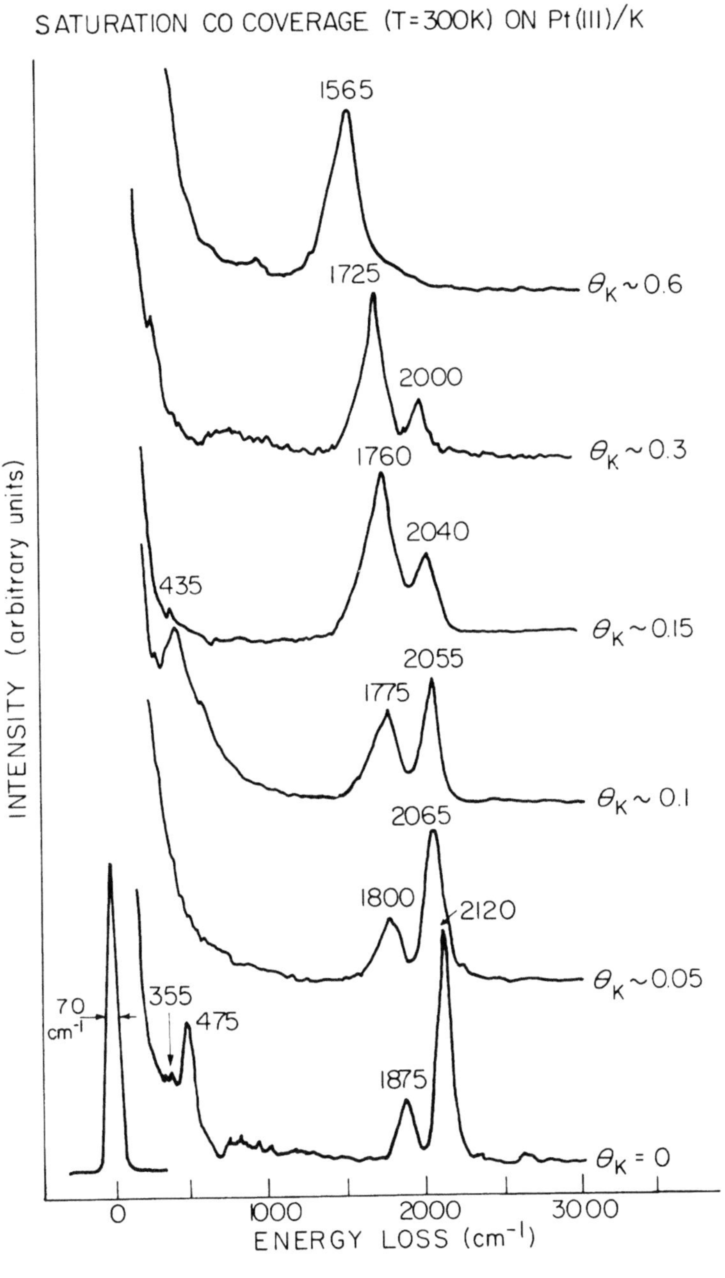

Fig. 18: Vibrational spectra of the saturation CO coverage chemisorbed on Pt(111) at 300K as a function of preadsorbed potassium.

studies showed (38) that the coadsorption of potassium and carbon monoxide on platinum surfaces decreased the CO stretching frequency by about 400 wave numbers as compared to the stretching frequency on the clean platinum, in the absence of potassium. This effect is shown in Figure 18. A decreased CO stretching frequency indicates a weakening of the CO bond and simultaneously it implies an increased metal-carbon bond strength. Thus, potassium seems to increase the bonding energy of CO to the surface, thereby improving the dissociation probability significantly. During ammonia synthesis the presence of potassium on iron surfaces increases the binding energy of dinitrogen and reduces the activation energy for its dissociation to produce atomic nitrogen on the surface (7). Since the dissociation of dinitrogen is the rate limiting step the ammonia synthesis rate is accelerated.

Another effect is the alkali metal catalyzed interaction of water vapor with carbon at the surface. It was reported recently (39) that when alkali hydroxides or carbonates are adsorbed on the surface of graphite they catalyze its reaction with water vapor. At low temperature in the range 200-250°C methane and carbon dioxide are produced. This reaction is an alternate route for the gasification of carbon that is usually carried out without a catalyst around 1000°C with water vapor to produce carbon monoxide and hydrogen. It is clear that by the low temperature reaction alkali hydroxides (potassium, lithium, etc.) act as catalysts for both the reduction of carbon to CH_4 and for its oxidation to CO_2 or CO. This reaction involves ionic intermediates, H^+ and OH^-, which form from water as catalyzed by the alkali hydroxides. These then, by a series of complex reactions that follow, can hydrogenate and oxidize the surface carbon. This low temperature reaction has an activation energy of about 11 kcal/mole and can be responsible for the removal of the carbonaceous deposit from various transition metal surfaces when water vapor is present.

7.3. Strong metal-support interaction

Control of the oxidation state of surface transition metal ions has been reported by Tauster and coworkers (40). By using relatively basic oxides, titanium oxide, zirconium oxide, etc. they find very different chemisorption and reactivity properties of transition metal ions. The influence of these refractory oxide supports on the catalytic activity was called "strong metals support interaction" (SMSI). Thus, using the support, higher oxidation state transition metal ions can be stabilized in addition to the presence of the same transition metals in the zero valent state.

7.4. Summary

Binary alloys (Au, Cu and Pt) have very different catalytic selectivities, improved rates and resistance to chemical degredation when compared with the one-component metal catalysts (Pt). Alkali metal additives to transition metals change the bonding of many adsorbates by charge transfer. In addition, alkali metal catalyzed reactions of water vapor with graphite to produce methane aids the cleaning of the metal catalyst surfaces during hydrocarbon reactions and the low temperature gasification of carbonaceous materials. Compound formation between the metal catalysts and the oxide support can be responsible for marked changes in the product distribution during catalytic reactions and maintaining the higher, desired oxidation state of the transition metal ions.

7.5. Questions that need theoretical and experimental scrunity

- What are the structural and electronic changes that occur upon alloying at each metal surface site (terrace, step, ledge) that are responsible for the change of bonding and reactivity of adsorbates?

- What is the mechanism of the water-carbon reaction that produces gaseous hydrocarbons and is catalyzed by alkali metals?

- What is the nature and conditions of compound formation between the metal and oxide support. Is the alloy catalyst a two phase or multiphase system that is produced from a single intermediate phase under reaction conditions? If so, is the composition of the multimetallic catalysts predictable from their phase diagram?

- Is there charge transfer between the insulator catalyst support (oxide) and the metal catalyst? If so, how is this charge transfer modifying the bonding of adsorbates?

- Are the structures of organic molecules modified when inside a zeolite cage? What are the mechanisms of alkylation, hydrogen transfer and molecular rearrangements in zeolites?

8. ACKNOWLEDGMENTS

This work was supported by the Director, Office of Energy Research, Office of Basic Energy Sciences, Materials & Chemical Sciences Division of the U.S. Department of Energy under Contract DE-AC03-76SF00098.

9. REFERENCES

1. Somorjai, G.A.: "Chemistry in Two Dimensions: Surfaces," Cornell University Press, 1981.

2. Blakely, D.W., Kozak, E., Sexton, B.A. and Somorjai, G.A.: 1976, J. Vac. Sci. & Technol. 13, pp. 1091.

3. Kahn, D.R., Petersen, E.E. and Somorjai, G.A.: 1974, J. Catal. 34, pp. 294.

4. Davis, S.M. and Somorjai, G.A.: 1980, J. Catal. 65, pp. 78.

5. Sexton, B.A. and Somorjai, G.A.: 1977, J. Catal. 46, pp. 167.

6. Spencer, N.D., Schoonmaker, R.C. and Somorjai, G.A.: 1981, Nature 294, pp. 643.

7. Engel, T. and Ertl, G.: 1979, Adv. Catal. 28, pp. 1-78.

8. Buchholz, J.C. and Lagally, M.G.: 1975, Phys. Rev. Lett. 35, pp. 442.

9. Spencer, N.D., Schoonmaker, R.C. and Somorjai, G.A.: 1982, J. Catal. 74, pp. 129.

10. Gillespie, W.D., Herz, R.K., Petersen, E.E. and Somorjai, G.A.: 1981, J. Catal. 70, pp. 147-159.

11. Toyoshima, I. and Somorjai, G.A.: 1979, Catal. Rev.-Sci. Eng. 19, pp. 105.

12. Ceyer, S.T. and Somorjai, G.A.: 1977, J. Catal. 57, pp. 361-371.

13. King, D.A.: 1975, Surface Sci. 47, pp. 384.

14. Ertl, G.: 1980, Catal. Rev.-Sci. Eng. 21, pp. 201.

15. Boudart, M.: 1981, Catal. Rev.-Sci. Eng. 23, pp. 1.

16. Gillespie, W.D., Herz, R.K., Petersen, E.E. and Somorjai, G.A.: 1981, J. Catal. 70, pp. 147; Davis, S.M., Zaera, F. and Somorjai, G.A.: 1982, J. Catal. (to be published).

17. Davis, S.M., Zaera, F. and Somorjai, G.A.: 1982, J. Am. Chem. Soc. (to be published).

18. Davis, S.M., Zaera, F. and Somorjai, G.A.: 1982, J. Catal. (to be published).

19. Lang, B., Joyner, R.W. and Somorjai, G.A.: 1972, Surf. Sci. 30, pp. 440; Van Hove, M.A. and Somorjai, G.A.: 1980, Surf. Sci. 92, pp. 489; Blakely, D.W. and Somorjai, G.A.: 1977, Surf. Sci. 65, pp. 419.

20. Davis, S.M., Zaera, F. and Somorjai, G.A.: 1982, J. of Cat. (to be published).

21. Salmeron, M. and Somorjai, G.A.: 1982, J. Phys. Chem. 86, pp. 341.

22. Davis, S.M., Gordon, B.E., Press, M. and Somorjai, G.A.: 1981, J. Vac. Sci. & Technol. 19, pp. 231.

23. Davis, S.M. and Somorjai, G.A.: 1981, J. Phys. Chem. (to be published).

24. Somorjai, G.A.: 1981, Proc. 9th Intl. Conf. on Atomic Spectroscopy, XXII CSI, Tokyo, Pure and Appl. Chem.

25. Dubois, L.H., Castner, D.G. and Somorjai, G.A.: 1980, J. Chem. Phys. 72, pp. 5234.

26. Bhasin, M.M., Bortley, W.J., Ellgen, P.C. and Wilson, T.P.: 1980, J. Catal. 54, pp. 120.

27. Castner, D.G., Blackader, R.L. and Somorjai, G.A.: 1980, J. Catal. 66, pp. 257.

28. Vannice, M.A.: 1975, J. Catal. 37, pp. 449.

29. Watson, P.R. and Somorjai, G.A.: 1982, J. Catal. 72, pp. 347.

30. Watson, P.R. and Somorjai, G.A.: 1982, J. Catal. 74, pp. 282-295.

31. Wagner, F.T. and Somorjai, G.A.: 1980, J. Am. Chem. Soc. 102, pp. 5494.

32. Ferrer, S. and Somorjai, G.A.: 1981, Phys. Chem. Lett. 85, pp. 1464.

33. Van Dammie, H. and Hall. W.K.: 1979, J. Am. Chem. Soc. 101, pp. 4373.

34. Ferrer, S. and Somorjai, G.A.: 1980, Surface Sci. 97, pp. L-304.

35. Rabo, J.A.: 1976, ed.: Zeolite Chemistry and Catalysis," ACS Monograph 171.

36. Sachtler, J.W.A., Van Hove, M.A., Biberian, J.P. and Somorjai, G.A.: 1981, Surface Sci. 110, pp. 19.

37. Sachtler, J.W.A., Biberian, J.P. and Somorjai, G.A.: 1981, Surface Sci. 110, pp. 43.

38. Garfunkel, E.L., Crowel, J.E. and Somorjai, G.A.: 1982, J. Phys. Chem. (to be published).

39. Cabrera, A. and Somorjai, G.A.: 1982, J. of Catal. 75, pp. 7.

40. Touster, S.J., Fung, S.C. and Gorten, R.L.: 1980, J. Am. Chem. Soc. 100, pp. 170.

NEW PERSPECTIVES IN SURFACE CRYSTALLOGRAPHY

J.B. Pendry
Blackett Laboratory, Imperial College, London SW7 2BZ, U.K.

The ideal balance between theoretical and experimental techniques occurs where comparable amounts of time and money are spent on each side of the problem. In LEED structure analysis this is true to within an order of magnitude for the sorts of structures currently being studied, but in the future we can already see a serious imbalance appearing. Experiments scale linearly in time with N, the number of beams (assuming that all are measured) and theory scales as N^3. The moral is that we should be working to increase the speed of computation and if possible to devise new methods that scale in time as N, so that a competitive position, once established, remains so for a wide class of experiments. Clearly N must be the optimum scaling because even the printing of all beams scales as N.

1. INTRODUCTION

The current generation of techniques is based on the layer approach in which calculations have a two-tier structure : calculation of scattering by a single layer followed by assembly of layers into a complete surface. The latter process is linear in the number of different layers and therefore is satisfactory by the linear criterion given above. Single layer scattering is more intractable and almost all current procedures scale as n^3 where n is the number of atoms per unit cell of the layer. Approximately speaking $N \propto n$ at a given energy. The exception to the n^3 rule is the reverse scattering (RS) method of Zimmer and Holland (1975) which scales at n^2 and is known to be competitive for large unit cells, but still falls short of the ideal scaling.

The main thrust of this paper is that the ideal scaling can be achieved. To anticipate the detailed arguments in condensed form : the extent of multiple scattering is limited by inelastic scattering. For current studies this is not a relevant observation because the multiple scattering path passes through many identical cells and there are techniques for effectively folding back the scattering into one cell by means of a rapidly performed lattice summation (e.g. Kambe (1967a,b)). Even if the

P.-O. Löwdin and B. Pullman (eds.), New Horizons of Quantum Chemistry, 339–350.

electron had infinite path length these techniques would always confine the calculation to describing the wavefunction within a single unit cell. However, the advantages to be gained from this folding operation cease to be critical once the unit cell size exceeds the longest multiple scattering path possible. In this asymptotic region the number of different multiple scattering paths scales linearly with the size of the unit cell and direct summation of the paths must eventually be preferred to the folding procedure. Folding scales at n^3 whereas direct summation scales as n, asymptotically. The problem is that asymptopia corresponds to very large n and unless an astute way of summing the multiple scattering paths is found, the linear techniques will never be feasible. Realisation of a calculation by an n-scaling technique represents the breaking of a barrier. Progress thereafter will be much easier because of the linear scaling.

It is not hard to find instances of systems that are impossibly time consuming with current technology. The classic example is the 7 x 7 Si (111) structure with 49 times the unit cell size of an unreconstructed Si (111) surface: calculations scale by a factor of 10^5. Another instance is that of steps and other defects which it would be useful to study with LEED in the way that HEED has been used to improve our understanding of bulk dislocations. Steps at surfaces can be active sites for catalysis and a means is needed for studying adsorption geomet at the step. Not least interesting of currently intractable problems is that of complex molecules at surfaces. To date only diatomic molecules have received a complete LEED analysis, but large organic molecules are important in surface chemistry. The current generation of programs limits us to molecules smaller than benzene and even this molecule presents difficulties. Then there are the surfaces of complex compounds How are polymer surfaces constituted? What lies on the exterior of a protein crystal? These are ambitious questions, but so too was the objective of protein crystallography in the early days of X ray diffraction. If the problem is sufficiently important a way can be foun

One conclusion to emerge from analysis is that experimental data-collect techniques need to be reconsidered for more complex surfaces. More data are needed as the number of parameters to be found increases, and fortunately the number of beams also increases proportionately. Especially crucial is the lateral complexity that a molecular adsorbate can have. Beams are selectively sensitive to surface geometry, as can b seen from a kinematic argument. The phase of diffraction into the $\underline{g}$th beam by an atom located at $\underline{R}$ is $\underline{g}.\underline{R}$, thus large $\underline{g}$ promotes good sensitivity to $\underline{R}$. Higher order beams are essential in the molecular case, and to achieve equal sensitivity to lateral and normal coordinates all beams should be measured.

Complex systems also favour the taking of as much data as possible at th same energy. θ and ϕ scans are to be preferred to energy scans because the calculational overheads per energy point grow large for complex systems. This is especially true of the asymptotic cluster method proposed in section 3 as a solution to the linearity problem.

The paper has the following plan. In section 2 a procedure is described that can be used with advantage to calculate for molecular adsorbates. In section 3 the asymptotic cluster method is proposed as a radical new approach to LEED, in which the solid is divided into a set of overlapping molecules, each containing a class of multiple scattering paths. Times scale as n in this procedure, and I believe that it is a feasible technique with current technology, given only a reasonable degree of symmetry in the surface. Finally, in section 4 there is some further discussion of the most advantageous means of data collection for interaction with the asymptotic cluster method.

2. THE MOLECULAR APPROACH

In a conventional LEED treatment of a molecular overlayer, no cognisance is taken of the molecular structure: the atoms of the molecule are assigned to the appropriate layer in the same way as bulk atoms. This increases the number of apparent variables by removing the constraint of molecular integrity, and can result in problems if two layers happen to be close together. An approach more in tune with the chemistry of the situation is to assign the molecule integral status by treating it as a non-spherical atom. Thus a new step in the lEED calculations emerges: that of assembling the component atoms into the molecule.

The basis for scattering by the molecule is a set of spherical waves centered on the molecule, just as for the atom, but because the molecule is generally larger than atoms more waves must be included in the basis set. The most obvious approach is to write down a set of multiple scattering equations based on waves centered on each of the component atoms, and transform to the single centre expansion only at the end of the calculation. For small molecules this is an appropriate way to proceed, but for larger molecules it is advantageous to identify a shell structure to the molecule, as is done in EXAFS calculations (Ashley and Doniach (1975), Lee and Pendry (1975)). There is an analogy with the division of a surface into layers: many of the same advantages follow. The first item in the calculation is scattering of spherical waves by a shell of atoms. There are two sorts of waves incident on the shell:

$$\sum_{\ell m} A^-_{\ell m}\, j_\ell(\kappa r) Y_{\ell m}(\underline{r}),\ \sum_{\ell m} A^+_{\ell m}\, h^{(1)}_\ell(\kappa r) Y_{\ell m}(\underline{r}) \tag{1}$$

from outside and inside respectively. These waves are expanded about the centres of the atoms in the shell.

$$j_\ell(\kappa r) Y_{\ell m}(\underline{r}) = \sum_{\ell'' m''} G^-_{\ell m, \ell'' m''}(\underline{R}_j) j_{\ell''}(\kappa|\underline{r}-\underline{R}_j|) Y_{\ell'' m''}(\underline{r}-\underline{R}_j) \tag{2}$$

$$h^{(1)}_\ell(\kappa r) Y_{\ell m}(\underline{r}) = \sum_{\ell'' m''} G^+_{\ell m, \ell'' m''}(\underline{R}_j) j_{\ell''}(\kappa|\underline{r}-\underline{R}_j|) Y_{\ell'' m''}(\underline{r}-\underline{R}_j) \tag{3}$$

where $$G^{\pm}_{\ell m,\ell''m''}(\underline{R}_j) = \sum_{\ell'm'} 4\pi\, i^{\ell-\ell'-\ell''}(-)^{m+m''} \begin{Bmatrix} h^{(1)}_{\ell'}(\kappa R_j) \\ j_{\ell'}(\kappa R_j) \end{Bmatrix}$$

$$\times\, Y_{\ell'-m'}(\underline{R}_j)\int Y_{\ell m}Y_{\ell'm'}Y_{\ell''-m''}d\Omega \tag{4}$$

Thus the wavefield incident on the jth atom is

$$\sum_{\ell''m''} a^{(0)}_{\ell''m''j}\, j_{\ell''}(\kappa|\underline{r}-\underline{R}_j|)Y_{\ell''m''}(\underline{r}-\underline{R}_j). \tag{5}$$

$$a^{(0)}_{\ell''m''j} = \sum_{\ell m} A^{+}_{\ell m}G^{+}_{\ell m,\ell''m''}(\underline{R}_j) + A^{-}_{\ell m}G^{-}_{\ell m,\ell''m''}(\underline{R}_j) \tag{6}$$

Multiple scattering takes place within the shell resulting in a total incident wavefield of $a_{\ell''m''j}$ for which the following self-consistent equation can be written

$$a_{LMj} = a^{(0)}_{LMj} + \sum_{\ell''m''k} a_{\ell''m''k}\, X_{\ell''m''k,LMj} \tag{7}$$

$$X_{\ell''m''k,LMj} = \tfrac{1}{2}\,[\exp(2i\delta_{\ell''}(k))-1]G^{+}_{\ell''m'',LM}(\underline{R}_j-\underline{R}_k)(1-\delta_{jk}). \tag{8}$$

hence the total outgoing wave is

$$\sum_{LM} \tfrac{1}{2}\,[\exp(2i\delta_L(j))-1]a_{LMj}\, h^{(1)}_L(\kappa|\underline{r}-\underline{R}_j|)\, Y_{LM}(\underline{r}-\underline{R}_j) \tag{9}$$

<u>Outside</u> the shell the scattered wavefield is

$$\sum_{\substack{LMj\\L'M'}} a_{LMj}\, \tfrac{1}{2}\,[\exp(2i\delta_L(j))-1]G^{-}_{LM,L'M'}(-\underline{R}_j)h^{(1)}_{L_1}(\kappa r)Y_{L'M'}(\underline{r}) \tag{10}$$

<u>inside</u> the shell the scattered wavefield is

$$\sum_{\substack{LMj\\L'M'}} a_{LMj}\, \tfrac{1}{2}\,[\exp(2i\delta_L(j))-1]G^{+}_{LM,L'M'}(-\underline{R}_j)j_{L'}(\kappa r)Y_{L'M'}(\underline{r}) \tag{11}$$

Substituting for a_{LMj} from (7), scattering matrices $T^{\pm\pm}$ can be identified, where the superscripts refer to the incident and scattered waves respectively, '+' and '−' referring to outward and inward moving waves:

$$T^{++}_{LM,L'M'} = \sum_{\substack{\ell m j \\ \ell' m' k}} G^{+}_{LM,\ell m}(\underline{R}_j)(1-X)^{-1}_{\ell m j,\ell' m' k}$$

$$\times \frac{1}{2} [\exp(2i\delta_{\ell'}(k))-1] G^{-}_{\ell' m' k,L'M'}(-\underline{R}_k) \qquad (12)$$

$$T^{-+}_{LM,L'M'} = \sum_{\substack{\ell m j \\ \ell' m' k}} G^{-}_{LM,\ell m}(\underline{R}_j)(1-X)^{-1}_{\ell m j,\ell' m' k}$$

$$\times \frac{1}{2} [\exp(2i\delta_{\ell'}(k))-1] G^{-}_{\ell' m' k,L'M'}(-\underline{R}_k) \qquad (13)$$

$$T^{--}_{LM,L'M'} = \sum_{\substack{\ell m j \\ \ell' m' k}} G^{-}_{LM,\ell m}(\underline{R}_j)(1-X)^{-1}_{\ell m j,\ell' m' k}$$

$$\times \frac{1}{2} [\exp(2i\delta_{\ell'}(k))-1] G^{+}_{\ell' m' k,L'M'}(-\underline{R}_k) \qquad (14)$$

$$T^{+-}_{LM,L'Mp} = \sum_{\substack{\ell m j \\ \ell' m' k}} G^{+}_{LM,\ell m}(\underline{R}_j)(1-X)^{-1}_{\ell m j,\ell' m' k}$$

$$\frac{1}{2} [\exp(2i\delta_{\ell'}(k))-1] G^{+}_{\ell' m' k,L'M'}(-\underline{R}_k) \qquad (15)$$

To calculate scattering by the whole molecule a procedure analogous to the layer doubling method can be followed. Starting with the central shell, scattering matrices T_1, the second shell scattering matrices T_2, is added to give a composite scattering T_{12}:

$$T^{-+}_{12} = T_2^{-+} + T_2^{--} T_1^{-+}(1-T_2^{+-} T_1^{-+})^{-1} T_2^{++} \qquad (16)$$

Each successive shell shell is put in place qnd the final scattering matrix is the non-spherical analogue of the atomic scattering matrix:

$$\frac{1}{2} [\exp(2i\delta_{\ell'})-1]\delta_{\ell\ell'\ mm'} \qquad (17)$$

and can be incorporated in existing layer programs with minimal modifications.

The number of partial waves needed to describe molecular scattering depends on the radius of the molecule, R_M, measured from the centre of the expansion. The centre should be chosen to minimize R_M.

$$\ell_{max} \simeq R_M \times \sqrt{2E} \qquad (18)$$

Typically for a small organic molecule such as benzone $\ell_{max.} \simeq 9$ at E = 4 Hartrees. One interesting aspect of this expansion is that it is more compact than a multi-centre expansion. A single centre expansion need only describe the wavefield on the surface of the molecule, thus the basic scales as R_M^2. A multi-centre expansion on the other hand describes the wavefield throughout the interior of the molecule and therefore scales as R_M^3 for spherical molecules of constant density. For planar molecules both bases are of comparable efficiency, and for a long-chain molecule the multi-centre expansion is the more efficient. An example of a diatomic molecule calculation may be found in Andersson and Pendry (1980).

An advantage of the molecular calculation is that any internal symmetry of the molecule can easily be built into the calculation, irrespective of angle of incidence which does notappear at this stage of the calculation. Clusters of enormous size (~100 atoms) can easily be handled if the configuration is of sufficiently high symmetry. Experiments made at constant energy, but sweeping through angular variables, enable the molecular scattering to be stored and re-used for each different angle. As the size of the molecules increases, the advantages of constant energy data-sets become greater.

3. THE ASYMPTOTIC CLUSTER METHOD

Current methods of calculation rapidly become impossibly time consuming as the 2D cell increases in complexity, because almost all methods scale as n^3, n being the number of atomsin the 2D cell. Yet a simple argument leads to the conclusion that the intrinsic complexity of multiple scattering saturates at a scaling proportional to n. Consider the number of scattering paths starting at a given atom and terminating the vacuum. This number is finite because of the finite inelastic scattering length, hence the total number of different paths is proportional to the number of different starting points, i.e. $\propto$ n. A means of calculating must be found that reflects this scaling law. A transition point can be expected, at which the new method becomes competitive.

The object is to establish the intrinsic structure that such schemes will possess, to contrast them with current schemes and to make estimates of feasibility. To this end the asymptotic cluster method is described below. It has the correct linear scaling with n but lacks the refinements that the ultimate theory will have.

Multiple scattering events are classified according to the first atom in the scattering chain. Thus A_j is the sum of all amplitudes starting at the jth atom and terminating in vacuo. The total amplitude is given by

$$A_{tot} = \sum_j A_j \tag{19}$$

The j-summation is limited in depth by the inelastic scattering length but is infinite parallel to the surface. Presently the summation will be reduced to a finite one. By tying the first scattering event to the jth atom all subsequent events are confined to a sphere, radius $\simeq \lambda_{inel}$, drawn about the jth atom. The problem can be solved by taking a cluster of atoms, i.e. a molecule, with the jth atom at its centre and solving the multiple scattering problem by the means discussed in the previous section. There is even a simplification: whereas in the general scattering problem external waves can hit any atom, in this instance only the central atom of the cluster is hit by external waves.

Let there be a wave incident on the surface:

$$\exp(i\underline{K}^{+}_{\underline{g}}.\underline{r}), \tag{20}$$

where

$$\underline{K}^{+}_{\underline{g}} = [(\underline{k}_{//} + \underline{g})_x, \ (\underline{k}_{//} + \underline{g})_y \ \pm \ (2E-2V_o-|\underline{k}_{//} + \underline{g}|^2)^{\frac{1}{2}}] \tag{21}$$

which can be expanded in spherical waves about the jth atom

$$\exp(i\underline{K}^{+}_{\underline{g}}.\underline{r}) = \sum_{\ell m} a_{j\ell m} \ j_\ell(\kappa|r-R_j|)Y_{\ell m}(\underline{r}-\underline{R}_j) \tag{22}$$

where

$$a_{j\ell m} = \exp(i\underline{K}^{+}_{\underline{g}}.\underline{R}_j) \ 4\pi \ i \ (-)^m \ Y_{\ell -m}(\underline{K}^{+}_{\underline{g}}) \tag{23}$$

The jth atom scatters these waves to give

$$\sum_{\ell m} a^{(0)}_{j\ell m} \ h^{(1)}_\ell \ (\kappa|\underline{r}-\underline{R}_j|)Y_{\ell m}(\underline{r}-\underline{R}_j) \tag{24}$$

where

$$a^{(0)}_{j\ell m} = \frac{1}{2} \ [\exp(2i\delta_\ell(j))-1]a_{j\ell m} \tag{25}$$

The interaction of the outgoing waves with the surrounding shells of atoms is treated sequentially. The spherical wave amplitude emerging from the nth shell are defined to be

$$a^{(n)}_{jLM} = \sum_{\ell m} a^{(0)}_{j\ell m} \; S^{(n)}_{j\ell m,LM} \tag{26}$$

and S can be defined by a recurrence relationship

$$S^{n+1} = S^{n}(1 - T^{+-}_{n+1} \; J^{-+}_{n})^{-1} \; T^{++}_{n+1} \tag{27}$$

where T_n are the reflection and transmission matrices of the nth shell, defined in equations (12)-(15); $S^{(0)}$ is defined to be the unit matrix; and J^{-+}_{n} is the reflection coefficient of the cluster comprising the first n shells plus the central atom:

$$J^{-+}_{o\ell m,\ell' m'} = \frac{1}{2}\,[\exp(2i\delta_{\ell}\;(j))-1]\,\delta_{\ell\ell'}\delta_{mm'}\;, \tag{28}$$

$$J^{-+}_{n+1} = T^{-+}_{n+1} + T^{--}_{n+1} \; J_{n}^{-+}(1-T^{+-}_{n+1} \; J^{+}_{n})^{-1} \; T^{++}_{n+1} \tag{29}$$

The total scattered wave is

$$\psi_S = \sum_{\substack{pj\\LM}} a^{(n)}_{jLM} \, h^{(1)}_{L} \; (\kappa|\underline{r}-\underline{R}_j-\underline{U}_p|)Y_{LM}(\underline{r}-\underline{R}_j-\underline{U}_p)\exp(i\underline{k}_{//} \cdot\underline{U}_p) \tag{30}$$

where $\underline{U}_p$ is a 2D lattice vector and the translation properties of the incident wavefield have been used. Expanding the scattered waves outside the crystal in plane waves gives

$$\psi_S = \sum_{g'jLM} \frac{2\pi \; i^{-L}}{\Omega\kappa K^{+}_{\underline{g}'z}} \; a^{(n)}_{jLM} \; \exp\;[i\underline{K}^{-}_{\underline{g}'}(\underline{r}-\underline{R}_j)]\,Y_{LM}(\underline{K}^{-}_{\underline{g}'}) \tag{31}$$

Substituting for $a^{(n)}_{jLM}$ from (26), (25) and (23) gives

$$\psi_S = \sum_{\underline{g}'} R_{\underline{g}'\underline{g}} \; \exp[i\underline{K}^{+}_{\underline{g}'}\cdot\underline{r}] \tag{32}$$

where

$$R_{\underline{g}'\underline{g}}(\underline{k}_{//}) = \sum_{\substack{j\ell m\\LM}} \frac{8\pi^2}{\Omega\kappa K^{+}_{\underline{g}'z}} \quad \exp[i(\underline{K}^{+}_{\underline{g}}-\underline{K}_{\underline{g}'})\cdot\underline{R}_j]$$

$$\times\ [i^{\ell}(-)^{m}\ Y_{\ell-m}(\underline{K}^{+}_{\underline{g}})]\ \frac{1}{2}\ [\exp(2i\delta_{\ell}\ (j))-1]S^{(n)}_{j\ell m,LM}$$

$$\times\ [i^{-L}\ Y_{LM}(\underline{K}^{-}_{\underline{g}'})] \tag{33}$$

Note that there is an asymmetry in S: many more values of LM than of ℓm will be needed because the LM expansion has to describe waves on the exterior of the cluster.

Alternatively a shell decomposition of the cluster need not be made. Multiple scattering would then be described by matrix 'X' of equation (B) and matrix R can now be written down by inspection:

$$R_{\underline{g}'\underline{g}}(\underline{k}_{//}) = \sum_{\substack{j\ell m\\ j'\ell'm'}} \exp\ [i(\underline{K}^{+}_{\underline{g}}.\underline{R}_{j}-\underline{K}^{-}_{\underline{g}}.\underline{R}_{j})]\frac{8\pi^2}{\Omega\kappa \underline{K}^{+}_{\underline{g}'z}}$$

$$\times\ [i^{\ell}(-)^{m}Y_{\ell-m}(\underline{K}^{+}_{\underline{g}})]\ (1-X)^{-1}_{\ell mj,\ell'm'j'}\ \frac{1}{2}\ [\exp(2i\delta_{\ell'}(j'))-1]$$

$$\times\ [i^{-\ell'}\ Y_{\ell'm'}(\underline{K}^{-}_{\underline{g}'})]. \tag{34}$$

This latter expression has the advantage that both the ℓm and $\ell'm'$ expansions are compact thus recalculation of R at a different value of $\underline{k}_{//}$ requires only that $Y_{\ell m}(\underline{K}_{\underline{g}})$ be evaluated for the small number of $\ell'm'$ values needed for a single atom. Probably the most economical procedure would be to calculate S in a shell model, and then transform to a multi-centre basis to find $(1-X)^{-1}$, which do not need to be recalculated for a new angle of incidence.

Both (33) and (34) are reminiscent of the kinematic scattering formula, which neglects multiple scattering:

$$R^{kin}_{\underline{g}'g} = \sum_{j\ell m} \frac{8\pi^2}{\Omega\kappa\ \underline{K}^{+}_{\underline{g}'z}}\ \exp[i(\underline{K}^{+}_{\underline{g}}-\underline{K}^{-}_{\underline{g}'}).\underline{R}_{j}]$$

$$\times\ [i^{\ell}(-)^{m}\ Y_{\ell-m}(\underline{K}^{+}_{\underline{g}})]\ \frac{1}{2}\ [\exp(2i\delta_{\ell}\ \ (j))-1]\ [i^{-\ell}Y_{\ell m}(\underline{K}^{-}_{\underline{g}'})]. \tag{35}$$

Equations (33) and (34) can be regarded as the kinematic formula plus multiple scattering corrections which appear as complications in the atomic scattering factor produced by subsequent transmission of the scattered wave through the other atoms.

All the remarks made in section 2 about symmetry hold here also and can be used to increase the speed of calculation. If the shell method is used to compute S, an obvious adaptation of the RFS method (Pendry, 1971) can be used to good effect. Waves are incident only on the central atom of the cluster which further increases the value of an RFS approach.

4. AN ASSESSMENT OF FLEXIBILITY

To estimate how time consuming the method is, take an energy

$$E = 2.5 \text{ H} \tag{36}$$

and an imaginary component of the potential

$$V_{oi} = -0.15 \text{ H} \tag{37}$$

so that

$$\lambda_{inel} \simeq \frac{\sqrt{2E}}{-V_{oi}} \simeq 15 \text{ a.u.} \tag{38}$$

To describe scattering by a molecule radius 15 a.u. would take partial waves up to

$$L_{max} \simeq \lambda_{inel} \times \sqrt{2E} \simeq 30 \tag{39}$$

or about 10^3 partial waves. This is a large but not impossible dimension for a matrix. If the surface had 4m symmetry these matrices would factorise into 8 matrices, each approximately 125 x 125. Current LEED programs invert matrices of this order to solve for layer scattering using the old methods. Alternatively if a multi-centre expansion were used the molecule would typically contain 100 atoms, each needing 25 partial waves, a total of 2,500 partial waves. Again, 4m symmetry would factorise these matrices into 8 matrices dimensions 300 x 300.

Demands on core size for the unsymmetrised single centre expansion would be dominated by the $10^3 \times 10^3$ scattering matrix which of course has 10^6 complex elements. However, once the inversion of this matrix is complete it is only necessary to store those elements of the inverse acting on partial waves originating from the centre of the cluster. These describe the waves around a single atom, therefore only $\simeq$ 25 partial waves would be needed and there are 2.5×10^4 complex elements for each inequivalent atom in the surface to be stored from one angle of incidence to another. Symmetry reduces the storage requirements to almost trivial amounts : $\simeq 10^4$ complex elements for one of the symmetrised scattering matrices, no more than one of which need be in core at a time.

One possible application of these asymptotic techniques is to a hybrid approach in which the fastest part of the current methods, that is calculation of inter layer scattering in a beam representation, is retained but the time consuming calculation of intra layer scattering is performed with the new method. The two-dimensionality makes the multi-centre expansion much more favorable. In a circle radius 15 a.u. we have

$$\simeq \frac{\pi \cdot 15^2}{20} \simeq 40 \text{ atoms}$$

each having 25 partial waves: a total of 10^3 unsymmetrised waves. Again assuming 4m symmetry yields 8 matrices of order 125×125 each.

The striking fact about the asymptotic cluster method is that most of the time consuming calculating is done only once per energy point. This implies that it is vary advantageous when using this scheme to have a lot of data available at the same energy: $I_{hk}(\theta,\phi)$ rather than $I_{hk}(E)$. Such data sets contain a surprisingly large amount of information. Typically a data set comprising a 5 H. range of 5 beams for 3 different angles of incidence, would be regarded as more than adequate. Peaks in I(E) data are spaced by

$$\Delta E \simeq 2|V_{oi}| \simeq 0.3 \text{ H}$$

therefore there are approximately

$$\frac{5 \times 5 \times 3}{0.3} = 250 \text{ bits}$$

of information in such a data set from a crude estimate. On the other hand in the $I(\theta,\phi)$ plot the minimum resolvable area of $\underline{k}_{//}$ space is typically

$$\Delta \underline{k}^2_{//} \simeq \frac{1}{\lambda^2_{incl}} \simeq \frac{V^2_{oi}}{2E} \simeq 0.005 \text{ a.u.}^2$$

The maximum area of $\underline{k}_{//}$ space accessible at 2.5 Hartrees in $2\pi E$, yielding

$$\frac{2\ E}{\Delta \underline{k}^2_{//}} \simeq 3 \times 10^3 \text{ bits per beam}$$

There are $\gtrsim$ 10 beams on the screen at 2.5 Hartrees giving a total of $\simeq 10^4$ bits of information. This is a vast amount of information; far more than is normally considered sufficient for an accurate and reliable determiantion. As the structure becomes more complex, more information will be required, but the number of beams scales in the same way as the information needed and $I(\theta,\phi)$ even at lower energies will always provide more than enough information.

A further argument supporting this form of data taking is that it provides the large and small g information necessary for good horixontal and vertical resolution. It also eliminates the need to know the energy variation of V_{or} and V_{oi}. Even the phase shifts themselves could be treated as variables to be determined, provided that only a small number of different atomic species was involved.

Because of the large amount of information available per energy, and the need to make a cluster calculation only once per energy, the criticality of the cluster calculation is reduced by a factor $\gtrsim 10^3$. It is possible to spend much more time inverting matrices (the time consuming step) in the cluster calculation.

Towards the Ultimate LEED Theory

Dimensions may be larger by a factor of $(10^3)^{1/3}$ and still give competitive times. Current calculations which may use 100x100 matrices can be expected to be roughly equivalent in time to the asymptotic cluster method using $10^3 \times 10^3$ matrices. Thus the cross over point for the new method would appear to have been reached in systems currently under study.

The conclusion is that the threshold of asymptopia has been passed especially for systems with some degree of symmetry. The possibility of progression with only a linear scaling law to systems of much more general interest than the limited number that can currently be studied is now open.

References

Andersson, S. and Pendry, J.B., 1980, J. Phys. C., 13, 3547-61.
Ashley, C.A. and Doniach, S., 1975, Phys. Rev., B11, pp 1279-88.
Kambe, K., 1967a, Z. Naturf., 22a, pp 322-30.
________, 1967b, Z. Naturf., 22a, pp 422-31.
Lee, P.A., and Pendry, J.B., 1975, Phys. Rev., B11, pp 2795-811.
Pendry, J.B., 1971, J. Phys. C, 4, 2514-23.
Zimmer, R.B. and Holland, B.W., 1975, J. Phys. C, 8, 2395-400.

INTRODUCTORY REMARKS TO SYMPOSIUM 5b: LARGE MOLECULES AND SOLID STATE THEORY.

Jaroslav KOUTECKÝ
Freie Universität Berlin
Institut für Physikalische Chemie
1000 Berlin 33, West-Germany

The quantum chemistry and the solid state theory represent two branches of quantum theory applied on two very near classes of objects which differ mainly in their sizes and kind of space order. The investigated problems are quite similar in both sciences but necessarily different methodical approaches and unlike customary scientific languages are the chief obstacles to the understanding between the groups of solid state theorists and of theoretical chemists.

The first plenary lecture of this Symposium by Albert Overhauser demonstrates the common difficulties of solid state theory and quantum chemistry in proper description of the electron correlation. It is perhaps worth mentioning that the analogues of charge-density-waves and spin-density-waves have been found when some molecular systems have been treated in one-electron approximation. Their connection with the real molecular properties is, of course, in this special case an open question . This example demonstrates a very frequent situation exhibiting the similarities and simultaneous differences in the problems which are dealt with in both sciences considered.

The connection between the properties of molecular semiconductors, metals and superconductors and the electronic structure of their constituent parts, i.e. molecules are dominant aspects of the second invited paper of Kim Carneiro. The concept of charge density waves, spin density waves, singlet and triplet superconductors in quasi-one-dimensional systems together with collective excitations are able to explain the properties of the molecular crystals with unusual electric conductivity.

In my opinion the theoretical research in the electronic structure of clusters of various sizes which are built up of atoms of various kind can serve as a future junction between the quantum chemistry and solid state theory. For this reason it seems to be appropriate to include few comments on this subject in these Introductory remarks.

The satisfactory experience with the quantum chemical methods obtained from the investigation of electronic structure of molecules helps tremendously in the studies of electronic properties of small clusters, mainly when the constituent atoms are light. The general inpact of cluster investigation involving a large variety of problem complexes is ob-

P.-O. Löwdin and B. Pullman (eds.), New Horizons of Quantum Chemistry, 351–356.

vious. They range from the properties of experimentally isolated metal clusters over the theory of crystal growth until the characterization of centers of catalytic activity. On the contrary, it is not obvious at all if it is possible to abstract from the obtained information on the structure of clusters at least some conjectures concerning the properties of corresponding very large systems. The simple enlargement of the cluster is a trivially right but in reality a not feasible answer to this question. Relatively large clusters which are already difficult to be treated with acceptably accurate and reliable methods have only a minority of atoms with the same number of nearest neighbors as have the bulk-atoms of the corresponding solid - in short the ratio surface/volume is too large.

I would like to emphasize here some quite general rules and regularities following from the theoretical work on alkali metal and earth-alkaline metal clusters [1-20]. They can help to suggest feasible ways how to extrapolate properly from the information on clusters for the better understanding of the properties of the solid built up from the same kind of atoms:

1) Among relatively very small Li and Na clusters the sections of the bcc and fcc crystal lattices can be found as very stable agglomerates. Their interatomic distances are very close to the interatomic distance in infinite crystals. The sections of the closed packed lattices in which Li crystallizes under 78° K are generally more stable than the sections from the bcc lattice. Moreover, the Li atom arrangements which can be considered as steps in pentagonal crystal growth are very stable. Similarily the Be clusters which are energetically preferred are the sections of the hcp lattice with the interatomic distances very near to the interatomic distances in Be crystal.

2) The correct prediction of approximate distances between the neighboring atoms is due to the same kind of binding in metal clusters and in metal crystals. In alkali metals the bonds are due to the overlaping of s-orbitals which is necessarily strongly stabilized by the participation of p-orbitals acting as polarization functions. Therefore, without the inclusion of the p-orbitals in the AO basis a correct prediction on the energetically favorable cluster shapes is completely impossible. On the other hand, the s-p hybridization is basic for the understanding of the bonds in the Be clusters. The s-p hybridization is less important for the other earth-alkaline metal clusters. The stability of the Mg_4 tetrahedron cluster results exclusively from the electron correlation. The inclusion of d-functions as polarization functions is indespensable for the proper description of the binding in this cluster.

3) The nature of binding in clusters explains also some properties found in the metal crystals. For example, the average

bond strength, expressed in terms of the binding energy per atom, does not change in alkali metals strongly along some deformation paths as long as no new overlaping between covalent atomic radii takes place. The "bond topology" is a more important factor than the smaller changes in "bond distances" for the cluster stability. This cluster property is due to the role of p-orbitals as polarization functions only and explains very naturally the softness of alkali metals. On the contrary, in the Be crystal the s-p hybridization and from it resulting pronounced directionality of bonds makes understandable the smaller deformability of this metal.

4) The compactness measured by the number of nearest neighbors of all atoms divided by the number of constituent atoms in the cluster is not necessarily the only dominant factor for the cluster stability. The other aspects of "bond" topology and "bond geometry" are also very important. For example, the existence of the Be atoms which posses only two "bonds" which form an acute mutual angle lowers the stability of the cluster in comparison with other Be clusters with the same number of atoms.

5) The spin properties are also the cluster features which strongly influence the cluster stability. The ground states of Li and Be clusters can have different multiplicity according to the cluster geometry. The energetically most stable smaller Li and Be clusters have always singlet ground state but very often small geometry changes connected with only small increase in energy give rise to the biradical features. It means that either lowest singlet and lowest triplet states are nearly degenerated or triplet state has quite lower energy than the lowest singlet state. Some Li clusters are singlet biradicals with a pair of unpaired electrons in the singlet ground state.

6) The behavior of the binding energy per atom with increasing number of atoms in clusters proves that the bond character in compact metals can be properly judged from the ab initio treatments of clusters. The more sophisticated theories which take into account the correlation effects predict the increase of binding energy per atom in clusters involving the atom species for which the bonds in diatomic molecules are very weak and very badly described in the one-electron approximation. The extrapolation of the binding energy per atom as function of the average number of bonds per atom yields an at least roughly right estimate of the sublimation heat of the corresponding crystal in the case of Li and Be. Because the "compactness" is not the only criterion for the cluster stability it is necessary for purposes of the mentioned extrapolation to consider the most stable clusters for the given number of atoms and the given average number of nearest neighbors.

7) The relative low stability of clusters with biradical features can cause higher reactivity of the cluster in the interaction with simple atoms and molecules than the comparable clusters exhibiting no biradical properties. Consequently, the cluster models for chemisorption and catalytic reactions should be carefully selected with special attention paid to the possible biradicaloid character of the model. Not enough transparent and seemingly inconsistent results of some cluster studies on the chemisorption using sophisticated methods are probably due just to the change in biradicaloid features of the clusters of different size which are employed to model the crystal surface.

8) From already above mentioned reasons the appearance of the electric conductivity of metals cannot be forced by the simple enlargement of the size of theoretically investigated clusters. Therefore, other signs of the metallic behavior must be sought for among the properties of relatively small clusters of metal atoms. The absence of the energy gap between the valence and conduction band is generally considered as a characteristic feature for metals and semiconductors with a zero energy gap in the framework of band theory. An existence of relatively small sections of the crystal lattice having triplet ground state with an energy which does not differ appreciably from the energies of lattice sections of comparable size with singlet multiplicity can be considered as a sign of developing metallic character. Similarily, the singlet cluster biradicals with relatively low energy and with approximately degenerated lowest singlet and triplet states can be regarded as symptomatic for metals. Such a behavior has been found in investigation of alkali and alkaline-earth metal clusters.

9) The proper application of the pseudopotential quantum chemical approaches yields parallel results to the results of a very sophisticated all-electron a priori methods. The treatment of medium size clusters with heavier atoms is therefore feasible.

The aim of these remarks is to encourage the research on clusters which can be used to check if the approximations and model notions employed in the solid state theory of particular solid do not miss some essential and fundamental features of the binding in this solid. The level of quantum chemical methods applicable for cluster calculations is sophisticated enough for this important purpose. Especially, the cluster models are indispensable tools for the description of surface properties of solids. These models should be compared and completed also with methods where the delocalization of electrons in larger parts of the solid is explicitely taken into account.

REFERENCES

[1] R.J. Messmer, in "The Nature of the Surface Chemical Bond", Eds.: T.N. Rhodin and G. Ertl(North-Holland, Amsterdam, 1979) p. 51.

[2] B. Kunz, in: Theory of Chemisorption, Topics in Current Physics 19, Ed. J.R. Smith (Springer, New York 1980) p. 115.

[3] D.W. Davies and G. del Conde, Chem. Phys. 12, 45 (1976).

[4] J.B. Brewington, C.F. Bender and H.F. Schaefer III, J. Chem. Phys. 64, 905 (1976).

[5] C.E. Dykstra, H.F. Schaefer III and W. Meyer, J. Chem. Phys. 65, 5141 (1976).

[6] K.D. Jordan and J. Simon, J. Chem. Phys. 67, 4027 (1977).

[7] C.W. Bauschlicher Jr., C.F. Bender, H.F. Schaefer III and P.S. Bagus, Chem. Phys. 15, 227 (1976).

[8] K. Hermann and P.S. Bagus, Phys. Rev. B17, 4082 (1977).

[9] J. Flad, H. Stoll and H. Preuss, J. Chem. Phys. 71, 3042 (1979).

[10] H.-O. Beckmann, J. Koutecký, P. Botschwina and W. Meyer, Chem. Phys. Lett. 67, 119 (1979).

[11] R.A. Whiteside, R. Krishnan, J.A. Pople, M.B. Krogh-Jespersen, P. von Ragué Schleyer and G. Wenke, J. Comp. Chem. 1, 307 (1980).

[12] H.-O. Beckmann, J. Koutecký and V. Bonačić-Koutecký, J. Chem. Phys. 73, 5182 (1980).

[13] H. Stoll, J. Flad, E. Golka and Th. Krüger, Surf. Sci. 106, 251 (1981).

[14] R.A. Chiles and C.E. Dykstra, J. Chem. Phys. 74, 4544 (1981).

[15] G. Pacchioni and P. Fantucci, J. Mol. Catalysis 12, 213 (1981).

[16] P. Fantucci, G. Pacchioni, R. Ugo and J. Fernandez-Rico, J. Mol. Catalysis 13, 379 (1981).

[17] R.A. Chiles, C.E. Dykstra and K.O. Jordan, J. Chem. Phys. 75, 1044 (1981).

[18] G. Pacchioni, H.-O. Beckmann and J. Koutecký, Chem. Phys. Lett. 87, 151 (1982).

[19] H.-O. Beckmann and J. Koutecký, Surf. Sci. 119, (1982).

[20] G. Pacchioni and J. Koutecký, Chem. Phys., submitted.

INSIGHTS IN MANY-ELECTRON THEORY FROM THE CHARGE DENSITY WAVE STRUCTURE OF POTASSIUM

Albert W. Overhauser
Department of Physics, Purdue University, West Lafayette, Indiana 47907, USA

Approximations frequently employed to simplify fundamental treatments of the theory of metals can lead to qualitatively incorrect conclusions. A striking example is the broken translational symmetry of the ground state of metallic potassium. Experimental confirmation of its charge-density-wave structure is reviewed. Such a structure cannot occur if local-density approximations to exchange and correlation are employed, or if Coulomb potentials are replaced by screened interactions.

1. INTRODUCTION

Many-electron theory becomes a most delicate problem in systems having a degenerate ground state. The Jahn-Teller effect comes to mind as an example. It hardly seems necessary to argue that an infinite degeneracy would provide an even greater challenge to theory. Metals fit this category since their finite density of states at the Fermi surface implies the existence of an infinite number of excited states within ε of the ground state (however small ε might be). Superconductivity, spin density waves, charge density waves, the Kondo effect, and the Mott metal-insulator transition point to a rich spectrum of intriguing phenomena of which, perhaps, many are yet undiscovered. That some have been given names does not mean they are well understood.

Consider the simplest metal of all: "jellium"--a favorite model for theorists, in that the positive-ion background is taken to be without structure. Electrons have charge and spin. So the fundamental response mechanisms are described by a (wave-vector-dependent) polarizability and spin susceptibility. Many theorists have tried to calculate these functions. Bloch[1] was the first to determine the enhancement of the spin susceptibility (over the Pauli value) resulting from exchange interactions. Wigner[2] pointed out that dynamic correlations reduce the magnitude of the enhancement. This latter contribution is a true many-body effect. (The exchange enhancement is a "mean-field" term and is known exactly.) A survey of theoretical "opinion" was compiled by Kushida et al.,[3] and is shown in Fig. 1. The lack of consensus can

P.-O. Löwdin and B. Pullman (eds.), New Horizons of Quantum Chemistry, 357–372.

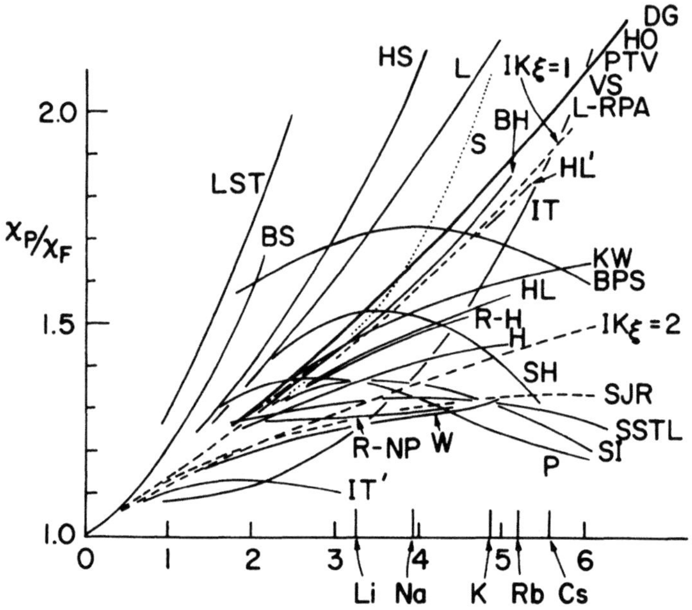

Fig. 1. Enhancement of the Pauli spin susceptibility for an electron gas, over that for free Fermions, versus electron separation parameter r_s. Values of r_s for the alkali metals are indicated. The labels on the curves refer to the theorists involved and may be interpreted using Ref. 3.

hardly be exaggerated. Clearly evidence from the real world is necessary to discipline our habits of thought and to dictate limits to our favorite approximations.

Fortunately two of the alkali metals, Na and K, come close to the theoretical ideal. The crystalline potential acting on the conduction electrons is so small that the degenerate Fermi sea is free-electron like. A common belief, supported by accurate deHaas-van Alphen measurements,[4] is that the Fermi surface deviates by only one or two parts per thousand from a perfect sphere. Nevertheless, many other measurements show that the translation symmetry of Na and K has been broken by a CDW instability, and that their Fermi surfaces are in fact multiply connected.

In the following sections we shall review some of the phenomena which reveal the CDW structure of K. Most research is carried out on K because Na crystals are destroyed by a Martensitic transformation near 35°K. Our attention will focus on recent experimental and theoretical research, which confirms the broken symmetry of the electronic ground state in a spectacular way.

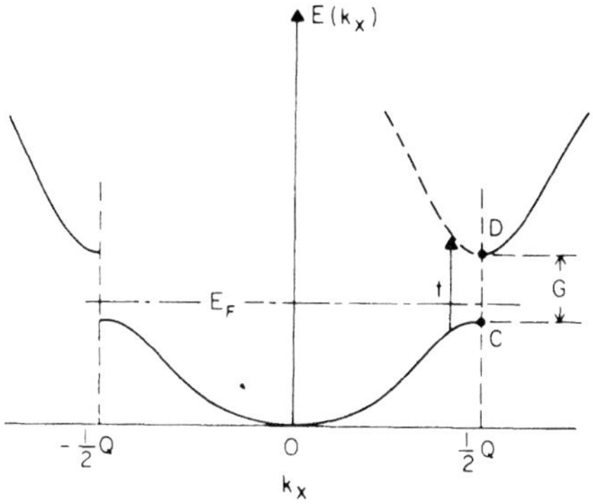

Fig. 2. Conduction-electron energy spectrum $E(k_x)$ for a CDW with $\vec{Q}$ along $\hat{x}$. The energy G, between states C and D is the magnitude of the CDW potential, Eq.(1). The optical transitions responsible for the Mayer-El Naby anomaly are indicated by the arrow t.

The relevance of these facts to many-electron theory will be discussed in Sec. 5. We note here that broken-symmetry states in simple metals were anticipated theoretically,[5] and that a microscopic theory of the correlation energy[6] shows a CDW structure, rather than a spin density wave, to be favored. Controversy arises from oversimplified accounts of many-electron effects, such as local-density treatments of exchange and correlation, or the use of screened interactions in energy calculations. Such approximations do not permit broken symmetries in simple metals. Appreciation of the ground state of K transcends, therefore, a mere account of the peculiarities of a particular metal.

2. OPTICAL ABSORPTION OF A CDW

An incommensurate CDW arises from, and causes, an extra periodic potential acting on the conduction electrons.

$$V(\vec{r}) = G \cos\vec{Q}\cdot\vec{r} \ . \tag{1}$$

The wavevector Q is approximately equal to the diameter $2k_F$ of the Fermi surface and, in general, is not an integral fraction of some reciprocal lattice vector. The presence of this potential in the one-electron Schrodinger equation causes energy gaps of magnitude G to appear in the energy spectrum $E(\vec{k})$ which, for jellium, would have been merely $\hbar^2k^2/2m$. The modified spectrum is shown in Fig. 2. The Fermi level E_F is shown to coincide with the lower edge of the CDW energy gap. New optical transitions from filled states (below E_F) to empty states (above) become allowed.[7] An important property of the new absorption is that it is uniaxial; it is proportional to $\cos^2(\hat{\varepsilon}\cdot\hat{Q})$, where $\hat{\varepsilon}$ is the polarization vector of the photon inside the metal. CDW optical transitions are vertical, in Fig. 2, and have a threshold equal to the gap G.

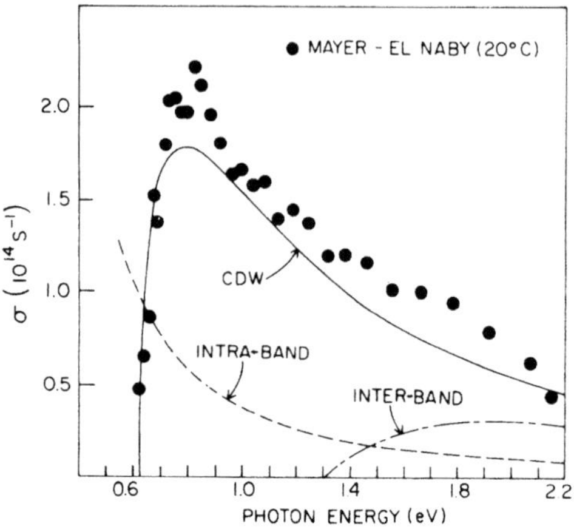

Fig. 3. Optical absorption of K. The intra-band Drude contribution has been subtracted from the data. Without a CDW structure only the inter-band contribution would remain. The solid curve is theoretical and assumes the presence of a CDW with energy gap G = 0.62 ev, Ref. 7.

The first evidence for a CDW structure in K came from the optical absorption measurements of Mayer and El Naby.[8] They found a strong, anomalous absorption having a threshold ~ 0.6 ev, well below the fundamental interband edge at 1.3 ev. Their work has been reproduced by Hietel[9] and by Harms,[10] all of whom worked on bulk K surfaces. Other workers, using evaporated films, have not observed the Mayer-El Naby anomaly; and this can be explained, because $\hat{Q}$ should be perpendicular to the surface of an evaporated film.[11] Inside the metal, then, $\hat{\varepsilon}$ is perpendicular to $\hat{Q}$ and the anomalous absorption strength should be ~ 0.

The Mayer-El Naby anomaly is shown in Fig. 3. It is independent of temperature and even persists into the liquid state. Its size and spectral shape are in excellent agreement with the theoretical spectrum,[7] for which the gap G is the only unknown parameter. A similar anomaly is found[9] in Na, for which the gap is ~ 1.2 ev.

The influence of the potential (1) causes the electronic wavefunctions to acquire extra momentum components.

$$\psi_{\vec{k}} \cong e^{i\vec{k}\cdot\vec{r}} + \alpha(\vec{k})e^{i(\vec{k}+\vec{Q})\cdot\vec{r}} + \beta(\vec{k})e^{i(\vec{k}-\vec{Q})\cdot\vec{r}}. \quad (3)$$

When one sums $|\psi_{\vec{k}}|^2$ over all occupied states, the total electron density is no longer uniform. It has a small sinusoidal modulation p.

$$n(\vec{r}) = n_o[1 - p \cos\vec{Q}\cdot\vec{r}]. \quad (4)$$

The Coulomb potential that would result from this charge modulation is compensated by a corresponding modulation in the positive-ion lattice.[6] The ions suffer a periodic displacement from their lattice sites $\vec{L}$:

$$\vec{u}(\vec{L}) = \vec{a}\ \sin\vec{Q}\cdot\vec{L}. \tag{5}$$

Because of this cancellation, which is almost perfect,[12] the CDW potential arises exclusively from exchange and correlation interactions. CDW's provide a critical test of many-electron theory since, for them, exchange and correlation is not a mere correction to a dominant Hartree potential.

3. OTHER CDW PHENOMENA IN K

Before discussing the recent, direct observation of open orbits in K, which show that its Fermi surface is multiply connected, we shall give a brief review of a number of anomalous properties studied during the last eighteen years. These also provide direct evidence for the broken symmetry; and give considerable information about other microscopic details.

3.1. Conduction-Electron Spin-Resonance Splitting

The g-factor for spin resonance (CESR) in a metal is a measure of the Fermi surface average of $g(\vec{k})$. This averaging occurs because electrons are scattered through ~ 10^4 states at the Fermi surface during one spin-lattice relaxation time. Nevertheless the CESR in K was found to be split into well resolved components[13] separated up to ~ 0.5 Oe in an applied field of 4000 Oe. Splittings up to 5 Oe were observed for H = 40,000 Oe;[14] so a g-factor splitting is involved. It is about 10% of the g shift caused by spin-orbit coupling. This splitting was explained quantitatively[15] by the Fermi surface distortion created by the CDW, which leads to a g-factor dependent on the angle between $\vec{H}$ and $\vec{Q}$. The splitting arises for samples subdivided into $\vec{Q}$ domains, and is proportional to the CDW energy gap G, Eq.(1). The same value, G = 0.6ev, obtained from the optical anomaly is required to fit the observed splitting.

3.2. Doppler-Shifted Cyclotron-Resonance Shift

Helicon waves (of wave vector q) generated by radio frequency excitation in a magnetic field $\vec{H}$ (normal to a K surface) can cause cyclotron resonance. This occurs when electrons traveling at the Fermi velocity, $v_F = \hbar k_F/m^*$, satisfy the doppler-shifted frequency condition:[16]

$$qv_F = eH_r/m^*c. \tag{6}$$

Since m* cancels in this equation, the resonance field H_r depends only on k_F, which in turn depends only on the lattice constant. For 10 MHz helicons,[17] the resonance field is 4% less than the 18 kOe value required by an energy band with spherical symmetry. This shift arises from the effect of the CDW on the $\vec{Q}$ component of conduction-electron velocity. It is explained quantitatively[18] by a CDW energy gap $G = 0.6$ ev.

3.3. Residual Resistivity Anisotropy

At helium temperature the electrical resistivity of a metal is dominated by electron scattering from impurities. Since these are point imperfections, the residual resistivity must then be isotropic (if the metal has cubic symmetry). However, measurements on single crystal spheres[19,20] have shown that, instead, the residual-resistivity tensor is a cigar-shaped ellipsoid with a major to minor axis ratio $\gtrsim 5$. The presence of the CDW energy gaps, Fig. 2, and the perturbed wavefunctions, Eq.(3), cause the resistivity parallel to $\vec{Q}$ to be about four times larger than perpendicular to $\vec{Q}$.[21] This ratio is reached if the CDW gap is $G = 0.6$ ev.

3.4. Linear Magnetoresistance

In large magnetic fields, such that the cyclotron frequency ω_c and scattering time τ satisfy $\omega_c\tau >> 1$, a metal (such as Na or K) should not exhibit a magnetoresistance if its Fermi surface is simply connected.[22] All workers have found instead that the resistance increases linearly in H without evidence of saturation, even for $\omega_c\tau \sim 400$. This can be explained (as an intrinsic property) only if the Fermi surface is multiply connected.[22] In Sec. 4 we shall show how the required open orbits arise from a CDW.

3.5. Induced Torque Anisotropy

If a spherical single crystal is placed in a rotating magnetic field, the induced currents create a torque; and for $\omega_c\tau >> 1$ this torque is proportional to the resistivity.[23] For a metal with a spherical Fermi surface the high-field torque should be isotropic. Measurements on ~ 200 specimens[19] of K revealed, however, a striking four-peaked pattern, even when the rotation axis was the three-fold, (111) axis. The anisotropy exhibited a range from 3:1 to 15:1. These observations were verified by Holroyd and Datars,[20] who found one sample with an anisotropy of 45:1 (at 23 kOe). This sample was spherical in shape to a precision of 2×10^{-3}. The torque anisotropies show that the symmetry of K (and also of Na) is less than cubic. The broken symmetry of a CDW is indicated.

3.6. The Oil Drop Effect

If small, spherical samples are supported in an induced-torque apparatus without being fastened with frozen oil, nearly isotropic torque patterns can be obtained.[20] However, if these same samples are remeasured after a drop of oil has been attached, large torque anisotropies are recovered. This phenomenon provides evidence for variations in $\vec{Q}$-domain texture (analogous to magnetic domains in a ferromagnet). An oil-free specimen may be subdivided into a random distribution of the twenty four possible orientations of $\vec{Q}$. It would then be macroscopically isotropic. A frozen oil drop creates a differential thermal stress which breaks the 24-fold degeneracy of $\vec{Q}$ orientations. This leads to a preferred texture (similar to the alignment of magnetic domains in a ferromagnet) and subsequently to an anisotropic induced-torque pattern.

3.7. Variability of the Residual Resistivity

The residual resistivity, caused by impurity scattering, varies from run to run on the same sample.[24] The changes can be in either direction depending on sample history, environment, and the rate of cooling from room to helium temperature. Sometimes the changes are as much as a factor of four. The residual resistivity anisotropy, Sec. 3.3, provides an explanation; since each run, beginning with a fresh cooldown from room temperature, permits a realignment of the $\vec{Q}$-domains, stimulated by thermal stress and altered surface conditions. The magnitude of the linear magnetoresistance, Sec. 3.4, also varies from run to run.

3.8. High-Field Hall Constant

In magnetic fields so high that $\omega_c\tau \gg 1$ the Hall coefficient must approach the constant value, $-1/nec$, where n is the conduction-electron density, known with a precision of 3×10^{-4} from the lattice constant. Different experimental techniques[25,26,27] all show violations, of the order of 3-8%, depending on crystal orientation. This phenomenon also indicates a multiply-connected Fermi surface.

3.9. Phason Heat Capacity Anomaly

Whenever there is a broken symmetry (involving a group of infinite order - e.g., the translation group) new collective excitations arise and have a frequency spectrum that goes to zero. For an incommensurate CDW these modes correspond to a dynamic phase modulation of the CDW.[28] They lead to a heat-capacity anomaly near 1°K.[29,30] This effect has been observed,[31] and has the expected location in temperature and magnitude.

3.10. Direct Observation of Electron-Phason Scattering

Pointcontact spectroscopy is a new technique (similar to that using superconducting tunnel junctions) for measuring inelastic scattering in normal metals.[32] The method has been applied to K,[33] and the observed spectrum above 2 mev is in good agreement with the theory for phonon scattering.[34] However a reproducible, sharp anomaly below 1 mev is also found which cannot be ascribed to ordinary acoustic phonons. Inelastic scattering of electrons by the new CDW excitations -

phasons - explains the anomaly quantitatively,[35] and without an adjustable parameter.

3.11. Difficulties

There are other phenomena not listed above - e.g., cyclotron phase resonance, ultrasonic attenuation, low-field Hall effect, etc. - which violate the expected behavior for a nearly-free electron model. However, there are also data which agree with the standard model and which have not yet been satisfactorily explained for a CDW structure. One of these is the apparent lack of an inhomogeneously broadened Knight shift in nuclear magnetic resonance.[36] Possibly very-long-range, indirect exchange interactions[37] may have something to do with this. The most important problem, however, is the non-observation of CDW effects in deHaas-van Alphen measurements.[4] A speculative explanation is that samples used have very small $\vec{Q}$-domain size - perhaps so small that the coherence volume of a Landau level contains 10^3- 10^4 domains. Interactions may lead then to an average deHaas-van Alphen frequency. Samples with larger $\vec{Q}$-domain size should reveal unusual effects. It should be noted, in this respect, that workers frequently discard data solely for the reason it doesn't behave "properly". In Cs, which also has an optical anomaly and a non-saturating magnetoresistance, up to 90% of the deHaas-van Alphen data is rejected.[38] Perhaps the discarded samples were in fact the better ones.

4. THE OPEN ORBITS OF POTASSIUM

The most spectacular evidence for a CDW structure in K is the observation of open-orbit peaks of magnetoresistance versus crystal orientation. Open orbits in momentum space (and in real space) occur whenever energy gaps truncate the Fermi surface so that some electrons cannot complete a closed cyclotron orbit. Instead the energy gaps cause a periodic Bragg reflection which forces those electrons to travel in (essentially) straight lines, even though $\omega_c\tau >> 1$.

4.1. Origin of Open Orbits

The existence of a CDW in K causes its Fermi surface to be cut by several pairs of energy-gap planes, as shown in Fig. 4. The fact that the positive ions form a discrete lattice, and not a smooth background, is the cause of (at least) two extra pairs, in addition to the main gaps. The heterodyne gaps arise from the beat periodicity between the lattice and the CDW. A static phason instability,[39] caused by the lack of commensuration, will also lead to a pair of small, subsidiary gaps, as shown.

In large magnetic fields the smaller gaps can be "broken down". That means that electrons can jump the gaps as if they were not there. Employment of all such options leads to five possible open orbit directions (in momentum space); and these are shown in Fig. 4.

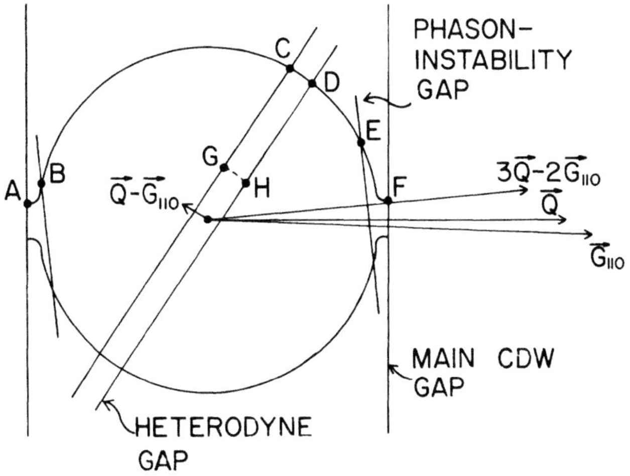

ORBIT	DIRECTION
$\overline{CD}$ or $\overline{AB}+\overline{EF}$	$\vec{Q}' \equiv \vec{G}_{110}-\vec{Q}$
$\overline{AF}$	$\vec{Q}$
$\overline{AC}+\overline{DF}$	$2\vec{Q}-\vec{G}_{110}$
$\overline{BE}$	$3\vec{Q}-2\vec{G}_{110}$
$\overline{BC}+\overline{DE}$	$4\vec{Q}-3\vec{G}_{110}$

Fig. 4. Fermi surface, energy-gap planes and open orbits for K with an incommensurate CDW. Magnetic breakdown is possible at the heterodyne gaps and the phason-instability gaps, but not at the main CDW gaps.

4.2. Theoretical Open-Orbit Spectrum

The optimum $\vec{Q}$ direction for a CDW does not coincide with a symmetry axis.[40] Consequently in a single crystal there will be twenty four equivalent axes. In general a macroscopic sample will be divided into $\vec{Q}$ domains. So the number of possible open-orbit directions is 5 x 24 = 120.

In order to calculate the magnetoresistance rotation spectrum for such a complicated system one must model the resistivity tensor for each $\vec{Q}$-domain, and then use effective medium theory[41] to treat the heterogeneous domain structure. This leads to patterns[42] such as those shown

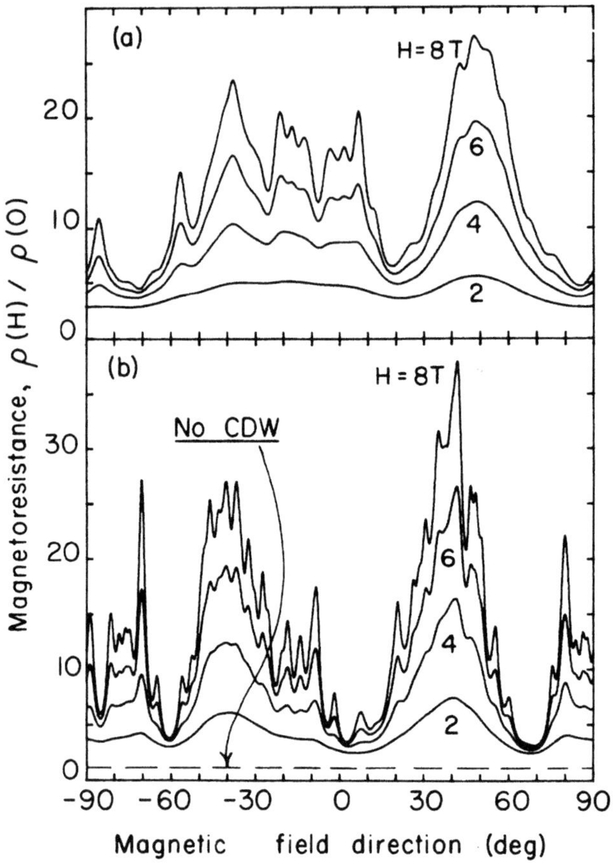

Fig. 5. Theoretical magnetoresistance rotation patterns calculated using effective medium theory. In (a) the $\vec{Q}$-domain size was taken to be 0.05 mm, and $\vec{H}$ was allowed to rotate in a (211) plane. In (b) the $\vec{Q}$-domain size was 0.12 mm, and $\vec{H}$ was in a (321) plane. Note the horizontal, dashed line in (b), which is the only allowed theoretical result for a spherical Fermi surface and no broken symmetry.

in Fig. 5. Sharp open-orbit structure appears for fields above 4T, if the $\vec{Q}$-domain size is taken to be ~ 0.1 mm. The patterns shown are to be compared with the horizontal line in Fig. 5b, which is the theoretical behavior (for all H) if the metal has a spherical Fermi surface.

4.3. Observed Open-Orbit Spectrum

Coulter and Datars have measured[43] magnetoresistance rotation patterns to 8.5 T by means of the induced torque technique. The results of two runs (typical of one hundred experiments) are shown in Fig. 6. There can be no doubt that the Fermi surface of K is multiply connected.

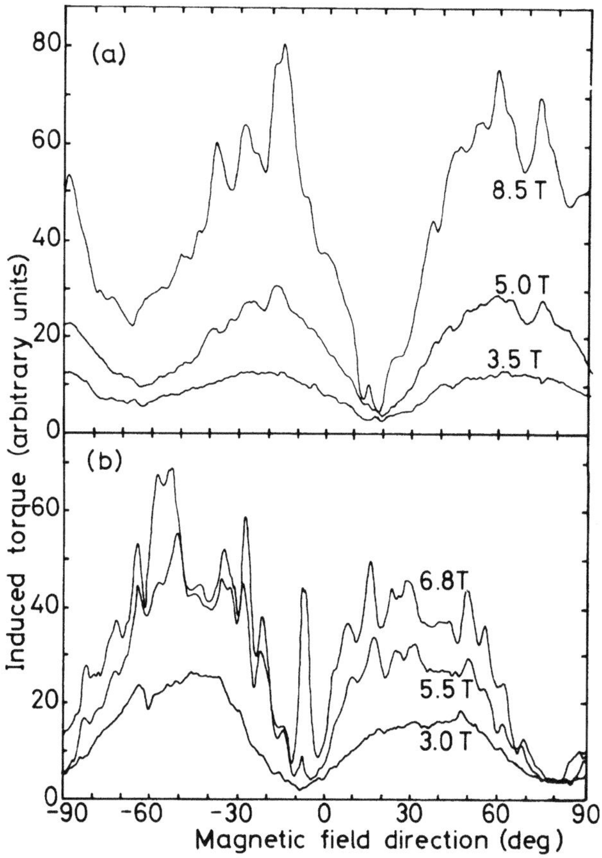

Fig. 6. Experimental magnetoresistance rotation patterns for single-crystal spheres of K. The data, taken by Coulter and Datars at 1.4°K, cannot be compared in detail with Fig. 5 since the volume fractions for each of the twenty four types of $\vec{Q}$-domain are unknown. The rotation planes are the same as in Fig. 5.

Further work at much higher fields is needed in order to determine the crystallographic directions of the various open orbits. Theoretically, some orbits will remain unresolved even at 24T.[42] Similar open-orbit patterns are also found in Na.

5. IMPLICATIONS FOR MANY-ELECTRON THEORY

Since the origin of CDW's in simple metals is understood microscopically,[5,6] the purpose of this section is to stress the limitations of several approximations frequently employed in many-electron theory. The failure (when using these) to foresee CDW structure, or to provide understanding, even after CDW's have been exhibited, should be kept in mind whenever one is tempted to use them.

5.1. Local-Density Approximations to Exchange and Correlation are not Predictive

The Slater approximation to the exchange potential[44] is

$$V_{XS} = -3e^2 k_F(\vec{r})/2\pi, \tag{7}$$

where $k_F(\vec{r})$ is the Fermi wave vector for a degenerate electron gas of local density $n(\vec{r})$.

$$k_F(\vec{r}) = [3\pi^2 n(\vec{r})]^{1/3}. \tag{8}$$

It has been argued[45] that Eq.(7) should be reduced by a factor 2/3. Many introduce a parameter α, so that the exchange (and correlation) potential is:

$$V_{XC} = \alpha V_{XS} \; . \tag{9}$$

It is easy to show that a CDW could never occur in K if this approximation were taken seriously.

Consider the periodic potential, Eq.(1), and let G = 0.6 ev and $Q = 2k_F$. A calculation of the fractional density modulation, p of Eq.(4), leads to $p \approx 0.17$. If this modulation is inserted into Eq.(8), the result for the amplitude of V_{XC}, Eq.(9) with $\alpha = 2/3$, is 0.2 ev. A local approximation is inconsistent by a factor of three. G would quickly iterate to zero.

Exchange potentials are highly non-local. This has been shown[46] by calculating the matrix elements of the exact exchange operator A for an electron gas with a small sinusoidal density modulation, $p \cos\vec{q}\cdot\vec{r}$. The off-diagonal matrix elements $\langle\vec{k}+\vec{q}|A|\vec{k}\rangle$, in Slater units, defined by Eq.(7), are illustrated in Fig. 7 for four values of $\vec{k}$ in the Fermi sea. The enormous disparity between the four curves exhibits a pathologic, non-local behavior. The singularity at $q = 2k_F$ for the $\vec{k}$ value denoted by Q, in the figure, is responsible for the SDW and CDW instability theorem.[5] When dynamic correlation potentials are included, the non-locality remains, though it is somewhat smaller.[47] A Kohn-Sham local-exchange potential corresponds to a horizontal line in Fig. 7 (at the ordinate 2/3) for all values of $\vec{k}$.

Na and K are the quantum systems which come closest to the idealization of a nearly-uniform electron gas, upon which all local-density functionals are based. Yet for these cases, where agreement ought to be best, the local-density method fails to reveal the qualitative nature of the ground-state charge density.

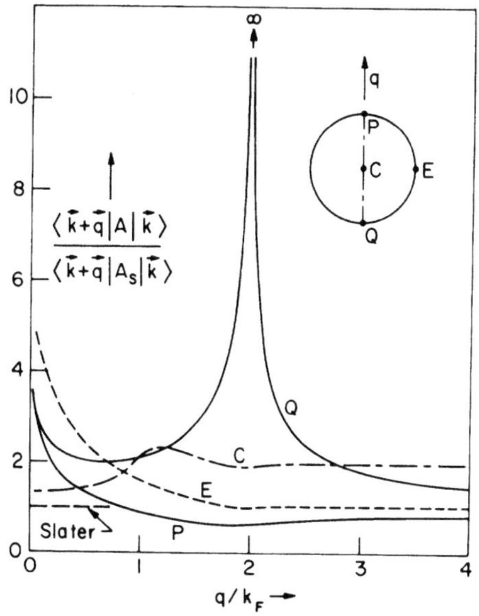

Fig. 7. Ratio of exact matrix elements of the exchange operator to those given by Slater's local-density approximation. Each curve is for the point $\vec{k}$ of the Fermi sea shown in the inset. The electron gas was taken to have a small sinusoidal density modulation, $\varepsilon \cos\vec{q}\cdot\vec{r}$. The matrix elements are in the (pure) momentum representation.

5.2. Screened Interactions are Dangerous

Since the work of Bohm and Pines[48] and the Fermi liquid theory of Landau,[49] it has been common to treat conduction electrons with a Fermion Hamiltonian employing screened Coulomb interactions. This is a valid approximation when electron-electron scattering processes are involved in transport problems. But such a model Hamiltonian cannot be used to investigate alternative ground-state structures. Calculations with dynamically screened interactions[50] have shown that any tendency towards SDW or CDW instability is suppressed. Screened interactions are qualitatively correct for treating spin-density excitations, but are qualitatively wrong for charge-density excitations (or instabilities).

5.3. The Coulomb Hole is Important

If one expresses the Coulomb energy for N electrons (in pure momentum states) by means of a Fourier analysis, the result is:

$$E_c^o = N\Sigma_{\vec{q}} \frac{2\pi e^2}{q^2} \quad + \qquad (10a)$$

$$- \frac{1}{2} \sum_{\vec{k},\vec{q}} \frac{4\pi e^2}{q^2} (n_{\vec{k}\uparrow} n_{\vec{k}-\vec{q}\uparrow} + n_{\vec{k}\downarrow} n_{\vec{k}-\vec{q}\downarrow}), \tag{10b}$$

where the n's are occupation numbers. The first term is just another way of writing the total rest mass, Nmc^2. The second term, (10b), is the exchange energy. The dielectric response of the electrons to the electric fields associated with the energy E_c^o can be represented by a dielectric function ε_q. Recall that the energy of a charged capacitor is reduced by ε if a dielectric is moved into the electric field; so also will the Coulomb energy of the N electrons be reduced. This reduction, the Coulomb energy part of the correlation energy, is:

$$\Delta E_c = N \sum_{\vec{q}} \frac{2\pi e^2}{q^2} \left(\frac{1}{\varepsilon_q} - 1\right) \tag{11a}$$

$$- \frac{1}{2} \sum_{\vec{k},\vec{q}} \frac{4\pi e^2}{q^2} \left(\frac{1}{\varepsilon_q} - 1\right) (n_{\vec{k}\uparrow} n_{\vec{k}-\vec{q}\uparrow} + n_{\vec{k}\downarrow} n_{\vec{k}-\vec{q}\downarrow}). \tag{11b}$$

ΔE_c is generally small compared to the exchange energy, (10b). However (10b) and (11b) can be combined; and their sum is the screened-exchange energy. The term (11a) is the Coulomb hole energy. It is much larger than screened exchange. It appears to be an innocuous constant, since it doesn't depend on the occupation numbers $n_{\vec{k}}$, and some workers treat it as such. But that is a serious error. If one cancels off the dynamic contributions of bare exchange (10b) by combining it with (11b), the dynamic effects of the Coulomb hole, (11a), become dominant. The Coulomb hole, from this point of view, is the driving force for a CDW ground state.

The more accurate approach to many-electron effects is to leave the two dynamic corrections (11a) and (11b) combined, for their sum is small. Then the large kinematic term (10b) can be calculated with precision. Unscreened exchange is, with this approach, the major cause of a CDW ground state, and ΔE_c provides a small reinforcement.[6]

REFERENCES

1. F. Bloch: 1929, Z. Physik 57, p. 545.
2. E. Wigner: 1938, Trans. Faraday Soc. 34, p. 678.
3. T. Kushida, J.C. Murphy and M. Hanabusa: 1976, Phys. Rev. B13, p. 5136.
4. D. Shoenberg and P.J. Stiles: 1964, Proc. Roy. Soc. A281, p. 62.
5. A.W. Overhauser: 1962, Phys. Rev. 128, p. 1437.
6. A.W. Overhauser: 1968, Phys. Rev. 167, p. 691.
7. A.W. Overhauser: 1964, Phys. Rev. Lett. 13, p. 190.
8. H. Mayer and M.H. El Naby: 1963, Z. Physik 174, p. 269.

9. B. Hietel and H. Mayer: 1973, Z. Physik 264, p. 21; and personal communication.
10. Peter Harms: 1972, "Dissertation, Technischen Universitat Clasthal", (unpublished).
11. A.W. Overhauser and N.R. Butler: 1976, Phys. Rev. B14, p. 3371.
12. G.F. Giuliani and A.W. Overhauser: 1981, Phys. Rev. B23, p. 3737.
13. W.M. Walsh, Jr., L.W. Rupp, Jr. and P.H. Schmidt: 1966, Phys. Rev. 142, p. 414.
14. G. Dunifer, personal communication.
15. A.W. Overhauser and A.M. de Graaf: 1968, Phys. Rev. 168, p. 763.
16. T. Kjeldaas, Jr.: 1959, Phys. Rev. 113, p. 1473.
17. P.A. Penz and T. Kushida: 1968, Phys. Rev. 176, p. 804.
18. A.W. Overhauser and Sergio Rodriguez: 1966, Phys. Rev. 141, p. 431.
19. J.A. Schaefer and J.A. Marcus: 1971, Phys. Rev. Lett. 27, p. 935.
20. F.W. Holroyd and W.R. Datars: 1975, Can. J. Phys. 53, p. 2517.
21. M.F. Bishop and A.W. Overhauser: 1978, Phys. Rev. B18, p. 2447.
22. I.M. Lifshitz, M. Ya Azbel and M.I. Kagonov: 1956, Zh. Eksp. Teor. Fiz. 31, p. 63 (English translation: 1957), Sov. Phys.-JETP 4, p. 41).
23. J.S. Lass and A.B. Pippard: 1970, J. Phys. E: J. Sci. Instrum. 3, p. 137.
24. H. Taub, R.L. Schmidt, B.W. Maxfield and R. Bowers: 1971, Phys. Rev. B4, p. 1134.
25. P.A. Penz: 1968, Phys. Rev. Lett. 20, p. 725.
26. D.E. Chimenti and B.W. Maxfield: 1973, Phys. Rev. B7, p. 3501.
27. S.A. Werner, T.K. Hunt and G.W. Ford: 1974, Solid St. Comm. 14, p. 1217.
28. A.W. Overhauser: 1971, Phys. Rev. B3, p. 3173.
29. M.L. Boriack and A.W. Overhauser: 1978, Phys. Rev. B18, p. 6454.
30. G.F. Giuliani and A.W. Overhauser: 1980, Phys. Rev. Lett. 45, p. 1335.
31. C.D. Amarasekara and P.H. Keesom: 1981, Phys. Rev. Lett. 47, p. 1311.
32. I.K. Yanson: 1974, Zh. Eksp. Teor. Fiz. 66, p. 1035 (English translation: 1974, Sov. Phys.-JETP 39, p. 506).
33. A.G.M. Jansen, J.H. van den Bosch, H. van Kempen, J.H.M. Ribot, P.H.H. Smeets and P. Wyder: 1980, J. Phys. F: Metal Phys. 10, p. 265.
34. M. Ashraf and J.C. Swihart: 1982, Phys. Rev. B25, p. 2094.
35. M. Ashraf and J.C. Swihart: 1982, Phys. Rev. Lett., in press.
36. D. Follstaedt and C.P. Slichter: 1976, Phys. Rev. B13, p. 1017.
37. G.F. Giuliani and A.W. Overhauser: 1982, Phys. Rev. B26, in press.
38. A.A. Gaertner and I.M. Templeton: 1977, J. Low Temp. Phys. 29, p. 205.
39. G.F. Giuliani and A.W. Overhauser: 1982, Phys. Rev. B26, in press.
40. G.F. Giuliani and A.W. Overhauser: 1979, Phys. Rev. B20, p. 1328.
41. M. Huberman and A.W. Overhauser: 1981, Phys. Rev. B23, p. 6294.
42. M. Huberman and A.W. Overhauser: 1982, Phys. Rev. B25, p. 2211.
43. P.G. Coulter and W.R. Datars: 1980, Phys. Rev. Lett. 45, p. 1021.
44. J.C. Slater: 1951, Phys. Rev. 81, p. 385.

45. W. Kohn and L.J. Sham: 1965, Phys. Rev. 140, p. A1133.
46. A.W. Overhauser: 1970, Phys. Rev. B2, p. 874.
47. K.J. Duff and A.W. Overhauser: 1972, Phys. Rev. B5, p. 2799.
48. D. Bohm and D. Pines: 1953, Phys. Rev. 85, p. 338.
49. L.D. Landau: 1956, Zh. Eksp. Teor. Fiz. 30, p. 1058 (English translation: 1956, Sov. Phys.-JETP 3, p. 920).
50. D.R. Hamann and A.W. Overhauser: 1966, Phys. Rev. 143, p. 183.

MOLECULAR SEMICONDUCTORS, METALS AND SUPERCONDUCTORS

Kim Carneiro
Physics Laboratory I
H.C. Ørsted Institute, University of Copenhagen
Universitetsparken 5, DK-2100 Copenhagen Ø
Denmark

Recent developments have demonstrated that molecular chain compounds may behave as semiconductors and metals and even show superconductivity. This is in accordance with theories of one dimensional conductors which indicate a very rich phase diagram. The competing instabilities are reviewed and the influences of correlations as well as one-dimensional fluctuations are described. Despite the fact that theory takes the molecular nature of real compounds into account in a very simple way, examples are shown where detailed understanding of the observed phenomena may be obtained.

INTRODUCTION

Although both theoretical models in one dimension as well as hypothetical metals built by molecules may at first sight seem hard to realize in the laboratory, the study of such objects as one-dimensional molecular conductors constitutes a rapidly increasing field of research.[1,2] The evolution is sketched in Figure 1, where the temperature dependent resistivity of some important compounds are shown. 20 years ago *molecular semiconductors* based on the acceptor molecule TCNQ were prepared with a room temperature conductivity of approximately 10 $\Omega^{-1}cm^{-1}$. An important step forward towards *molecular metals* was made 10 years ago when the matching donor molecule TTF was prepared, since the charge transfer salt TTF-TCNQ displayed a room temperature conductivity of 1000 $\Omega^{-1}cm^{-1}$ as well as metallic behaviour down to 54 kelvin, where the metal-insulator transition takes place, owing to a charge density wave instability as anticipated by Peierls. During subsequent years both donor and acceptor molecules were modified with the purpose of increasing the conductivity even further, often by the successful application of pressure. In 1980 the molecule TMTSF was used to prepare molecular conductors which at low temperatures conducted as well as copper at room temperature, and soon thereafter the study of organic or *molecular superconductors* began.

Because the *molecular conductors* are built by planar molecules, which in the crystalline state form conducting stacks perpendicular to the molecular planes, they are essentially one-dimensional conductors.

P.-O. Löwdin and B. Pullman (eds.), New Horizons of Quantum Chemistry, 373–393.

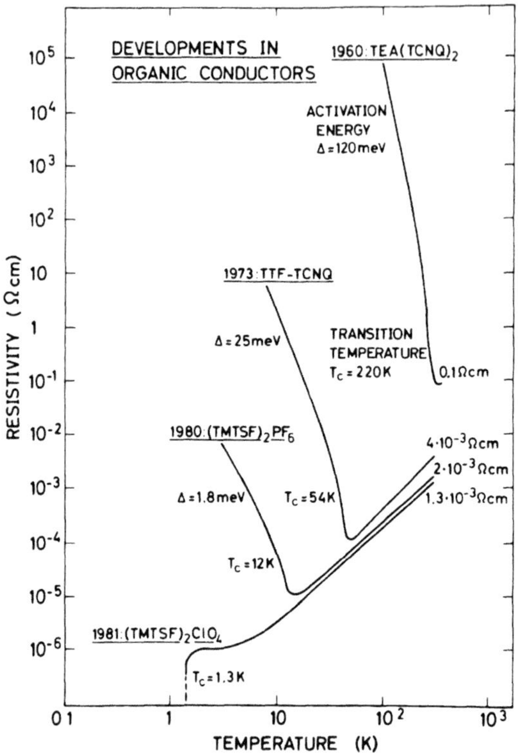

Figure 1. Developments in molecular conductors. The room temperature conductivity, the transition temperature and the activation energy are shown for some typical compounds.

In this regard they are similar to highly *conducting platinum-chain salts* which amongst chemists have been known for almost 150 years, as well as to the recently developed *conducting polymers*. Hence, the intricacies of one dimensional theoretical models are of crucial importance for the understanding of these three classes of compounds.

ONE DIMENSIONAL GROUND STATES

Peierls[3] originally considered a one-dimensional conductor and reached the conclusion that it was "unlikely to exist", since an insulator would always have lower energy. His argument is indicated in Figure 2. By opening a gap in the electronic spectrum at the Fermi level ε_F electrostatic energy is lowered at the expence of the elastic energy necessary to deform the regular ionic lattice with a period corresponding to $2k_F$. In three dimensional metals the same electron-phonon interaction leads to a small anomaly in the phonon spectrum;[4] but in one dimension it leads to a genuine instability, so that the ground state of the hypothetical metal will always be a charge density wave

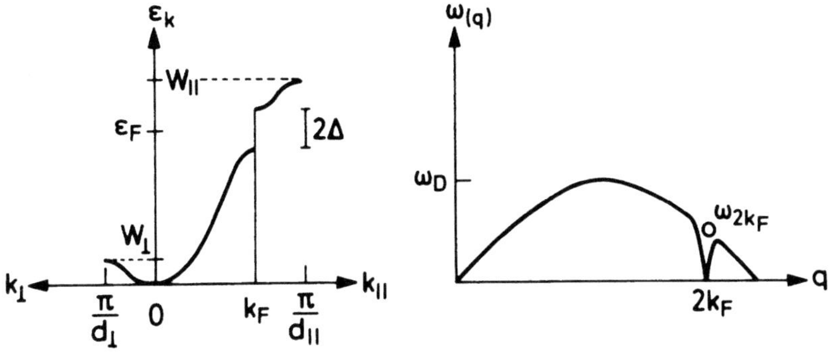

Figure 2. (a) Electronic band structure ε_k for a quasi-one-dimensional conductor in the tight binding approximation. Band widths are $W_{||}$ and $W_{\perp}$ in the parallel and perpendicular directions respectively. The gap 2Δ at the Fermi level (k_F, ε_F) is shown. (b) The phonons of a one-dimensional conductor, exhibiting a softening at $q = 2k_F$ according to Peierls. ω_{2k_F} is the unperturbed phonon frequency at $q = 2k_F$.

insulator with a gap Δ. Using a specific Hamiltonian due to Fröhlich[5] one gets at zero temperature:

$$\Delta = \alpha\varepsilon_F \, e^{-1/\lambda} \tag{1}$$

where α depends on the band structure,[6,7] and λ is a dimensionalless electron-phonon coupling constant. At finite temperatures the gap decreases, and within the mean field approximation one gets for the transition temperature:

$$T_P^{MF} = \Delta/(1.76\ k_B) \quad , \tag{2}$$

i.e. the usual BCS-relation.[8] This transition into the CDW-state is often referred to as the Peierls transition.

Considering electron-electron interactions Bychkov *et al.*[9] pointed out that the CDW was only one out of four possible ground states in a one dimensional electron gas. Dependent upon the forward and backward scattering amplitudes (called g_2 and g_1 respectively) one gets a spin density wave (SDW) which is also insulating, together with two superconducting states, namely singlet (SS) and triplet (TS) superconductivity. All four ground states may be described in a way similar to the Peierls instability by separate gaps Δ which as in (1) depend on scale energies (αE_F in (1))and coupling constants(λ in (1)). And since (2) is assumed to be valid,the biggest gap is the instability with the highest transition temperature, i.e. the ground state corresponding to the choice of g_1 and g_2. Of course three dimensional systems also show SDW and SS; but similar to the case of the CDW instability one dimension displays the instabilities and the competition between them

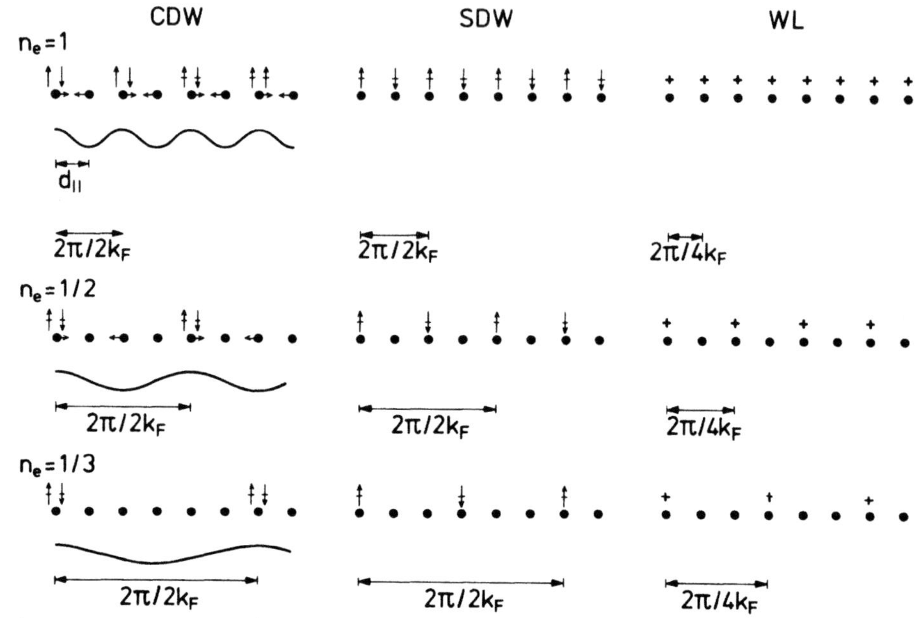

Figure 3. Electronic configurations of the three insulating ground states of a one-dimensional conductor. CDW: charge density wave. SDW: spin density wave. WL: Wigner lattice. The ions, shown by solid circles (●), are separated by $d_{||}$. The electrons (+) are arranged with periodicities indicated. For CDW the solid line shows the $2k_F$-lattice distortions. The configurations are shown for the three typical charge trans

most clearly. The fact that $(TMTSF)_2X$ compounds show all four states except TS has spurred an increased theoretical activity along these lines.[10,11]

Considering the superconducting instability specifically in molecular crystals Little pointed out that the intramolecular vibrations could in principle give rize to high temperature superconductivity.[12] In the BCS theory for ordinary metals the scale energy is related to the Debye frequency ω_D which is of the order of 300 cm^{-1} ; but if a reasonably large λ could be achieved for electron-vibron coupling then the much higher frequency of the internal modes would give a higher transition temperature. The model was later extended to consider an excitonic superconductor with an even higher T_c.[13]

So far we have assumed spin degeneracy as well as the existence of a transfer integral $t_{||}$ giving rize to a band width $W_{||} = 4t_{||}$. However, electron correlations are known to be important in molecules and cannot *a priori* be neglected in molecular conductors. If the Coulomb interaction U dominates, i.e. $U >> 4t_{||}$, then the ground state is a Wigner lattice (WL) according to the solution of the Hubbard Hamiltonian.[14] In this case the electrons act effectively as spinless

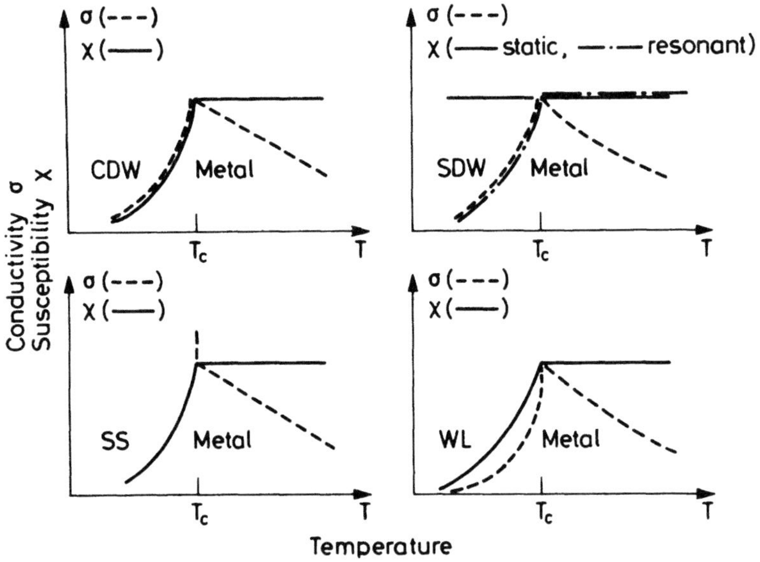

Figure 4. Conductivity $\sigma_{||}$ and magnetic susceptibility $\chi_{||}$ of a one-dimensional conductor, shown for four different ground states. SS: Singlet conductivity. In the SDW only the resonant susceptibility becomes activated below T_c, and in the WL the activation energy for susceptibility is one third of that for conduction.

fermions with a periodicity corresponding to $4k_F$.[15] Also the competition between $2k_F$ instabilities and the Hubbard transition has been considered for $U \approx 4t_{||}$ which often seem to be realistic in molecular semiconductors; but in this case the situation is less clear.[16]

The three insulating states CDW, SDW and WL are shown in Figure 3. It is important to notice that the details of the behaviour depend on the number of electrons n_e (or holes n_h) per ion. The Fermi vector is given by

$$k_F = \frac{1}{2} n_e \pi / d_{||} = (1 - \frac{1}{2} n_h) \pi / d_{||} \quad . \tag{3}$$

The differences in electronic structure between the different ground states is reflected in the physical properties, e.g. the conductivity $\sigma_{||}$ and the magnetic susceptibility $\chi_{||}$, which in principle are enough to distinguish between the various phases. Both $\sigma_{||}$ and $\chi_{||}$ reflects the gap Δ and are most often activated as demonstrated in figure 4. In the SDW-case there is an important difference between the behaviour of the resonant susceptibility (as measured e.g. by

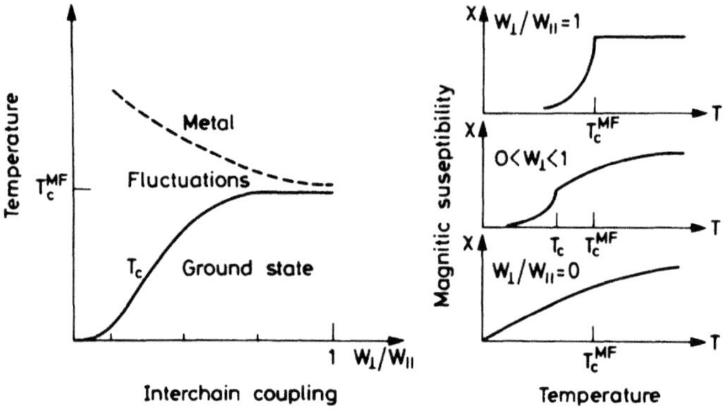

Figure 5. Phase transitions in a quasi-one-dimensional system. In a three dimensional system ($W_\perp/W_{||}=1$) mean field theory is essentially correct and the transition occurs at T_c. In a one-dimensional system no phase transition takes place when $T>0$; but T_c determines the range of fluctuations. In a quasi-one-dimensional system the phase transition occurs at T_{3D} which depends on $W_\perp/W_{||}$. (a) Phasediagram. (b) The behaviour of the magnetic susceptibility.

electron spin resonance) and the static susceptibility.[17] In the WL-case the activation energy for $\chi_{||}$ is Δ/π.[18]

Already Fröhlich pointed out that collective excitations may have a profound effect on the low temperature properties when he proposed the sliding CDW as a mechanism for superconductivity.[5] In recent field theoretical models such nonlinear excitations are identified as solitons or polarons and they modify the simple picture outlined here.[19,20]

THE ROLE OF FLUCTUATIONS

Despite the fact that theories predict a variety of interesting instabilities for a one-dimensional conductor, these will strictly speaking not give rize to phase transitions at finite temperatures.This is due to the argument by Landau that in one dimension the ground state is reached only at $T = 0$,[21] since at finite temperatures the existence of domains will prevent long range order. Another way of stating this is to notice that the entropy term in the free energy will dominate except at $T = 0$. Therefore, the instabilities described above will, if the system is really one-dimensional, cause critical fluctuations to occur. This is of course also the case in three dimensional systems, but here the critical temperature region is confined to typically $T_c \pm 10\%$. In one dimension $T_c = 0$ and the critical region is extended well above T_c^{MF} (cf. (2)) so that this temperature is to be taken as a scale temperature for the onset of critical fluctuations.[22]

However, molecular conductors are not one-dimensional, they are strongly anisotropic or quasi-one-dimensional; and this brings back a finite transition temperature. In a crystal with conducting chains this is conceived in the following way: The one-dimensional instability gives rize to ordered domains along each chain, and at low enough temperatures the domains will have a size (or correlation length $\xi_{||}$) which allow an ever so small coupling transverse to the chains to be effective. Then the system is three dimensional and a usual transition takes place. One way of introducing interactions between chains is to consider a transverse transfer integral $t_{\perp}$ giving a transverse band width $W_{\perp} = 4t_{\perp}$ and to consider the suppression of T_c with respect to T_c^{MF} as a function of $W_{\perp}/W_{||}$.[23] The situation is schematically shown in Figure 5, together with the effect of fluctuations in a measurement of the susceptibility χ .

In general the role of fluctuations in approaching each of the ground states are different. (For specific results see e.g. ref.23 for the Peierls instability and ref. 24 for SS). Hence one may observe fluctuations stemming from one instability before the transition due to another instability takes place. Such effects have been reported for TTF-TCNQ and related salts, where superconducting fluctuations may show up in $\sigma_{||}$ at temperatures above the Peierls transition. Similarly SDW fluctuations seem to be present in $(TMTSF)_2C\ell O_4$ above the superconducting transition. Fluctuations add an interesting (and complicated feature to the study of molecular conductors.

REAL COMPOUNDS

Practical insight into the richness of physical behaviour of one-dimensional models is provided by existence of extended orbitals in molecules. If such orbitals form the basis of a (partially filled) conduction band in the solid state then we may get a molecular conductor, and in Figure 6 is shown some characteristic compounds where such bands exist. Table 1 gives values for the bandwidths $W_{||}$ as well as a signature of the one-dimensionality $\sigma_{||}/\sigma_{\perp}$. In view of the above discussion of the importance of Coulomb correlations U is also shown.

In KCP($K_2Pt(CN)_4Br_{0.3}\cdot 3H_2O$) the overlap occurs in the Pt-d_{z^2} orbital and the partial band filling is obtained by partial oxidation of Pt^{+2} . Conducting platinum chains have been reported in the chemical literature long time ago,[25] but they were first studied systematically by Krogmann in the late sixties.[26] New Compounds are still being made, both based on tetracyano-platinate $[Pt(CN)_4]^{-2}$ and bis-oxalato-platinate $[Pt(C_2O_4)]^{-2}$.[27,28] As a natural extension of the latter brief reports have been made on conducting bis-squarato- and bis-croconato-platinates, $[Pt(C_4O_4)]^{-2}$ and $[Pt(C_5O_5)]^{-2}$ respectively.[29,30] Solid state measurements indicate that the electronic overlap along the chain is so large that a free electron model is appropriate for the conduction band, and there is no evidence for a finite U . These and other simplifying factors make platinum compounds excellent model compounds for the Peierls instability.[31]

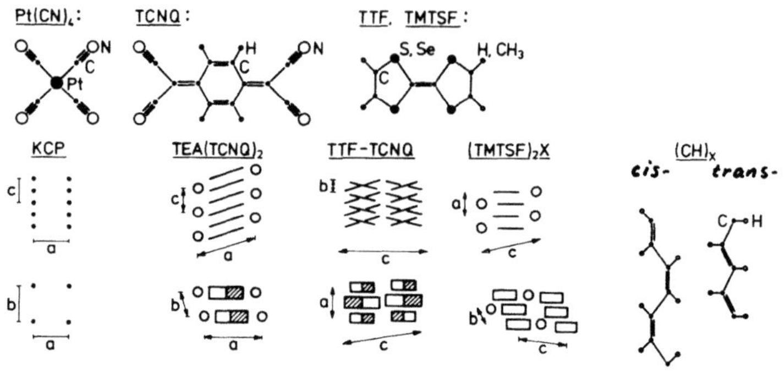

Figure 6. Stacking molecules and schematical crystal structures for some quasi-one-dimensional conductors.

The relation between the electronic structure of the individual Pt-ligand-species and the solid state properties is not understood in detail and in general only rather crude band structure calculations have been made.[32-36] However, the free electron behaviour in KCP has been explained.[37] A point which is not understood is the accessible range of bond lengths with metallic resonance. Empirically one finds $d_{||}$ ranging from 2.80 Å (just the bond length in Pt-metal) to 2.96 Å, whereas all salts with bond lengths bigger than 3.0 Å have so far been non-conducting. Calculations, however, show nothing magic around 3 Å.[38,59] This is interesting because the possible materialization of the coupling between conduction electrons and excitons requires $d_{||} \approx 3.5$ Å in order to allow for large ligands. Also why the number of electrons n_e is fixed in each salt and why $1.6 < n < 1.75$ for all salts is basically not understood.

In TEA(TCNQ)$_2$ (triethylammonium-bis-tetracyanoquino-dimethane), the second salt shown in Figure 6, the overlap occurs between the delocalized π-orbitals of the planar TCNQ-molecules which on the average assumes the charge $n_e = \frac{1}{2}$. M(TCNQ)$_2$ salts were studied extensively in the sixties employing different cations M^+, ranging from spherical such as K^+ and Cs^+ and large organic ones such as $TEA^+(NH(C_2H_5)_3)$ to polar cations such as $NMP^+(C_{12}H_8N_2CH_3)$[39,40] The typical crystallographic arrangement is shown in Figure 6. The TCNQ-planes are tilted with respect to the conducting axis and often the cations result in an appreciable dimerizati on. It is also obvious that the transfer integrals are quite different in the two directions perpendicular to the chain as evidenced by two distinctly different conductivity ratios.

In $TCNQ^{-\frac{1}{2}}$ salts the magnitude of $W_{||}$ and U is not entirely clea[r.] Based on charge transfer bands in crystalline optical spectra one gets

Compound	$W_{\|\|}$(eV)	U(eV)	$\sigma_{\|\|}/\sigma_{\perp}$
$K_2Pt(CN)_4Br_{0.3}\cdot 3H_2O$(KCP)	4.5(a)	0	10.000
$TEA(TCNQ)_2$	0.75(a) 0.4(b)	≈1	160,2850
TTF-TCNQ	0.20, 0.44(b)	small	300
$(TMTSF)_2X$	1(a)	≃0	200,30.000
CH_x	10.4(b)	disputed	-

Table 1. Electronic bandwidth $W_{||}(=4t_{||})$ and on-site correlation energies U for some quasi-one-dimensional conductors.(a) experimental values from reflectivity measurements. (b) theoretical values from band structure calculations. The estimates of U are described in the text. The degree of one dimensionality is indicated by the conductivity anisotropy $\sigma_{||}/\sigma_{\perp}$. The two bandwidths in TTF-TCNQ are for TTF and TCNQ.

$U \approx 1eV$[41] and the activation energies for $\chi_{||}$ and $\sigma_{||}$ (cf. above) point towards $U/W_{||} >> 1$.[42] However, the analysis of a well-defined plasma frequency for IR-reflectivity via the relation

$$\omega_P^2 = 4\pi\rho_M n_e \, e^2/m^* \qquad (4)$$

yields an effective mass m^* corresponding to $W_{||} = 0.75$ eV. ρ_M is the number density of conducting molecules. This puts $TEA(TCNQ)_2$ into the region of $U \simeq W_{||}$, where theories are difficult and where some experiments may reflect small U behaviour (the CDW instability) whereas others reflect large U (the Hubbard instability).

Ab initio calculations have been carried out for the TCNQ molecule, giving detailed insight into its electronic structure, and demonstrating its high electron affinity.[43] Charge transfer integrals calculated between dimers[44,45] and trimers[45] are in reasonable agreement with experimental values based on (4).

The big leap in room temperature conductivity and the preservation of a metallic state down to low temperatures created big excitement when these properties were first reported in TTF-TCNQ.[46] Several new features appear when TTF (tetrathiofulvalene) forms crystals with TCNQ as opposed to the 1:2 TCNQ salts. In particular the following was considered in the search of explanations for the improved metallic behaviour with respect to $M(TCNQ)_2$ salts:

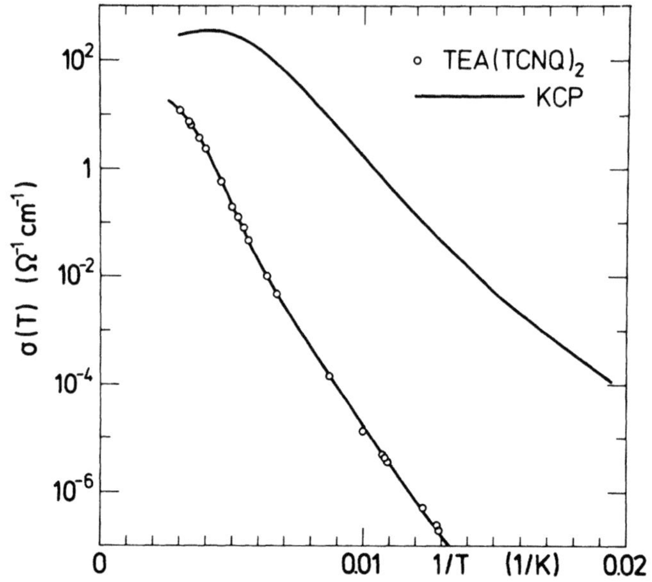

Figure 7. Temperature dependent conductivity $\sigma_{||}(T)$ for $TEA(TCNQ)_2$ and KCP.

i) *The charge transfer is 0.59.* This corresponds to Fermi vector which is incommensurate with the lattice as opposed to the 1:2 salts. Since at least the half-filled band will be extra susceptible to the Peierls instability because of umklapp scattering (and 1:2 salts *are* half-filled in the presence of dimerization) commensurability was thought to be "bad for conductivity". However, since the $(TMTSF)_2X$ family consists of 2:1 salts this agreement is at least of limited validity.

ii) *The donor stack conducts.* The fact that π-orbitals delocalize both both along the TTF and the TCNQ stack will of course increase the conductivity. But this direct effect is relatively small in TTF-TCNQ at high temperatures,[47] and the TTF bandwidth is small compared to TCNQ as shown in Table 1.[44,45] However, when Se is substituted for S to give TSF-compounds, these seem to be dominated by the characteristics of the donor molecule.

iii) *U is decreased.* If the indirect effect of TTF on TCNQ is to enhance the delocalisation on the latter then correlation effects could decrease and the conductivity increase. Indeed the activation energies for susceptibility and for conductivity are the same, demonstrating that U is not large compared to $W_{||}$. But even at room temperature the finite size of U is apparent from a $4k_F$ distortion of the lattice as measured by X-rays.[48]

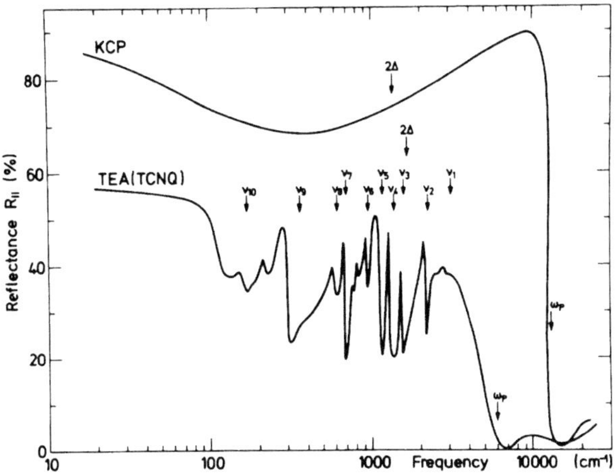

Figure 8. Reflectance along the conducting axis of $TEA(TCNQ)_2$ and KCP. The perpendicular reflectivity is typically 5% and structureless. The derived plasma frequencies ω_p are shown together with the positions 2Δ where excitations across the Peierls gap set in. In $TEA(TCNQ)_2$ the 10 a_g vibrational modes of the TCNQ molecule (ν_1-ν_{10}) give a pronounced structure in $R_{||}$.

iv) *The herring-bone crystal structure.* TTF-TCNQ crystallizes in an interesting structure and apart from the question of why it does that, it is important to understand the effect of the structure on the transfer integrals in particular perpendicular to the conducting chains.[44]

During the decade after the occurrence of TTF-TCNQ the search for compounds with higher conductivity took the form of an intimate interplay between chemistry and physics. Molecules were modified and if conducting crystals would form, their physical properties would be related to the molecular and crystalline properties to sort out the relative importance of i)-iv) mentioned above. Systematic studies were difficult because changing one "chemical parameter" without changing all the others simultaneously were most often not possible; but here the application of pressure added much to the systematic information gained.[49] And as the transition temperature for the metal-insulator transition became lower and lower, superconductivity appeared.

The first and so far the only compounds exhibiting superconductivity are the $(TMTSF)_2X$ salts based on tetramethyl-tetraselenafulvalene and small anions such as PF_6^-, $C\ell O_4^-$, ReO_4^- and NO_3^-.[50] By being 2:1 salts they demonstrate that the problem of commensurability is not crucial (cf. i)). The TMTSF molecule is larger than TTF which according to ii) should decrease U and no evidence for $U > 0$ has been found. The band width is very large which may be attributed to the fact that the molecular planes are perpendicular to the conducting axis and not tilted. This has

Parameter Unit	$d_{\|\|}$ (Å)	F $\pi/d_{\|\|}$	ε_F eV	Δ (meV)	λ	T_c (K)	η
KCP	2.89	0.85	3.25	72	0.30	100	0.02
$TEA(TCNQ)_2$	3.28	0.25	0.15	120	0.40	220	0.18

Table 2. Parameters relevant for the Peierls instability in KCP and $TEA(TCNQ)_2$.

a big effect on the transfer integrals.[45] Hence the features ii)-iv) are important in the design of molecular conductors, whereas the role of i) remains somewhat unclear.

The last compound shown in Figure 6 is the polyacetylene $(CH)_x$ which has become the prototype polymeric conductor. Here π-electron delocalize along the polymer chain; but only upon doping does $(CH)_x$ conduct. This may be thought of as stemming from the fact that pure $(CH)_x$ is a Peierls semiconductor at room temperature and that impurity levels in the gap are necessary for conduction.[51] At a few percent doping $(CH)_x$ have room temperature conductivities of the same order as the organic charge-transfer salts. Similar results are obtained for other polymers such as polypyrole (C_4NH) and polyparaphenylene $(C_6H_4Y)_x$.[52,53]

The lack of crystal structure constraints on the degree of doping makes the polymers excellent candidates for varying n_e in the models above. However, the techniques for doping and alignments of the microscopic chains are still rather complicated.[54,55]

Band structure calculations have recently been used to investigate the interesting high temperature properties of pure and doped $(CH)_x$ both in its trans and its cis modification, but the role of correlations is not settled.[56] Ovchinnikov *et al.* have argued for the existence of strong correlation effects with reference to organic molecules,[57] whereas others have suggested that U is small like in π-bonding of inorganic systems.[58]

Below we will look into some specific examples of how the behaviour of quasi-one-dimensional molecular conductors may be brought into contact with the mathematical models described above.

MOLECULAR SEMICONDUCTORS

As an example of the instabilities in linear conductors we consider how the Peierls instability manifestsitself in the platinum chain conduc-

		KCP	$TEA(TCNQ)_2$
From phonons	λ_{ph}	0.31	0.1
From vibrons	λ_{vib}	0.00	0.5
	$\lambda_{ph} + \lambda_{vib}$	0.31	0.6
From activation energy	λ	0.30	0.4

Table 3. Electron phonon coupling constants λ in KCP and $TEA(TCNQ)_2$.

tor KCP and in the organic conductor $TEA(TCNQ)_2$. The behaviour of the two compounds is in some respects very similar, but they differ in the important aspect that in $TEA(TCNQ)_2$ the intra-molecular vibrations play an important role in establishing the low temperature semiconducting (or insulating) state, whereas in KCP only external modes contribute to the coupling constant λ . To see this we determine λ in two ways namely from the macroscopic properties of the low temperatures as well as from the microscopic properties at high temperatures.

KCP undergoes a Peierls transition at T_c = 100 K but the transition is so broad that the $2k_F$ lattice distortion is seen by X-rays even at room temperature . In this way k_F is measured and its value is given in Table 2. The conductivity (see Figure 7) is activated at low temperatures and shows structure in the broad transition region. The activation energy Δ is shown in Table 2. As mentioned above the conduction electrons are free in KCP and therefore ε_F may be calculated from k_F . Then λ may be calculated from the formula corresponding to (1) in the CDW case:[7]

$$\Delta = 8 \frac{1-k_F d_{||}/\pi}{1+k_F d_{||}/\pi} \varepsilon_F e^{-1/\lambda} \quad . \tag{5}$$

The alternative way of calculationg λ is from the microscopic properties at high temperatures. In principle this means well above T_c^{MF} which from (2) may be estimated from Δ to be 475 K, i.e. inaccessible because KCP decomposes just above room temperature. However, accurate values for the important parameters may be obtained allowing the determination of λ from:[31]

$$\lambda_{ph} = \frac{\varepsilon_F}{8} \frac{\hbar}{2M} \left(\frac{2\pi}{d_{||}}\right)^2 \frac{1}{\hbar^2 \omega_0^2} \quad , \tag{6}$$

where M is the $Pt(CN)_4$-mass and ω_0 is the unperturbed phonon-frequency indicated in Figure 1. This gives λ=0.31 for KCP. Since (6) describes the contribution from phonons only, the fact that $\lambda = \lambda_{ph}$

shows that in KCP electron-phonon coupling is dominant in the Peierls transition.

$TEA(TCNQ)_2$ is more difficult to analyse unambiguosly, but it is however illuminating to derive numbers and compare them to KCP. Focusing on the features relating to a Peierls distortion (i.e. neglecting U and $4k_F$) we get a transition temperature $T_c = 220$ K and $\Delta = 120$ meV. $k_F = 0.25\ \pi/d_{||}$ as given by stoichiometry, and for the guarterfilled band (1) gets modified to:[7]

$$\Delta = 2\ W_{||}\ e^{-1/\lambda} \quad , \qquad (7)$$

yielding the $\lambda = 0.40$. However, an estimate of λ_{ph} from external modes gives approximately 0.1.[44,47,60] This shows that electron-phonon coupling plays a minor role in establishing the Peierls gap and indicates that coupling between conduction electrons and internal modes is of importance.

This coupling was independently pointed out by Rice *et al.*[60] from an analysis of the infrared reflectivity from $TEA(TCNQ)_2$,[42] shown also for KCP in Figure 8.[61] Both compounds show structure stemming from the plasma edge at ω_P and of the excitations across the gap 2Δ, but further the reflectivity of $TEA(TCNQ)_2$ has distinct features at frequencies which correspond to the 10 a_g modes of the TCNQ-molecule. This may be used to derive the electron-vibron coupling constants and their sum λ_{vib} is 0.51 ± 0.1 . In Table 3 we compare $\lambda_{ph}+\lambda_{vib}$ to λ derived from conductivity. The agreement is very good for KCP showing that the Peierls ground state here can be well accounted for by a relatively simple microscopic model. In $TEA(TCNQ)_2$ phonons cannot explain the increase in λ with respect to KCP; but on the other hand the electron-vibron model seems to overestimate the dc conductivity gap.

From the Δ's one may calculate the mean field transition temperature to be 475 K and 790 K for KCP and $TEA(TCNQ)_2$ respectively, and in both cases do the actual transitions occur at much lower T_c's. This is consistent both with the broad transition region demonstrating the role of fluctuations as well as with the large conductivity anisotropy. In terms of the bandwidth anisotropy $\eta = W_\perp/W_{||}$, this may be found from the theory of Horowitz *et al.*:[23]

$$T_c/T_c^{MF} = \exp\{-1.4\ \frac{\Delta}{W_\perp}\} \quad , \qquad (8)$$

and the results are shown in Table 2. One clearly sees that platinum salts are more one-dimensional than the organic conductors.

Our analysis also demonstrate that not only is $U \approx W_{||}$ in organic conductors. One also finds $W_\perp \approx \Delta \approx \varepsilon_F$, three quantities which theory usually assumes to be well separated. Nothing is simple in organic conductors!

MOLECULAR METALS

The metallic state of quasi-one-dimensional conductors remained a hypothetical high temperature state of the CDW-semiconductor until TTF-TCNQ allowed the study of this synthetic metal over an extended temperature region. Molecular metals turned out to be interesting in their own right, since indeed their behaviour demonstrates that electronic transport is changed much more drastically than one would estimate from the bare effect of the small bandwidths compared to normal metals (cf. Table 1).

In the metallic region one often finds that the measured conductivity has the following temperature dependence:

$$\sigma_{||}(T) \propto T^{-2.3} \tag{9}$$

as opposed to $\sigma_{||}(T) \propto T^{-1}$ of pure ordinary metals. This dramatic behaviour of molecular metals, giving rize to giant conductivity peaks close to the Peierls transition, has been attributed to superconducting fluctuations.[46] Since such fluctuations should show op only along the chain the increase of the conductivity anisotropy $\sigma_{||}/\sigma_{\perp}$ with decreasing temperature observed in several compounds is another evidence for collective effects at high temperatures.[62] This of course implies that the scale temperature T_c^{MF} is high for superconductivity, but only the observation of a high real transition temperature will unambiguously settle this question.

From our discussion of the Peierls instability in KCP and $TEA(TCNQ)_2$ one might expect that also in TTF-TCNQ the fluctuations would smear out the transition at $T_c = 54$ K and dominate the resistivity at rather high temperatures. Yet the transition appears sharp. This is, however, consistent with the fact that T_c is less depressed with respect to T_c^{MF} in TTF-TCNQ than the two other compounds. Inserting $\Delta = 25$ meV from low temperature conductivity into (2) one gets $T_c^{MF} \approx 160$ K . Hence both the low T_c (= 54 K), the relatively high value of T_c/T_c^{MF} (cf. (8)) and the conductivity anisotropy $\sigma_{||}/\sigma_{\perp}$ indicate that fluctuations due to the CDW should set in only relatively close to T_c . But of course this does not exclude that superconducting fluctuations could occur at high temperatures.

It is, however, not necessary to invoke collective effects to account for the unusual temperature dependence of the conductivity, since both the molecular degrees of freedom as well as the big lattice anharmonicity, both modify the electronic transport. It was early pointed out that molecular rotations (librons) would be a natural explanation for (9) since the conductivity in this case has the following temperature dependence:[63]

$$\sigma_{||}^{lib}(T) \propto \omega_{lib}^{4} T^{-2} \tag{10}$$

Here ω_{lib} is the libron frequency which owing to the anharmonic interactions (corresponding to a big Grüneisen parameter) would change enough to account for the difference in exponent between (10) and (9). However, although scattering from translational phonons has the temperature dependence

$$\sigma^{ph}(T) \propto \omega_{ph}^{2}\, T^{-1} \tag{11}$$

at constant volume, still the observed conductivity may be quantitatively accounted for even if phonon scattering is dominant.[47] The reason why both (10) and (11) may be modified to agree with experiments lies in the uncertainty in the assessment of the volume and temperature dependence of the involved frequencies.[47,64]

Both the libron and the phonon theory assumes that transport in molecular metals may be conceived of as in a metal with a well-defined conduction band. However, even this assumption literally is violated since the electronic mean free path at room temperature is of the order of inter-molecular spacings. Therefore one might expect the electrons to localize, in particular considering the one-dimensional nature of the conductivity.[64,65] Neither localization nor the other crucial problem, namely the role of correlations, have been treated quantitatively with respect to transport in molecular metals. As mentioned, despite the concern about its validity, band transport theories explain the observed conductivity very well.

MOLECULAR SUPERCONDUCTORS

When superconductivity was first observed in $(TMTSF)_2PF_6$ under pressure[66] and later in the absence of applied pressure in $(TMTSF)_2ClO_4$,[67] a new dimension was added to the study of molecular conductors. In Figure 8 we compare the results for $(TMTSF)_2ClO_4$ with Hg , as observed by Kammerlingh-Onnes in 1912.[68] Clearly the first synthetic superconductor is only a poor reproduction of the first natural one, both in terms of the magnitude of the transition temperature and the sharpness of the transition. However, superconductivity in the $(TMTSF)_2X$ compounds do exhibit several extraordinary features. And of course the mere existence of superconductivity makes TMTSF a very special molecule.

Dependent of the anion X,$(TMTSF)_2X$ exhibit three different ground states,[50] namely CDW when X = ReO_4 with T_c = 180 K, SDW when X = PF_6, AsF_6 and NO_3, $T_c \approx 12$ K and SS when ClO_4 with $T_c \approx 1.3$ K. When pressure is applied the metal-insulator transition temperatures (CDW or SDW) decreases quickly, and at pressures of 6-12 kbar superconductivity is observed ($T_c \approx 1$ K), with the important exception of X = NO_3 . The superconducting transition temperature decreases gradually with increasing pressure. It is possible to classify the compounds and their behaviour (CDW, SDW or SS) in terms of symmetry of the anion;[50] but the variation in the Se-network of the crystals may also have a profound

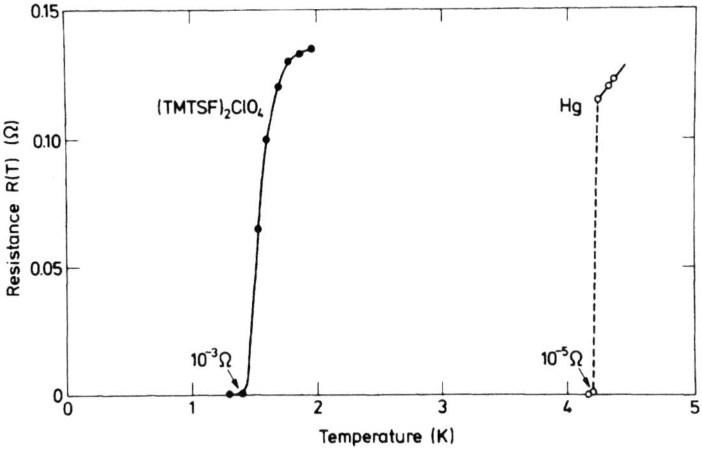

Figur 9. Superconductivity in Hg and $(TMTSF)_2ClO_4$.

effect on the different properties of $(TMTSF)_2X$ compounds.[69]

Several observations indicate that at low temperatures $(TMTSF)_2X$ compounds are two-dimensional rather than one-dimensional. Crystallographic investigations show short Se-Se contacts within the TMTSF-sheets, (the $\vec{a}\vec{b}$-plane) and at low temperatures there is evidence of a plasma-edge both in the conducting direction as well as along $\vec{b}$.[70] Both results indicate a large $t_\perp$ in the $\vec{b}$ direction. Critical magnetic fields in the superconducting phase also show less difference between the $\vec{a}$ and $\vec{b}$ direction than between these two and the $\vec{c}$ direction.[71] Finally relation between Δ and T_c for the insulating states compare well with (2) indicating that fluctuations are not very important.

Nevertheless, other observations indicate strong superconducting fluctuations. Tunnelling spectroscopy, designed to measure the superconducting indicate $\Delta = 1.8$ meV and $\Delta = 4$ meV for $(TMTSF)_2PF_6$ (under pressure) and $(TMTSF)_2ClO_4$ respectively.[72] From (2) this corresponds to T_c^{MF} much larger than the observed $T_c \approx 1$ K , and indeed the experiments show evidence of a fluctuating gap well above T_c. Also far infrared transmission experiments have recently been shown to have a minimum, consistent with $\Delta = 4$ meV in $(TMTSF)_2ClO_4$.[73,74]

So although organic superconductors have only been studied experimentally for two years, they have already demonstrated a controversial behaviour posing fundamental questions to our understanding of the classical field of superconductivity.

SUMMARY

In this paper I have tried to review and compare the physical properties of one-dimensional models with those of molecular conductors, and indeed the richness in behaviour of the one class may be understood in the context of the other.

Large planar molecules with delocalized π-electronic orbitals may be stacked to give a sufficient overlap perpendicular to the planes of the molecules to give high conductivity. The bandwidth range from 0.5 to 1 eV. Electronic correlation energies (cf. Table 2) are difficult to estimate quantitatively and despite the fact that correlations are traditionally large in organic molecules there is growing evidence that they play a minor role in highly conducting molecular compounds.

By analyzing examples of the Peierls instability, conductivity in the metallic phases and finally superconductivity I have demonstrated that the molecular nature influences the properties of molecular conductors in a variety of ways, which in certain cases have been quantitatively accounted for. In other cases apparent inconsistencies between different experimental results raise questions that force us to reconsider our view upon several aspects of solid state physics.

Conducting platinum chain compounds are formed based on overlap of d_{z^2} orbitals, when planar ligands are stacked. Here conduction electrons are very nearly free, corresponding to large bandwidths, and correlation effects are negligible. In Pt-chain conductors the specific molecular nature seems to be relatively unimportant, and these compounds are very good model compounds for the Peierls instability.

In conducting polymers π-electrons delocalize along the polymeric chain resulting into very large bandwidths. However, conductivity is achieved only by adding impurities so that high conductivity is associated with strong disorder effects.

ACKNOWLEDGEMENTS

It is a pleasure to acknowledge fruitful discussions on the organic conductors with K. Bechgaard and C.S. Jacobsen and the close collaboration with A.E. Underhill and J.M. Williams on the platinum salts. This study was supported in part by NATO (Grant #016.81)

REFERENCES

1. For recent reviews see for example: *The International Conference on Low Dimensional Synthetic Metals*: 1981, Chemica Scripta 17, pp.1-230; and
2. *Proceedings of the International Conference on Low-Dimensional Conductors:* 1982, Mol.Cryst.Liq.Cryst. 79, part A-E.
3. Peierls, R.E., *Quantum Theory of Solids* (Oxford at the Clarendon): 1954, pp.107-108.
4. Kohn, W.: 1959, Phys.Rev.Lett. 2, p.393.
5. Fröhlich, H.: 1954, Proc.R.Soc. (London) A 223, p.296.
6. Rice, M.J., and Strässler, S.: 1973, Solid State Commun. 13, pp.125-128.
7. Nielsen, J.B., and Carneiro, K.: 1980, Solid State Commun. 33, pp.1097-
8. Bardeen, J., Cooper L.N., and Schrieffer, J.R.: 1957, Phys.Rev. 108, p.1175.
9. Bychkov, Yu. A., Gorkov, L.P., and Dzylozhinskii, J.E.:1966, Sov. Phys.JETP 23, p.489.
10. Horovitz, B., Gutfreund, H., and Weger, M.: 1982, ref.2,pp.591-604, and: 1981, Solid State Commun. 39, p.541.
11. Barisic, S., Emery, V.J., Brazovskii, S., and Bruinsma, R.: 1982, in ref.2.
12. Little, W.A.: 1964, Phys.Rev. A134, p.1416.
13. Gutfreund, H., and Little, W.A., *Chemistry and Physics of One-Dimensional Metals* (Plenum Press, ed. H.J. Keller): 1977, p.279.
14. Hubbard, J.: 1963, 1963 and 1969, Proc.Roy.Soc.(London) A 276, p.238, A 277, p.237, and A 281, p.401.
15. Bernasconi, J., Rice, M.J., Schneider, W.R., and Strässler, S.: 1975, Phys.Rev. B 12, p.1090.
16. Huizinga, S., and Kommandeur, J.: 1981, Chemica Scripta 17, pp.59-60.
17. Scott, J.C.: 1982, Mol.Cryst.Liq.Cryst. 79, pp.405-416.
18. Takayama, H., Lin-Liu, and Maki, K.: 1980, Phys.Rev. B 21, p.2388.
19. Brazovskii, S, and Kirova, N.: 1981, Chemica Scripta 17, pp.171-172.
20. Bishop, A.R., Campbell, D.K., and Fesser, K.: 1982, Mol.Cryst.Liq. Cryst. 77, pp.253-264.
21. Landau, L.D., Lifshitz, E.M.: 1969, *Statistical Physics* (Addison-Wesley).
22. Scalapino, D.J., Sears, M., and Terrell, R.A.: 1972, Phys.Rev.B 6, p.3409.
23. Horovitz, B., Gutfreund, H., and Weger, M.: 1975, Phys.Rev. B 12, pp.3174-3285.
24. Schultz, H.J.: 1982, Mol.Cryst.Liq.Cryst. 79, pp.199-212.
25. Döbereiner, J.W.: 1833, Pogg.Ann. 28, p.180.
26. Krogmann, K.: 1969, Angew.Chem.Int.Ed.Engl. 8, p.35.
27. For a recent review see: Williams, J.M., Schultz, A.J., Underhill, A.E., and Carneiro, K.: 1982, *Extended Linear Chain Compounds* (Plenum, ed. J.A. Miller) pp.73-118.
28. Underhill, A.E., Watkins, D.M., Williams, J.M., and Carneiro, K.: *ibid* pp.119-156.
29. Toftlund, H.: 1979, J.C.S.Chem.Comm.,pp.837-838.

30. Toftlund, H.: unpublished results.
31. Carneiro, K.: 1982, Mol.Cryst.Liq.Cryst. 79, part C.
32. Fritsche, L., Rafat-Mehr, M.: 1975, Lecture Notes in Physics, 34, pp.97-107.
33. Whitmore, D.M.: 1974, Phys.Lett. 50 A, pp.55-56.
34. Bullett, D.W.: 1978, Solid State Commun. 27, p.467.
35. Miller, J.S.: 1976, Inorg.Chem. 15, p.2357.
36. Reis, A.H. Jr., and Peterson, S.W.: 1978, Ann.N.Y.Acad.Sci.313, p.560.
37. Messmer, R.P., and Salahub, D.R.: 1975, Phys.Rev.Lett. 35,pp.533-536.
38. Abarbanel, A.M.: 1975, Ann.Phys. 91, pp.356-365.
39. Melby, L.R., Harder, R.J., Hertler, W.R., Mahler, W., Bensen, R.E., and Mochel, W.E.: 1962, J.An.Chem.Soc. 84, p.3374.
40. Berlinsky, A.J.: 1976, Contemp.Phys.17, pp.331-354.
41. Soos, Z.G., and Klein, D.J.: 1976, *Treatise on Solid State Chemistry* (ed. N.B. Hannay, Plenum Press) 3, pp.679-767.
42. Farges, J.P.: 1980, *The Physics and Chemistry of Low Dimensional Solids* (ed. L. Alcacer, NATO ASI, series C, 56, D. Reidel) pp.223-232
43. Johansen, H.: 1975, Int.J.Quant.Chem.IX, pp.459-469.
44. Berlinsky, A.J.: 1976, Solid State Commun.19, pp.1165-1168.
45. Hermann, F., Salahub, D.R., and Messmer, R.P.: , Phys.Rev.B 16, pp.2453-2465.
46. Coleman, L.B., Cohen, J.J., Sandman, D.J., Yamagishi, F.G., Garito, A.F., Heeger, A.J.: 1973, Solid State Commun. 12, pp.1125-1131.
47. Conwell, E.M.: 1981, Phys.Rev.B 22, pp.1761-1780.
48. Khanna, S.K., Pouget, J.P., Comès, R., Garito, A.F., Heeger, A.J.: 1977, Phys.Rev. B 16, pp.1468-1479.
49. Jérome, D., and Weger, M.: 1977, *Chemistry and Physics of One-Dimensional Metals* (ed. H.J. Keller, NATO ASI, series B, Plenum) pp.341-367.
50. Bechgaard, K.: 1982, Mol.Cryst.Liq.Cryst. 79, pp.1-13.
51. Heeger, A.J., MacDiarmid, A.G.: 1982, Mol.Cryst.Liq.Cryst. 77,pp.1-24
52. Diaz, A.: 1982, Chemica Scripta 17, pp.145-148.
53. Clarke, T.C., Kanazawa, K.K., Lee, V.Y., Rabolt, J.F., Reynolds, J.R. Street, G.B.: 1981, Chemica Scripta 17, pp.161-167.
54. Shirakawa, H., and Ikeda, S.: 1971, Polymer J. 2, p.231.
55. Luttinger, L.B.: 1962, J.Org.Chem. 27, p.1591.
56. Grant, P.M., and Batra, I.P.: 1979, Solid State Commun. 29,pp.225-228
57, Ovchinnikov, A.A., Ukrainskii, I.I., and Kventsel, G.V.: 1973, Sov.Phys.Uspekhi 15, p.575.
58. Grant, P.M., and Batra, I.P.: 1981, unpublished results.
59. Wangbo, M.-H., Hoffmann, R.: 1978, J.Amer.Chem.Soc. 6093-6098.
60. Rice, M.J., Pietronero, L., and Brüesch, P.: 1977, Solid State Commun. 21, pp.757-760.
61. Brüesch, P., Strässler, S., and Zeller, H.R.: 1975, Phys.Rev. B 12, p.219.
62. Jérome, D.: 1980, *The Physics and Chemistry of Low Dimensional Solids* (ed. L. Alcacer, NATO ASI, Series C, Reidel) pp.123-142.
63. Gutfreund, H., and Weger, M.: 1977, Phys.Rev. B 16, p.1753.
64. Weger, M.: 1980, *The Physics and Chemistry of Low Dimensional Solids* (ed. L. Alcacer, NATO ASI, Series C, Reidel) pp.77-100.

65. Mott, N.F., and Troose, W.D.: 1961, Adv.Phys. 10, p.107.
66. Jérome, D., Mazaud, A., Ribault, M., and Bechgaard, K.: 1980, J.Phys.Lett. 41, p.95.
67. Bechgaard, K., Carneiro, K., Olsen, M., Rasmussen, F.B., Jacobsen, C.S.: 1981, Phys.Rev.Lett. 46, pp.852-856.
68. Kammerlingh-Onnes, H.: 1911, Acad.van Wetenschappen (Amsterdam) 14, p.113, p.818.
69. Wudl, F.: 1982, Mol.Cryst.Liq.Cryst. 79, pp.67-78.
70. Jacobsen, C.S., Tanner, C.S., and Bechgaard, K.: 1981, Phys.Rev. Lett. 46, pp.1142-1145.
71. Greene, R.L., Haen, P., Huang, S.Z., Engler, E.M., Choi, M.Y., and Chaikin, P.M.: 1982, Mol.Cryst.Liq.Cryst.79, pp.183-197.
72. More, C., Roger, G., Sorbier, J.P., Jérome, D., Ribault, M., and Bechgaard, K.: 1981, J.Phys.Lett. 42, p.313.
73. Thomas, G.A., Kwo, J., Rosenbaum, T.F., and Wudl., F.: 1982, Bull. Amer.Phys.Soc. 27, pp.150-151.
74. Na, H.K., Timusk, T., Delricu, J.M., Jérome, D., Bechgaard, K., Fabre, J.M.: 1982, *ibid* p.151.

INTRODUCTION

Bernard Pullman

The scope and the diversity of the topics which are going to be covered by the Symposium on Quantum Biology and Pharmacology of this 4th International Congress of Quantum Chemistry are particularly wide, this large-scale range of problems considered being still more evident if we take into account the communications to be presented at the associated invited posters session. This impressive dimension of this research field is not astonishing if we consider the rapid development of quantum biology during the last years, a development which certainly owes much to the two regular annual meetings in which its current progresses are being presented and discussed : the Sanibel Symposia and the Jerusalem Symposia. Neither can this Symposium here be considered separately from the Satellite Symposium on Quantum Chemistry in Biology to be held next week at Lidingo, organized by Inga Fischer-Hjalmars and Anita Henriksson-Englo.

It has become a regular and most profitable tradition of the Sanibel and the Jerusalem Symposia to have experimentalists invited and closely associated with the presentation and discussion of problems in modern molecular and quantum biology. It is with great pleasure that I observe a step in this direction also at this biology Symposium of this Quantum Chemistry Congress. The organizers certainly could not have made a better choice in inaugurating what I hope will become a tradition in this respect than inviting Prof. Youri Ovchinnikov, Director of the Shemyakin Institute of Bioorganic Chemistry in Moscow and Vice-President of the Academy of Sciences of the Soviet Union. Professor Ovchinnikov has made many fundamental contributions to molecular biology, all at the front line of this rapidly developing discipline, but his most celebrated and known work is connected with the structure and mechanism of functionning of ionophores. It is about problems related to ion transfer accross membrane that he will talk to us today but you may remark that this time his lecture will not be about the oligopeptide or depsipeptide type of ionophores but on membrane proteins as ion channels. How closely his interests are to those of quantum biologists may be seen in the fact that Mme Pullman will give a lecture, not here but at the satellite Symposium at Lidingo, on "Theoretical studies on cation transport systems".

P.-O. Löwdin and B. Pullman (eds.), New Horizons of Quantum Chemistry, 395–397.

Our next speaker Professor Peter Schuster, whom we all knew for a long time as an expert on hydrogen bond, will talk to us on the evolution of molecules, a most fascinating problem for, I would say, everybody. I myself had a proof of this fascination nearly 20 years ago. It originated in a short stay that I made in Uppsala, in 1962. I felt very relaxed, as I always do when I am here, and wrote a note entitled "Electron Delocalization and Biochemical Evolution". In this article I proposed an explanation on the involvement of certain types of molecules in the origin of life, for instance for the fact that say, porphyrins, rather complex molecules, are found by paleobiology, to have been there at the very eve of life, some 3 billion years ago and have been utilized by nature for such a variety of function as oxygen transport in haemoglobin, electron transport in cytochromes or photosynthesis in chloroplasts, while it is obvious that this common squeleton is certainly not the best possible one for all these functions. I published this note in Nature and received a number of request for reprints which constitutes the absolute record of the number ever requested for any other of my publications. It is a proof I think of the fascination that "the beginnings" exerce on all of us. We all know of the remarkable, renovating contribution that Professor Eigen has made to this topic. Professor Schuster has joined the Eigen school, so to say, and has become one of its prominent members. It is with great curiosity that we are waiting for his message.

The last two lectures by Madame Pullman and Dr. Clementi, both too well known to the participants to need an introduction, will be devoted to nucleic acids. The discovery of the molecular or rather macromolecular structure of the nucleic acids was at the origin of modern molecular biology and as I have related in my Reminiscences at the 6th Sanibel Symposium on Quantum Biology it inaugurated also the development of Quantum Biochemistry. The nucleic acids remain still at the very center of experimental and theoretical molecular biology and it is thus not astonishing that they receive this treat at our symposium. As a matter of fact some of the most recent developments have given a new dimension to the study of these biopolymers. These developments are connected with the recent discoveries that these acids may exist in a number of biologically significant allomorphic forms, which differ among themselves in geometrical parameters and, what seems to be particularly important, in conformation. This polymorphism may be due to external factors (humidity, salt concentration etc) or, as discovered recently, induced by variations in the base sequence, a situation which stresses the great interest of studying both the intrinsic properties of these fundamental biological substrates and the influence on these properties of the various environmental factors to which they may be submitted. Moreover, the dependence of the polymorphism on base sequence leads to the consideration of a possible regional heterogeneity (microheterogeneity) of native DNA as a function of the order of the base pairs along its helical axis. Such microheterogeneity could be of particular significance for the specificity of interaction or association of the nucleic acids with external reactants and this line of research is actively developed today with respect to such important types of reactants as carcinogens, mutagens, antitumor agents, antibiotics etc. Some of the less regular nucleic acids like tRNAs are in themselves a diversified such system and Madame Pullman will present to you the result of studies on a representative of this class of compounds, $tRNA^{Phe}$, which is probably the most complex single molecule studied till today by the method of quantum chemistry. I myself was hoping to deal with the problem of DNA from the point of view of its polymorphism and microheterogeneity at the Satellite Symposium at Lidingo but important last minute obligations prevent me unfortunately from being there.

I hope that I have fullfield my duty, prescribed by the organizers, to talk to you for 15 minutes and I am now going to leave the stage for our invited speakers.

MEMBRANE PROTEINS AS ION CHANNELS

Yu.A.Ovchinnikov
Shemyakin Institute of Bioorganic Chemistry, USSR
Academy of Sciences, Moscow, USSR

Membrane proteins perform many vitally important functions in cells. The general characteristic of all these proteins is their influence on transport of various effectors, in particular, on ion transport, as a rule, in response to an outer action. For the last two decades we, at the Shemyakin Institute, were engaged in studying the problems of ion transport (1-3). For ionophores, a classical representative of which is valinomycin, the molecule conformation was established and its potential possibilities were elucidated that allowed us to understand the mechanism of action of these compounds. Nowadays more and more information is available on the properties of gramicidin A, the best studied model channel former. Investigations of ion channels are carried out in many laboratories throughout the world by various methods: biochemical, physical, electrophysiological, etc., however, the results obtained are not always in accordance. Their isolation in a pure state is a serious problem and here highly specific effectors are of undoubt benefit. They are also applied in studying the action of membrane channel formers; on the other hand, using such specific ligands, one can succeed in obtaining information on the channel structure.

Various aspects of structural organization of ion channels are discussed in this paper, summarizing the data on sodium channels of electrically excitable membranes, acetylcholine receptor as well as two retinal-containing proteins, bacteriorhodopsin from halophilic microorganisms and visual rhodopsin, obtained in the Shemyakin Institute.

Generation and transmission of nerve impulses are tightly connected with fast sodium channel functioning. These transmembrane ion-conducting systems are well characterized by various electrophysiological methods, but you find practically no data on their molecular organization. Sodium channels are proposed to be complicated protein complexes with various functional components: selective filter and gating mechanism (4). Research of molecular organization of sodium

P.-O. Löwdin and B. Pullman (eds.), New Horizons of Quantum Chemistry, 399–415.

channels by conventional methods is difficult because of their very low content in electrically excitable membranes and high lability in a solubilized state. Therefore, the use of various neurotoxins capable of modifying some channel functions is practically the only effective way for their investigation. There are at least three groups of neurotoxins differently affecting functioning of fast sodium channels (5) Tetrodotoxin and saxitoxin block the transmembrane current of sodium ions and are most probably bound to the ion-conducting portion of channels. The group of alkaloid toxins: batrachotoxin, veratridine, aconitine and grayanotoxins activate sodium channels and change their ion selectivity. Apparently these neurotoxins also interact with channel components performing the transport of sodium ions. Toxins of scorpions and sea anemones affect inactivation of fast sodium channels and apparently,they are bound to the channel portion regulating the work of the ion-conducting component. Thus, the study of the receptors of various neurotoxins which affect functioning of sodium channels allowed us to approach understanding of its molecular nature.
Investigation and isolation of toxin membrane receptors involve obtaining their radioactive derivatives suitable for the radio-ligand analysis. To obtain a tetrodotoxin derivative of highly specific radioactivity the method of catalytic reduction of unsaturated substrates with gaseous tritium was used (6). After periodate oxidation the formed ketoderivative of the toxin was treated with HCN for obtaining the corresponding cyanohydrin (Fig.1). Then catalytic reduction of the

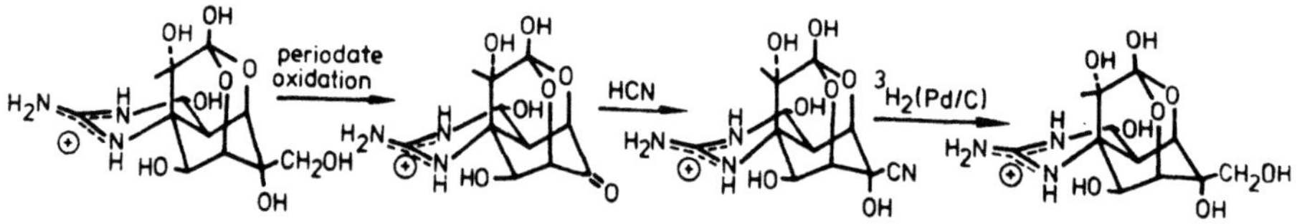

Figure 1. Scheme for obtaining radioactive tetrodotoxin derivative.

modified neurotoxin with gaseous tritium in the presence of 5% Pd/C was carried out. In this case the toxin analog containing an aldehyde group was apparently formed as the main product that was transformed into d,l form of tetrodotoxin at the additional reduction. According to the data obtained specific radioactivity of the tetrodotoxin preparation was 10 Curi/mmol.
Axon crab membranes and rat brain synaptosomes were used for purification of tetrodotoxin receptors. At sedimentation in

sucrose density gradient tetrodotoxin receptors of the axonal membrane solubilized with Lubrol PX were established to be present in two fractions with sedimentation coefficients 9,5 and 16S. Receptor components of the axonal membrane are not bound to concanavaline and wheat germ lectin, that probably points to the absence of carbohydrate components specific to these lectins in their molecules, but does not exclude the glycoprotein nature of receptors.

Upon fractionation of axonal membranes solubilized with Lubrol PX, gel-chromatography, sedimentation in sucrose concentration gradient and ion exchange chromatography on DEAE-Sepharose the correlation was found between tetrodotoxin receptors and the proteins of molecular weights 53 000, 77 000, 85 000 and 230 000 (7). However, for unequivocal identification of the receptors their complete purification and stabilization in a solubilized state is necessary.

To isolate tetrodotoxin receptors from rat brain membranes with solubilized Lubrol PX use was made of subsequent chromatography on DEAE-Sepharose, WGA-Sepharose and Sepharose CL-6B as well as sedimentation in sucrose concentration gradient. We succeeded in the 340-fold purification of toxin receptors and obtaining the membrane fraction binding more than 500 pmol of tetrodotoxin per 1 mg of the protein. Its sedimentation coefficient is about 11S. By means of electrophoretic analysis it was established that the proteins of molecular weights 53 000, 76 000, 90 000 and 250 000 were predominantly contained in this fraction.

The comparative analysis of the receptor components of tetrodotoxin and neurotoxins from the scorpion venom was of considerable interest. We used neurotoxin M_{10} from the scorpion *Buthus eupeus* venom (Fig.2) (8). Investigation of the scor-

10
Val-Arg-Asp-Gly-Tyr-Ile-Ala-Asp-Asp-Lys-Asp-Cys-

20
Ala-Tyr-Phe-Cys-Gly-Arg-Asn-Ala-Tyr-Cys-Asp-Glu-

30
Glu-Cys-Lys-Lys-Gly-Ala-Glu-Ser-Gly-Lys-Cys-Trp-

40
Tyr-Ala-Gly-Gln-Tyr-Gly-Asn-Ala-Cys-Trp-Cys-Tyr-

50 60
Lys-Leu-Pro-Asp-Trp-Val-Pro-Ile-Lys-Gln-Lys-Val-

Ser-Gly-Lys-Cys-Asn

Figure 2. Amino acid sequence of neurotoxin M_{10} from the scorpion *Buthus eupeus* venom.

pion toxin receptors is known to be considerably difficult as its reception depends on the value of the membrane potential and because of the low content of sodium channels in electrically excitable membranes (9). To elucidate the nature of the neurotoxin M_{10} receptors its radioactive and photosensitive analog preserving biological activity and suitable for covalent modification of the receptor was used (10). The photosensitive label in modified neurotoxin M_{10} analog was located on lysine 57 situated in the direct vicinity of the characteristic region of invariant and hydrophobic amino acid residues of the toxin molecule and, therefore, the covalent bond should be formed predominantly with receptor components of electrically excitable membranes. At photoactivation the modified analog of the neurotoxin specifically binds two protein components of electrically excitable membranes from rat brain, crab axones and neuroblastoma cells, molecular weights of these proteins being 76 000 and 51 000 (11). Thus, protein components of these molecular weights probably compose the membrane system performing potential-dependent transport of sodium ions. One can hope that further studying of various neurotoxin receptors makes possible determining the structure and peculiarities of functioning of fast sodium channels in detail.

Nicotinic acetylcholine receptor (AchR) of the postsynaptic membrane is a well characterized membrane receptor (12-17). It can exist in dimeric (∿500 kD) and monomeric (∿ 250 kD) forms, the latter consists of 5 subunits of molecular weights from 40 to 65 kD (stoichiometry $\alpha_2\beta\gamma\delta$). The N-terminal sequences of these subunits are described in literature (15). The experiments with liposomes and lipid bilayers showed that the complex of 5 subunits isolated in an individual state is capable of reproducing the basic property of the intact receptor; i.e. acetylcholine binding is accompanied with ion permeability changes. Investigations of proteolysis showed that all the subunits are transmembrane proteins.

In spite of rapid progress many problems remain unsolved: we still do not know the groupings responsible for the binding of the neurotransmitter and induced changes in ion permeability and mutual arrangement of AchR subunits. Spatial relationships of the receptor site (the site where the binding of acetylcholine and such bioeffectors as cobra venom neurotoxins occurs) and the so called channel component (which is blocked with histrionicotoxin and phencyclidine) are also unclear.

Neurotoxins from the cobra venom (18) deserve special attention. Finding of the source in which AchR content is extremely high, electric organs of electric rays,as well as application of affinity sorbents with bound cobra venom neurotoxins influenced much the success in the receptor isolation in a pure state. Investigation of the topography of AchR and neurotoxin interacting sites could promote not only better understanding of the mechanism of a neurotoxin action but also shed light on the organization of the AchR active site.

The necessary data were obtained using spin and fluorescence labels incorporated in different sites of neurotoxins of long and short types. Binding of the corresponding derivatives was followed by EPR and fluorescence methods (19,20). In case of neurotoxins from *Naja naja oxiana* the results of spectral methods were compared with those of formation of photoinduced cross-linkings between AchR and a series of p-azidobenzoyl derivatives (21).

To obtain the maximal possible number of monolabeled derivatives the reaction was carried out in the presence of guanidinium hydrochloride at the deficiency of dansylchloride or N-hydroxysuccinimidyl esters of spin labels. The obtained mixture was separated by ion exchange chromatography and location of labels in each derivative was carried out by the methods of protein chemistry. We succeeded in obtaining and isolation of 6 monosubstituted photoactivable derivatives using the reaction with N-hydroxysuccinimide ester of p-azidobenzoic acid (22). The labels (ε-amino groups of Lys 15, Lys 25, Lys 26, Lys 44, Lys 46 and α-amino group of Leu 1) were localized by the direct method. So, in the dark azide and disulfide bonds of the toxin derivatives were simultaneously reduced with dithiotreitol (thus we avoided intramolecular cross-linking), then carboxymethylation was carried out followed by tryptic digest and separation of the obtained peptides in the light.

Binding studies were performed with AchR preparations solubilized in Triton X-100 (isolation by affinity chromatography on neurotoxin - Sepharose 4B, the AchR activity ranged from 7-10 nmol of α-toxin binding sites per 1 mg of the receptor protein). For the Lys 46 dansylated NT-II derivative one could observe an increase in fluorescence and a blue shift of the emission maximum upon its binding to AchR (Table 1).

Table 1. Fluorescence of dansylated neurotoxin II derivatives bound to acetylcholine receptor.

Labeled residue	Emission maximum (nm) free	AchR/ bound	Relative fluorescence intensity (AchR bound/free) λex, nm 330	298
Leu 1	560	560	1,0	1,3
Lys 15	556	555	1,0	1,3
Lys 25	555	548	1,3	1,6
Lys 26	563	544	1,5	2,6
Lys 46	570	545	3,0	5,3

These changes favoured the dansyl label incorporation into a more hydrophobic environment of the receptor upon the complex formation. Fluorescence increasing is larger when excitation is at 298 nm, that gives evidence of energy migration from AchR triptophan(s) into dansyl group of the neurotoxin. Similar spectral changes but smaller then those of Lys 46 derivatives were observed for neurotoxins having dansyl groups at Lys 25 or Lys 26 (Table 1). Binding of the NT-II derivatives dansylated at Leu I or Lys 15, to AchR can be detected by the energy transfer, however, in case of λex = 330 nm emission spectra were not changed, i.e. this AchR derivative does not affect the dansyl group microenvironment. Since in case of Lys 46, Lys 26 and Lys 25 derivatives the experimentally determined values of fluorescence polarization are lower than those calculated according to the Perrin equation, one can conclude that even coming into contact with AchR the dansyl groups enter into spacious pockets and retain a considerable intrinsic mobility.

At titration of AchR with various concentrations of dansylated neurotoxins the 2:1 stoichiometry of the neurotoxin-AchR complexes and their dissociation constants were determined (19, 21). By displacement of dansylneurotoxins the dissociation constants for native NT-II and monospin labeled derivatives (see below) were found. The obtained results showed that the NT-II monoderivatives having sufficiently high toxicity and preserving the conformation of the original neurotoxin (20, 21, 23) are suitable objects for studying the neurotoxin-receptor topography.

The interaction of a series of the NT-II monoderivatives (spin labels at Leu 1, Glu 2, Lys 15, Lys 25, Lys 26,His 31, Lys 44 and Lys 46) as well as two spin labeled derivatives of the long type neurotoxins with AchR were investigated by EPR. The signals of the latter compounds in the EPR spectra were shown to undergo some changes at the binding to AchR(Fig. 3).

The analysis of the label motion (independent on the movement of the neurotoxin or its complex with AchR) is quite complex, however, the presence of a large series of derivatives of the same protein permits to come to some qualitative conclusions: labels in some positions (e.g. Lys 44) are not restricted in their motion at the complex formation, whereas for others (e.g. Gln 2, Lys 26 or His 31) these restrictions are clearly defined. However, even these restrictions preserve the difinite freedom of the label motion. That follows from the comparison of their correlation times with those for the toxin-AchR complex (19, 21). (This conclusion is in accordance with the above mentioned data on fluorescence polarization).

By means of the method of paramagnetic probes (24) we followed the changes in accessibility of spin labels to the outer medium at formation of the complexes with AchR by spin-labeled toxins. Table 2 shows that depending on the spin label position in the neurotoxin molecule and the charge of the

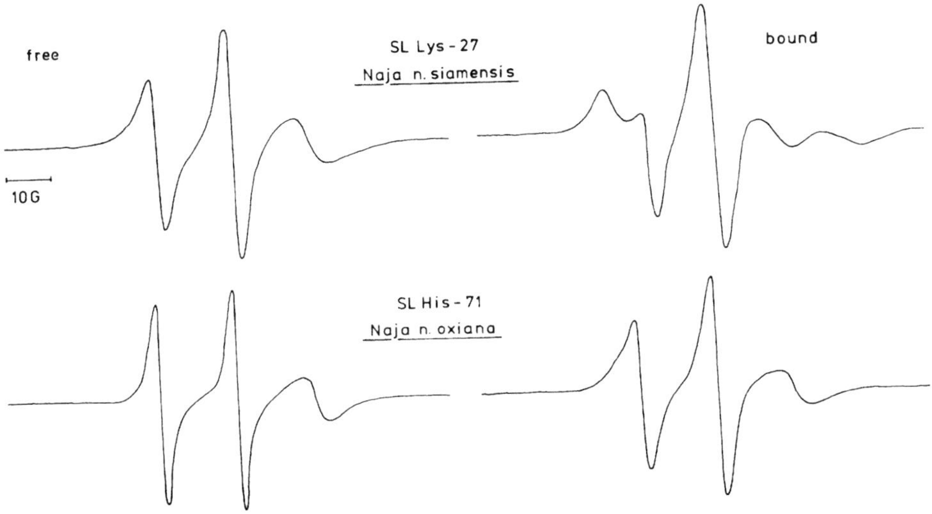

Figure 3. EPR spectra of spin-labeled neurotoxins (numeration of residues according to the homologous series (18)) AchR-$3\cdot10^{-6}$M, Triton X-100 - 0,025%, pH7, 4 , 21°C.

Table 2. Spin label accessibility to paramagnetic probes

Labeled residue	Paramagnetic probe			
	$K_3Fe(CN)_6$ $k_w/2k_p^*$		$Ni(CH_3COO)_2$ $k_w/2k_p$	
	free	bound	free	bound
Leu I	1,5	2,3	1,2	7,5
Glu 2	0,5	>4	1,3	1,7
Lys 15	1,3	4,4	0,9	1,1
Lys 25	2,3	4,2	1,5	2,8
Lys 26	2,9	>10	1,5	6,2
His 31	0,5	>10	1,0	2,4
Lys 44	1,0	1,2	1,2	0,8
Lys 46	1,1	10	0,9	7,0
His 71	0,7	0,6	1,6	1,2
Lys 27	0,8	4,6	3,3	>10

k_w is the rate constant of spin exchange between the probe and the free spin label in solution, k_p is the rate constant of spin exchange between the probe and the spin label bound to the protein (24).

paramagnetic probe the label accessibility changes in different ways: accessibility of the label at Lys 44 remains practically the same and the label at Lys 26 is screened from both probes. Fig.4 shows the scheme for the topography of the AchR-neurotoxin-II binding obtained from EPR and fluorescence data.

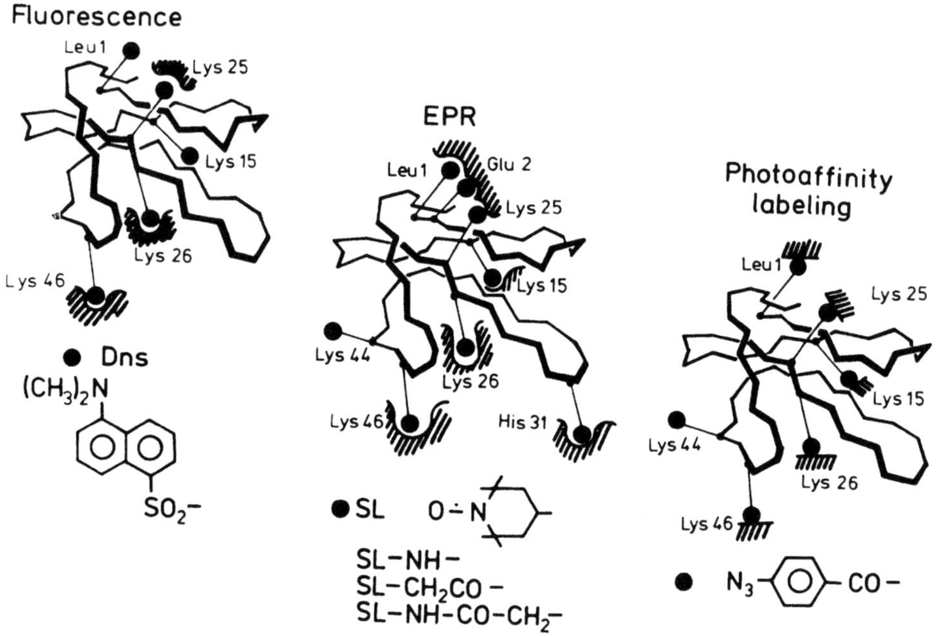

Figure 4. Topography of the AchR-neurotoxin binding sites. A circle near each residue corresponds to the relative NT-II monoderivative. The regions near the labels, which undergo the direct AchR influence, are hatched.

EPR data for NT-II and long neurotoxins are in good accordance (binding of the label at Lys 27 and no label binding in the C-terminal fragment located at the opposite side of the central loop (25)).

For molecule positions Lys 26 and Lys 46 EPR and fluorescence data confirm the contacting of the corresponding labels with AchR. On the other hand AchR does not affect the dansyl group at Leu 1, however, for the spin label at this position one can observe the limited lability and decreasing accessibility to Ni^{+2} ions. The data on the AchR interaction with the photoactivable NT-II derivatives shed light on this apparent descrepancy.

All 6 p-azidobenzoyl derivatives of NT-II are effectively

bound to AchR in the dark (22) that follows from their ability to compete with (^3H)-acetylated NT-II or toxin 3 from *Naja naja siamensis*. The similar way was used while studying formation of photoinduced cross-linkings with AchR. Control tests showed that under the chosen conditions spectral characteristics of the receptor are not substantially changed and its activity does not decrease.
The derivative labeled at Lys 44 does not form photoinduced cross-linkings with AchR, whereas all other derivatives (labels at Leu I, Lys 15, Lys 25, Lys 26 and Lys 46) form covalent complexes with AchR of extremely high efficacy (70-80%)(22). Thus, the obtained data are in good accordance with the EPR results (Fig.4). Apparently the regions of multiple contacts are located between AchR and neurotoxin that is a result of interactions of spatially proximated neurotoxin groupings with AchR (e.g. Lys 26, His 31 and Lys 46). AchR affects the label incorporated at one of these groupings due to undisturbed neighbouring contacts. On the other hand single contacts also exist which are probably of less functional role as in case of Leu 1 and Lys 15 residues: labels at these positions can contact AchR surface or lie outside depending on their nature.
The complementary data on photoaffinity modification and spectroscopic approaches show that a considerable portion of the neurotoxin surface - fragments of at least three disulfide-confined loops are involved in the AchR highly specific binding. Besides, we succeeded in obtaining new data on AchR site where neurotoxins are bound. Spectral investigations revealed the presence of tryptophan residues as well as positively and negatively charged groups in this active site (21).Judging from conformational rigidity of NT-II (stability to pH changes, medium polarity or modification of amino groups (23)) we concluded that at the binding of the neurotoxin to AchR it preserves its spatial structure and at the first approximation it can be considered as a mould of the corresponding AchR site. Thus the latter should be sufficiently extensive. Taking into account the distances between binding groups of the toxin (20, 21) and the diameter of the AchR molecule on the extracellular site of the membrane (where neurotoxins are bound) (13-17) it is evident that two neurotoxin molecules binding to AchR cannot be accomodated within two α-subunits but should contact neighbouring subunits. Really long neurotoxins were observed to be cross-linked with several subunits (26 - 29), however the contacting site of the neurotoxin was not identified in any case. Establishment of the subunits at which each label is bound in different regions of the neurotoxin surface (for this purpose we obtained radioactive photoactivable NT-II derivatives) permits not only to determine its orientation in the AchR active site but also to choose among the known variants (13,14,16,17) of mutual arrangement of subunits. In turn it is

a necessary stage in elucidating the topography of the receptor and AchR channel sites.

Modern biophysics pays a special attention to studying light transformation by photosensitive bacteria and photoreceptor cells. In spite of the considerable success recently achieved in understanding the primary photochemical transformations in bacteriorhodopsin and visual rhodopsin many problems concerning their functioning are still unclear (30,31) Bacteriorhodopsin is one of the best studied membrane proteins capable of active transport of hydrogen ions across purple membranes of halophilic microorganisms. Visual rhodopsin, like bacteriorhodopsin, generates a large electrical potential utilizing the energy of the absorbed light quanta (32). However, the role of the electric potential in visual excitation is still unknown in detail.

Henderson and Unwin succeeded in determining the molecular architecture of bacteriorhodopsin with 7Å resolution (33) by combination of electron microscopy and diffraction methods. According to them a protein molecule consists of 7 α-helical rods spanning the entire width of the membrane perpendicularly to its surface. This model (resolution 7Å) is a starting point to develop more detailed study of the three-dimensional structure of bacteriorhodopsin. However, to elucidate the groups which participate in functioning of the protein providing the transfer of hydrogen ions across the cell membrane, one should know the protein structure and its polypeptide chain arrangement in the membrane.

Thus, determination of the bacteriorhodopsin primary structure became the next step in the study of its three-dimensional structure and mechanism of action.

That appeared to be quite a difficult task due to the extreme hydrophobicity of the protein itself and its fragments formed as a result of the molecule cleavage by various chemical methods. Structural analysis of bacteriorhodopsin was based on studying the peptide fragments cleaved off from the protein molecule at methionine, tryptophan, tyrosine and aspartic acid residues. In the course of the peptide structure study use was made of the methods of the automated sequence analysis (34). The amino acid sequence of bacteriorhodopsin was established, it was shown that the protein molecule consists of 248 amino acid residues (35-37), but does not contain cystine, cysteine and histidine residues. Analyzing the primary structure of bacteriorhodopsin clearly defined hydrophobic regions of the polypeptide chain were found. These are sequences 38-41, 102-107, 157-166, 231-248 that compose about 7% of all amino acids in the protein molecule. Hydrophilic sites of the protein are supposed to be located on the membrane surface and some of them link α-helical rods. That was confirmed when studied the action of proteolytic enzymes on bacteriorhodopsin integrated in the native purple membrane (37).

Such an approach allowed us to show unequivocally that the above mentioned hydrophilic regions of the protein are accessible to proteolytic enzymes and consequently are situated near the membrane surface as connections of α-helical rods. On the basis of these results as well as the Henderson data (31) the model of the bacteriorhodopsin molecule organization in the native purple membrane was proposed (Fig. 5).

According to this model the polypeptide chain of the protein is arranged in the membrane as 7 α-helical rods, each of them consists on the average of 28-30 amino acid residues. The rods are linked with peptide fragments of 6-8 amino acid residues. The N- and C-terminal residues of the protein are located on the opposite sites of the membrane. The majority of the charged residues (lysine, arginine, glutamic acid, aspartic acid) are situated near the membrane surface where they contact an aqueous medium. The clear assymetry is observed in location of these residues on the outer and inner surfaces of the membrane. The considerable amount of these amino acids (about 20 residues) is located on the cytoplasmic membrane surface and only 6 residues are situated on the outer surface. It is also noteworthy that some charged protein groups are at quite long distances from the membrane surface (Arg 32, Asp 35, Asp 115, Lys 172, Arg 175, Asp 211) and formation of ion paires can be a factor stabilizing the rigid protein conformation in the membrane. Besides these these residues along with OH groups of Ser, Tyr, Trp can be a part of an ion-conducting pathway from the cytoplasmic space to an aldimine bond, formed by Lys 216 and retinal (38, 39) and further to the outer membrane surface. The sufficient number of hydrophilic amino acids in the membrane moiety of bacteriorhodopsin confirm the above assumption and their preferred localization on the inner sites of α-helical rods (40).

The structural analysis of bacteriorhodopsin stimulated studying of another photosensitive membrane protein - visual rhodopsin. According to the experimental data lately accumulated we concluded that a rhodopsin molecule fulfills the role of a modulator of the cyclic nucleotide level (41).

To determine the amino acid sequence of visual rhodopsin and to elucidate its polypeptide chain structural organization in the photoreceptor membrane we used the approach that had been applied for the bacteriorhodopsin study (34). As in case of bacteriorhodopsin the data on cyanogen bromide cleavage at methionine residues provided the main structural information concerning visual rhodopsin. The complete amino acid sequence was reconstructed by the analysis of overlapped methionine-containing peptides, obtained after cleavage of the protein with trypsin, chymotrypsin and protease from *Staphylococcus aureus* (42).

The analysis of the C-terminal thermolytic fragment of the

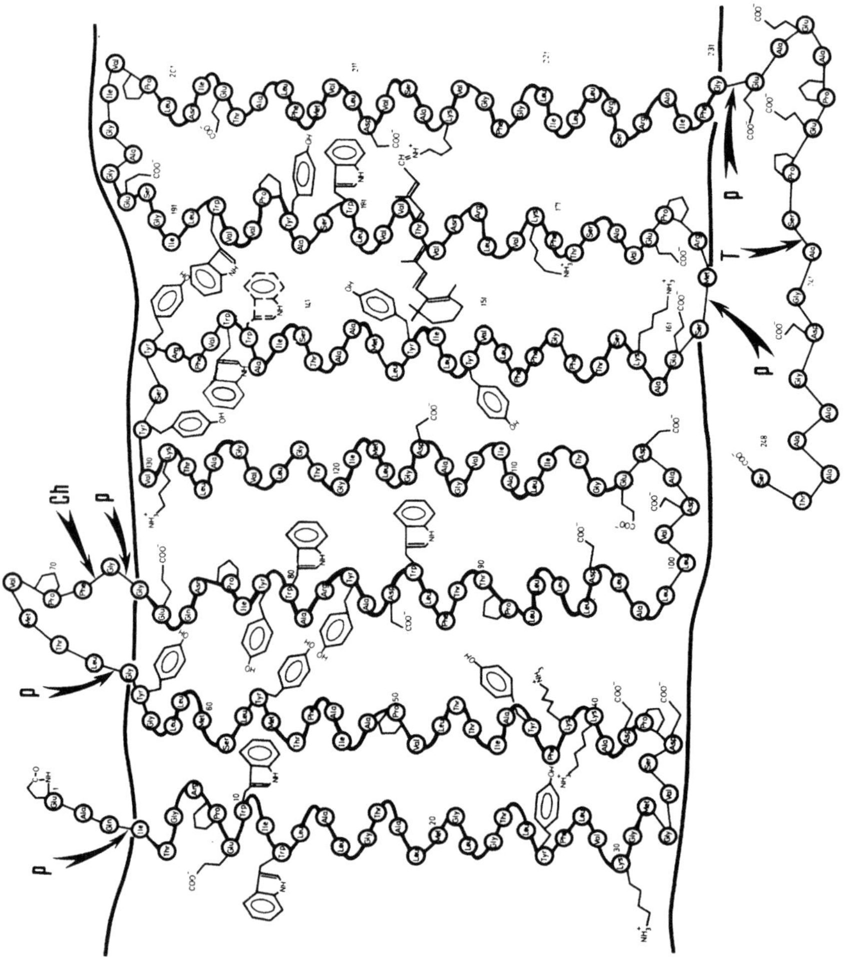

Figure 5. Packing of bacteriorhodopsin molecule in the purple membrane.

protein (240-326) with covalently linked retinal showed that the ε-amino group of Lys-295 is the site of chromophore linking.

The primary structure of visual rhodopsin is characterized by availability of some quite long hydrophobic sites broken off by short hydrophilic fragments in the polypeptide chain. To localize hydrophilic segments in the native molecule the photoreceptor membranes were treated with proteolytic enzymes. The action of thermolysine, papain and protease from *Staphylococcus aureus* results in two membrane-bound fragments, F_1 and F_2, of molecular weights 25 and 12 kD, respectively. Preparations of the membrane treated with these enzymes preserve spectral characteristics typical of the native membranes. Fragment F_1 is the N-terminal part of the rhodopsin molecule and contains covalently bound polysaccharide chains, whereas fragment F_2 belongs to the C-terminal region of the protein molecule and includes Lys 295 responsible for chromophore binding. The noncovalent complex of two fragments dissociated only upon illumination. Determination of the amino acid sequence of fragment F_2 showed that all the used enzymes attacked the protein molecule at the same site of the polypeptide chain located on the cytoplasmic membrane surface: bonds 240-241 (papain), 239-240 (thermolysine) and 238-239 (protease from *Staphylococcus aureus*). One should note that formation of fragments F_1 and F_2 is accompanied with the cleavage of the short C-terminal fragment (325-347) and the peptide (235-240) from the protein.

Treatment of native membranes with chymotrypsin resulted in cleavage of another bond (146-147) in the rhodopsin molecule.

Thus, protein sites (240-241), (239-240), (238-239) and (146-147) appeared to be accessible to the action of protease from the cytoplasmic surface of photoreceptor membranes.

To discover the protein sites in the intradisc space native membranes were subjected to freezing and thawing and the obtained preparations were treated with chymotrypsin. Then peptides (1-10), (11-13). (14-24), (25-30) which belong to the N-terminal fragment located in the intradisc space were found in a supernatant. At the papain action on inside-out membranes another bond (186-187) was cleaved.

The analysis of the action of proteases on the photoreceptor disc membranes of native and inside-out orientation showed that at least 30 amino acid residues from the N-terminus of the protein and bond (186-187) are located in the intradisc space and protein sites (146-147), (238-239), (239-240), (240-241) are on the outer surface of the membrane.

The membrane moiety exposed into the membrane contains 210 amino acid residues that follows from the action of proteolytic enzymes on the visual rhodopsin molecule integrated in the native photoreceptor membrane.

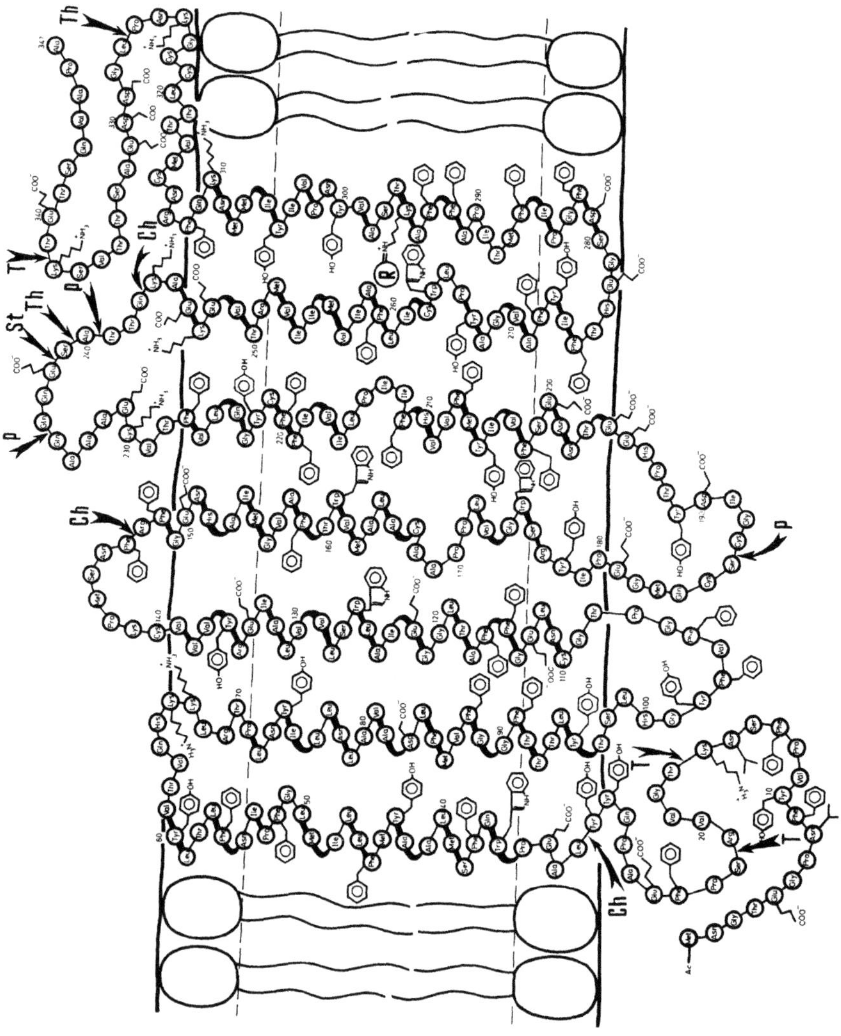

Figure 6. Packing of visual rhodopsin molecule in the photoreceptor membrane.

The analysis of distribution of hydrophobic and hydrophilic amino acids along the protein chain and proteolysis allowed us to assume that the polypeptide chain of rhodopsin is packed in the membrane as 7 α-helical rods spanning its width (Fig. 6).
Possible similarity of the packing models for rhodopsin and bacteriorhodopsin polypeptide chains most likely resembles a general principle of the structural organization of these two membrane proteins participating in cell transport processes.
Recent progress in the studies of membrane proteins, which can be phenomenologically considered as ion channels, have been briefly discussed in the paper. These proteins differ greatly in their molecular weights (∿25 kD for bacteriorhodopsin and ∿250 kD for acetylcholine receptor) and structural organization.
Bacteriorhodopsin is the best characterized protein: its primary structure and polypeptide chain arrangement in the membrane are known, by means of diffraction method its spatial organization has been elucidated in general, the data on the role of amino acid residues of the protein and structural elements of the chromophore are available. Apparently, in the future there will be a possibility to compare the hypotheses concerning the proton transport via the groupings which occupy a definite position in the protein primary structure and are located in a definite membrane region. The approaches, which appeared to be fruitful in case of bacteriorhodopsin, could be useful upon the study of complicated subunit components - ion channels of chemo- and electrically excitable membranes.

References

1. Yu.A.Ovchinnikov, V.T.Ivanov, and A.M.Shkrob, Membrane Active Complexones, BBA Library (Elsevier, Amsterdam, 1974), Vol.12.
2. Yu.A.Ovchinnikov, Eur.J.Biochem.94, 321(1979).
3. Yu.A.Ovchinnikov, Int.J.Quantum Chem.20, 614 (1981).
4. B.Hille, Biophys.J.22, 283 (1978).
5. T.Narahashi, Physiol.Rev. 54, 813 (1974).
6. V.A.Kovalenko, V.N.Pashkov, E.V.Grishin, Yu.A.Ovchinnikov, V.P.Shevchenko,and N.F.Myasoedov, Bioorg.Khim. (USSR) 8, 719 (1982).
7. V.A.Kovalenko, V.N.Pashkov, and E.V.Grishin, Bioorg.Khim. (USSR) 7, 1828 (1981).
8. E.V.Grishin, Int.J.Quantum Chem. 19, 291 (1981).
9. G.N.Mozhayeva, A.P.Naumov, E.D.Nosyreva, and E.V.Grishin, Biochim.Biophys.Acta 597, 587 (1980).
10. E.V.Grishin, N.M.Soldatov, Yu.A.Ovchinnikov, G.N.Mozhaeva, A.P.Naumov, A.N.Zubov, and B.Ch.Nisman, Bioorg.Khim.

(USSR), 6, 724 (1980).
11. E.V.Grishin, N.M.Soldatov, and Yu.A.Ovchinnikov, Bioorg. Khim. (USSR) 6, 914 (1980).
12. T.Heidmann, and J.-P.Changeux, Ann.Rev.Biochem. 47, 317 (1978).
13. A.Karlin in the Cell Surface and Neuronal Function, C.W. Cotman, G.Paste, G.L.Nicolson, Eds., (Elsevier/North Holland, New York 1979), p.191.
14. E.A.Barnard, J.Lamprecht, M.Lo, E.Nockes, K.Simikawa, J.Cavanagh, and J.O.Dolly, in Synaptic Constituents in Health and Disease, M.Brzin, D.Sket, and H.Bachelard, Eds. (Mladinska Knjiga-Pergamon, Ljubljana, 1980), p.224.
15. M.A.Raftery, M.W.Hunkapiller, C.D.Strader, and L.E.Hood, Science 208, 1454 (1981).
16. F.Hucho, Trends Biochem.Sci. 6, 242 (1981).
17. J.Kistler, R.M.Stroud, M.W.Klymkowsky, R.A.Lalancette, and R.H.Fairclough, Biophys.J. 37, 371 (1982).
18. E.Karlsson, in Handbook of Experimantal Pharmacology, C.Y.Lee, Ed., (Springer-Verlag, Berlin, 1979), p.159.
19. V.I.Tsetlin, E.Karlsson, A.S.Arseniev, Yu.N.Utkin, A.M. Surin, V.S.Pashkov, K.A.Pluzhnikov, V.T.Ivanov, V.F. Bystrov, and Yu.A.Ovchinnikov, FEBS Lett. 106, 47 (1979).
20. V.T.Ivanov, V.I.Tsetlin, E.Karlsson, A.S.Arseniev, Yu.N. Utkin, V.S.Pashkov, A.M.Surin, K.A.Pluzhnikov, and V.F. Bystrov, in Natural Toxins, D.Eaker, T.Wadstrom, Eds. (Pergamon Press, Oxford and New York, 1980), p.523.
21. V.I.Tsetlin, E.Karlsson, Yu.N.Utkin, K.A.Pluzhnikov, A.M.Surin, V.I.Kondakov, V.F.Bystrov, V.T.Ivanov, and Yu.A.Ovchinnikov, Toxikon 20, 83 (1982).
22. K.A.Pluzhnikov, A.A.Karelin, Yu.N.Utkin, V.I.Tsetlin, and V.T.Ivanov, Bioorg.Khim. (USSR) 8, 850 (1982).
23. V.I.Tsetlin, A.S.Arseniev, Yu.N.Utkin, A.Z.Gurevich, L.B.Senyavina, V.F.Bystrov, V.T.Ivanov, and Yu.A.Ovchinnikov, Eur.J.Biochem. 94, 337 (1979).
24. G.I.Likhtenstein, Spin Labeling Methods in Molecular Biology (John Wiley, New York, 1976).
25. M.D.Walkinshaw, W.Saenger, and A.Maelicke, Proc.Natl. Acad.Sci. USA 77, 2400 (1980).
26. V.Witzeman, and M.A.Raftery, Biochem.Biophys.Res.Commun. 85, 623 (1978).
27. F.Hucho, FEBS Lett. 103, 27 (1979).
28. N.Natanson, and Z.Hall, J.Biol.Chem. 255, 1698 (1980).
29. R.E.Oswald, and J.-P.Changeux, FEBS Lett. 139, 225, (1982).
30. W.Stoeckenius,Acc.Chem.Res. 10, 337 (1980).
31. R.Uhl, E.W.Abrahamson, Chem.Rev. 81, 291 (1981).
32. V.I.Bolshakov, G.R.Kalamkarov, M.A.Ostrovsky, Dokl.Acad. Nauk SSSR 248, 1485 (1979).
33. R.Henderson, P.N.T.Unwin, Nature 257, 28 (1975).
34. N.G.Abdulaev, Yu.A.Ovchinnikov, Methods in Enzymol. 88, 723 (1982).

35. Yu.A.Ovchinnikov, N.G.Abdulaev, M.Yu.Feigina, A.V.Kiselev, N.A.Lobanov, FEBS Lett. 100, 219 (1979).
36. G.H.Khorana, G.E.Gerber, W.C.Herlihy, C.P.Gray, R.I. Anderegg, K.Nihei, K.Biemann, Proc.Natl.Acad.Sci. USA 76, 5046 (1979).
37. Yu.A.Ovchinnikov, N.G.Abdulaev, M.Yu.Feigina, I.D.Artamonov, A.S.Zolotarev, M.B.Kostina, A.S.Bogochuk, A.I. Miroshnikov, V.I.Martinov, A.B.Kudelin, FEBS Lett.(1982) in Press.
38. H.D.Lemke, D.Osterheld, FEBS Lett. 128, 255 (1981).
39. E.Miller, A.H.Johnson,M.Akhtar,FEBS Lett. 130, 187 (1981).
40. D.M.Engelman, G.Zaccai, Proc.Natl.Acad.Sci. USA 77, 5894 (1980).
41. W.H.Miller, C.D.Nicol, Nature 280, 64 (1979).
42. Yu.A.Ovchinnikov, N.G.Abdulaev, M.Yu.Feigina, I.D.Artamonov, A.S.Zolotarev, A.S.Bogochuk, M.B.Kostina, V.I.Martinov, A.I.Miroshnikov, A.B.Kudelin, Bioorg.Khim. (1982) in Press.

EVOLUTION OF MOLECULES

Peter SCHUSTER
Institut für Theoretische Chemie und Strahlenchemie
Universität Wien, Währingerstraße 17, A 1090 Wien

Capability of self-replication is a unique feature of polynucleotides. Their properties represent the ultimate molecular basis of biochemical and biological evolution. The kinetics of self-replication can be approached theoretically by means of ordinary differential equations (O.D.E.). Depending on the nature of the mechanism of replication we are dealing with O.D.E.'s which have different properties. Template induced replication leads to competition between polynucleotide sequences. Stable mutant distributions called "quasispecies" are formed, when the mechanism of replication is accurate enough. The two most important features of Darwinian evolution variability through mutations and optimization of the replication rate through selection can be observed at the molecular level already. By means of higher order terms competition can be suppressed provided some conditions are fulfilled. Then, the replicating elements cooperate. A rigorous proof has been presented which relates cooperation to the formation of a closed loop of catalytic interactions. Such a closed positive feedback loop has been called a "hypercycle". Cooperation of replicating elements allows to develop new properties of the system. These new properties are established on the expenses of global optimization. Second order catalysis hinders the system to approach the global optimum of the rate of replication.

1. INTRODUCTION

Polynucleotides, RNA or DNA, are the only known molecules which have an intrinsic capability of self-replication. This property is a direct consequence of their molecular structure which is characterized by two basic and indispensible features:
(1) the complementary of bases in Watson-Crick base pairs, A=U(T) and G≡C, and
(2) the possibility to form double-helices which have a uniform, repetitive backbone structure.
Every base in this structure is uniquely defined by that in the

P.-O. Löwdin and B. Pullman (eds.), New Horizons of Quantum Chemistry, 417–437.

opposite or vis-a-vis position. Thus, there is a uniquely defined negative for every given polynucleotide.

The molecular geometry of the famous double helix provides the basis for complementary and for direct replication. Complementary replication starts from a given polynucleotide, the plus strand. A negative copy is synthesized sequentially, i.e. step by step or base by base from the 5' to the 3' end. The negative strand acts as a template for the synthesis of a new plus strand molecule. Complementary replication is an example for an autocatalytic process with a unique intermediate, the minus strand. Examples of complementary replication are viral RNA synthesis, the mechanism of which has been studied extensively (Biebricher et al., 1981 and 1982) as well as template induced, enzyme free oligo-ribonucleotide synthesis (Lohrmann et al., 1980).

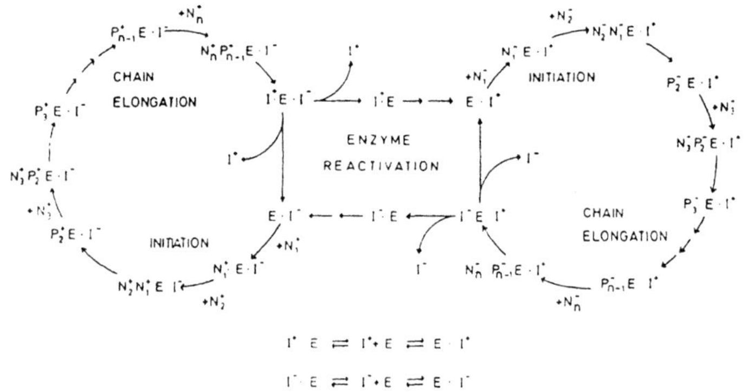

Figure 1. A cyclic mechanism for bacteriophage RNA replication in vitro as used by Biebricher et al. (1981) to interpret the kinetic results obtained with the Qß system. E, I^+ and I^- represent free Qß polymerase and the plus and minus strand of the RNA to be replicated. $N_1^+, N_2^+, \dots, N_n^+$ and $N_1^-, N_2^-, \dots, N_n^-$ are the nucleoside triphosphates in the sequence as they appear in the plus and minus strands respectively. The chain of the newly synthesized molecule grows always from the 5'- to the 3'-end. $P_2^+, P_3^+, \dots, P_{n-1}^+$ and $P_2^-, P_3^-, \dots, P_{n-1}^-$ are used as symbols for the growing chains. A nucleation length of two bases is assumed. Enzyme reactivation is necessary because the two polynucleotides are not in suitable positions to restart polymerization: the newly synthesized strand is bound with the correct end (3') but in the wrong site (synthetic site) whereas the old strand is sitting in the correct (reading) site but bound with the wrong end (5'). The many step mechanism thus consists of three distinguishable processes: initiation of RNA synthesis chain propagation and reactivation of the enzyme.

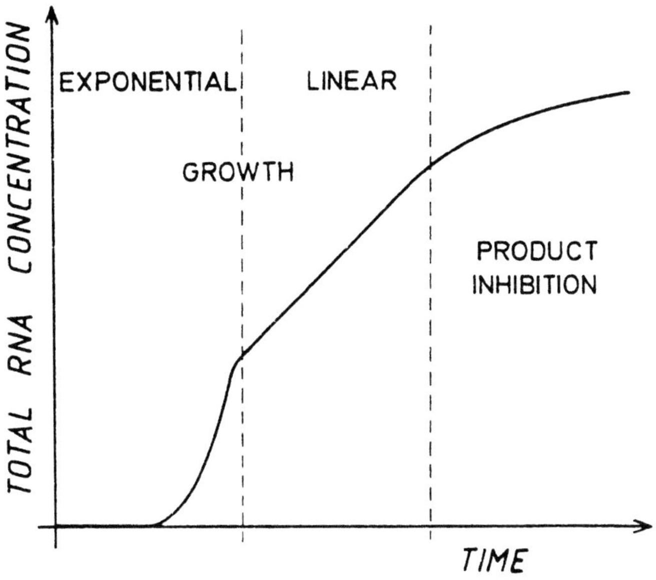

Figure 2. RNA synthesis in the test tube. A characteristic experimentally recorded curve shows three phases of growth: exponential growth at low template concentration, linear growth above saturation of the enzyme by polynucleotides and further levelling off at still higher concentrations of template when enzyme reactivation becomes the rate determing step.

Direct replication occurs with DNA. The molecular mechanism is complicated and involves about ten different enzymes. The local topology of the process is summarized in the so-called replication fork (Kornberg, 1980). Despite the highly complex molecular mechanism RNA and DNA replication obeys the law of simple autocatalysis under suitable conditions (Biebricher et al., 1981). An example is shown below.
Chemical kinetics of ordinary, i.e. non-autocatalytic, chemical reactions has been investigated thoroughly by various techniques. In particular we mention the formal mathematical approach by Feinberg (1977) and Horn. Although, their basic formalism is very general, the important theorems are only valid for non-autocatalytic reaction steps. Despite various efforts to develop a fully general theory of chemical kinetics - as one of the most promising approaches we mention the work by Clarke (1980) - no really satisfactory and generally enough treatment of autocatalysis has been presented yet. The reason for this lack is easy to understand: autocatalytic processes lead to highly non-linear differential equations and these are hard to analyze by the conventional techniques. In this contribution we shall try to describe an approach towards autocatalysis and self-replication which is characterized by the demand of smooth extensibility towards biochemical and biological evolution.

2. AUTOCATALYSIS, COMPETITION AND SELECTION

The basic process we consider is a simple bimolecular autocatalytic reaction

$$A + B \underset{f_2}{\overset{f_1}{\rightleftharpoons}} 2\,A \qquad (1)$$

Such a reaction has two characteristic phases: a phase of exponential growth far off equilibrium and a phase of relaxation towards equilibrium:

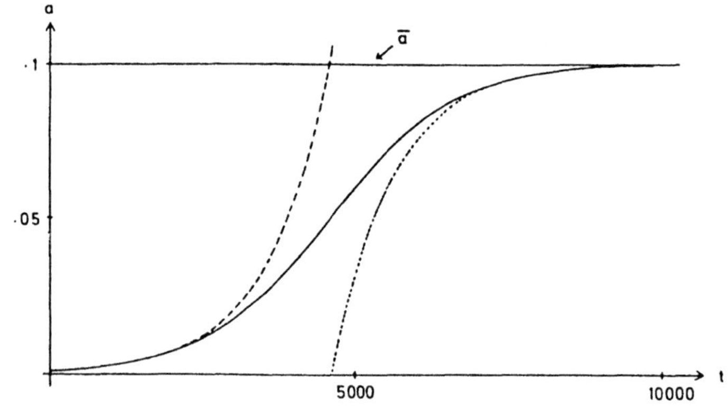

Figure 3. A characteristic solution curve of equation (1) at buffered concentration of B, $b=b_o=\bar{b}$. The concentration of the autocatalyst is denoted by a . $\bar{a}$ is the equilibrium concentration. Pure exponential growth and relaxation to equilibrium are shown as broken lines. The constants chosen are $f_1=1[t^{-1}c^{-1}]$, $f_2=0.01[t^{-1}c^{-1}]$ and $\bar{b}=0.001[c]$ in arbitrary time $[t]$ and concentration $[c]$ units.

In order to keep the process within the interesting non-linear range, i.e. the range of exponential growth we have to introduce some coupling to an external source. There are different ways to approach this goal:
(1) periodic renewal of the material consumed by serial transfer experiments,
(2) continuous renewal of material in flow reactors and
(3) coupling of the autocatalytic reaction to an energy driven recycling process. This is what actually happens in the biosphere, where almost all energy comes from photochemistry of sunlight and is used via photosynthesis of plants.

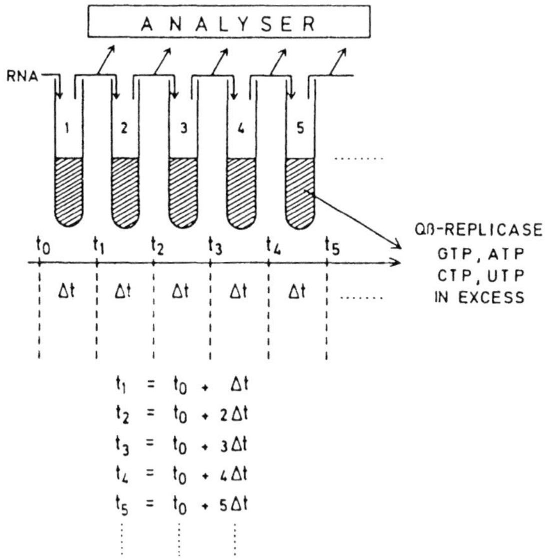

Figure 4. The technique of serial transfer experiments (Spiegelmann 1971). RNA, in particular RNA of the bacteriophage Qß, grows in a medium which contains the enzyme Qß replicase as well as the four nucleoside triphosphates ATP, UTP, GTP and CTP in excess. After a time Δt a sample is taken out of the test tube. Part of it is analyzed, part of it is transferred into a new medium. The procedure is repeated after time intervals of Δt. The conditions are chosen such that the RNA is growing in the exponential phase.

We consider the third case at first and analyze the three-step mechanism (2). For the sake of simplicity we assume all reactions to be irreversible - this is essential only for the recycling reaction (2c); g(E) indicates that this rate is under external control (Schuster, 1981b):

$$A + B \xrightarrow{f} 2A \qquad (2a)$$

$$A \xrightarrow{d} C \qquad (2b)$$

$$C \xrightarrow{g(E)} B \qquad (2c)$$

Since there is no material flow into the system, we have the conservation law: $a+b+c=c_0$. We denote concentrations by small letters: $[A] = a$, $[B] = b$ etc.

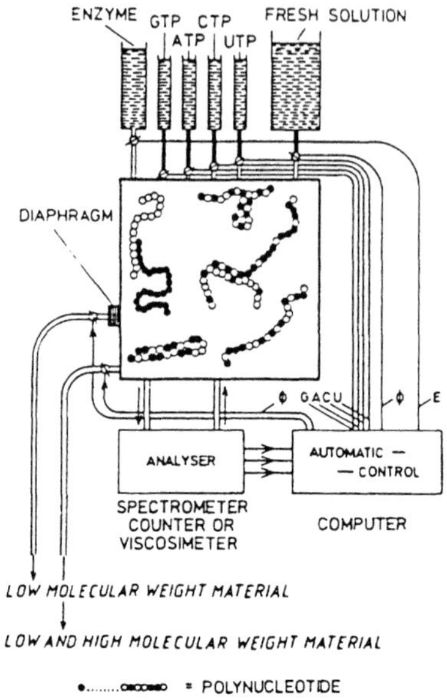

Figure 5. The dialysis reactor for evolution experiments. This kind of flow reactor consists of a reaction vessels which allows for temperature and pressure control. Its walls are impermeable to polynucleotides. Energy rich material is poured from the environment into the reactor. The degradation products are removed steadily. Material transport is adjusted in such a way that the concentrations of monomers is constant in the reactor. A dilution flux ϕ is installed in order to remove the excess of polynucleotides produced by multiplication. Thus the sum of the concentrations $[I_1] + [I_2] + \ldots + [I_N] = \Sigma x_i = c$ may be controlled by the flux ϕ. Under "constant organization" ϕ is adjusted such that the concentration $c=c_o$ is constant. The regulation of ϕ requires internal control which may be achieved by analysis of the solution and data processing by a computer as indicated above.

The steady states of the differential equation corresponding to the mechanism (2) can be obtained by straightforward analysis (Table 1). We obtain two solutions P_1 and P_2, stability of which follows from linearization. The steady state P_1 is asymptotically stable iff $c_o<d/f$ and P_2 is asymptotically stable iff $c_o>d/f$. The positions of P_1 and P_2 as functions of the total concentration are shown below.

Mass action kinetics applied to the kinetics of equation (2) leads to the differential equation

$$\dot{a} = (fb-d)a$$
$$\dot{b} = gc-f.b\ a$$
$$\dot{c} = da - gc,$$

which has only two degrees of freedom because mechanism (2) is restricted by mass conservation: $a+b+c=c_o$ or $\dot{a}+\dot{b}+\dot{c}=0$. The total concentration of the system is denoted by c_o. The conditions for steady states $\dot{a}=0$, $\dot{b}=0$ and $\dot{c}=0$ are fulfilled by two solutions, state (1)

$$\bar{P}_1:\ \bar{a}=0,\ \bar{b}=c_o,\ \bar{c}=0$$

and state (2)

$$\bar{P}_2:\ \bar{a}=\frac{g}{g+d}(c_o-\frac{d}{f}),\ \bar{b}=\frac{d}{f},\ \bar{c}=\frac{d}{g+d}(c_o-\frac{d}{f}).$$

Stability analysis can be performed by linearization. According to two degrees of freedom we obtain two eigenvalues of the Jacobian matrix w_1 and w_2. In particular we find

$0<c_o<\frac{d}{f}$:	$\bar{P}_1$, $w_1^{(1)}<0$	and	$w_2^{(1)}<0$
	$\bar{P}_2$, $w_1^{(2)}<0$	and	$w_2^{(2)}>0$
$\frac{d}{f}<c_o$:	$\bar{P}_1$, $w_1^{(1)}<0$	and	$w_2^{(1)}>0$
	$\bar{P}_2$, $w_1^{(2)}<0$	and	$w_2^{(2)}>0$

Hence, $\bar{P}_1$ is stable below and $\bar{P}_2$ above the critical concentration $c_{cr}=d/f$.

Table 1. Analysis of the steady states of mechanism (2)

Two aspects are relevant for our further considerations:
(1) the existence of a non-vanishing stationary concentration of the autocatalyst A requires a total concentration c_o which exceeds the critical threshold value $(c_o)_{cr}=d/f$ and
(2) above this critical concentration the autocatalytic reaction proceeds at the maximum rate. All B above threshold is converted into A and C, and hence, B remains at the threshold value. The autocatalytic process exhausts the available reservoir as far as possible.

Next, we consider two autocatalysts, I_1 and I_2, living on common resources. We model such a situation by the reaction mechanism (3):

$$B + I_1 \xrightarrow{f_1} 2I_1 \tag{3a}$$
$$B + I_2 \xrightarrow{f_2} 2I_2 \tag{3b}$$
$$I_1 \xrightarrow{d_1} C \tag{3c}$$
$$I_2 \xrightarrow{d_2} C \tag{3d}$$
$$C \xrightarrow{g(E)} B \tag{3e}$$

Again we have a mass conservation law: $x_1+x_2+b+c=c_o$ (From now on it is convenient to denote autocatalys s or polynucleotide sequences by I_i and their concentrations by $[I_i] = x_i$; $i=1,2,\ldots,n$).

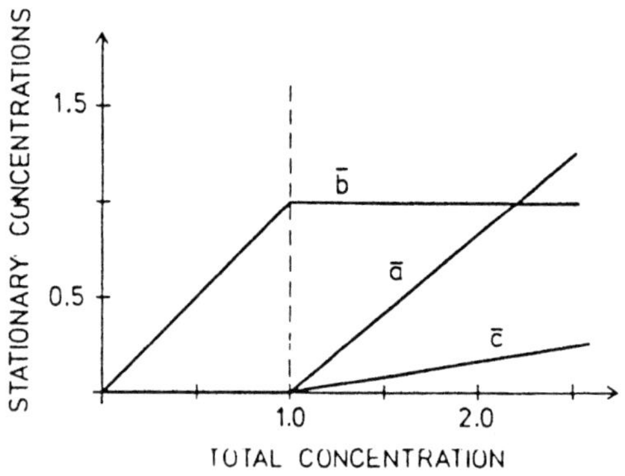

Figure 6. The staedy states of mechanism (2). We observe a simple type of bifurcation at the critical concentration $c_{cr}=d/f$. At concentrations $c_o<c_{cr}$, $\bar{P}_1$ is stable. Neither the autocatalyst A nor the degradation product C are present in non-zero concentrations. Above threshold the steady state $\bar{P}_2$ is stable at which all three reactants are present.

The analysis of mechanism (3) is straight forward (table 2).

We find three mutually exclusive steady states P_1, P_2 and P_3. By mutually exclusive we indicate that only one of the three states can be asymptotically stable for a given value of c_o. At P_1 both autocatalysts are at concentration zero. At P_2 I_1 is present and I_2 vanishes and at P_3 the inverse is true: only I_2 is present at non-zero concentration.

Mass action kinetics applied to the kinetics of mechanism (3) leads to the differential equation

$$\dot{x}_1 = (f_1 b - d_1)x_1$$
$$\dot{x}_2 = (f_2 b - d_2)x_2$$
$$\dot{b} = gc - (f_1 x_1 + f_2 x_2)b$$
$$\dot{c} = d_1 x_1 + d_2 x_2 - gc.$$

Again we have mass conservation $x_1+x_2+b+c=c_o$ or $\dot{x}_1+\dot{x}_2+\dot{b}+\dot{c}=0$ The differential equation thus has three degrees of freedom.

The conditions for the steady states $\dot{x}_1=0$, $\dot{x}_2=0$, $\dot{b}=0$ and $\dot{c}=0$ are fulfilled by three solutions, state (1)

$$\overline{P}_1:\ \overline{x}_1=0,\ \overline{x}_2=0,\ \overline{b}=c_o,\ \overline{c}=0,$$

state (2)

$$\overline{P}_2:\ \overline{x}_1 = \frac{g}{g+d_1}\left(c_o - \frac{d_1}{f_1}\right);\ \overline{x}_2=0,\ \overline{b} = \frac{d_1}{f_1},\ \overline{c} = \frac{d_1}{g+d_1}\left(c_o - \frac{d_1}{f_1}\right).$$

and state (3)

$$\overline{P}_3:\ \overline{x}_1< 0,\ \overline{x}_2= \frac{g}{g+d_2}\left(c_o - \frac{d_2}{f_2}\right),\ \overline{b}= \frac{d_2}{f_2},\ \overline{c}= \frac{d_2}{g+d_2}\left(c_o - \frac{d_2}{f_2}\right).$$

Stability analysis can be performed by linearization. The three eigenvalues of the Jacobian matrix for state (1) are

$$\overline{P}_1:\ w_1^{(1)}=f_1c_o-d,\quad w_2^{(1)}=f_2c_o-d_2 \quad \text{and} \quad w_3^{(1)}= -g.$$

Hence, $\overline{P}_1$ is stable in the range $0<c_o<\min(d_1/f_1, d_2/f_2)$. Above this critical concentration $\overline{P}_2$ or $\overline{P}_3$ is stable depending on the relative value of d_1/f_1 and d_2/f_2.

Table 2. Analysis of the steady states of mechanism (3).

In order to decide which steady state P_2 or P_3 is approached by the system we calculate the time derivative of the ratio of concentrations:

$$\left(\frac{x_1}{x_2}\right)' = \frac{\dot{x}_1 x_2 - x_1 \dot{x}_2}{x_2^2} = \left(\frac{x_1}{x_2}\right)\{(f_1 b - d_1) - (f_2 b - d_2)\} =$$

$$= \left(\frac{x_1}{x_2}\right)\{(f_1 - f_2)b - (d_1 - d_2)\} \tag{4}$$

There are two clear-cut situations:
(1) $f_1>f_2$ and $d_1 \leqslant d_2$: $(x_1/x_2)'>0$ provided $b>0$ and hence I_2 vanishes.

I_1 is selected.

(2) $f_1<f_2$ and $d_1\geq d_2$: $(x_1/x_2)'<0$ provided b>0 and the opposite occurs: I_1 vanishes and I_2 is selected.

The cases $f_1>f_2$ and $d_1>d_2$ as well as $f_1<f_2$ and $d_1<d_2$ are more tricky but selection does occur.

Apart from the kineticly degenerate case $f_1=f_2$ and $d_1=d_2$ we thus find that autocatalysts living on a common source compete and selection according to the rate constants of replication and decompositions takes place. One characteristic example is shown below.

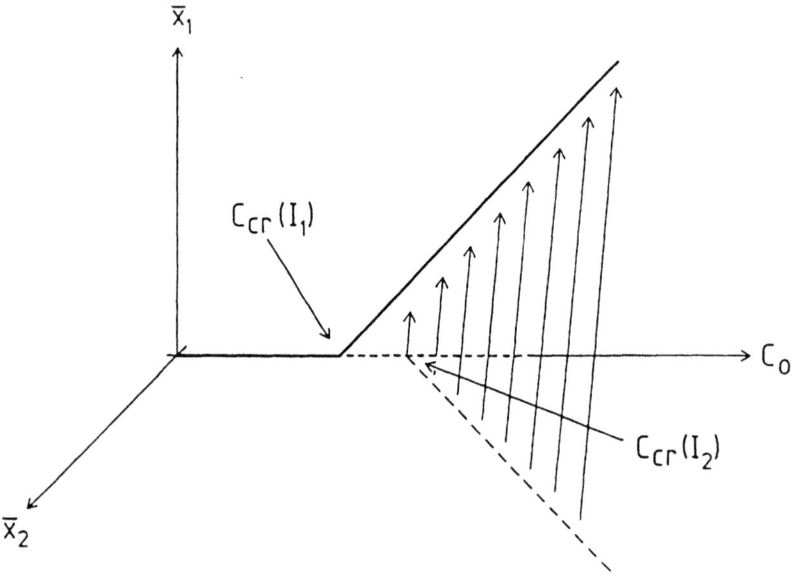

Figure 7. Steady states and bifurcation points in a system of competing autocatalysts as described by mechanism (3). We observe three steady states, the zero state $\bar{P}_1$ and two reactive states $\bar{P}_2$ and $\bar{P}_3$. Note, that the steady state which branches off the zero state first (here $\bar{P}_2$ at the critical concentration $c_o=c_{cr}(I_1)=d_1/f_1$) is always stable whereas the other reactive state (here $\bar{P}_3$) remains unstable. The stable reactive state corresponds to $x_2=0$. Hence the autocatalysts compete and selection of I_1 takes place.

3. THE REPLICATOR EQUATION

In this section we generalize our system of competing autocatalysts. We consider n replicating elements and appropriately use the expression "replicator" (Dawkins, 1976 and 1982) for it. In particular these replicators are polynucleotides. The n replicators live on a common source. This source sustains a certain total concentration of replicators

$$c(t) = \sum_i x_i ; \quad x_i = [I_i] ; \; i=1,\ldots,n \tag{5}$$

Thus, the function c(t) depends on the kinetics of replication and on some external program superimposed. We may express this by the following differential equation which is non-autonomous (Eigen and Schuster, 1979):

$$\dot{x}_i = \Gamma_i(\vec{x}) - \frac{x_i}{c}\phi \; ; \; i=1,\ldots,n; \; \vec{x}=(x_1,\ldots,x_n) \tag{6}$$

The growth function $\Gamma_i(\vec{x})$ describes the mechanism of replication in kinetic terms. According to the laws of mass action it can be developed into a power series in $\vec{x}$:

$$\Gamma_i(\vec{x}) = k^{(i)} + \sum_j k_j^{(i)} x_j + \sum\sum_{jk} k_{jk}^{(i)} x_j x_k + \ldots \tag{7}$$

Each term in (7) corresponds to a certain class of reaction: spontaneous formation is of zero order in $\vec{x}$, template induced replication is of first order, template induced and catalyzed reaction, finally, is of second order in $\vec{x}$.

The external program operating on the system is subsumed under the constraint ϕ. As we see from equation (6) this constraint is assumed to be unspecific: it operates on all replicators in the same way like a dilution flux in a flow reactor. There is a strict relation between total concentration and contraint:

$$\frac{dc}{dt} = \sum_i \dot{x}_i = \sum_i \Gamma_i(\vec{x}) - \phi \tag{8a}$$

or

$$c(t) = c(0) + \int_o^t \left(\sum_i \Gamma_i - \phi\right) d\tau \tag{8b}$$

For special constraints, called constant organization (Eigen, 1971), the total concentration becomes constant:

$$c = \text{CONST.} = c_o : \quad \phi = \sum_i \Gamma_i(\vec{x}) \tag{9}$$

This stationary state is stable provided

$$\left[\sum_i \Gamma_i(\vec{x})\right]_{c=c_o} > 0 \tag{10}$$

is fulfilled. Under constant organization equation (6) is of the much simpler form:

$$\dot{x}_i = \Gamma_i(\vec{x}) - \frac{x_i}{c_o} - \sum_j \Gamma_j(\vec{x}); \quad \sum_i x_i = c_o; \; i=1,\ldots,n \tag{11}$$

We shall call (11) the stationary replicator equation provided $\Gamma_i(\vec{x})$ refers to some replication kinetics.

In an often realized and important case the non-stationary and the stationary equation are related by a simple transformation. We introduce "internal coordinates" or relative concentrations:

$$\vec{\zeta} = \frac{1}{c}\vec{x} \quad \text{and hence} \quad \sum_i \zeta_i = 1 \tag{12}$$

Insertion of (12) into (6) leads to the equation

$$\dot{\zeta}_i = \frac{1}{c}\{\Gamma_i(\vec{x}) - \zeta_i \sum_j \Gamma_j(\vec{x})\}; \; i=1,\ldots,n \tag{13}$$

The transformation is not complete yet since we have a mixture of old and new variable in (13). It can be completed, however, in the important special case when the growth functions are homogeneous in $\vec{x}$:

$$\Gamma_i(\vec{x}) = \Gamma_i(c\vec{\zeta}) = c^{\lambda}\Gamma_i(\vec{\zeta}) \tag{14}$$

(λ herein is the degree of homogeneity). Combining (13) and (14) we find:

$$\dot{\zeta}_i = c^{\lambda-1}\{\Gamma_i(\vec{\zeta}) - \zeta_i \sum_j \Gamma_j(\vec{\zeta})\}; \; i=1,\ldots,n \tag{15}$$

Apart from a phase factor, $c^{\lambda-1}$, which is always positive and, hence, leaves the trajectories invariant, the nonstationary equation (6) formulated in relative concentrations is identical with the stationary equation for $c_o=1$. For homogeneous growth functions a knowledge of the stationary solutions is sufficient to describe the evolution of the system under any external constraint. The ultimate reason for this simple behaviour is to be seen in the fact that the phase portrait of equation (11) is independent of c_o in this case.

In the following three sections we shall consider always cases with homogeneous growth functions and therefore assume $\sum_i x_i = 1$ without loosing generality.

4. SELECTION AND OPTIMIZATION

We come back to our simple initial example of competing autocatalysts or replicators and consider it once again under the conditions of the stationary replicator equation. We put

$$\Gamma_i = (f_i - d_i)x_i = w_i x_i; \; i=1,\ldots,n \tag{16}$$

and obtain ($\sum_i x_i=1$):

$$\dot{x}_i = (w_i-\phi)x_i;\ i=1,\dots,n \tag{17}$$

The constraint or mean reproductive success of the replicators, ϕ , in this example is simply the ensemble average of the net production $w_i=f_i-d_i$:

$$\phi = \sum_i w_i x_i = E\{w\} \tag{18}$$

A straight forward calculation shows that again selection takes place: the replicator with the maximum net production is selected:

$$I_m:\ w_m = \max\{w_i;\ i=1,\dots,n\};\ \lim_{t\to\infty} x_m=1 \tag{19}$$

At the same time ϕ approaches a maximum value:

$$\frac{d\phi}{dt} = \sum_i w_i\dot{x}_i = \sum_i w_i^2 x_i-(\sum_i w_i x_i)^2 =$$

$$= E\{w^2\} -(E\{w\})^2 = E\{(w-E\{w\})^2\}\geqslant 0 \tag{20}$$

A concrete numerical example is shown below. What we observe actually is the basic mechanism of Darwin's natural selection. The prerequisite was competition between replicators growing at a first order autocatalytic growth law. In case a more efficient mutant enters the system it will instantaneously replace the previously selected species.

5. REPLICATION WITH ERRORS - THE QUASISPECIES

No physical process can occur with absolute accuracy. In reality we are always dealing with a certain frequency of copying errors. Copying from time to time gives rise to new sequences of polynucleotides. We call these erroneous copies mutants of the original sequence. Template induced replication with errors has been studied extensively in the literature (Eigen, 1971; Thompson and McBride, 1974; Jones et al.,1976; Eigen and Schuster, 1979; Epstein and Eigen, 1979; Küppers, 1980; Swetina and Schuster, 1982). Therefore we present here only the origin of the problem as well as the most relevant results. We admit errors in the replication process and therefore account for interconversions between different replicators through the replication process, e.g. I_i is not only synthesized through replication on its own template but appears also in an error copy of I_j:

$$I_i + \sum_\lambda \nu_{i\lambda}B_\lambda \xrightarrow{f_iQ_{ii}} 2\, I_i \tag{21a}$$

$$I_j + \sum_\lambda \nu_{i\lambda}B_\lambda \xrightarrow{f_jQ_{ij}} I_i+I_j \tag{21b}$$

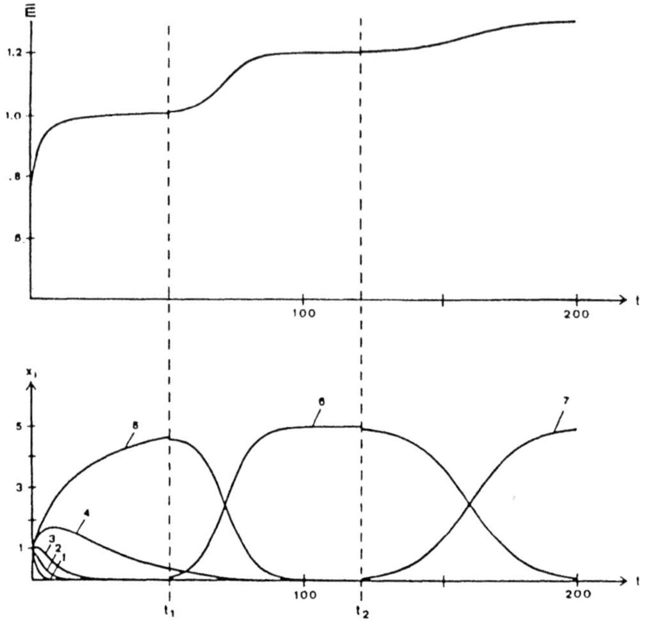

Figure 8. Selection in an ensemble of polynucleotides under the constraint of constant organization. We present solution curves of the differential equation (17): $\dot{x}_i = x_i(w_i - \bar{E})$, $i=1,2,..$ with $w_1=0.3$ $[t^{-1}]$, $w_2=0.65[t^{-1}]$, $w_3=0.82$ $[t^{-1}]$, $w_4=0.96[$ $w_5=1.01[t^{-1}]$, $w_6=1.2[t^{-1}]$ and $w_7=1.3[t^{-1}]$ and $\bar{E}=\Sigma w_i x_i/\Sigma x_i$. The initial conditions are: $x_1(0)=x_2(0)=\ldots x_5(0)=1$ and $x_6(0)=x_7(0)$ The mean excess productivity $\bar{E}$ starts from an initial value of $\bar{E}=0.748$ and increases steadily as the population becomes homogeneous: $\bar{E}\to 1.01$ and $x_1, x_2, x_3, x_4 \to 0$ as $x_5 \to 5$. One easily recognizes the case for the increase of $\bar{E}$: the less efficientl growing polynucleotides are eliminated. They disappear in the same sequence as their excess productivity w_i increases, namel 1,2...5. At $t=t_1=50$ $[t]$ we observe a fluctuation $x_6=0\to x_6=\delta$ the appearance of I_6 as a favourable mutant leads to further increase in $\bar{E}$ which now approaches the value $\bar{E}=1.2$ as I_5 disappears when the population becomes homogeneous again: $x_1=\ldots x_5=0$, $x_6=1$. The same story happens when a new more efficient mutant appears at $t=t_2=120$ $[t]$. Now I_6 is replaced by I_7 and $\bar{E}$ approaches the temporary optimum $\bar{E}=1.3$. This example illustrates well the nature of the selection process and the role of the mean excess productivity $\bar{E}$ as the quantity to be optimized.

By $\Sigma\nu_{i\lambda}B_\lambda$ we denote the low molecular weight material which is necessar to synthesize I_i. As before we denote the rates of synthesis on the templates I_i and I_j by f_i and f_j respectively. The factors Q_{ii} and Q_{ij}

denote the frequencies of correct and erroneous replication products. Accordingly we have the conservation relation:

$$\sum_i Q_{ij} = 1 \qquad (22)$$

From the mechanism (21) and the corresponding decomposition reaction of the replicators which are the same as before we obtain the differential equation

$$\dot{x}_i = (f_i Q_{ii} - d_i - \phi)x_i + \sum_j f_j Q_{ij};\ i,j=1,\ldots,n \qquad (23)$$

Solutions of equation (23) can be grouped into two qualitatively different situations:

(1) The replication process is accurate enough. This means that the quality factor of the selected species is above an error threshold which is determined by the rate constants f_i and d_i, $i=1,\ldots,n$:

$$Q_{mm} > Q_{min}$$

Selection takes place as in the previous case of correct replication but instead of a single replicator a whole distribution of replicators is selected. This distribution consists of a master sequence I_m which is accompanied by a whole family of closely related mutants. The whole ensemble, master sequence plus mutant distribution, has been called a "quasi-species" in analogy to the concept of an actual biological species. As in the case of correct replication the mean net production ϕ is optimized during selection (Jones, 1978).

(2) The accuracy of replication is below the critical limit. There is no influence of the rate constants of replication on the distribution of replicators. They are present only according to their statistical weights in the replication process. No selection takes place.

The critical accuracy sets a limit to the length of polynucleotides which can be replicated with sufficient accuracy to guarantee stable transfer of information from generation to generation. This limit is determined by the accuracy of the replication process. In nature three examples have been studied with sufficient detail so that reliable estimates on the accuracy of replication are possible: enzyme-free template induced RNA-synthesis, enzyme catalyzed RNA-synthesis by virus specific enzymes and DNA-replication in procaryotes. In table 3 we compare the results from the error limit with the lengths of these polynucleotides in nature. Interestingly, we find that particularly in case of simple viruses nature operates close to the error limits. These results have been confirmed independently by an analysis of the sequence distributions in some RNA viruses: Bacteriophage Qß (Domingo et al., 1976, 1978; Batschelet et al., 1976), foot-and-mouth disease virus (Domingo et al., 1980) and influenza virus (Ortin et al., 1980; Fields and Winter, 1981).

Single digit accuracy $\bar{q}$	Error rate per digit $1-\bar{q}$	Superiority σ	Maximum digit content ν_{max}	Biological examl[e]
0.95	5.10^{-2}	2	14	Enzyme-free RNA replication
		4	28	
		10	46	
		20	60	
0.99	1.10^{-2}	2	4	t-RNA precursor
		4	139	
		10	230	$\nu \sim 80$
		20	300	
0.9995	5.10^{-4}	2	1386	single stranded RNA replication via specific replicases Phage Qβ, ν =450[0]
		4	2772	
		10	4606	
		20	5991	
0.999999	1.10^{-6}	2	$0.7.10^6$	DNA replication via polymerases and proof-readi[ng]
		4	$1.4.10^6$	
		10	$2.3.10^6$	
		20	$3.0.10^6$	
0.9999999	1.10^{-7}	2	$6.9.10^6$	E.coli, $\nu=4.10^6$
		4	$13.9.10^6$	
		10	$23.0.10^6$	
		20	$30.0.10^6$	

Table 3. Single digit accuracy and lengths of genetic information. By ν we denote the number of digits, i.e. the number of bases in the polynucleotide. The quality factor Q is given by $\bar{q}^{\nu}$, where $\bar{q}$ denotes the single digit accuracy. This is the frequency of the incorporation of the correct base. Hence, $(1-\bar{q})$ is the error rate per digit. The superiority σ of the master sequence is a function of the rate constants f_i and d_i; $i=1,\ldots,n$ (See Eigen and Schuster, 1979).

6. COOPERATION OF REPLICATORS

Replicators following a simple rate law of replication as given in (16) compete and either a single molecule or a mutant distribution is selected. A system with a given replication mechanism and therefore also with a given accuracy of the replication process is limited in the amount of information that can be transferred from one generation to the other. Polynucleotides exceeding a certain length are not replicated faithfully and are of no use for evolution. On the other hand, we know that replicating elements cooperate under certain conditions and it is straight forward to ask for the simplest mechanism that can suppress competition. So we are looking for a functional relation between replication rates and concentrations which turns a set of competitor into an organized system, the elements of which are the previously independent replicators.

Here, we consider exclusively the mathematical problem. The physical background has been discussed in several papers (Eigen, 1971; Eigen and Schuster, 1979 and 1982; Schuster 1981a, Eigen at al., 1981). In case we apply mass action kinetics the analysis simplest. We only need to take into account the next term of the series expansion (7). For the sake of simplicity we neglect mutation terms a are confronted with the differential equation (24).

$$\dot{x}_i = (\sum_j k_{ij}x_j - \phi)x_i; \quad i,j=1,\ldots,n \tag{24}$$

with

$$\phi = \sum_{i,j} k_{ij}x_i x_j$$

Equation (24) has been studied in great detail (Hofbauer et al.,1980) A special example of (24) which, however, is not relevant for our purpose here is the selection equation of population genetics (Fisher, 1930). For our present purpose we require two general results. Cooperation between self-replicating elements is observed in systems in which the elements are connected by a positive feedback loop of catalytic actions. In graph theory such a closed loop is called a Hamiltonian arc. In systems of low dimension the existence of a Hamiltonian arc is a necessary condition for cooperation.

Independently of the dimension n another general theorem of cooperation could be derived (Schuster et al., 1979; Hofbauer et al.,1981): the replicators of a system show cooperative behaviour if the network of catalytic action consists exclusively of such a positive feedback loop (e.g. figure 9, A). In the context of the rate constants in equation (24) this means that only particular off-diagonal elements of the matrix $K=\{k_{ij}\}$ may be different from zero:

$$K = \begin{pmatrix} 0 & 0 & \ldots\ldots\ldots\ldots 0 & k_{12} \\ k_{21} & 0 & \ldots\ldots\ldots\ldots 0 & 0 \\ 0 & k_{32} & \ldots\ldots\ldots\ldots 0 & 0 \\ & & & \\ 0 & 0 & \ldots\ldots\ldots k_{n,n-1} & 0 \end{pmatrix}$$

Under this condition equation (24) simplifies further ($k_{i,i-1}=K_i$; all indices understood modulo n: n+1=1; 0=n):

$$\dot{x}_i = x_i(K_i x_{i-1} - \phi); \quad i=1,\ldots,n \tag{25}$$

A system which is described by an equation of type (25) is called an elementary hypercycle (Eigen and Schuster, 1979).

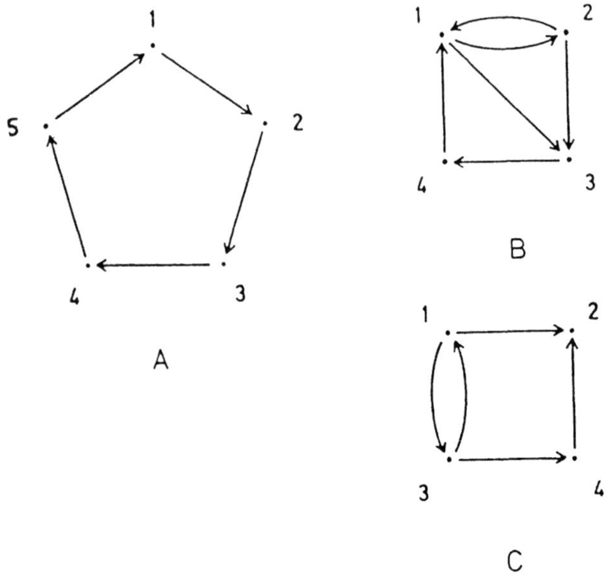

Figure 9. Hamiltonian arcs and cooperation: We present three examples of graphs representing special cases of equation (24). All diagonal terms are assumed to be zero ($k_{ii}=0$, $i=1,\ldots,n$). This is no restriction in generality since equation (24) is invariant towards additive constants in the columns of the matrix K: $k_{ij} \rightarrow k_{ij}+\delta_j$. Hence we can substract the diagonal terms: $k_{ij} \rightarrow k_{ij}-k_{ii}$ (Hofbauer et al., 1980). We consider the non-hyperbolic case, this is the case of exclusively non-negative off diagonal terms. We draw a directed arc $j \rightarrow i$ for every non zero rate constant k_{ij}. Characteristic examples are A: a graph which consists of a Hamiltonian arc only and corresponds to an elementary hypercycle, B: a graph which contains a Hamiltonian arc and C: a graph which does not contain a Hamiltonian arc.

For a physical interpretation of this theorem of cooperation we consider the simple dimensional case n=2.

Cooperation requires that mutual catalytic enhancement exceeds self-enhancement. In terms of rate constants of equation (24) this requirement implies $k_{12}>k_{22}$ and $k_{21}>k_{11}$. Higher dimensional systems (n>2) have to meet additional requirements: mutual enhancement has to be such that it sustains a closed catalytic feedback loop. Due to this functional catalytic relationships the system is transformed from a competitive ensemble into an organized unit: relative concentrations of the members of the cycle are stable against fluctuations. Another important feature concerns genetic information on properties related to it. The simultaneous presence of many (n) replicators increases the coding capacity of the

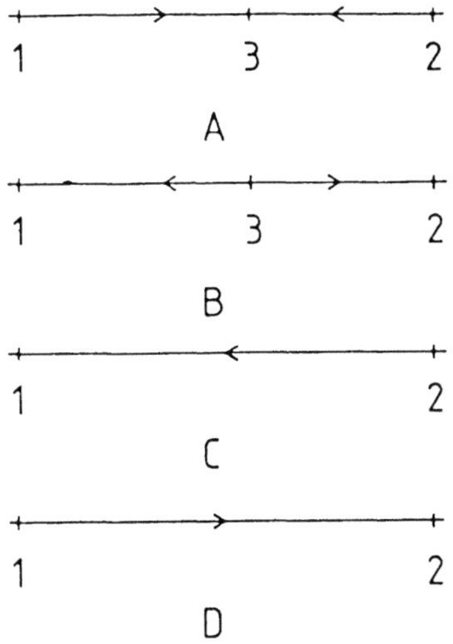

Figure 10. Dynamical systems with second order catalysis (equation 24) and n=2. We distinguish four cases with different qualitative behaviour, A: $k_{11}<k_{21}$ and $k_{22}<k_{12}$ leads to cooperation and hypercyclic stabilization of the ensemble (I_1,I_2), B: $k_{11}>k_{21}$ and $k_{22}>k_{12}$ leads to competition between I_1 and I_2, C: $k_{11}>k_{21}$ and $k_{22}<k_{12}$ yields selection of I_1 and D: $k_{11}<k_{21}$ and $k_{22}>k_{12}$, finally, yields selection of I_2.

system and opens up new possibilites for further development.

Finally, it seems also interesting to ask for the role of the mean net production ϕ in a second order catalytic system like the one described by equation (24). We consider again the simple case with n=2. A straight forward calculation shows that ϕ does not approach a maximum value in a system with hypercyclic organization: the point of maximum ϕ does not coincide with the stable distribution of the two replicators I_1 and I_2:

$$\phi = \phi_{max}: x_1 = x_2 = \frac{1}{2} \quad \text{but} \quad \lim_{t\to\infty} x_1 = \bar{x}_1 = \frac{k_1}{k_1+k_2} .$$

Thus we come to the conclusion that the entire systems have to pay for the functional stabilization by a reduced mean net production. This result that the mean net production ϕ is not optimized is quite general. In some special cases, optimization of ϕ may nevertheless occur. We mention one important case: when the matrix of rate constants is symmetric, $k_{ij}=k_{ji}$, ϕ approaches a (not necessarily global) optimum value. This is the case in the already mentioned selection equation (Fisher, 1930).

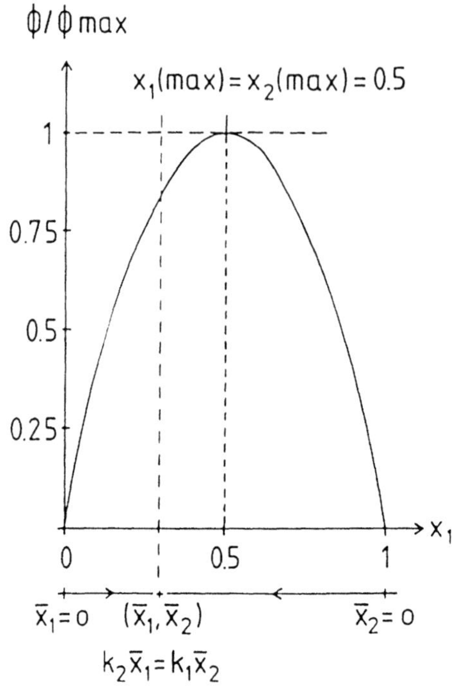

Figure 11. The evolution of ϕ in an elementary hypercycle corresponding to equation (25) with n=2. The dynamical system is characterized by an asymptotically stable fixed point determined by $k_2\bar{x}_1=k_1\bar{x}_2$ $(x_1+x_2=1)$. This steady state does not coincide with the maximum reproductive success: $\phi_{max}\rightarrow x_1=x_2=0.5$.

7. CONCLUDING REMARKS

Competition of replicators growing at a first order rate law and cooperation leading to functional organization observed with some systems with second order rates are two properties of replicating molecules which are not encountered outside the domain of autocatalytic processes. These properties are major factors shaping the dynamics of early evolution (Eigen and Schuster, 1982). Other concepts, like formation of compartments and consecutive development of cell division, development of metabolic pathways and utilization of light in photosynthesis presumably were later steps requiring already an organized ensemble of molecules capable of replication and protein synthesis through translation. The importance of the two concepts mentioned here are, however, not restricted to the evolution of molecules. Closely related problems, translatable into the same mathematical equations occur in many other stages of biological evolution.

REFERENCES

Batschelet, E., Domingo, E. and Weissmann, C. (1976) Gene 1, 127
Biebricher, C.K., Eigen, M. and Luce, R.(1981),J.Mol.Biol. 148,369 and 391
Biebricher, C.K., Diekmann, S. and Luce, R. (1982) J.Mol.Biol. 154, 629
Clarke, B.L. (1980) Adv.Chem.Phys. 43, 1
Dawkins, R.(1976) The selfish Gene. Oxford Univ.Press, Oxford
Dawkins,R.(1982). The extended phenotype. W.H. Freeman, Oxford
Domingo, E., Flavell, R.A. and Weissmann, C.(1976) Gene 1,3
Domingo, E., Sabo, D., Taniguchi, T. and Weissman,C.(1978) Cell 13, 735
Domingo, E., Davilla, M. and Ortin, J. (1980) Gene 11, 333
Eigen, M. (1971) Naturwissenschaften 58, 465
Eigen, M and Schuster, P.(1979) The Hypercycle, Springer Verlag, Berlin
Eigen, M. and Schuster, P. (1982), J.Mol.Evol., in press
Eigen, M., Gardiner, W., Schuster, P. and Winkler-Oswatitsch, R., (1981) Sci.Am. 244 (4), 88
Epstein, I.R. and Eigen, M. (1979) Biophys.Chem. 10, 153
Feinberg, M. (1977) Mathematical aspects of mass action kinetics. In Chemical reactor theory (L.Lapidus, N.R.Amundson, eds.)pp.1-78 Prentice Hall Inc., Englewood Cliffs, N.J.
Fields, S. and Winter, G.(1981) Gene 15, 207
Fisher, R.A. (1930) The genetic theory of natural selection, Clarendon Press, Oxford
Hofbauer, J., Schuster, P. and Sigmund, K. and Wolff, R., (1980) SIAM J.Appl.Math. 38, 282
Hofbauer, J., Schuster, P. and Sigmund, K. (1981) J.Math.Biol. 11,155
Jones, B.L. (1978) J.Math,Biol. 6, 169
Jones, B.L., Enns, R.H. and Ragnekar, S.S. (1976) Bull.Math.Biol. 38, 12
Kornberg, A.(1980) DNA Replication. W.H.Freeman, San Francisco
Küppers, B.O.(1980) Bull.Math.Biol. 41, 803
Lohrmann, R., Bridson, P.K., and Orgel, L.E. (1980) Science 208, 1464
Ortin, J., Najera, R., Lopez, C., Davilla, M. and Domingo, E.(1980) Gene 11, 319
Schuster, P.(1981a) Prebiotic Evolution. In Biochemical Evolution (H.Gutfreund, ed.) pp.15-87, Cambridge University Press, Cambridge, U.K.
Schuster, P.(1981b) Selection and evolution in molecular systems. In Nonlinear Phenomena in Physics and Biology (R.H.Enns, B.L.Jones, R.M.Miura and S.S.Rangnekar, eds.) pp.485-548. Plenum Publ.Co., New York.
Schuster, P., Sigmund, K. and Wolff, R. (1979) J.Diff.Equs. 32, 357
Spiegelman,S. (1971) Quart,Rev.Biophys. 4, 213
Swetina, J. and Schuster, P. (1982), Biophys.Chem., in press
Thompson, C.J. and McBride, J.L.(1974) Math. Biosc.21, 127

THE MCM (MOST COMPLICATED MOLECULE) IN TODAY'S QUANTUM CHEMISTRY : $tRNA^{Phe}$. MOLECULAR POTENTIAL VERSUS MOLECULAR FIELD.

Richard LAVERY and Alberte PULLMAN.
Institut de Biologie Physico-Chimique, 13, rue Pierre et Marie Curie - 75005 Paris - France.

ABSTRACT

The distribution of the electrostatic field of yeast $tRNA^{Phe}$ is calculated on an envelope surrounding the macromolecule and compared with the corresponding distribution of the electrostatic potential. It is demonstrated that, whereas the strongest potentials of the macromolecule are concentrated at the center of its three dimensional structure, the strongest fields are found all along its phosphodiester backbone, associated with the anionic oxygens of the phosphates. It is further shown that the binding of counterions to the macromolecule has very different effects on the resulting potentials and fields.

INTRODUCTION

The polyanionic character of the nucleic acids, due to their ionised phosphate groups, and the fact that these macromolecules often interact with species which are positively charged in their active form, suggests that the electrostatic properties of the nucleic acids can play an important role in their reactive behaviour. This has been confirmed in a recent series of publications dealing with the electrostatic potential of various conformations of double-stranded DNA (1-4, and references therein) and of yeast $tRNA^{Phe}$ (5-7 and references therein). These potentials, when combined with a measure of the steric accessibility of the site under attack (1, 2, 8), have notably enabled the explanation of many experimental data concerning the reactivity of the nucleic acids towards charged electrophiles.

Very recently, we have suggested an extension of these studies of reactivity by the calculation of electrostatic fields (9-11) : whereas the potential is a satisfactory guide to what is felt by a charged species in the vicinity of the macromolecule (the energy of a point charge being the magnitude of the charge multiplied by the local potential), the field is an equally good guide to what is felt by an approaching dipolar species (the energy of a point dipole being the scalar

P.-O. Löwdin and B. Pullman (eds.), New Horizons of Quantum Chemistry, 439–451.

product of the dipole moment with the local field). By the term dipolar species we imply those molecules for which a single center multipole expansion is dominated by the dipole term. A very important species in this class is the water molecule, which leads us to hope that the study of the fields of the nucleic acids will yield useful information on the nature of their hydration.

Our recent studies concerning DNA in various conformations have shown that the distributions of the potential and of the field surrounding this macromolecule are quite different and that, moreover, the binding of counterions influences these two electrostatic properties in quite different ways (9-10). In the present paper we will present a comparison of the potential and field distributions for one particular type of nucleic acid, transfer RNA, specifically yeast $tRNA^{Phe}$ (12). This molecule, which is made of 76 nucleotides (roughly 2500 atoms) presents a considerably more complex structure than the double helical DNAs and, as our previous calculations of potential have shown, its three-dimensional form obtained by X-ray crystal studies (13-16 and references therein), has a very important influence on its external electrostatic properties.

METHOD

Let us recall that the potential and the field of a molecule are defined by :

$$V\,(P) = \sum_{\alpha} \frac{Z_\alpha}{|\underline{r}_{\alpha P}|} - \int \frac{\rho(i)}{|\underline{r}_{iP}|}\, d\tau_i$$

$$\underline{E}\,(P) = \sum_{\alpha} \frac{Z_\alpha\, \underline{r}_{\alpha P}}{|\underline{r}_{\alpha P}|^3} - \int \frac{\rho(i)\, \underline{r}_{iP}}{|\underline{r}_{iP}|^3}\, d\tau_i$$

where V(P) is the potential and $\underline{E}$(P) the field at point P, due to the nuclei of charge Z_α, distant by $\overline{r}_{\alpha P}$ and to the electronic distribution $\rho(i)$ whose volume element $d\tau_i$ is at a distance r_{iP}.

A detailed description of the technique that we employ to calculate the potential or field of a macromolecule has been given in our previous publications, to which we refer the reader (3, 4, 9). In essence, the procedure consists of subdividing the macromolecule into appropriate subunits which are small enough to have their _ab initio_ wavefunctions evaluated. In the case of the nucleic acids these subunits are the phosphates, the sugars and the bases. Once the wavefunctions of these subunits are known, the potential or field that they create may be calculated from the corresponding electron density distributions. The potential or field of the macromolecule is then obtained by the superposition of the contributions of all the subunits appropriately positioned in space.

In $tRNA^{Phe}$, the conformations of each of its 76 phosphate and sugar groups are different, in contrast to the regularity of the backbone components of the helical DNAs. In order to avoid the calculation of a wavefunction for each of these subunits, they are arranged into groups having closely related conformations, 10 such groups for the phosphates and 9 for the sugars, all the members of each such group subsequently being represented by the wavefunction of one representative member (5-7). This procedure permits the electrostatic properties of all the subunits to be calculated sufficiently accurately, while allowing very considerable savings in computation.

In order to calculate these electrostatic properties rapidly and over large areas it is, moreover, useful to replace the exact electron distribution of each subunit by a multicenter, multipole expansion. The expansion we employ, termed OMTP (17, 18), (which expands individually each overlap distribution), yields accurate potentials and fields down to a distance of 2 Å from the constituent atoms of the unit and enables very considerable computational savings. Very close to the subunit it is, nevertheless, necessary to calculate its electrostatic properties from the exact electron distribution.

There are many ways of presenting the potentials or fields obtained for a macromolecule, for example in planes or lines traversing its structure, or at points close to its reactive sites, etc. The representation that we shall employ in the present publication, and which has proven particularly useful for giving an overall view involves the use of a surface envelope (7, 19). This envelope is formed around the macromolecule by the overlapping of spheres centered on each of its atoms, these spheres having radii proportional to the van der Waals radii of the atoms concerned. The macromolecule is then viewed through a planar "window" on which is drawn a grid of points. From each point a perpendicular is projected onto the surface, the potential or field is computed at the point of contact and the value is marked at the corresponding point on the window. Shadings are used to indicate the magnitude of the potential or field (vide infra). Quite generally, we employ a factor of 1.7 for the envelope radii, which enables the surface electrostatic properties to be calculated exclusively with the use of the multipole expansions of the subunits (the smallest sphere, that of hydrogen with a van der Waals radius of 1.2 Å, then having an envelope radius of 2 Å). This envelope is also well adapted to describing the behavior of reactive species approaching the macromolecule at distances where they would just begin to interact strongly with its specific reactive sites (2 Å - 4 Å). Note that in the case of the surface fields, we can represent graphically both the intensity and the direction of this vectorial quantity on the surface envelope.

The geometry that we have employed for $tRNA^{Phe}$ is that of the orthorhombic crystal due to Sussman and coworkers (20). In this crystalline form the macromolecule is associated with various counterions : magnesium, sodium and spermines. As the positions of the magnesium ions are well resolved (14, 21), we shall use them in the present discussion as a preliminary test of the effect of counterions on the electrostatic properties of the macromolecule.

The units that we shall use for the potentials are volts and for the fields, volts/Å. This enables an easier comparison of the two quantities than the units we previously employed for potential (which, in fact, correspond to the potential energy of a proton), namely, kcal/mole. The conversion between kcal/mole and volt can be performed by simply dividing by 23.06.

RESULTS AND DISCUSSION

In discussing the surface electrostatic properties of $tRNA^{Phe}$ we have limited ourselves to two views of the macromolecule, faces

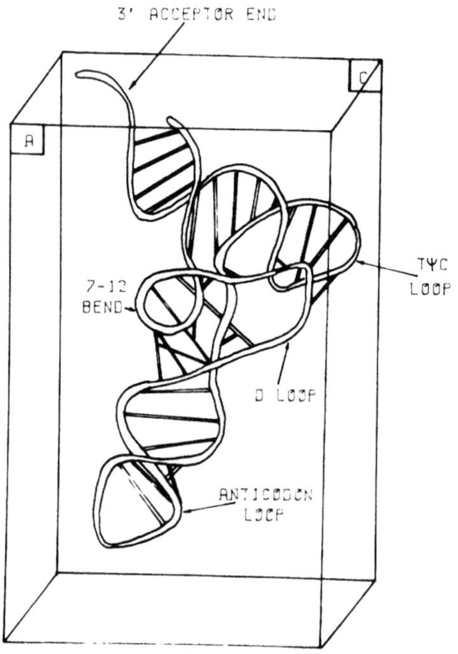

FIGURE 1. Diagrammatic representation of $tRNA^{Phe}$ and of the windows viewing faces A and C of the macromolecule.

TABLE 1. Shading used for the surface envelope potentials and fields.

SHADING	POTENTIAL (VOLTS)	FIELD (VOLTS/Å)	
		UNSCREENED	SCREENED
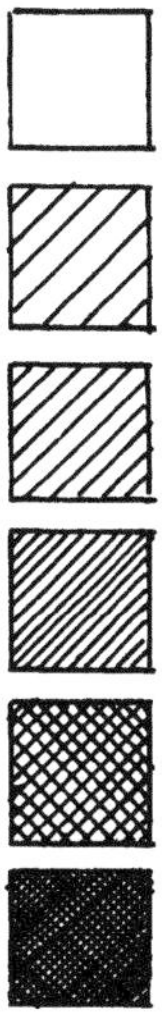	- 21.00	0.00	0.00
	↓	↓	↓
	- 28.08	0.65	1.28
	↓	↓	↓
	- 35.17	1.30	2.57
	↓	↓	↓
	- 42.25	1.95	3.85
	↓	↓	↓
	- 49.33	2.60	5.13
	↓	↓	↓
	- 56.42	3.25	6.42
	↓	↓	↓
	- 63.50	3.90	7.70

termed A and C in fig. 1 which correspond to the large sides of its "L-shaped" three-dimensional form (7). The windows viewing these faces, in the plane of which the surface potentials and fields will be presented, and a schematic representation of $tRNA^{Phe}$ are given in figure 1. The surface properties will be indicated by various degrees of shading, darker regions indicating either more negative potentials or stronger field intensities. The conventions employed for these shadings are given in table 1.

The results on face A, are presented in figure 2. This figure contains, in the center, two diagrams of the macromolecule, viewed through window A, the upper (2b) indicating all the non-hydrogen atoms and the lower (2e), somewhat easier to interpret, containing only those atoms belonging to the sugar-phosphate backbone. On both these diagrams the positions of the four magnesium ions located in the orthorhombic crystal (20) are indicated by black spots. On the left of these central diagrams are the potential (fig. 2a) and the field (fig. 2d) for $tRNA^{Phe}$. On the right, are the potential (fig. 2c) and the field (fig. 2f) for the macromolecule in the presence of the four bound counterions. Each of the subdivisions on the potential and field graphics represent one

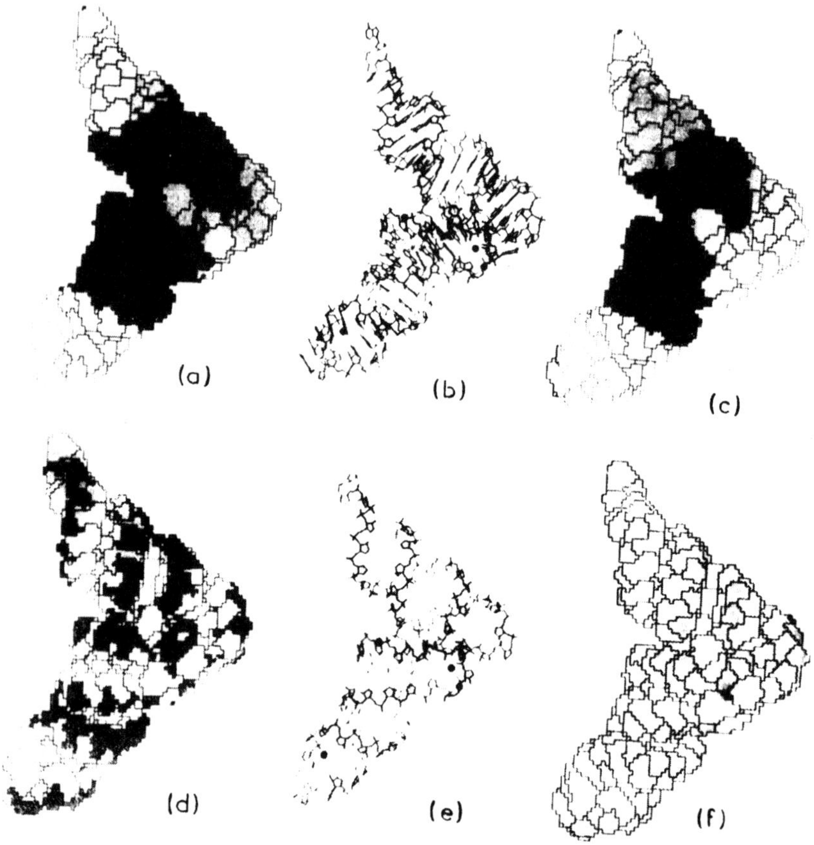

FIGURE 2. Surface potentials and fields on the face A of $tRNA^{Phe}$.
(a) Potential without 4 Mg^{2+}
(b) Diagram of $tRNA^{Phe}$
(c) Potential with 4 Mg^{2+}
(d) Field without 4 Mg^{2+}
(e) Diagram of the $tRNA^{Phe}$ backbone
(f) Field with 4 Mg^{2+}

subunit of the macromolecule visible in the surface envelope. Note finally, from table 1, that the shadings of the potential for the isolated or screened tRNA are identical, but those for the fields are not, the shadings of the fields for the tRNA covering almost twice the range of those for the unscreened molecule.

The potentials (2a) on face A, show a very important concentration of the most negative values in the center of the molecule. In the inner angle of the three-dimensional form of $tRNA^{Phe}$, close to the bend in the phosphodiester backbone formed by the phosphates 7-12, the potentials are about 40 volts stronger than at the extremities, notably, the anticodon loop, the acceptor end and the TΨC and D loops. This is due to the concentration of the anionic phosphate groups in the center of $tRNA^{Phe}$, which, because of the long range of the electrostatic potentials (which for a simple charge, decline with distance as $1/r$) leads to very important superposition effects.

The field intensities (2d) present a very different picture. Since the field of the subunits of the macromolecule falls off much more rapidly with distance (that for a simple charge varying as $1/r^2$), the superposition effects can be expected to be attenuated. This is indeed the case and figure 2d shows strong field intensities over the entire surface envelope, wherever the anionic oxygens of its phosphate groups point toward this surface. The fields of the macromolecule are thus much more closely related to those of its subunits than are the macromolecular potentials (the isolated anionic phosphate presenting maximal surface fields of roughly 1.9 volt/Å, considerably stronger than those of the isolated bases or sugars, roughly 1.2 volt/Å, see also reference 11). Moreover, in figure 2d, it can be seen that the fields in the grooves of the double helical sections of $tRNA^{Phe}$ are generally much weaker than those on the bordering backbones, which is the reverse of the usual distribution of potential. This is particularly clear for the deep groove in the anticodon stem and for that at the lower end of the acceptor stem, which present the weakest field intensities of the entire face A.

The effect of the inclusion of the four magnesium ions into the calculations can be deduced from the comparison of figure 2a with figure 2c and figure 2d with figure 2f. Note that only one of the four cations is bound on face A of the macromolecule, that which, in the D loop, is closest to the center of the structure. Nevertheless an overall change in the surface potentials is visible (fig. 2c), with a contraction of the strongest potentials around the center of the macromolecule and an appreciable decrease in their magnitude (of up to 8 volts). The D loop and the anticodon loop have both become notably less negative due to the magnesiums bound in these regions, but the center of the molecule remains the most negative zone, despite the presence of a further magnesium ion bound in the 7-12 bend. The effect of this ion alone is clearly insufficient to overcome the very strong influence of the centrally concentrated phosphates.

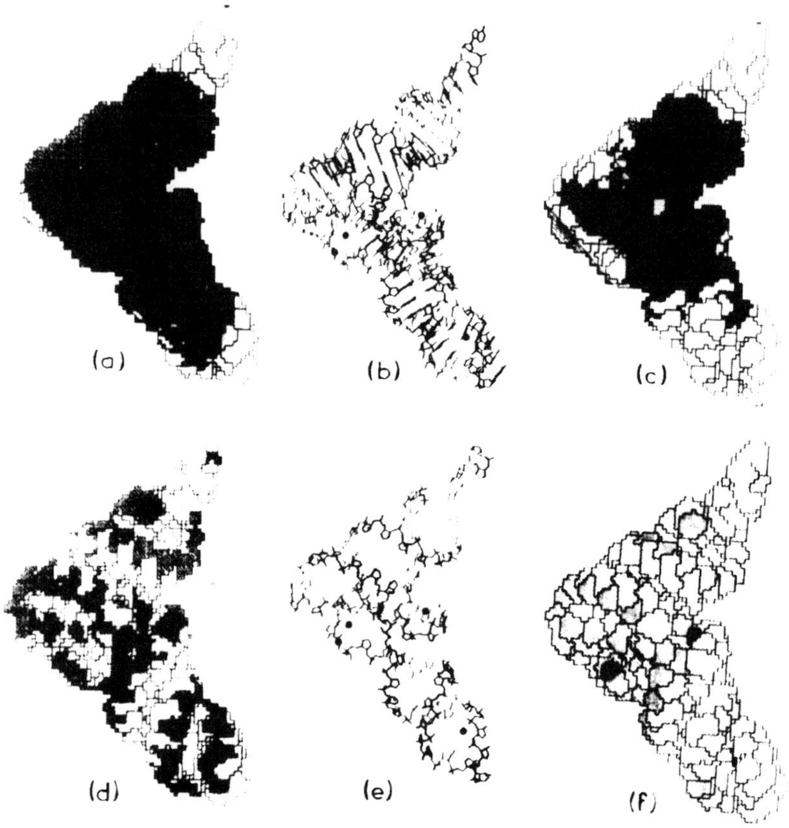

FIGURE 3. Surface potentials and fields on the face C of $tRNA^{Phe}$
(a) Potential without 4 Mg^{2+}
(b) Diagram of $tRNA^{Phe}$
(c) Potential with 4 Mg^{2+}
(d) Field without 4 Mg^{2+}
(e) Diagram of the $tRNA^{Phe}$ backbone
(f) Field with 4 Mg^{2+}

The effect on the field distribution is very different. The ion bound on face A, in the D loop, produces very strong fields in a very restricted zone around its binding position (fig. 2f), the maximal intensity being almost twice that calculated for the unscreened macromolecule. This can be understood in terms of the ion pair formed by the cationic magnesium and its associated anionic phosphates. Such a close approach of opposing charges leads to very large potential gradients (9), or in other words, fields. Because of the change of scale in shading the fields, the rest of the surface might appear to be associated with very weak values, but, in fact, the values, other than those around the visible ion pair, remain almost unchanged, due to the short-range character of the field.

We now turn to face C of $tRNA^{Phe}$, the results for which are presented in figure 3. The organization of this figure is the same as that of figure 2, with two central diagrams of the macromolecule (3b and 3e), the unscreened surface potential (3a) and field (3d) on the left and the screened potential (3c) and field (3f) on the right. The positions of the magnesiums are once again marked by black spots.

The distribution of the potential on the surface of the isolated macromolecule is globally similar to that on the opposing face with an important concentration of the most negative values in its center. There is, however, apart from the deepest zone on the 7-12 bend, a second zone of very strong potential in the center of the D loop and a slightly weaker, but well defined zone in a pocket formed by the curved anticodon loop. As we have shown before (7), these zones correlate rather well with the observed binding sites of the magnesium ions in the crystal.

Turning to the field distribution, we observe again no central concentration but, as for face A, strong fields are found wherever the phosphate anionic oxygens are close to the surface envelope. Intervening weak zones can be observed in the grooves of the double helical stems and in the inside of the anticodon, the D and the TΨC loops.

When the effect of the four magnesiums are added to the calculations, a contraction of the deepest potential zone can again be observed (fig. 3c), most notably withdrawing from the anticodon loop and the edges of the D and TΨC loops. There is also an overall weakening of potential, more important than that observed for face A (of up to 16 volts). As a result, whereas for the isolated macromolecule, the deepest potential was on face C (- 119 volts versus - 117 volts on face A), after addition of the four magnesiums, three of which are bound on this face, the potential minimum is found on face A (- 109 volts versus - 103 volts on face C).

As concerns the field distribution on face C in the presence of the magnesiums (fig. 3f), three zones of particularly strong intensity are observed, corresponding to the three visible magnesium binding positions in the anticodon loop, in the D loop and in the 7-12 bend.

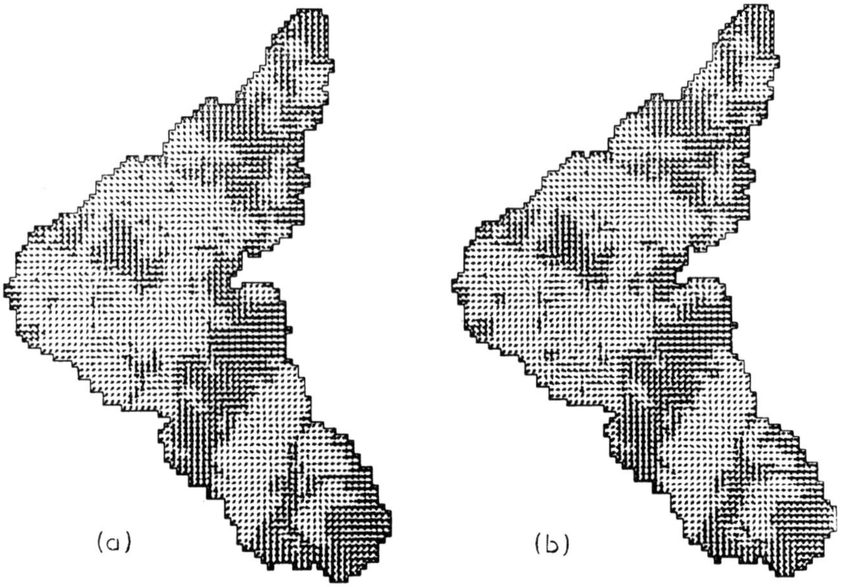

FIGURE 4. Surface field directions on the face C of $tRNA^{Phe}$ (the symbols : Δ , ⧖ and ⏀ indicate fields pointing outwards, by more than 30°, inwards, by more than 30° or tangential within ± 30°, with respect to the surface envelope).
(a) without 4 Mg^{2+}
(b) with 4 Mg^{2+}

Finally we show, in figure 4, the directions associated with the surface fields for the case of face C, with and without the presence of the four Mg^{2+} ions. These directions are indicated by three symbols, Δ for fields which point outwards from the surface envelope by more than 30°, ⧖ for fields which point inwards by more than 30° and ⏀ for fields pointing within ± 30° from the tangent to the surface envelope. Each of these symbols is also oriented to indicate the direction of the field component tangential to the surface envelope.

In the absence of the Mg^{2+} ions (fig. 4a) it can be remarked that the field direction is almost never outwards from the macromolecule This is understandable due to its strong negative charge. In several zones the field points inwards. These zones correspond rather closely to the strongest field intensities in fig. 3d and occur over the sections of the tRNA backbone where the phosphate anionic oxygens are pointed most directly towards the surface envelope. The field direction in the grooves of the double helical stems and in the interior of the loops are mostly more or less tangential to the surface envelope, but occasionally indicate the presence of an electronegative atom of a base close to the surface by their radial orientation around its position.

The effect of adding the 4 Mg^{2+} ions can be seen in fig. 4b. There are localized changes in field direction around each of the three cations bound on this face of $tRNA^{Phe}$, the field generally changing its direction from inward pointing to roughly tangential to the surface envelope. These ions are, however, not strong enough to overcome the influence of the 76 phosphate groups and cause outward pointing fields, with the exception of a very small zone in the 7-12 bend. The field directions over the rest of the surface remain virtually unchanged.

It is also interesting to remark that the maximal field intensities in $tRNA^{Phe}$, both before and after screening (3.85 volts/Å and 7.62 volts/Å respectively), are considerably greater than those calculated for double helical DNA's, isolated or screened (the largest values being those of A-DNA, 2.94 volts/Å and 3.5 volts/Å respectively) (10). In the case of the unscreened macromolecules, this may be attributed to the distances of approach between neighbouring phosphate groups, which, in the center of $tRNA^{Phe}$ are considerably shorter than in any of the DNA's we have studied. For the screened macromolecules the difference is due to the nature of the counterions employed, Mg^{2+} for tRNA and Na^+ for DNA. It is clear that the divalent magnesium will produce very much stronger field intensities in its neighbourhood than the monovalent sodium.

CONCLUSIONS

If it is recalled that the potentials of a macromolecule may be used as a guide to the behavior of an approaching charged species, whereas the field may be used as a guide to the behavior of dipolar species, and notably of the water molecule, the dramatic differences in the potential and field distributions around $tRNA^{Phe}$ may be expected to be of considerable importance.

Thus for a reaction occuring in aqueous solution, the solvent and any charged attacking species will be subjected to different influences from the macromolecule. In particular, the central concentration of potential may be contrasted with the distribution of the field, which is strong all along the phosphodiester backbone. This may be important for understanding reactions in the case of a possible competition between solvent and reactive species at a given reactive site.

Moreover, we have shown that the presence of counterions, should have remarkably different effects on solvent binding and electrophilic attacks, since whereas they produce a weakening of the potential over the entire surface of the macromolecule, they leave the field distribution almost unchanged except in very limited zones around the ion binding positions where it is strongly enhanced. Work is in fact continuing in studies of specific reactions, taking into account both solvent and counterions, in view of obtaining a realistic modelling of an interaction in solution.

REFERENCES

(1) Pullman, B., Lavery, R. and Pullman, A. : Eur. J. Biochem., in press.

(2) Pullman, B., Pullman, A. and Lavery, R. : in "The Living State" ed. R.H. Mishra, Wiley-Interscience, in press.

(3) Pullman, A. and Pullman, B. : 1981, Quart. Rev. Biophys. 14, pp. 289-380.

(4) Pullman, A. and Pullman, B. : in "Chemical Applications of Atomic and Molecular Electrostatic Potentials", 1981, ed. P. Politzer and D.G. Truhlar, Plenum, N.Y., pp. 381-405.

(5) Lavery, R., Pullman, A., Pullman, B. and de Oliveira, M. : 1980, Nucl. Acid. Res. 8, pp. 5095-5111.

(6) Lavery, R., Corbin, S. and Pullman, A. : 1981, Int. J. Quant. Chem., Quant. Biol. Symp. 8, pp. 171-183.

(7) Lavery, R., Pullman, A. and Corbin, S. : 1981, in "Biomolecular Stereodynamics", Vol. I, ed. R.H. Sarma, Adenine Press, N.Y., pp. 185-193.

(8) Lavery, R., Pullman, A. and Pullman, B. : 1981, Int. J. Quant. Chem. 20, pp. 49-62.

(9) Lavery, R., Pullman, A. and Pullman, B. : Theoret. Chim. Acta, in press.

(10) Lavery, R. and Pullman, B. : Nucl. Acid. Res., in press.

(11) Lavery, R., Pullman, A. and Pullman, B. : in preparation.

(12) Transfer RNA's are adaptor molecules which selectively transport a given amino acid and place it at the appropriate spot on the ribosomal matrix where the biosynthesis of proteins occurs. $tRNA^{Phe}$ is specific for phenylalanine.

(13) Quigley, G.J., Teeter, M.M. and Rich, A. : 1978, Proc. Natl. Acad. Sci. (USA) 75, pp. 64-68.

(14) Sussman, J.L., Holbrook, S.R., Warrant, R.W., Church, G.M. and Kim, S.-H. : 1978, J. Mol. Biol. 123, pp. 607-630.

(15) Jack, A., Ladner, J.E. and Klug, A. : 1976, J. Mol. Biol. 108, pp. 619-649.

(16) Stout, C.D., Misuno, H., Rubin, J., Brennam, T., Rao, S.T. and Sundaralingam, M. : 1976, Nucl. Acid. Res. 3, pp. 1111-1123.

(17) Port, G.N.J. and Pullman, A. : 1970, FEBS Letters 31, pp. 70-73.

(18) Goldblum, A., Perahia, D. and Pullman, A. : 1979, Int. J. Quant. Chem. 15, pp. 121-129.

(19) Lavery, R. and Pullman, B. : 1981, Int. J. Quant. Chem. 20, pp. 259-272.

(20) We have employed a set of coordinates further refined from those of reference 14, communicated to us by Dr. S.-H. Kim.
(21) Holbrook, S.R., Sussman, J.L., Warrant, R.W., Church, G.M. and Kim, S.-H. : 1977, Nucl. Acid. Res. 4, pp. 2811-2820.

Acknowledgements: This work was supported by the National Foundation for Cancer Research (U.S.A.).

INDEX OF SUBJECTS

GPSR Compliance

The European Union's (EU) General Product Safety Regulation (GPSR) is a set of rules that requires consumer products to be safe and our obligations to ensure this.

If you have any concerns about our products, you can contact us on ProductSafety@springernature.com

In case Publisher is established outside the EU, the EU authorized representative is:

Springer Nature Customer Service Center GmbH
Europaplatz 3
69115 Heidelberg, Germany

Batch number: 10815974

Printed by Printforce, the Netherlands